A TEXT BOOK

OF

GEOMORPHOLOGY

A TEXT BOOK OF GEOMORPHOLOGY

P. Dayal

M.A. (Aligarh), Ph. D. (London)

Simon Research Fellow, University of Manchester, Formerly University Professor and Head of the Department of Geography, Patna University and Vice-Chancellor, Magadh University, Bodh- Gaya.

Visiting Professor, North-Eastern Hill University, Shillong.

Professor Emeritus, Patna University.

RAJESH PUBLICATIONS

NEW DELHI - 110 002

A TEXT BOOK OF GEOMORPHOLOGY

ISBN 10 : 81-85891-56-7
ISBN 13 : 978-81-85891-56-9

Published by :
Sanjay Gupta
RAJESH PUBLICATIONS
1, Ansari Road, Darya Ganj,
New Delhi - 110002
Ph.: 011-43084725
rajeshpublications@gmail.com
www.rajeshpublications.com

Typesetting by :
Ritu Pandey
New Delhi- 110002

Printed in India at :
R.K. Offset Process, Delhi - 95

Third Revised Edition : 2005
Fourth Revised Edition : 2007
Fifth Revised Edition : 2010
Sixth Edition : 2011
Seventh Revised & Enlarged Edition : 2015
Reprint : 2016
Eighth Edition : 2017
Ninth Edition : 2019
Tenth Edition : 2021
Reprint : 2023

Preface to the First Edition

When my book on Geomorphology in Hindi entitled *Bhu-Akriti Vigyan* was published seven years ago, many colleagues in the different universities suggested that there was need for a similar book in English for the geography students of Indian universities. Originally it was intended to bring out an English version (if not exactly an English translation) of *Bhu-Akriti Vigyan*. But it was soon realized that it would be more realistic to use this opportunity to revise the book thoroughly and incorporate new materials. This process has taken nearly six years and what has emerged is a wholly new book, different in content and organization and more comprehensive in scope than the original book. Most of it was written between 1983 to 1986 while the author was a Visiting Professor in the Department of Geography, School of Environmental Studies, North-Eastern Hill University, Shillong. The library facilities at NEHU and the warm and congenial atmosphere provided by the then Vice-Chancellor, Dr. B.D. Sharma made the task an extremely pleasant one.

The book essentially follows the pattern of courses prescribed in Indian universities and includes a rather odd chapter on the origin of the earth which still finds place in the course of study of most Indian universities. On the whole an attempt has been made to strike a balance between structural and erosional Geomorphology, and the new concepts of plate tectonics, sea-floor spreading and continental drift and mountain building which have revolutionised geomorphological thinking have been given adequate coverage. Simultaneously, an attempt has been made to incorporate the latest findings in fluvial, glacial, periglacial, karst and coastal and ocean floor geomorphology together with planation surfaces and the evolution of slopes. Throughout the emphasis has been on the complexity of landforms and the interaction of many factors and processes. A summary of the basic concepts and conflicting views has also been provided. The book concludes with a fairly comprehensive chapter on Applied Geomorphology and its ramifications. An important omission is a chapter on Quantitative Geomorphology which the author has been constrained to omit primarily because he felt that he would not be able to do justice to it. Luckily, there are a number of good books available on the subject for use by students.

My colleagues and students in the Department of Geography, Patna University where I have spent the best part of my teaching career have contributed in various ways to the making of this book. Dr. L. N. Ram, Dr. S.N. Prasad, Dr. S. Adhikari, Sri K. Paswan, Dr. Ms., Urvija Shanker and Dr. Satish Sinha have particularly helped with library work. I have also benefitted from the ideas of and discussions with Dr. A.B. Mukherjee of Panjab University, Dr. R.K. Rai of NEHU, Shillong, Dr. R.P. Singh of Magadh University, Dr. E. Ahmad of Ranchi University, Dr. K.R. Dikshit of Poona University and Dr. L.R. Singh of the Pant Social Science Institute, Allahabad, Most of the maps and diagrams have been prepared by Sri Surendra Lal Das, Draftsman, Department of Geography, and Dr. Rajendra Prasad, Reader in Geography (Formerly Cartographer), Patna University. The author, of course, owes a deep debt of gratitude to the various well known writers on the subject whose works he has freely consulted and used.

It is hoped that the book will fulfill a long-felt need and will be found useful for students beginning their advanced study of the subject in Indian universities. Every effort has been made to include as many examples from India as possible which have been culled from widely scattered sources.

Patna, January 1990 **P. Dayal**

Preface to the Second Edition

The warm reception that the book has received has made it necessary to bring out a revised edition earlier than expected. In this edition printing imperfections have been removed and some minor corrections have been made. Chapters 16 and 20 have been revised to incorporate new materials on river regimes, channel morphology and pediplanation. A number of new diagrams have been added and the bibliography has been updated. It is hoped that this revised edition will be found quite useful by the university students as well as by those taking the administrative service examinations conducted by the Union and State Public Service Commissions.

July 25, 1994 **P. Dayal**

Preface to the Third Edition

The third edition is being published by **Rajesh Publications** New Delhi. The book has been thoroughly revised and updated and several new additions have been made to the Bibliography.

January, 2005 **P. Dayal**

Preface to the Seventh Edition

In this edition a new chapter on Materials of the Earth's Crust - Rocks and Minerals has been added and forms chapter six of the book. Some additions have been made in the bibliography and important topics discussed in new chapter have been included in the new index. The final proof has been read by Dr. V.N.P. Sinha for which I am obliged to him. I thank him for help in many other ways.

January, 2015 **P. Dayal**

Some Opinions

By far the most comprehensive text-book on Geomorphology authored by an Indian scholar. Designed for the graduate courses in Indian Universities. Has a blend of traditional themes and the recent developments. Well illustrated with a useful index. Chapters on periglacial landforms, coastal landforms and geomorphology of ocean floor, basic concepts and applied geomorphology. Painstakingly written and the treatment very comprehensive. Both the students and the teachers will find it easy to read and understand.

Transactions of the Institute of Indian Geographers
Vol. 13 (2), 1991

Extremely useful for M.Sc. Geography students and those appearing at State Service Examinations. High quality material. Author and publisher deserve special compliments.

Dept. of Geography
Sri Venkateswara University
in a Communication to Publishers, 21.2.92

Written primarily keeping in view the graduate and post-graduate syllabi of Geomorphology in Indian Universities. Covers a wide range of geographical phenomenon. Basic concepts and modern views on geomorphology have been incorporated with application of geomorphology in day to day life as well as in allied disciplines. The book serves as a readymade reference for students and teachers.

Annals, National Association of Geographers, India
Vol. 12, No. 1 & 2, December 1992

An exhaustive title incorporating all new developments in Geomorpholgy. Characterized by scholarship, command over language and efficient communication skill. Makes the subject interesting and lively.

S.S. Chib
Head, Dept. of Geography
University of Jammu

This text-book, painstakingly written by an eminent scholar of geography with long teaching & administrative experience in several Indian Universties is unique for its extensive coverage and wealth of information. No other textbook on this subject published in India can claim to encompass so many varied topics within the limitation of a single volume and so modest a price. Written primarily for the undergraduate and graduate students of earth sciences, especially of geography, in universities from all over India, this book is particularly useful to those preparing for service examinations. Easy to read and understand and profusely illustrated, it is a good textbook, and will be of great help to Indian students and teachers.

Everyman's Science
Vol. 34, No1 1999
Journal of Indian Science Congress Association

CONTENTS

Chapter 1

Meaning, Scope and Development of Geomorphology

Meaning of Geomorphology

Geomorphology is an important branch of Physical Geography. It is concerned with the scientific study of the origin and evolution of the relief features of the earth. The literal meaning of Geomorphology is a discourse on the form of the earth. It has been derived from the Greek words *ge* meaning earth, *morphe* meaning form and *logos* meaning discourse. Thus geomorphology is a systematic and organised description and analysis of the landforms of the earth. Here the word landform has to be understood in its widest connotation and includes not merely the micro features of the land but also the major relief features such as continents and ocean basins, mountains, plateaus and plains, and a discussion of their origin is equally relevant. This is possible only if we have knowledge of the structure of the earth, the nature of rocks, diastrophism, volcanism, plate tectonics, continental drift as well as the processes of weathering and erosion, because all these factors influence, fashion and change the relief features of the earth. Thus the study of landforms has both a structural and an erosional dimension, and both the aspects have to be considered for a proper understanding of the evolution and distribution of landforms.

Both Worcester and Von Engeln use the term geomorphology in this extended sense. According to Worcester "Geomorphology is the interpretative description of the relief features of the earth. In other words geomorphology is the science that describes the surface of the lithosphere, explains its origin and interprets its history".[1] Von Engeln too does not consider geomorphology to be merely the science of land forms but also includes within its fold the shape of

1. Worcester, P.G.—*A Textbook of Geomorphology*, Van Nostrand, East-West Ed., 1965, p.3.

the entire earth, its configuration and the disposition of its largest units.[1]

According to Strahler, geomorphology is an analysis of the origin and evolution of earth features. Geomorphology does not study merely the physical, chemical and biological processes affecting the evolution of land forms but also structure of the earth's crust, the geological processes as well as the climatic influences, because it is the combined influence of all these factors that determines the land forms.[2] According to him it is possible to recognize two major types of land forms (i) Initial land forms which owe their origin to diastrophism and vulcanism and are the product of crustal deformation, and (ii) Sequential land forms which result from the action of exogenetic or external forces on the initial land forms through weathering, erosion and mass movements. In areas where indogenetic or internal forces have remained active till recent date through earth movements and volcanic action, we come across complex mountain ranges and high plateaus. On the other hand, in areas where the external forces have been operating uninterruptedly for a long time the surface is relatively low and flat marked by a variety of lesser topographic features.

Strahler is further of the view that although historically geomorphology has dealt largely with the surface of the continents, it may be extended to include the relief features of the continental shelves and ocean basins.[3] Thornbury is also of the same view. According to him geomorphology is the science of land forms including the submarine topography.[4] Bloom puts it more emphatically when he says that 'geomorphologists must not let their world end at low tide'.[5]

Bloom defines geomorphology as " systematic description and analysis of landscapes and the processes that change them"[6] He emphasizes that such a study must include both structure and process, as the origin and evolution of landscape can only be

---

1. Von Engeln, O.D.— *Geomorphology*, Macmillan, New York, 1960.p.1.
2. Strahler, A.N.- *The Earth Sciences*, Harper International Edition, 1968, p. 452
3. Ibid
4. Thornbury, W.D.-*Principles of Geomorphology*,Wiley International Ed., 1984, p.1
5. Bloom, A.L.—*Geomorphology : a systematic analysis of late Cenozoic landforms*, Prentice-Hall of India, New Delhi, 1979, p.xv
6. Ibid., p.1

understood with reference to their geological history and the processes that have shaped the earth's surface through time.

Sparks in his book *Geomorphology* gives a rather narrow definition of the subject. According to him, "essentially it is the study of the evolution of land forms, especially land forms produced by the processes of erosion... The origin of the major features of the earth's crust is usually not considered to be within the sphere of geomorphology."[1] In his view geomorphology need not include a study of mountain building, earthquakes, volcanoes, the distribution and origin of continents and oceans. Geologists have been studying these subjects but their study requires such a deep knowledge of mathematics and physics that it is not possible for geologists and geographers to do justice to these subjects. He quotes Davis in support of his view who has stated that while geology studies the evolution of structure, geography studies the wearing down of the structure. Few scholars, however, would agree with the restrictive and narrow definition of geomorphology given by Sparks. Other eminent British physical geographers like Steers and Wooldridge do not subscribe to this view.[2] Wooldridge considers both geophysical topics and erosional geomorphology as the rightful domain of geomorphology and emphasizes that both these aspects should be studied by geographers.

According to the German geologist Machatschek, geomorphology is the study of the physical processes fashioning the form of the solid surface of the earth as well as the resultant land forms.[3] Australian physical geographer Ollier states in unmistakable terms that geomorphology includes the study of both continent-ocean relationships and the origin, evolution and characteristics of lesser land forms. According to him geomorphology has assumed new dimension when the earth's geomorphic history and features have to be analysed in the context of such dynamic geo-physical processes as sea-floor spreading, plate tectonics and continental drift.[4]

---

1. Sparks, B.W.—*Geomorphology*, Longman, London, 2nd ed, 1972, p.1
2. Steers, J.A.—*The Unstable Earth: Some Recent Views in Geomorphology*, Methuen, 1973.
   Wooldrige, S.W. and Morgon, R.S.- *The Physical Basis of Geography, An Outline of Geomorphology*, Longmans, London, 1948.
3. Machatschek, F.- *Geomorphology*, Longman, 1969.
4. Ollier, C.D.-*Tectonics and Landforms*, Longman, 1981, p.2.

Relation of Geomorphology with Geography and Geology

There has been some difference of opinion among scholars as to whether geomorphology should be considered a branch of geography or geology. According to Lobeck geomorphology or the study of landforms is a branch of geology.[1] Similarly Thornbury is of the view that geomorphology is primarily geology.[2] On the other hand, Wooldridge and Morgan consider geomorphology to be a legitimate and important branch of geography. Sparks too is of the same view and states that although a background knowledge of geology is essential for the student of geomorphology, geomorphology is more closely related to geography than to geology.[3]

William Morris Davis, the father of American geomorphology considered geomorphology to be essentially a branch of geography. In his opinion the subject matter of geology is more related to time sequence and the past, while the view point of geomorphology is essentially geographical as it gives priority to the existing patterns.

In reality both geographers and geologists have contributed to the development of geomorphology, and their contribution, though varying in nature and quantity in the different countries, have been essentially complementary to each other. Geologists in their studies have been mainly emphasizing the influence of geological history and nature of rocks in the evolution of landscape, though some of them have also shown interest in the activities of the tectonic or gradational processes. On the other hand, geographers have laid greater emphasis on the relationship between the nature of land forms and the processes that have shaped them, and more recently they have also tried to study the influence of human activities on land forms. In fact, the field of geomorphology has been cultivated both by geographers and geologists and in the process the subject has been influenced by and gained from the methods and view points of both. While the association of geomorphology with both geography and geology (as also with some other disciplines) has been extremely fruitful and useful, it has to be admitted that till recently both geologists and geographers, and more particularly the former, have given only a peripheral position to the analysis of landforms in their studies. This is an area where co-operation

1. Lobeck, A.K.-*Geomorphology*, McGraw Hill, 1939, p.3.
2. Thornbury. W.D.-Op. cit., p. 143.
3. Sparks, B.W.-op. cit, pp.4-5.

between geology and geography is not only fruitful and desirable but is in fact essential. Dury and Bloom have rightly remarked that geomorphology is a part of two sciences, and geomorphologists may regard themselves as either geographers or geologists or preferably both.[1]

Scope of Geomorphology

From the above discussion it is clear that the essential subject matter of geomorphology is the systematic analysis of the origin, evolution and present pattern of the earth's relief features. We may divide this study into three main sub-divisions, though it has to be recognised that these sub-divisions are so inter-dependent and interrelated that this division may appear some what artificial. These major sub-divisions are *structure, process, and classification and evolution of the various landforms*. Structure, process and evolution of landforms are all dynamic. Their interactions are therefore highly complex and the nature and intensity of the interactions change with time and place.

Structure is primarily determined by internal forces which include plate tectonics, diastrophism and volcanic action. Structure includes the constitution of the earth's crust, the disposition of the continents and ocean basins, the nature of rocks, the stratigraphic arrangement of rocks and their tectonic displacement and deformation, the building up of mountains, folds and faults and so on.

Process is essentially related to external forces. The most universal of these forces is gravity which regulates the downward movement of materials on the surface of the earth. Sometimes the influence of gravity is direct and apparent, as we can see in the formation of talus but by and large it exerts its influence indirectly through the medium of running water, glaciers and solifluction. Through this process debris are transported from higher to lower levels resulting ultimately in the reduction of relief. In addition to gravitational movements, tangential movements are also important in the atmosphere and hydrosphere, and these can be observed in winds, currents, waves and tides. It is obvious that from the point of view of geomorphology gravitational movements are far more

1. Bloom, A.L.- op. cit. p.2. Dury, G.H.-Some recent views on the nature, location, needs, and potential of geomorphology, *Professional Geographer*, 24, 1972, pp,199-202

important and influential than the tangential movements.

The nature and intensity of the geomorphic processes are influenced by various factors and they vary from region to region. The most influential of these factors are climate and the associated vegetation. Climate to great extent determines the process. By climate here we mean not merely the present climate but also the past climates, as many landscapes bear the imprint of more than one set of climate-controlled processes. Process oriented geomorphology emphasizes the shaping of landforms in response to various climatic conditions and also takes into consideration the climatic influences of the past.

In the end landforms are classified according to their genesis, and their *evolution* is analysed with reference to the dynamic structure and the dynamic processes. It is obvious that the mechanisms of causality are extremely complex in geomorphology, and one action may lead to several reactions which may be contrary to one another. It is, therefore, extremely important to have a comprehensive and historical perspective in the analysis of landform evolution. Both structure and process are important in geomorphic evolution but an account of the interaction of surface processes with geological structures does not provide an adequate explanation of landforms unless we include in the explanation the length of time the process has been operating which will give us an idea of the relative stage of development.[1]

It would thus appear that the subject matter of geomorphology includes a consideration of both structure and process as well as a systematic analysis of the evolution of land forms through time, although the primary aim of the subject is a study of the present day landscape, its patterns and its evolution.

Relief Features of Different Order

Salisbury[2] introduced the concept of hierarchy of relief features, and following him we may divide the relief features of the earth into three major classes on the basis of their size. This classification is useful because "each successive level includes features of successively

1. "Acknowledgement that landscapes are not immutable leads to the immediate conclusion that if change is in progress, something (structure) is altered by something (process) to a definable extent (stage) or for a definite interval (time)". Bloom A.L.—op. cit, p.5.
2. Salisbury, R.D.—*Physiography*, 3rd ed., Henry Holt, New York, 1919.

smaller size, greater unity of origin and more transient durability."[1]

(i) **Relief features of the first order**—This includes the two major earth features— the continental platforms and the ocean basins. At the junction of these two are the continental margins. To a very great extent the arrangement of continents and oceans would appear to be fairly stable but in the long geological history of the earth we come across evidences of many changes in their form, extension and geographical location.

(ii) **Relief features of the second order**— Superimposed on the continents and ocean basins are the relief features of the second order and include the plains, plateaus and mountains on the continental surface as well as the mid-oceanic ridges, oceanic plains and trenches on the ocean floor together with major features of the continental margins. The relief features of the second order owe their origin mainly to endogenetic or internal forces, especially diastrophism and volcanicity, although some plains and plateaus may also be formed by erosion. They are essentially constructional features and provide the basis for the development of the relief features of the third order.

(iii) **Relief features of the third order**— The lesser features which develop on the second order relief features are included in the relief features of the third order. Valleys, canyons, waterfalls, beaches, eskers, drumlins, sand-dunes and such other smaller land forms are examples of third order relief features. They have been essentially carved by external or exogenetic forces through the process of erosion. They are therefore referred to as destructional landforms. The agencies of denudation such as weathering, rivers, winds, glaciers, waves etc. act rather slowly and it has taken them thousands or millions of years to carve out the lesser features found on the surface of the earth today. A study of the relief features of the third order occupies a lot of space in geomorphology and the degradational as well as the aggradational features produced by these agencies and processes are analysed in great detail.

Development of Modern Geomorphology[2]

Modern Geomorphology is said to have its beginning in the

1. Bloom, A.L.—op. cit, p.14.
2. This section is based on Thornbury, W.D.-op.cit pp 4-13 See Also P. Chorley, R.J., Dunne, A.J. and Beckinsale, R.P.-*The History of the Study of Landforms*, John Wiley, 1964, New York

works of James Hutton (1726-1797) and he is considered to be the originator of modern geomorphic ideas. But before him, between the 15th to 18th centuries a number of scholars had put forward ideas which were closely akin to modern geomorphology. In this connection, we may first mention the name of Leonardo da Vinci (1452-1519). He pointed out that the rivers themselves carve their own valleys through erosion and that the rivers obtained their debris by erosion of one part of the land surface, transported them and then deposited them in another area. Buffoon, a French scholar (1707-1788) stated that the rivers were a powerful agent of erosion, and their capacity to erode was so great that they could destroy land surface and ultimately reduce them to sea-level by prolonged erosion. He also criticized the religious superstitious beliefs of the time and opined that the age of the earth could not be measured in thousands of years. Targioni-Tozetti of Italy (1712-1784) also recognized the erosive power of streams. He was of the view that the irregularities in the courses of rivers were related to the differences in the nature of rocks. Where the rivers pass though an area of relatively soft rocks, their valleys are wide, but in areas of hard and resistant rocks the river valleys are narrow. Thus he was the first to suggest the concept of differential erosion resulting from differences in structure and rocks.

Geotthard of France(1715-1786) has referred to the lowering of mountains as a result of river erosion. He has also pointed out that all the debris derived from river erosion are not deposited in the sea, but a large part of it is deposited along the banks of rivers giving rise to flood-plains. He, however, thought that the ocean was a more powerful agent of erosion than the rivers, and in this connection he referred to the rapid erosion of the Chalk Cliffs of Northern France by the sea. He also tried to explain the formation of several hills in Central France from volcanic action.

Another French scholar, Desmarest (1725-1815) has shown with reference to specific examples that the valleys through which the rivers of Central France flow have been in fact carved by them. He was the first to try to understand the evolution of landscape in successive gradual stages.

The area of study of De Saussure of Switzerland (1740-1799) was the Alps. He laid great stress on the power of streams to erode the mountains and put forward the view that the river valleys in the Alps mountains are the creation of the rivers which flow through them. Besides erosion by rivers, he also acknowledged the erosive action of glaciers. He collected a number of field observations which was used later by Hutton to support his theory.

We thus find that the above scholars laid the foundations of geomorphology through their studies and views. James Hutton made full use of the materials and thoughts contained in their works in the formulation of his theory and has acknowledged the same. Although James Hutton is considered to be the father of modern geomorphology, the view of his contemporary M.U. Lomonosov (1711-1765) also deserve mention. Lomonosov is known as the founder of modern Russian geography and geology. He was the first to suggest that the landscapes are the result of interaction between internal and external forces and that these processes are universal in nature.

James Hutton (1726-1797)—James Hutton was born in Scotland in 1726. He was the first to realize the immensity and significance of geological time. In 1785 he presented a paper before the Royal Society of Edinburgh in which he propounded his epoch making "Theory of the Earth'. This theory is better known as the doctrine of uniformitarianism. It is based on the hypothesis that the present is the key of the past. In other words, with the help of the present day landscape and the processes operating on them today, it is possible to get an idea of the processes that operated in the past and the changes that they produced. The changes that have taken place on the surface of the earth in the remote past were not entirely different from the changes that are taking place today, and were, in fact, very similar. The physical laws which are active today in the evolution of landscape were also operative in the past. He was impressed by both the physical and chemical processes of erosion, and with the help of convincing examples showed that the hills and mountains of today were not permanent features and have been chiselled and fashioned by exactly those very slow erosion processes which are active today. Rivers erode the land, transport the alluvial sediments and in the end deposit them in the sea. It is through this process that the sedimentary rocks of the earth were formed in the past. He realized that the vast thickness of the older sedimentary rocks is proof of the fact that the processes of erosion and deposition have been in operation on the earth's surface for immeasurably long periods of time. He refers to both fluvial and marine erosion but lays more emphasis on erosion by rivers and the development of river valleys. In fact, some of the basic modern concepts about erosion are to be found in the theory of Hutton. It has therefore been rightly said that Hutton was much ahead of his times.

Hutton believed in the orderliness of nature and its obedience to certain laws, and was opposed to the doctrine of catastrophism—the myth of successive destructions of the face of the earth by violent and supernatural cataclysms, then prevalent among the Christian priests. He stated in clear terms that the past history of our globe must be explained by what can be seen to be happening now. Hutton's paper on "Theory of the Earth; or an Investigation of the Law observable in the Composition, Dissolution and Restoration of Land upon the Globe" was first published in the Transactions of the Royal Society of Edinburgh in 1788. Later, in 1795 it was published in the form of a book entitled "*Theory of the Earth with Proofs and Illustrations*", the theory being supported by new proofs and illustrations. Hutton's views received scant attention in his life time. After the death of Hutton, his friend John Playfair (1748-1819), a Professor of Mathematics, elucidated his view in clear and beautiful language and published them in a book entitled "*Illustrations of the Huttonian Theory of the Earth*" in 1802. This had the effect of popularizing his views and led to its much wider acceptance. Later Sir Charles Lyell called Hutton's theory Uniformitarianism and he became its great exponent and supporter.

After Hutton, the development of geomorphology took place in different directions in Europe and North America in the wake of new ideas and principles and it would be convenient to consider the development in the two continents separately.

Developments in Europe after Hutton

As has been pointed earlier, Sir Charles Lyell (1797-1875) fully supported the principle of unformitarianism and gave it a prominent place in the various editions of his book '*Principles of Geology*'. Through this book Hutton's views received wide publicity and acceptance. However, it appears from the eleventh (1872) and subsequent editions of Lyell's book that his faith in the erosive capacity of running water had weakened and he had started giving much greater importance to marine denudation.

The Concept of Ice Age— An important discovery of the European geologists in the 19th century was the existence of Pleistocene Ice Age in Northern Europe. They discovered enough evidence which showed that this entire area was covered by icesheet during the Pleistocene period. The main credit for the establishment of this idea is usually given to Louis Agassiz (1807-1873), but besides Agassiz there were several other scholars who had found evidences of glaciation in the different parts of Europe and they were of the view that the alpine galciers were more extensive in the past than they are today. In this connection the names of Playfair (1815), Venetz of Switzerland (1821,1829), Esmark of Norway (1824), Bernhardi of Germany (1832) and Jean de Charpentian of Switzerland (1834) deserve special mention. In fact Charpentian was mainly responsible for getting Agassiz interested in glaciation, and it was only after his travels with the former that Agassiz expressed his support for the view of Charpentian. In 1837 Agassiz read a paper before the Helvetic Society in which he propounded the theory of a great ice age. He published a summary of his views in 1840. Charpentian did not like that Agassiz should have published his paper before him. Charpentian too published his theory of continental glaciation in 1841. There may be some hesitation in calling Agassiz the originator of this theory, but there is no doubt that it was largely due to his untiring efforts that the people finally came to accept the truth of the great ice age. By 1870 the theory was fully established, and people had started talking in terms of several glacial periods. In 1894 James Geikie, the famous Scottish geologist published his book '*The Great Ice Age*' in which he divided the Pleistocene Ice Age into several glacial periods interrupted by relatively warmer interglacial periods. This was confirmed by Penck and Bruckner in 1909 by their investigations in the Alps which led to the identification of four glacial periods or stages separated by three interglacial periods in the Great Ice Age.

Marine Erosion— In the 19th century considerable work was also done in the area of marine erosion by the European geologists. Special mention may be made in this connection of Sir Andrew Ramsey (1814-91), Baron Ferdinand Von Richthofen (1833-1905) and Sir Charles Lyell (1797-1875). Ramsey described the plains built by marine erosion in Wales and S.W. England. Based on his travels in China, Richthofen emphasized the work of winds in marine erosion as well as in the formation of loess. Throughout his life Charles Lyell continued to emphasize the great importance of marine erosion, and it was largely due to the prestige that he enjoyed that people took a long time to realize that erosion by running water was actually far more important than erosion by the sea. Ramsey started giving far more prominence to water erosion than marine erosion in his later years, particularly after 1860.

River Erosion— British geologist, George Greenwood was a supporter of river erosion. In 1857 he put forward the view that the depth to which a river could cut its valley depended upon the location of resistant rocks in the valley bottom further downstream. Thus he laid the foundation of the concept of local or temporary base level of erosion in the course of a river. In 1862 Jukes in his paper on the rivers of Southern Ireland divided rivers into two important classes— (a) transverse streams which cut across geological structure, and (b) longitudinal streams, which flow parallel to the structure or flow along the zone of weaker rocks in the direction of the strike of the rocks. He held the view that the transverse streams develop earlier than the longitudinal streams, and the latter should be treated as subsequent streams. He also referred to the phenomenon of river piracy. These views are now accepted as some of the fundamental principles in the development of river systems.

Toward the latter part of the 19th century a number of books relating to geomorphology appeared. In 1869 Peschel published a compilation of the views regarding the evolution of landforms. In 1886 Richthofen too presented these principles in a more comprehensive form. In 1894 Arthur Penck published a genetic discussion of landscapes. It can therefore be safely stated that geomorphology had established itself as a branch of geology by the end of the 19th century in Europe, though the word geomorphology had not come into vogue and instead the word physiography was commonly used.

Developments in North America

In North America the maximum development of geomorphology took place between 1875 to 1900, that is, in the last quarter of the 19th century. This is known as the heroic age in

American geomorphology, as it was during this period that the most important geomorphological theories were propounded. The names of Major J.W. Powell (1834-1902), G.K. Gilbert (1843-1918) and C.E. Dutton (1841-1912) are particularly noteworthy, as these were the people who showed the way and it is basically on the foundation of the concepts formulated by them that later on W.M. Davis built his theory of the geomorphic cycle.

Major J.W. Powell (1834-1902) — Powell is considered to be the founder of the American school of geomorphology. On the basis of his observations and investigations in the Colorado plateau and Yuinta mountains in Western U.S.A. he put forward a number of very valuable concepts. He suggested that landforms should be classified on the basis of geological structure. He suggested two bases for the classification of river valleys— structure and origin. The first classification can be based on the relationship between the rock strata and the river valleys formed on them. On the basis of origin, he divided the river valleys into antecedent, consequent and superimposed. Powell was the first to recognize and propound the concept of the base level of erosion of rivers. He indicated that there is an ultimate limit below which the rivers cannot erode, and this limit is determined by the sea level. He put forward his views in the following words : "We may consider the level of the sea to be a grand base level, below which the dry lands cannot be eroded; but we may also have, for local and temporary purposes, other base levels of erosion, which are the levels of the beds of the principal streams which carry away the products of erosion... Where such a stream crosses a series of rocks in its course, some of which are hard, and others soft, the harder beds form a series of temporary dams, above which the corrasion of the channel through the softer beds is checked, and thus we may have a series of base levels of erosion, below which the rocks on either side of river, though exceedingly friable, cannot be degraded."

Powell also recognized that after reaching the base level of erosion the land surface would be reduced to a featureless flat plain. He did not make use of any terminology for this ultimate land surface. It was this kind of reduced land surface which was later named peneplain by Davis. In the words of Powell, " What I have called base level would, it fact, be an imaginary surface, inclining slightly in all its parts towards the lower end of the principal stream draining the area through which the level is supposed to extend or having the inclination of its parts varied in direction as determined by tributary streams". Thus Powell conceived that if the process of erosion continues uninterruptedly for a long period of time then ultimately the land surface would be reduced to a level plain only slightly above the sea level. Thus we owe the concept of peneplain to Powell, though he had not actually used this term.

After observing the unconfirmities in the rocks of the Grand Canyon of Colorado he declared that these unconfirmities were indicative of periods of land erosion in the geological past. He indicated the geomorphic differences between the erosional scarps and the fault scarps caused by displacement of rocks. He also showed familiarity with the concept of migration of river divides, although Gilbert has dwelt on its significance at great length.

G.K. Gilbert (1843-1918)— Gilbert has been called the first real geomorphologist of North America. He has made a detailed analysis of the processes of subaerial erosion and more especially of the changes in the river valleys in the course of erosion. He has laid stress on the importance of lateral erosion in the development of valleys. He was the first person to establish the relationship between the load of a river, its volume, velocity and gradient, based on quantitative studies. He tried to explain the erosional landforms developed in an area of intrusive igneous rocks through his study of the Henry mountains of Utah. He showed that the stream flowing on the steeper slope of an asymmetrical ridge performs more erosion than the stream flowing on the gentler slope, with the result that the water divide keeps on migrating from the more active stream towards the less active stream. He called this the law of unequal slopes.

C.E. Dutton (1841-1912)— The name of Dutton is usually associated with the theory of isostasy. He was the first person to use the term isostasy in 1889. But he is better known for his penetrating analysis of the different landforms. He furnished several evidences in support of a long period of erosion before the formation of the present canyons in the Colorado plateau, which he called the period of the great denudation. He furnished several examples from the Colorado plateau in support of the base level concept which were later used by Davis in the formulation of his theory of peneplanation.

W.M. Davis (1850-1934)— The development of American geomorphology bears a very strong imprint of William Morris Davis. If Hutton and Playfair are considered to be the founders of geomorphology, then Davis is undoubtedly the father of American geomorphology. Davis was a great definer, analyst, expositionist and systematizer. Before him the views of American geomorphologists were rather scattered. He pieced together the scientific views of Powell and Gilbert on Western U.S.A. and gave them the form of an organised and systematic science. Further, he introduced the genetic method in the study of landforms and thus gave a new direction and dynamism to geomorphology. No scholar till today has written as much on geomorphology as Davis has done. Between 1880 and 1934 he published nearly 400 research papers and books. His book entitled '*Geographical Essays*' was published in 1909, but his first book containing a systematic analysis of the principles of geomorphology was published in 1912 in the German language.

Davis propounded several new theories and concepts, but his best known and most powerful theory is the one relating to geomorphic cycle. According to this theory, in the evolution of landscape, the landforms pass through a logical order or sequence of development with the help of which it is possible to identify the stage of their development. This sequence was termed youth, maturity and old age by Davis. In the evolution of any landscape, together with the stage, geological structure and process are also extremely important. Thus according to him, landscape is a function of structure, process and stage, He showed that geographical cycle or geomorphic cycle is that interval of time during which an uplifted landmass is changed into a featureless lowlying plain under the influence of the processes of erosion. He used the term peneplain for the featureless surface developed at the end of the cycle of erosion. The cycle of erosion starts immediately after the uplift of a landmass, and after having passed through the stages of youth, maturity and old age, the uplifted landmass ultimately is reduced to the base level of erosion and changes into a peneplain.

Between 1920-30 W. Penck of Germany opposed the theory of the cycle of erosion propounded by Davis. In the evolution of landscape based on the work of running water, Davis had assumed that first of all there is a relatively rapid uplift of a landmass, and after that there is a long period of stillstand during which the geomorphic cycle is completed. Penck and his supporters were of the view that this is not a normal sequence. In fact, normally the uplift is slow at first and afterwards it is so rapid that it prevents the landscape from undergoing through those stages whose penultimate result is a featureless lowlying surface.

The Davisian concept was based on the work of running water which according to him was the process of normal erosion. The concept of normal erosion met with opposition both in Germany and the Soviet Union. The scientists of these countries in their travels, had noticed marked differences in the landscape of the different climatic regions and they were not quite convinced that the Davisian principle could be applied in all circumstances. Later on, Davis also talked of abnormal erosion and besides the normal cycle of erosion based on the work of running water, put forward the concepts of glacial and arid erosion cycles.

No satisfactory alternative to the geomorphic cycle of Davis has been proposed so far. None the less, it would be true to say that not many people are inclined to agree with the concept of stage proposed by Davis and are not quite prepared to accept it in that form. They believe in the orderly development of landforms and also accept the importance of structure and process in their evolution, but they are sceptical about the concepts of cycle and stage. Inspite of the objections which have been raised against the views of Davis, we have to admit that there is no other person whose stamp on geomorphology is more than that of Davis. Simultaneously, it is important to remember that the development of the science of geomorphology is not dependent upon the works and views of only a handful of great personalities, but innumerable workers working in relatively small areas have contributed their share to this development.

Recent Trends in Geomorphology

Till about the early 1950's the Davisian cycle formed a paradigm within which geomorphological problems could be organised. After that follows a period of persistent questioning of the Davisian model and a gradual change from the temporal mode of explanation and the inductive scientific approach to process-response systems approach based on the use of the deductive scientific method. These and other related recent developments and trends are discussed below:

(i) ***Increasing co-operation between geographers and geologists***— Davis considered geomorphology to be a part of geography. His argument was that geomorphology is basically concerned with the study of the present ; it should therefore be considered to be a part of geography and not of geology which is concerned with the study of the past. On the other hand, during the past few decades the development in geomorphology has been more geological than geographical. This has happened primarily due to two factors: (a) the geomorphic studies have been found to be useful for the different branches of geology and there has been increasing application of geomorphology to the study of engineering geology, ground water geology and soil science, and (b) American geographers have taken comparatively less interest in the study of physical geography and have been emphasizing the human aspects of geography in their studies. In Europe, and more particularly in Britain, Germany, Russia and Poland the trend has been just the opposite, and geomorphology has been increasingly looked upon as an important branch of geography and several geographers have made notable contributions to the development of the subject. It has to be remembered that this is an area where the co-operation of both geographers and geologists is both desirable and fruitful and there is need for maintaining a proper balance between the geological and the geographical aspects of geomorphology. Fortunately, there is increasing evidence of this realization in the more recent works.

(ii) ***Growth of Structural Geomorphology***— An important development has been that modern approaches to the study of geomorphology have become much more broad-based. Apart from the physical laws governing the operation of the agents of denudation, the materials upon which these agents act have come to be more closely studied in terms of their physical properties and relations. The creation of new crustal structures by volcanicity and tectonic activity, crustal behaviour and deformation have increasingly become of vital concern and interest to geomorphologists. This field of study has come to be termed as structural geomorphology. With the concept of sea-floor speading and plate tectonics physiography of the ocean basins has currently become an integral part of geomorphology and has been vigorously pursued. As a consequence the study of geomorphology has come to have a global as well as a regional perspective.

(iii) ***Increasing criticism of the Davisian model***—The last forty years have witnesed increasing criticism of the views of W.M. Davis, especially those relating to his concept of cycle of erosion and its attendant concern with denudation chronology. In arid climates the ultimate landscape resulting from the action of water is considered to be pediplain and not peneplain as suggested by Davis. It cannot be said that the concept of pediplain proposed by L.C. King has been accepted as an alternative concept, but it has been given an important position in the evolution of landscape in the arid regions. Further, people are not inclined to accept the view that there are different periods of land uplift and land erosion. On the other hand, there is a continuous interaction between the tectonic movements and the agents of erosion so that the landscapes are in a continuous process of evolution. The mountain building movements are not sudden and short-lived episodes but prolonged manifestation staggered throughout geological time. According to Tricart," if this is so, the very foundations of the cycle of erosion crumble. There is no longer tectogensis followed by morphogenesis, but continuous interaction between tectonic movements and the agencies of erosion"[1]

People are not prepared to accept the cyclic concept in the evolution of landscape though they agree that there is evolution in which there are oscillations but there is no return to the initital stage. "If we withdraw the cyclic element involved in stage and look upon it simply as step in an evolution which does not lead back to the starting point, and if we add to this evolution, and also to the process a morphoclimatic element, then the trilogy is still a worthwhile concept.... Davis's mistake was to regard only the stage as cyclic; he completely ignored the possibility of changes in *process*"[2] The division of landscapes into young, mature and old at once prejudices the explanation of the landscape since it is

1. Tricart, J.—*Structural Geomorphology*, Longman, English Translation, 1974, p.19.
2. Ibid.

assumed that a continuous process of erosion was involved with each part of the landscape passing through each stage of development. In fact time itself is seen as the explanatory variable, but this could be of value only if the process producing change was known and the mechanism producing change acted continuously to allow the landscape to pass through the necessary sequence of stage. Within the Davisian model the processes operating and the changes in them were not understood in detail, and it often proved difficult to locate landscapes into an exact stage. It has gradually come to be realised that what was important was not that change was occuring but rather what were the processes inducing change, at what rate and with what effect on the form and evolution of the landscape concerned.[1]

(iv) *Growth of Climatic geomorphology*—Another important development has been the growth of climatic geomorphology according to which the landscapes which develop in the different climates are different, and the processes of erosion differ significantly in the different climatic regions. Climate exerts its influence also through the natural vegetation because every region has its own distinctive natural vegetation. Based on this the concept of morpho-climatic systems has evolved which determine the processes of landform evolution and give rise to a distinctive ensemble of landforms.

The principal proponents of climatic geomorphology are Peltier, Birot, Budel, and Tricart and Cailleux. Peltier (1950) has identified nine morphogenetic regions—glacial, periglacial, boreal, maritime, selva, moderate or the normal of Davis, savanna, semiarid and arid.[2] These regions are based on the assumption that a specific climate resulted in distinctive geomorphic processes resulting in a climate-process system in each morphogenetic region. P. Birot (1960) has discussed the cycle of erosion in different climates. Their general contention is that the normal cycle is so oversimplified a concept and for existing conditions so unreal and abnormal that present landforms "must be interpreted as a product of complex evolution, as a mixture of newly developing agents and of survivals in course of removal at varying rates."

1. Goudie, Andrew et al (ed)—*Geomorphological Techniques*, edited for the British Geomorphological Group, George Allen and Unwin, London , 1981, p.3
2. Peltier, L.C.—The geographic cycle in periglacial regions as it is related to climatic geomorphology. *Ann. Assoc. Am. Geog.* 40, 1950, pp 214-36

French geomorphologists J. Tricart and A. Cailleux in their book *Introduction to Climatic Geomorphology*, have put forward a strong plea for climatic geomorphology.[1] Their view is that relief is in part dependent on climate and that in addition to their direct action climatic influences on landforms work principally through vegetation and soils. They also propose a division of the world into major morpho-climatic zones and state that morpho-climatic adjustment or equilibrium is reached in a given region when the land forms are predominantly determined by a morphogenic system that is dependent on climatic factors.

J. Budel of Germany is more emphatic in his views and considers climate to be more important than structure in geomorphic evolution. This is so, because according to Budel every climatic zone has its own specific complex of dynamic processes and the significance of the structural features of the earth's crust changes in relation to these dynamic processes. In different climates, for example, the morphological hardness of rocks differs. Limestone is morphologically a soft rock in the permafrost area of polar regions, but in other regions it offers extreme resistance to surface degradation as rain water is carried into subterranean channels along joints and there is little surface run-off. Granite in the tropics under conditions of deep chemical weathering, is a soft rock, but it is extremely hard in most non-tropical areas. Everywhere climate determines the mode of exogenic processes which actively shape relief. In contrast the structural features represent varying passive phenomena. Further, according to Budel while the structural features are haphazard in their distribution over the earth, climatic variation is systematic and organized in the great climatic zones over the entire earth. Thus taking into account both the total complex of current exogenic processes based on climate and the distinctive landforms produced in every climatic zone by these processes, he divides the earth's surface into five climato-morphological zones[2]:—

1. Zone of glaciated areas ie polar regions and high mountain ranges.

---

1. Tricart J. and Cailleux A.—*Introduction to Climatic Geomorphology*, Longman, London, 1972 (Translated by Jonge)
2. Budel, J.—Geomorphology in the *Encyclopedia of Geomorphology*, edited by R.W. Fairbridge, Reinhold, 1968.

2. Zone of pronounced valley formation ie parts of sub-polar regions at present ice-free, mostly with permafrost.

3. Extra-tropical (ie temperate zone of valley formation). This includes most of the mid-latitude regions and is today characterized by very moderately active processes and usually strongly marked fossil features of cold glacial periods.

4. Sub-tropical zone of pediment and valley formation ie transition zone between 3 and 5., internally highly differentiated.

5. Tropical zone of planation surface formation ie the humid, particularly the seasonally humid tropical regions where deep chemical weathering and surface sheet-wash are both important.

According to Budel complete adaptation of the entire relief to the characteristic complex of exogenic processes in any morphological zone takes some ten million years. During this period climatic changes may occur so that a new complex of morphogentic processes become established leading to a different relief peculiar to itself. The relicts of the former landscape, however, persist for a long time. In the mid-latitudes several such changes have taken place during last 10-12 million years. Consequently we find many ancient relief features owing their existence to earlier process complexes under different climates in addition to the modern features resulting from the morphogenetic processes of today. Thus we observe several landform generations corresponding to the several climatic changes. To identify them precisely is the aim of climatogenetic geomorphology which adds time as a factor. While the polar, mid-latitude and sub-tropical regions show a complexity of climatic relief generations, the humid tropics appear to be an exception where warm humid conditions have existed for many millions of years and sheet-wash processes have dominated the picture with little variation from the late Tertiary (Pliocene) through the Quaternary period to the present time. Further, a large part of the humid tropics are characterized not only by climatic stability but also by structural stability. It is therefore argued that given relatively stable structure and climate, the geomorphological processes must have operated long enough and over a wide enough area to give rise to distinctive land forms in parts of humid tropics.

Thus climatic geomorphology postulates that different climates, by influencing processes, develop distintive assemblage of landforms, and climate which controls all dynamic processes takes precedence over all other components including structural components. While the relationship between land forms and climate is significant, it has to be admitted that much more field empirical evidence is needed to substantiate the idea of distinctive climatogenetic landforms. In its present state much of the work would seem to suffer from overgeneralization and subjectivity. A major difficulty is that most landscapes are polygenetic in nature at the present time and show the influence of two or more climatic types. This complicates geomorphological analysis because present forms are not adjusted in many cases to the processes acting on them at present.

The supporters of climatic geomorphology are quite critical of the Davisian concepts, especially his concept of normal cycle as this is based on limited observations in temperate mid-latitudes. W.M. Davis, of course, realized the importance of climate in the shaping of glacial and arid landscapes. He, however, called the conditions in these zones as accidents which was not quite justified as they cover large areas especially when we include periglacial and semiarid climates where the climatogeomorphic conditions are very similar.

(v) *Growth of Applied Geomorphology*—Another newly developed branch of geomorphology is applied geomorphology in which the practical applications of geomorphology in day to day life are examined. Geomorphological knowledge is now being increasingly used in engineering projects, in the determination of mineral resources, in the construction of aerodromes, roads and dams, in flood-control and irrigation projects, in regional planning and development, in soil conservation and land use and in the establishment of human settlements. Thus a very close relationship is now being developed between the landforms and human activities, and this has added a new dimension and dynamism to the subject and has brought it closer to geography.

(vi) ***Rise of Quantitative Geomorphology***—Closely related to the growth of applied geomorphology is the emphasis on field observation and laboratory experiments and the rise of Quantitative Geomorphology with the help of which it is possible to measure and establish the geomorphic processes and laws with greater precision and accuracy. In some countries experimental laboratories have been set up in which the hydraulic laws of waves, currents and rivers are studied and determined, and several field-stations have been established to make detailed observation and study of the operation of the natural processes. So far as the use of quantitative methods in geomorphology is concerned, this is a common tendency observable in all scientific subjects. This is a desirable trend and has resulted in greater precision in geomorphic studies. Since the early fiftes geomorphology has been generally progressing towards a more scientific foundation through its concern with measurement and scientific analysis of data with special reference to practical problems and the adoption of statistical analysis as a means of testing hypotheses. Implicit in the quantitative approach is the use of the deductive scientific method.

Thus a theoretical approach to landscape form and development has replaced the explanation of unique landscape features which was dominant till about 1950 when the concept of Davisian cycle held sway. This new approach is dependent on the formulation of an idealised view or model of reality. Such models may then be tested either to verify that they are true reflection of the real world or if this is not the case, they can be revised and improved.

The new quantitative geomorphology has been very much concerned with understanding of the relevant geomorphic processes and the rates at which such processes operate. From knowledge gained about process activity, it has been possible to theorise on the landforms which should be produced by such processes. 'One major result of the process study has been the relegation of time to the position of a parameter to be measured rather than as a process in its own right. Another major result of the change in geomorphological emphasis has been a reduction in the spatial and temporal scales within which lanforms are now considered. The scale of landforms being examined has been reduced, partly because process and form are best related at small scales and partly because over the larger scales with which the denudational chronologists were concerned process operation is not invariate as Davis envisaged'.[1]

Field observation is extremely important in geomorphology but qualitative observation is not enough and even the recording of static features in the landscape is being increasingly done quantitatively. A word of caution is however necessary. While making use of mathematics and statistics, one has to ensure that the subject does not become intractable and obscure and geomorphology does not become subordinate and subservient to the mathematical techniques. Moreover, one has to constantly remember that the geomorphic characteristics of an area are influenced by extremely complex processes and are not always amenable to mathematical and statistical analysis. This should not be understood to mean that the application of statistical techniques should be discouraged in geomorphology but that there is need for great care and caution, and deep analytical understanding about the problems and errors involved in the use of statistical laws in geomorphology is called for.

Questions

1. Discuss critically the meaning and scope of Geomorphology.
2. Give a concise account of the development of Geomorphology in *either* Europe *or* North America.
3. Evaluate the contribution of any *two* of the following to the development of geomorphic thought:-

   Hutton, Davis, Powell, Agassiz.
4. Present a critical account of the scope of Geomorphology and its recent trends.
5. What are the main geomorphic concepts of Davis? What are the objections which have been advanced against them ?
6. Examine critically the recent trends in geomorphology with special reference to climatic and quantitative geomorphology.

---

1. Goudie, Andrew et al (ed)—*Geomorphological Techniques*, George Allen and Unwin, London, 1981,p.4

Chapter 2

Origin of the Earth

The Solar System

We know that the earth is a member of the solar system. Sun, the principal member of the system is located in the centre, and revolving round the sun are the nine planets of which our earth is one. These planets can be divided into two groups— Terrestrial or the Inner planets and the Outer planets. The *Inner Planets* are nearer the sun and include Mercury, Venus, Earth and Mars. All four are relatively small and they are located at relatively short distances with respect to one another as well as with respect to the sun. The remaining five planets constitute the *Outer Planets*—Jupiter, Saturn, Uranus, Neptune and Pluto. They are located not only far away from the sun but also with respect to one another. With the exception of Pluto, they are also very much larger in size than the inner planets. Pluto which is 5,870 million kms away from the earth and 7,300 million kms away from the sun, is almost equal in size to the earth. The remaining four outer planets are referred to as the *Great Planets* on account of their large size.

In addition to the above nine planets there are several minor planets called *Planetoids* or *Asteroids*. They exceed 40,000 in number and are mostly located between Mars and Jupiter. The diameter of the largest among the asteroids is only 768 kms, but the great majority are much smaller than this. These asteroids also revolve round the sun like the planets.

All the planets including the asteroids revolve round the sun in the same direction and the orbit of each is an ellipse. With the exception of pluto, the direction of rotation of the remaining planets is also the same. These facts seem to suggest that possibly all the planets of the solar system have originated in a similar way. With the exception of mercury, venus and pluto, the remaining six planets have satellites or *moons* which revolve round them. The two largest planets—jupiter and saturn have 15 and 23 moons respectively. These two planets are located in the middle and as we move away from them in either direction, the number of moons attached to the planets becomes less but the size of the moons increases relative to the size of the planets. Mars has two moons and the earth has one. Similarly uranus has five moons and neptune has two. The above fact is considered significant in understanding the origin of the system.

Besides the planets, the solar system also contains *comets*, which move round the sun but their orbit is extremely long and eccentric. Sometimes they come very close to the sun and at other times they are very far away from the sun. The orbital period of most comets is more than ten thousand years, but there are some, such as Halley's comet, whose orbital period is only 40 to 80 years. The comet has a broom like appearance. Its head called coma is extremely bright and the long broom shaped tail juts out of it. The comet is composed of extremely light materials such as dust and other gaseous materials, and the nucleus of the head is composed of solid matter and frozen gases.

Resembling the comets to some extent are the *meteors* which also exist in the solar system. These are infinitesimal particles of matter which travel at great speed through space and some times also enter the earth's atmosphere. When they enter the earth's atmosphere, they appear as shooting stars in the sky. When they are outside the earth's atmosphere, they are called *meteoroids*, but when they can be seen shining inside the earth's atmosphere they are referred to as *meteors*. Sometimes meteoric dust also falls in large quantities on the earth through the atmosphere. The meteorites and the meteoric dust give us an idea of the wandering cosmic matter or raw material of the universe. They also help us in understanding the internal structure of the earth.

Origin of the Earth and the Solar System

Man has been curious about the origin of the earth and the solar system from ancient times and various hypotheses have been put forward from time to time on the basis of imagination, speculation as well as observation. None of these, however, can be considered to be wholly satisfactory or widely acceptable. There appears to be, however, unanimity on one point regarding the origin of the solar system and that is that all the planets of the solar system appear to have originated broadly at the same time and in a similar way, as the directions of both their rotation and revolution round the sun are the same. But there are several other questions for which no satisfactory answers are available. Sun is composed mainly of hydrogen and helium, but the core and the mantle of the earth are composed mainly of ferrous and silicate minerals and lacking in hydrogen. There is predominance of methane, ammonia and hydrogen in the bigger planets but the smaller planets are all deficient in these. Why is there such a marked difference in the composition of the bigger and the smaller planets? Can we conclude from this that the planets have been formed by the separation of materials in fragments from the sun? Were the earth and the other smaller planets sometime in a wholly molten condition? Why is the speed of rotation and revolution of the planets so much faster in comparison to the sun's rotation? In the absence of any reliable evidence, no satisfactory answers are available to these highly exciting and interesting questions.

Based on the scientific studies and evidences, some conclusions have been arrived at in respect of the origin of the earth and the solar system about which there is not much difference of opinion. For instance, it is now generally believed that the stars of our and other galaxies came into existence about 7,000 million years ago and possibly the earth came into existence about 4,500 to 5,000 million years ago. We may divide the scientific views regarding the origin of the solar system broadly into two classes:

(a) *Monistic hypotheses or Parent hypotheses* in which the origin of the solar system is believed to have taken place from a single star or nebula. This process is called the evolutionary process of the origin of the solar system. The chief protagonists of this school of thought are Kant, Laplace, Lockyer, Hoyle, Weizsaker and Kuiper.

(b) *Dualistic hypotheses or Bi-Parental hypotheses* in which the planets of the solar system are believed to have originated as a result of the coming together of two stars. In this school of thought we may include the Planetesimal hypothesis of Chamberlin and Moulton, the Tidal Hypothesis of Jeffreys, the Binary Star theory of Russell and a few others.

The Gaseous Hypothesis of Kant

The German philosopher Immanuel Kant first proposed his

Gaseous Hypothesis in 1755 which had its basis in Newton's law of gravitation. According to Kant, in the beginning hard, cold and motionless particles of primordial matter lay scattered in space. Due to mutual gravitational attraction these particles started colliding and falling on one another. The collision generated heat as well as rotation. In course of time, through this process the original cold and motionless cloud of matter was converted into a hot and rotating nebula. With further increase in temperature this nebula began to rotate with such great speed that buldge started developing in its middle part as a result of centrifugal force. When the centrifugal force became still more intense, the central buldge separated from the nebula in the form of a ring which changed into a planet in course of time. Nine successive rings came out of the nebula in this manner and they ultimately formed nine planets. The remaining part of the original nebula formed the sun. Thus according to Kant the earth has originated by the aggregation and solidification of the materials which separated from the nebula in the form of a ring on account of centrifugal force. Through repetition of the same process circular rings also separated from the planets resulting in the formation of satellites. It is in this manner that, according to Kant, the entire solar system has originated.

Kant's hypothesis has been found to be against mathematical laws. Kant's assumption that the mutual collision of particles resulted in the generation of rotational movements in the nebula is against the principle of the conservation of angular momentum. The second assumption of Kant that the rotation of the nebula increased in speed with increase in its size also goes against the above law, because the speed decreases with increase in size, and when the speed increases, size will decrease. Thus the basic assumptions on which Kant built his hypothesis have been found to be untenable. The importance of his hypothesis, however, lies in the fact that this was the first hypothesis based on Newton's Law of gravitation and was the forerunner of the famous Nebular hypothesis proposed by the French mathematician Laplace in 1796.

Nebular Hypothesis of Laplace

As stated above, Laplace presented his views regarding the origin of the earth's in 1796. His views had great resemblance with those of Kant, but in all probability he had no knowledge about Kant's publication. His views were primarily influenced by two facts : (a) existence of nebula in the universe, and (b) the presence of a ring round the planet saturn. Laplace starts with the assumption that there existed in the universe a massive hot and rotating nebula. This nebula went on decreasing in size as a result of constant cooling. He makes no mention of how this nebula originated, but by his imagining the existence of a nebula, he saves himself from the criticism levelled against Kant's hypothesis about collision of primordial matter generating heat and movement.

As a result of the rapid rotational movement of the nebula, its outer part started losing heat by radiation. Cooling led to contraction of the nebula and decrease in its volume. The reduction in volume resulted in increase in its speed of rotation. The increased rotation led to increase of centrifugal force, until a stage was reached when there was an exact balance between the centrifugal force and the gravitational force at the equator of the rotating nebula. When this stage was reached the ring of material near the equator would have no weight and this part therefore started buldging out. This upper part could not stay with the still cooling and shrinking central nucleus and separated from the nebula in the form of a ring and started revolving round the nebula in the direction of the latter's rotation. By repetition of this process, a number of rings separated from the parent nebula and each ring aggregated to form a planet after condensation. Planet after planet were born in this manner, each forming a spheroidal mass of heated gas which later on solidified. Laplace accounted for the formation of the seven planets then known through this process, and further repetition of the same process led to the birth of the satellites of the planets. The central residual mass of the nebula remained as the sun.

Several modifications of the nebular hypothesis have been proposed, the most notable among these being the Meteoritic Hypothesis of Lockyer. Lockyer suggested that the original nebula was an aggregation of meteorites in which there was continuous supply of incandescent gas through constant collision. But many of the objections which have been raised against the theory of Laplace are equally valid against this modified version.

The hypothesis of Laplace had the merit of being simple, and with its help it was possible to explain the uniformity in the direction of rotation and revolution of all the planets and the satellites (with the exception of a few satellites). Further, his hypothesis could also explain why the orbits of the different planets were roughly circular and in the same plane. According to this hypothesis, the earth and the other planets were first in a gaseous state, and later on they passed through a molten stage and finally the solid crust was formed. The increase of temperature in the earth's interior as well as volcanicity were in favour of this hypothesis. In the opinion of many scholars, the earth appears to have passed through a phase of molten stage in its history.

For all these reasons, the hypothesis of Laplace continued to hold ground for more than a hundred years, but several objections started being raised against it by the beginning of the twentieth century and the theory came to be seriously doubted in the light of new scientific facts. The main objections which have been raised against this hypothesis may be stated as follows :—

(i) British physicist James Clerk Maxwell proved with the help of his investigations that if we assume the total mass of all the planets to be a ring around the sun, then this mass is so small that it was impossible for it to condense into planets under the influence of gravitational attraction.

(ii) The second important objection is related to the distribution of angular momentum in the solar system[1]. At the moment 98% of the total angular momentum of the solar system is present in the planets and only 2% is found in the sun. If the hypothesis of Laplace is correct then the sun should have several times its present angular momentum and its speed of rotation should be 200 times its present speed. Further it has been found that total angular momentum of the solar system which should be equal to the angular momentum of the original nebula, is not adequate enough for the disintegration and formation of rings by excess of rotation of the nebula.

(iii) It has not been pointed out by Laplace and his supporters as to how the gaseous rings collected into rounded planets. According to the law of dynamics it is more probable that the rings would break into smaller and roughly equal parts and several small planets would be formed in the same orbit.

(iv) It is more probable that on account of the small degree of cohesion between the particles of the nebula, the formation of the rings would

1. *Angular Momentum =mass x velocity x radius.*

be a continuous and not an intermittent process. The contraction of the nebula would also be continuous and not intermittent.

(v) If the rings were composed of extremely hot gases, then the chances are that they would escape from the gravitational field of the sun and thus the gaseous rings would be lost in space.

(vi) According to this hypothesis all the satellites should revolve in the direction of their parent planets, but the direction of revolution of some of the satellites of uranus, neptune, jupiter and saturn are retrograde to the normal direction of the revolution of the planets.

The Planetesimal Hypothesis of Chamberlin and Moulton

This hypothesis was proposed in 1905 by Professor T.C. Chamberlin and F.R.Moulton of the University of Chicago. Through this hypothesis they tried to explain not only the origin of the earth but also the origin of the atmosphere and the oceans. Contrary to the views then prevailing, they thought that the planets have not originated from a nebula the remnant of which is the sun, but that they originated as a result of the coming together of the sun and another star. Masses of gas erupt from the sun at periodic intervals which are called prominences. According to them, as another big star happened to pass close to the sun, the severe tidal distortion due to the approach of the star together with the eruptive tendency inherent in the sun, caused a disruption of the sun's mass. The tidal disruption led to the ejection of jets of solar matter to great distances, the passing star drawing them forward in the line of its own motion. The total mass of gaseous matter that was ejected from the sun was much greater than the total mass of the existing planets, but a large part of it fell back on to the sun's surface. The portion that remained out of the sun, condensed from the gaseous state into tiny solid particles called planetesimals. The planetesimals developed a nucleus round which the scattered tiny planetesimals collected, and thus through the aggregation of innumerable planetesimals around separate nuclei the different planets were formed. According to this hypothesis it is not necessary to assume that the earth was ever in a molten state. The earth has been formed as a result of the aggregation of planetesimals round a nucleus. As the planetesimals collected round the nucleus, the size of the earth went on increasing, first rapidly and then slowly, until it reached its present size and form. The heat in the earth's interior was the result of the mutual collision of the planetesimals and their aggregation during the above period of development. As a result of the increase of pressure in the earth's interior there was reorganization of the planetesimals and the interior came to have higher density.

Possibly there was no atmosphere in the early stages of the earth's formation. But as the size of the earth increased with the aggregation of the planetesimals, the atmospheric elements also started developing. The atmosphere may have originated in two ways. Firstly, as the planet forming planetesimals aggregated, water vapour, carbon dioxide, nitrogen and other gases became availabe and volcanic eruption supplied oxygen. Secondly, when the earth's size became sufficiently large, free atmospheric molecules were attracted towards the earth because of its gravitational attraction. Thus the atmospheric elements were drawn from both internal and external sources.

The source of the hydrosphere was also planetesmals, and very likely it began forming first below the surface and then slowly came up to the surface. Further, water was also obtained from the condensation of atmospheric vapour. This water started collecting in the depressions caused by volcanic eruptions. In the beginning the ocean existed in the form of several lakes and then it was extended gradually to form the oceans. The materials eroded from the land by leaching and weathering started collecting in the water and the landmass also became lighter. On the other hand, the areas covered by water were further depressed as a result of the weight of water. Through this process the area of the hydrosphere went on increasing and the depth of the oceans also increased.

In the course of evolution of the earth the pressure in its interior went on increasing resulting in increasing temperature. Volcanic activity started from the internal molten reservoirs and the first Archaean igneous rocks were formed. Chamberlin has envisaged three principal stages of the earth's evolution :-

(a) Period of Planetesimal accretion,

(b) Period of dominant vulcanism, and

(c) The actual geological period.

Although the basic assumptions of this hypothesis namely, the formation of planetesimals from the gaseous mass and the formation of the planets by planetesimal accretion have been considered important concepts, various objections have been raised against this hypothesis some of which are listed below:-

(i) The basis of this hypothesis is the passing star close to the sun at a sufficiently high speed. In order that matter may be drawn out from the sun, the passing star must have come within a distance of the sun not more than the diameters of the two stars themselves. This seems to be an extremely unlikely event. Further, their closing velocity must have been at least 3,000 miles per second. This figure is considered unreasonable because it exceeds the escape velocity of the galaxy.

(ii) The hypothesis has been built on a catastrophic and extraordinary event. It is unreasonable to base any concrete scientific hypothesis on such an event.

(iii) It the planets have been formed by planetesimal accretion, then they should possess much less angular momentum than they actually have.

(iv) According to Jeffreys the nucleus of the planet was incapable of attracting the atmospheric elements. Even if there were some gases and water in the nucleus, it was not possible for them to come out once they had aggregated with the planetesimals. Thus in his view the hypothesis presents a wrong picture of the origin of the atmosphere and is unable to solve the other problems relating to the origin of the earth.

The Tidal Hypothesis of James Jeans and Harold Jeffreys

The British astronomer Sir James Jeans propounded his tidal hypothesis of the origin of the solar system in 1919 and Sir Harold Jeffreys introduced some important modifications in it in 1926. The names of both Jeans and Jeffreys are, therefore, associated with the tidal hypothesis. The hypothesis does not take into account the possible eruptive tendencies in the sun. According to this hypothesis the solar system has been formed as a result of the coming together of the sun and another passing star. A huge star, many times bigger than the sun, happened to pass close to the sun. The gravitational attraction of the passing star had a powerful tidal effect on the sun and through tidal action a part of the sun was raised in the form of a long cigar-like gaseous filament, which ultimately broke loose and separated from the sun on account of the high gravitational attraction of the star. The outer fragments of this filament were lost in space, its inner part fell back upon the surface of the sun, and the middle part condensed and compacted into sections like the beads of a garland to form the nine planets and started revolving round the sun. The filament, according to Jeans, was unstable lengthwise and soon broke up into several sections, each forming a distinct aggregation of matter later to develop by cooling and contraction into a planet. Jeffreys suggested, however, that the aggregations may come into being shortly after the ejection of the corresponding solar material. In other words, the filament may not be strictly regarded as continuous but rather as a series of condensations whose masses and spatial distribution are shown by him to be roughly in accordance with the present position and characteristics

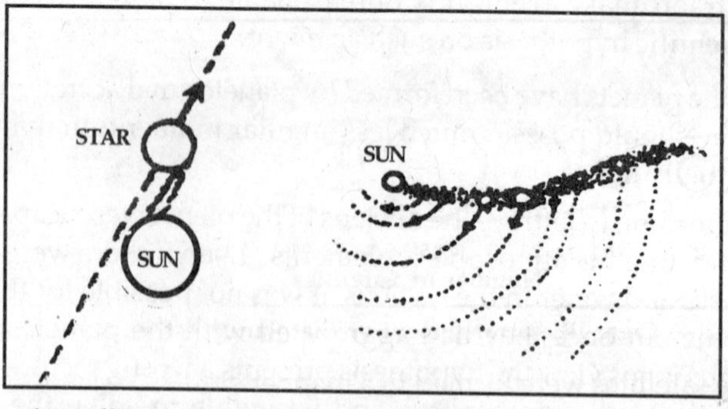

Fig. 1 — According to Jeans and Jeffreys, under the influence of a passing star a gaseous filament broke away from the sun, which later broke into several sections and developed into planets.

of the planets in the solar system. By a repetition of the same process, under the influence of the gravitational attraction of the sun and possibly also of the passing star, some materials further separated from the planets and condensed to form the satellites. Broadly speaking, the size of the planet determined the number of its satellites.

It is possible to explain a number of characteristics of the solar system with the help of this hypothesis:-

(i) The tidal filament ejected out of the sun should be thickest in the middle and should taper towards both the ends, as its one end would be influenced by the gravitational attraction of the sun while the other end will be under the influence of the gravitational attraction of the passing star. Consequently, when the planets are formed after the breaking up of the filament into sections, the larger planets will be formed in the middle and the smaller ones towards the two ends. The actual arrangement of the planets of the solar system is in conformity with this hypothesis. The two largest planets, jupiter and saturn, are situated in the middle and as we go away from them in either direction the size of the planets becomes smaller. Starting from the side of the sun, the first planet is mercury which is the smallest and after that we have venus and earth which get bigger in that order. Beyond the earth is mars which, however is smaller than the earth. But broadly speaking, the hypothesis is able to explain the sequences in the size of the planets. Pluto was discovered after this hypothesis was propounded, but the size of pluto is also in conformity with the hypothesis.

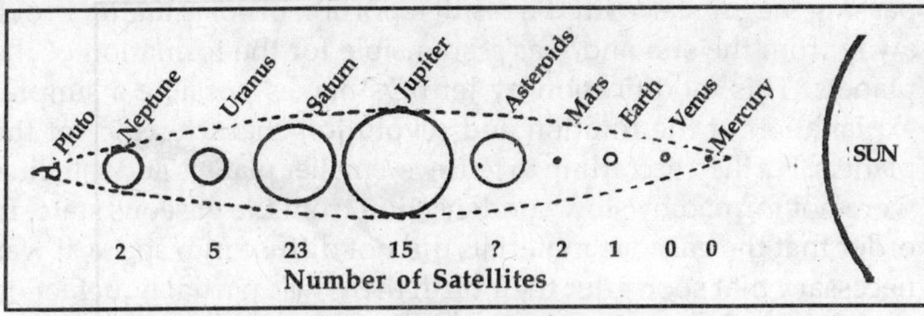

Fig. 2 — Cigar-like arrangement of the planets in the solar system

(ii) The satellites were formed in a manner similar to the formation of the planets by ejection of tidal material from the sun. Consequently we find a similar sequence in size in the case of satellites as we find in the planets. In other words, if one planet has several satellites, then the biggest satellite is located in the middle and the smaller ones are towards the edges.

(iii) According to this hypothesis the bigger planets remained in the gaseous state for a longer period because they took a longer time to cool on account of their larger size. This explains why more satellites were formed from the bigger planets, although the size of the satellites was relatively small in comparison to the parent planet. The middle sized planets gave rise to fewer satellites but the size of these satellite was relatively large in comparison to the parent planet. The planets at the far ends cooled quickly because of their small size and therefore did not give rise to any satellite.

This sequence is maintained in the satellite of the different planets. Jupiter and saturn, the two largest planets have several small satellites, while small planets like mercury and venus have no satellites. Thus the hypothesis is able to explain not only the size and arrangement of planets but also the size and number of the satellites.

(iv) According to this hypothesis, the orbits of the planets should not synchronize with the orbit of the sun, but should be inclined at a definite angle on account of the gravitational attraction of the sun. In fact, the orbits of all the planets are actually inclined at different angles. The declination of the earth is 23.5^0, while that of mercury is 7^0.

Modification of the Hypothesis by Jeffreys

Jeffreys postulated an actual collision of the sun with the passing bigger star with the result a portion of solar matter broke away from the sun and was responsible for the formation of the planets. This modification by Jeffreys makes possible a simpler explanation of the rotation and revolution characteristics of the planets. Further, according to Jeffreys, smaller planets and satellites were not formed by slow condensation from the gaseous state. In order that the gaseous materials did not diffuse into space, it was necessary that soon after their birth there was partial liquefaction or solidification so that they could hold together. In the case of these bodies a liquid core must immediately have been formed by cooling partly by expansion of the gases and partly by radiation. The earth is supposed to have cooled down until it was completely liquid and thereafter to have solidified through loss of heat by radiation. In the process of solidification its materials became zoned in the order of their densities. In the opinion of Jeffreys all these changes were accomplished within relatively short period of time. When the earth was in a liquid state, the heat supplied by convection to its surface kept pace with loss by radiation, but after formation of the crust, heat could be conveyed to the surface only by the slow process of conduction, with the result that the temperature at the surface would fall rapidly. Whatever vapour was there in the primitive atmosphere would condense after the formation of the solid crust and the normal process of erosion would have been thus initiated.

Criticism of the hypothesis of Jeans and Jeffreys

The tidal hypothesis of Jeans and Jeffreys enjoyed wide acceptance for some time, but more recently several objections have been raised against it:-

(i) Levin (1958) put forward the view that the stars lie at such vast distances from each other in the space that the possibility of their coming close together or their mutual encounter can easily be ruled out.

(ii) Where did the big star which was responsible for tidal ejection of solar material go ? No satisfactory answer is available to this question.

(iii) The gaseous matter which would be ejected from the sun would be so hot and there would be so much momentum in the gaseous particles that they would soon be lost in space.

(iv) If the planets have been formed from solar matter, then the sun and the different planets should all be composed of similar materials. But in fact, this is not so. There is predominance of lighter elements (hydrogen and helium) in the sun, while the planets are composed mostly of heavier elements like iron, silicon, aluminium etc.

(v) N.N. Parisky has tried to show by mathematical calculations that this hypothesis is unable to explain the distance between the sun and the different planets. American astronomer Russell has also pointed out that on the basis of this hypothesis the planets should be much closer to the sun but in fact they are much farther apart. He has great doubts if the matter ejected from the sun could at all spread over such great distances and result in the formation of the planets.

The Binary Star Hypothesis of H. N. Russell

The principal weakness of the hypotheses that we have considered so far, is that they are unable to provide a satisfactory explanation for the great distance of the planets from the sun and their circular orbits round the sun. This difficulty can be removed if it is assumed that the materials which formed the planets were far removed from the sun from the very beginning,

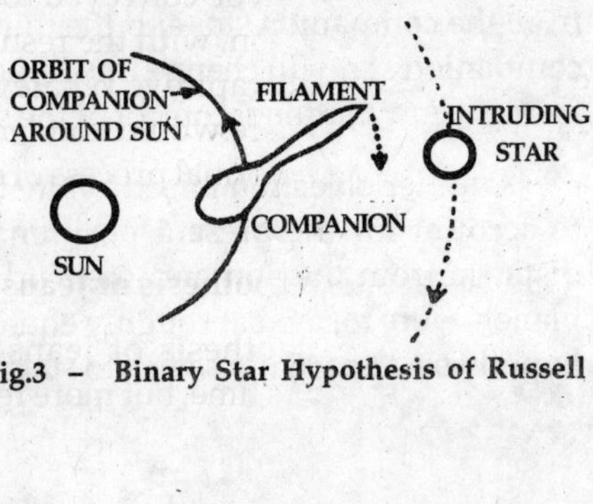

Fig.3 – Binary Star Hypothesis of Russell

or in other words, if it is assumed that the planets have not been born from the sun. Accordingly, H.N. Russell suggested that there was another companion star of the sun and the two together formed a binary star or a twin-star system. A third star happened to pass close to the companion star of the sun, and this resulted in the ejection of gaseous matter from the latter in the form of a filament which ultimately separated from it. The planets were formed from this gaseous filament of matter in course of time. In the beginning the planets were closer together and the satellites owe their birth to the mutual gravitational attraction between them. This third star was too far away from the sun to have any impact on the latter.

The suggestion that the primitive sun was a binary star cannot be dismissed as mere imagination, because at least 10% of the stars in the universe are binary stars. In fact in the opinion of some scholars the number of binary stars is probably 30% of the total. This hypothesis helps us to explain the great distances of the planets from the sun as well as their high angular momentum. But the greatest weakness of the hypothesis is its inability to account for the removal of the sun's companion from its control and for the retention of the tidal filament which is later supposed to condense into planets and revolve round the sun. R.A. Lyttleton (1936) has supported this hypothesis and has tried to show by mathematical calculations that this is possible under certain circumstances. If it is assumed that (a) the mass of the sun, the companion star as well as the third star was roughly equal, (b) the distance of the companion star from the sun was 1,700 million miles, (c) the speed of revolution of the companion star round the sun was 6 miles per second and the speed of the third star was at least 20 miles per second, and (d) the third star had come within a distance of approximately 3 to 4 million miles of the companion star. Under the above stated conditions, it is possible that tidal filaments will be ejected both from the companion star and the third star, the elliptic orbit of the companion star will change into hyperbole orbit and it will escape from the gravitational control of the sun.

Another objection raised against this hypothesis, is its failure to account for the present position of the planets, that is, their distance from the sun and their orbit, if it is assumed that the planets were formed at roughly equal distance from the sun as the hypothesis apparently seems to suggest.

The Fission Hypothesis of Ross Gun

The process by which binary stars may originate has been used by Ross Gun to explain the origin of the solar system. A twin-star or binary star system is created by the division of a star by fission when as a result of its cooling and contraction there is increase in the speed of its rotation. According to Ross Gun there was a chance encounter between a rapidly rotating star and another star when the former was in a state of fission. Consequently the tidal effect of the other star upon the disintegrating star was profound and the tidal material broke into several parts and started revolving around the first star (i.e. the sun) as planets.

The principal objection against this hypothesis is that it appears extremely improbable that the encounter between the two stars should take place exactly at the time when one was in a state of fission, as ordinarily the meeting of two stars is in itself a rare and extraordinary event.

The Cepheid Hypothesis of A.C. Banerji

Professor A.C. Banerji of India proposed in 1942 that the sun and the planets of the solar system have originated from a cepheid. Cepheid is a special kind of star which is 5 to 20 times bigger in size than the sun and in which there is a regular pulsation. According to Banerji another star came close to a cepheid. Due to the attraction of this star the pulsation of the cepheid increased and it became unstable resulting in the ejection of enormous amounts of tidal material from the cepheid which ultimately condensed to form the sun and its planets. The velocity of ejection was so great that the sun and its planets were freed from the gravitational control of the cepheid, and the revolution of the planets is related to the other star which in course of time receded away into space. This hypothesis is unable to explain a number of characteristics of the solar system, nor has an exposition of this hypothesis been presented in sufficient detail.

The Nova Hypothesis of F. Hoyle and R.A. Lyttleton

In 1939 British astronomer Hoyle put forward a new hypothesis according to which the planets of the solar system have originated from the explosion of a supernova. According to this hypothesis the sun had another companion star which was much bigger than the sun, and the two together constituted a binary system. There was a terrific explosion in the companion star. Such explosions do happen in space. It is estimated that in our galaxy i.e. the Milky way, about twenty stars suddenly explode with brilliant light and suddenly they become abnormally luminous. Such stars with short period abnormal luminosity are called nova, and when the explosion and light are extremely intense, they are called supernova. The explosion in the nova or supernova is possibly related to atomic reactions. When there is explosion, enormous amounts of gaseous matter are thrown out of the nova/ supernova and its expansion becomes so great that suddenly its luminosity increases by hundreds of thousands. When the ejected gases become cold, the luminosity of the star decreases and it becomes less bright.

In the opinion of Hoyle and Lyttleton the planets of the solar system have originated from the expanding gases obtained from the explosion of a supernova. The expansion of the exploded gases was more in the direction of the sun as a result of which this gaseous matter came within the gravitational attraction of the sun, and later on it condensed to form the planets and the satellites in course of time. As a reaction to the explosion the supernova recoiled and started receding away from the sun and ultimately separated. This hypothesis is able to explain the high angular momentum of the planets, but is unable to offer any satisfactory explanation for the rotational movement of the planets as also for the origin of the satellites.

The Electromagnetic Hypothesis of Hannes Alfven

Most scientists have taken the help of gravitational attraction or tidal force to explain the origin of the planets of the solar system. In 1942 Alfven tried to explain the origin of the solar system by reference to electromagnetism. Like the earth, there is a magnetic field surrounding the sun. According to Alfven, in the early days the magnetic attraction of the sun was many times more than its gravitational attraction. In his view, the sun's rotation was also faster than at present. On account of the gravitational attraction of the sun, the atoms of a nebula present in space were attracted towards the sun. These atoms became surrounded by extensive clouds of ionized atoms whose extent was roughly equal to the present extent of all the planets taken together. According to the laws of motion for charged particles in a magnetic field, the ionized atoms collected in the equatorial plane of the sun at a distance approximately equal to the present distance of jupiter and saturn from the sun, and started revolving round the sun. On account of the revolving atomic matter round it, the speed of rotation of the sun decreased. It is this atomic matter which in course of time has condensed to form the different planets. The magnetic field of the planets attracted other atoms which resulted in the formation of the satellites. This hypothesis has not been developed in all its details, and more particularly it fails to explain the origin of the inner planets, namely mercury, venus, earth and mars.

The Inter-Stellar Dust Hypothesis of Otto Schimdt

In 1943 the Soviet scientist Otto Schimdt proposed a new hypothesis for the origin of the solar system, basing himself on the earlier views of Kant and Laplace. Large quantities of gases and dust particles are found scattered in the universe. Schimdt is of the opinion that in the beginning the sun was able to attract some gases and dust particles on account of its gravitation. This sheet of cloud of dust and gases started revolving round the sun. In the initial stage the cloud particles of different sizes were revolving round the sun in an irregular and unorganized manner. But subsequently the heavier particles collected near the bottom of the cloud heap and the cloud cover took the form of a vast flat saucer or plate. Due to mutual collision the particles started consolidating by condensation. On consolidation they formed the embryos of the planets and later on took the form of asteroids. These asteroids started revolving round the sun within the dust disc in the direction of the latter. The asteroids collected and absorbed the scattered matter by gravitational attraction which increased their size and they turned into planets. Even after the formation of the planets, sufficient material remained in space in attenuated (immature) condition which started revolving round the planets and in due course condensed to form the satellites.

Schimdt has taken the help of mathematics and logic in presenting his hypothesis. (i) According to him the planets have originated not from the sun but from clouds of dust and gases. It is therefore not unnatural to find marked difference in the angular velocity of the sun and the planets. Further, the difference in the angular velocities of the planets themselves can be explained with reference to the differences in the angular velocities of the dust and gaseous particles at the time of their condensation. (ii) The inner planets of the solar system are composed of heavier elements (silica, iron, aluminium etc.) while there is predominance of lighter elements (hydrogen, helium, methen etc.) in the outer planets. According to Schimdt, when in the beginning the gas and dust particles collected round the sun in the form of a saucer, it was not possible for the sun's rays to go far after penetrating the saucer. In other words, the inner side of the saucer was hotter, and here the planets with heavier elements were formed. On the outer side of the saucer where the temperature was lower, the lighter gaseous materials condensed to form the planets on account of the cold, (iii) Schimdt has shown that the distance between the different planets is in accordance with statistical laws, because bodies of different size and speed will consolidate only at certain fixed distances in their revolution round the sun. (iv) The orbit of the planets of the solar system is circular. The reason for this is that the different particles after collision will attain their average motion. Having attained their average motion, the orbits of the planets would be similar and they will move in the same direction.

Although it is possible to solve a numebr of problems with the help of Schimdt's hypothesis, it has not been fully accepted. One major objection against this hypothesis which has been pointed out is that it is not able to explain how the sun came to attract the clouds of gas and dust in the first instance.

The Nebular -Cloud Hypothesis of Dr. Von Weizsacker

German physicist Karl Von Weizsacker presented a hypothesis in 1943 which very much resembles the hypothesis of Laplace described earlier. According to this hypothesis, the origin of solar system is due to the condensation of fine particles of dense interstellar clouds of gases and dust of a nebula which surrounded the sun. Weizsacker stated that it was possible to meet the objections raised against the hypothesis of Kant and Laplace on the basis of new scientific knowledge. We know that interstellar space is filled with mixtures of gases and minute dust particles, the chemical composition of which is the same as that of the sun or of the other stars. This interstellar matter is composed mostly of hydrogen and helium gases mixed with very small quantities of extremely fine dust particles. When the sun was formed by condensation of interstellar matter, a large portion of this interstellar matter remained as a giant rotating envelope or cloud round the sun. Alternatively, it is also possible that the sun entered into an extensive nebular cloud. This nebular cloud which was composed mainly of hydrogen, helium and dust particles of various elements started revolving round the sun like a ring, under the sun's influence. Large masses or lumps of matter were formed as a result of the mutual collision of the dust particles present in the nebular cloud, which ultimately became the planets. After the initiation of this process the large lumps or bodies will easily attract the smaller particles and the size of these bodies will increase by accretion. Weizsacker showed that the planets were formed in this manner in about 100 million years. So long as their size went on increasing by accretion of new matter, the planets were hot, but when the fall-out of interstellar dust particles stopped their process of growth also came to an end. Their upper layers cooled by radiation resulting in the formation of the solid crust.

In explaining the formation of the planets at varying distances from the sun, Weizsacker has compared the materials revolving round the sun with a garland in which there were several big and small circular eddies. The diameter of this nebular disc or disc like envelope surrounding the sun was approximately equal to the present diameter of the solar system and its thickness was about 200- 300 million miles. Through his hypothesis he has tried to give a satisfactory explanation of the varying distance of the different planets as also the formation of the different satellites, but his whole analysis is highly technical and complex.

The principal merit of this hypothesis lies in the fact that it contains a higher degree of probability and logic than the other hypotheses, as even at the present time we find nebular clouds in abundance in space. Further, if this mode of formation of the solar system is accepted, then there should be other planetary systems as well in the milky way, as the passage of a star through a nebula is an ordinary and normal event.

It may be pointed out that the hypotheses of Otto Schimdt and Weizsacker though different, contain remarkable similarity of views, and both these hypotheses have generated a new interest in the nebular hypothesis of Laplace proposed towards the end of the 18th century.

The Protoplanet Hypothesis of Gerald Kuiper

American astronomer Gerald Kuiper suggested certain

amendments to the hypothesis of Weizsacker and in so doing put forward his own protoplanet hypothesis in 1951. In his opinion, the condensation in the nebulous ring round the primordial sun took place earlier than the condensation of the sun. In other words, the planets were created earlier and the present form of energy producing and light giving sun came to develop later. There was no difference in the composition of the nebula and the primordial sun. Both consisted primarily of hydrogen and helium with admixture of not more than 1 to 2% of heavier elements. As the nebula contracted, it became less and less steller and divided into a number of separate clouds or protoplanets. In the case of each original planet, the solid particles collected in the centre round which there existed an extensive envelope of gases. The protoplanets were of different sizes but they were much bigger than the present planets.

According to this hypothesis, as the proto-planets contracted the satellites were also formed close to the planets through a similar process. The rotation of the original nebula was in an anti-clock wise direction. That is why this is the direction of revolution of all the planets. In the beginning the shape of the proto-planets was elongated on account of the tidal attraction of the sun and on the protoplanets their long axis kept pointed towards the sun. This made the direction of rotation the same as the direction of revolution. In other words their periods of rotation and revolution were once equal[1].

In the meantime the central mass was slowly condensing and developing into a star (the sun). Due to atomic reactions tremendous heat was generated in its interior parts and intense radiation started from the sun. When the sun became luminous, the solar wind of radiation and ejected particles drove away into space the remaining part of the nebula as well as the lighter gases of the proto-planets, and the remaining heavier parts of the proto-planets completely disappeared from the nearer planets, with the result that their mass and density became more. But only a part of the hydrogen-helium gases could be blown off the surface of jupiter, saturn and the outer planets. Thus their density is relatively lower and especially the two largest planets-Jupiter and Saturn have the lowest density. Gradually the size and mass of the proto-planets decreased due to

1. Ordway, R.J.-*Earth Science*, Van Nostrand, East-West ed., Indian Reprint, 1971, p. 438

contraction and their speed of rotation increased. As a result of decrease in their mass, their gravitational attraction also decreased and their satellites moved further away from them.

We thus find that at the present time the basic assumptions of Kant and Laplace are once again being accepted. It is now believed that the origin of the solar system is not the result of some special accident or catastrophe but is due to the systematic contraction and condensation of a solar nebula. "Since about 1943, there has been a swing back to theories of Laplacian type, in which the planets formed from a rotating gaseous nebula or dust cloud either around the sun and perhaps derived from it or else parental to the sun which formed during the condensation processes.....The objection to Laplacian theories that the sün's angular momentum is too small has now been overcome... In the original gas cloud any viscous in the gases would have slowed down the rotation at the interior of the nebula and accelerated the movement at the edge. Magnetic forces could have achieved the same result[1].

As opposed to Laplace, however, the original nebula is believed to be relatively cold. Famous physicist and Nobel-Laureate Harold C. Urey has supported the view of Kuiper on the basis of his own findings.[2]

Present Position and the Big Bang Theory

It is now widely accepted that the sun, planets and their satellites formed from a cloud of gas and dust, contracting under its own gravity. The original cloud had some degree of rotation. So as the centre contracted, the conservation of angular momentum compelled the rest of the cloud to form the flattened disc, the Solar Nebula, rotating in the same plane as the Sun. Nucleosynthesis had begun by the time this stage was completed and the sun (the central part of the disc) shone brightly. Material was then lost to interstellar space and condensation also occurred, producing the solid bodies of the solar system. As a result the density of the nebula decreased, heat radiated more easily and cooling occurred. The planets condensed and aggregated from the nebular material. The

1. Gass I.G., Smith, P.J. and Wilson, R.C.L.(editors)—*Understanding the Earth*, 2nd Edition, ELBS, 1973,p. 66
2. Urely, H.C.,—The origin of the Earth, *Scientific American*, off print 833, October 1952, Readings in Earth Science, Vol I, Indian Reprint, 1975, p.239-44.

accretion of the planets could have been either slow or rapid and was completed about 4.5 billion years ago. This is taken as the age of the solar system and the earth. Inclination of the orbits favours condensation for the satellites of the planets, though the retrograde orbit of Jupiter's largest satellite (Triton) suggests capture from somewhere in the outer solar system.

The solar system is a very small fraction of the universe. The universe contains millions of galaxies, each galaxy has thousands of millions of stars and each star has a number of planets around it. The universe is not fixed but expanding, as the galaxies are running away from one another at high speed as the frontiers of the universe go on expanding. Formerly the edge of the universe was believed to be 2 to 4 billion light years away but with the discovery of quassars (black dots) it is now believed to be 8 to 10 billion light years away. From the expanding universe and the outward velocities of galaxies, it has been conceived that at a finite point in time all matter in the universe was packed into a single gigantic dense primordial body or fireball from which in an apocalyptic explosion about 15 billion years ago, fragments were sent flying outward to form everything in the universe as we know it today - the galaxies, stars, planets and satellites etc. Thus everything in the universe emerged from a single point called singularity about 15 billion years ago. This is called the big bang theory according to which all matter in the universe was created by this explosion at a fixed time and subsequently developed into galaxies, stars, planets and satellites by an evolutionary process. This theory was postulated in 1950's to 1960's and has been validated by authentic evidences received from COBE (Cosmic Background Explorer). Space is full of clouds some of which is left over from the Big Bang and others given out by exploding stars. Cosmologists believe that after the Big Bang the universe cooled rapidly to allow atomic particles to form larger chunks of matter which eventually became stars, galaxies, and planets etc. But the cause of the explosion is not known nor is there any evidence of what preceded the Big Bang and the great mystery of creation still remains elusive.

Questions

1. Give a concise account of three principal hypotheses put forward to explain the origin of the solar system and the earth.
2. Discuss critically any three of the following :-

(i). Nebular Hypothesis of Laplace,
(ii). Tidal Hypothesis of Jeans and Jeffreys,
(iii). The Binary Star Hypothesis of Russell,
(iv). The inter-Stellar Dust Hypothesis of Otto Schimdt,
(v). The Nebular cloud Hypothesis of Kuiper,
(vi). The Big Bang Theory

3. 'Since about 1943, there has been a swing back to theories of Laplacian type". Discuss this statement with reference to the theories that have been put forward regarding the origin of the solar system.

Chapter 3

Geological History of the Earth

Age of the Earth

According to the current methods of radio-metric dating, the age of the earth is put at between 4.5 and 5 billion years. The great age of the earth is, however, a relatively recent discovery. James Hutton and others who believed in the principle of uniformitarianism knew that the earth was very old but they had no method of knowing its exact age. Several methods were devised which were in use till about the beginning of the 20th century. These were based on the rate of sedimentation, salinity of the oceans, age of fossils, the rate of cooling of the earth and the tidal force of the moon. Based on the law of superposition of sedimentary strata and the age of fossils, various estimates of the earth's age were made by different scientists which ranged from 3 million to 150 million years. Towards the end of the 19th century, basing himself on the quantity of salt that would be deposited in the ocean each year from the land and the total amount of salt present in the oceans, John Joly calculated the age of the earth at about 90 million years. In 1897 Lord Kelvin, a famous physicist, on the assumption that the earth was originally molten and had cooled to its present condition, put the age of the earth at not more than 100 million years, and possibly much less. Nuclear fusion and radioactivity had not then been discovered, and Kelvin believed that the sun's output of energy was of a conventional nature and could illuminante the earth only for a few tens of million years. He believed that the sun was much hotter in the past and would become much cooler in the future and therefore the earth could support organisms for a period of only 30 to 40 million years.

In 1904 Rutherford for the first time took the help of radioactive substances in determining the age of the earth. The discovery of radioactivity provided a reliable means of calculating the ages of rocks and minerals which contain radio-active isotopes (elements).

The method is called radiometric dating. Radiometric dating is reliable because a radio-active element decays to a daughter element (isotope) at a constant and known rate that is referred to as the half rate, the time required for one-half of the originally present (parent) element to decay to the daughter element. Knowing the decay rate, measurement of the ratios of the parent and daughter elements provides the age. Uranium and thorium are the two principal radio-active elements. In a fixed time they disintegrate and decay and change into lead. Similarly, potassium changes by decy into argon in a fixed period (Half-life 1.3 billion years). As stated earlier radio-active isotopes decay is constant and unaffected by any chemical or physical agents. Till about 30 years ago the age of the earth calculated on this basis was placed at about 3,000 million years. But with more precise scientific knowledge in recent years, it has been possible to make more accurate calculations. Account has also been taken of the relative percentage of lead isotopes in ancient rocks and meteorites, and the age of the earth is now placed at around 4,600 million years. It is believed that chances of error in this calculation are very small, and this age is therefore now widely accepted.

Geological History of the Earth

Radiometric dating has made possible the specific or absolute dating of rock units which represent various events in the earth's distant past. But prior to the development of radiometric dating, people depended on relative dating by taking into account the law of superposition according to which in an undisturbed sequence of sedimentary rocks, each bed is older than the one above it and younger than the one below. Where igneous intrusions or faults cut through other rocks, they are considered to be younger than the rocks they cut. Layers of rocks are said to be *conformable* when they are found to have been deposited without interruption. But where a complete sequence of rocks representing the entire period is not available and there are breaks in the rock record, they are called *unconfirmities*. Help was also taken of the fossils or remains of organisms preserved in the rocks from the geological past, by applying the principles of faunal succession and organic evolution. Fossils also provided indications of past environments. By adding up evidence from these diverse sources, it has been possible to build the geological calender or the geological history of the earth using methods of relative dating. It is only recently that absolute dates have been added to the calender.

The geological history of the earth covers such a vast span of time that in order to comprehend it in the correct perspective it has been subdivided into units of varying magnitude. The largest divisions of geological time are called *eras*. Three eras are recognised--the *Palaeozoic* (ancient life), the *Mesozoic* (middle life) and *Cenozoic* (recent life). As is clear from their names, they signify profound world wide changes in life forms. Each of the eras is subdivided into smaller time units known as *periods*. The Palaeozoic has seven periods, the Mesozoic three periods and the Cenozoic two periods. Each period is characterized by less profound changes in life forms. Each of the twelve periods is further sub-divided into *epochs*, and each epoch into *ages*. The terms used for the corresponding stratigraphical rock units of formations are group, system, series and stages respectively.

It will be observed that the Palaeozoic era begins some 600 million years ago. The enormously long period of more than 4000 million years prior to the Cambrian (the first period of the Palaeozoic era) is simply referred to as the Pre-cambrian with no further subdivision, as our knowledge about those early times is very limited. This is so, because prior to the Cambrian, very simple life forms such as algae, bacteria, fungi, worms and sponges have not left adequate fossil record. Further, the bulk of the Pre-cambrian rocks are highly complex metamorphic rocks which in many cases have been badly eroded or are covered by rocks of more recent date.

On the basis of the above divisions a standard geological time scale has been prepared which is accepted throughout the world and with reference to which the geological history of the earth can be studied. As the major units of the standard geological time-table were delineated by workers in Western Europe and Great Britain many of the terms used are named after European place names. But they are used in all countries for age determination, correlation and comparative studies.

Division of Geological Time into Eras

We have seen above that geological time has been divided into four major divisions called eras — Pre-Cambrian, Palaeozoic, Mesozoic and Cainozoic or Cenozoic. The oldest and by far the longest is Pre-Cambrian which covers 90% of geological time but our knowledge about this era is very meagre.

Our knowledge about the remaining three eras starting with about 600 million years ago is relatively more. The Palaeozoic era lasted some 345 million years and the principal form of life consisted of marine invertebrates, fishes and amphibians. This was followed by the Mesozoic which lasted for about 160 million years and at this time dinosaurs and other reptiles were the principal animals. The Cainozoic era is the most recent and includes the history of the earth during the past 65 million years when mammals and broad-leaved trees had become important. Cainozoic era is divided into two periods—Tertiary and Quaternary, but some scientists treat them as two different eras, as man appeared on the surface of the earth in the Quaternary period.

Pre-Cambrian Era

This is the oldest era of geological history. The duration of this era is from the beginning of the earth nearly 4.6 billion years or more till about 0.6 billion years ago. In other words, Pre-Cambrian alone compasses 90% of all geological time. The Pre-Cambrian rocks are mostly crystalline igneous and metamorphic rocks which are extremely complex. The oldest of the Pre-Cambrian rocks are said to belong to the Archaean period. These are wholly azoic and contain no trace of life whatsoever. The rocks are crystalline and there is predominance of gneisses and schists, and whatever sedimentary rocks occur are all thoroughly metamorphosed. The Peninsular plateau of India is composed for the most part of Archaean rocks which include the rocks of the Dharwar system which were originally sedimentaries but have been so thoroughly metamorphosed that they are indistinguishable from the basement gneisses and schists.

Just above the Archaean system are the Late Pre-Cambrian or Proterozoic rocks. These rocks are mostly sedimentary where one can find traces of the earliest known forms of life. The rocks of the Cuddapah system and the Vindhyan system in India belong to this period. In the Cuddapah system which consists of marine sedimentaries, there are no fossils. In the Vindhyan system as well, apart from stray algal plant remains and primitive brachiopods, no clear-cut evidence of the existence of life has been found.

The rocks of the Pre-Cambrian era cover extensive areas in all the continents and generally form areas of low or moderate relief on account of the denudation of ages. These are called continental shields. They are highly metamorphosed, folded and faulted, but since the Pre-Cambrian times they have remained largely unaffected by earth movements and constitute stable blocks. Examples of such stable blocks are the Canadian shield in North America, the Baltic shield in northern Europe, Angara shield in mid-Siberia, practically the whole of Africa which is the most extensive and the highest shield, the shield of Penninsular India, the Brazilian plateau in South America, and the plateaus of western Australia and the Antarctica.

In the later sedimentary rocks of the Pre-Cambrian era we find traces of algae, sponges and jelly fish, but no definite fossil remains have been found, as the fauna of those days had not developed shell or skeletal material. But it would not be possibly correct to conclude that no life had developed at that stage as no fossil evidence is available, because in the era immediately after this i.e. in the Palaeozoic era we come across abundance of fossils. This indicates that the process of the evolution of life had probably begun much earlier.

The Pre-Cambrian rocks are extremely rich in metallic materials, and almost all the important occurrences of iron ore, gold, copper manganese, uranium, chromium, lead, zinc, and mica in the world are found in the rocks of the Pre-Cambrian era. But due to the high degree of metamorphism that these rocks have undergone as well as absence of life and vegetation in those early days, they are poor in the fossil fuels like coal, petroleum and gas.

Palaeozoic era

The Palaeozoic era started some 550 million years ago and continued till about 250 million years ago. This era is divided into six periods of which the oldest is the Cambrian followed by the Ordovician, Silurian, Devonian, Carboniferous and the Permian.

By the beginning of the Palaeozoic era the continental shields had become stable and deposition had started taking place in the geosynclines along their margins. The processes of deposition and consequent sinking in the geosynclines continued till about the Silurian period, and during this time there was also folding and volcanic activity at intervals during the Ordovician and Silurian periods. But the mountain building activity intensified during the Devonian period, and there are several evidences of orogenesis during the Devonian and Permian periods. During the Upper Palaeozoic the extensive Hercynian mountain system, only isolated

The geological time-scale

Eras	*Periods and Systems*	*Time from the beginning of the period to the present (in millions of years)*	*Distinctive Life*
Cainozoic Kainos=recent zoe=life(recent life	**Quaternary**		
	Recent or Holocene } Epochs	0.01	Modern Man
	Pleistocene (Glacial) } Epochs	1.8-2.5	Stone-age man
	Tertiary		
	Pliocene	5.3	Variety of mammals inc.
	Miocene } Epochs	24	elephants ; flowering
	Oligocene } Epochs	34	plants and broad
	Ecocene } Epochs	55	leaved trees
	Paleocene } Epochs	65	
Mesozoic Mesos=Middle (Mediaeval life)	Cretaceous	142	Extinction of Dinosaurs
			Age of Reptiles
	Jurassic	205	(Dinosuars) and Birds.
	Triassic	248	Reptiles & dinosaurs appear, first corals.
Palaeozoic Palaios=ancient (arcient life.)	**Upper Palaeozoic**		
	Permian	290	Age of Amphibians
	Carboniferous	354	wide spread forest
	Devonian	417	Age of Fishes
	Lower Palaeozoic		
	Silurian	443	Age of Marine
	Ordovician	495	Invertebrates
	Cambrian	540-560	Abundant fossils first appear
Pre- cambrian	**Late Pre-Cambrian** or Proterozoic (earlier)	2500 ? (not defind)	Scanty remains of primitive invertebrates and earliest known forms of life.
	Earlier Pre-Cambrian or Archaeon (Primeaval)	4000 ? not defined	Almost lifeless.
	Eozoic (dawn)	4560 "	

Origin of the continental crust nearly 4600 million years ago.

relics of which are present to-day, was formed in Europe.

During this era the south pole was possibly somewhere in the South Atlantic Ocean and the north pole in the North Pacific Ocean, and the equator passed through South Australia, Arabian Peninsula, Middle Europe and eastern North America. We find coral limestones in the Devonian period in North America, Europe and South-east Asia and Australia which indicates a hot climate. On the other hand, in the Devonian and Permian periods there are evidences of relatively cold climates in South America, South Africa and India. In the Carboniferous and early Permian periods North America, South Africa and Europe were clothed with forests of fern which were responsible for formation of the huge coal deposits of these regions. There are evidences to show that the climate became arid or semi-arid in the Permian period, and during this period the famous salt deposits of North America and Europe (Germany) were deposited. But in India (as well as in other parts of the Gondwanaland) the climate became warmer after the Upper Carboniferous glaciation, and it is in the rocks of the Permian period that we find the rich coal deposits of Damuda series in India.

Remains of different organisms are found in the marine sediments from the beginning of Palaeozoic i.e. from the Cambrian period. First the invertebrates and then the vertebrates evolved. There is an abundance of fishes in the Silurian and Devonian periods; amphibians developed during the Carboniferous and Permian periods and the reptiles made their first appearance.

In the Palaeozoic sedimentary rocks and more especially in the Carboniferous rocks we find coal, petroleum and gas, and the Silurian and Permian rocks contain potash, sodium and other salts.

Mesozoic Era

The Mesozoic era is of a much shorter duration than the Palaeozoic and is divided into three periods—Triassic, Jurassic and Cretaceous. This era is best known as the age of dinosaurs and reptiles. It is during this era that birds make their appearance in Jurassic and the small mammals begin to appear in the Triassic. There was also development of coniferous and broad-leaved trees during this era which provided necessary food for the birds and animals.

In the Triassic sedimentation was proceeding in the geosynclines of North and South Americas. Jurassic was also a relatively quiet period, but there was extension of the geosynclines and because of the formation of geanticline in the Alpine geosyncline it was divided into two parts. The Tethys sea which had come into existence by the end of the Palaeozoic era, continued to exist till the Cretaceous without any significant deformation. But by the close of the Mesozoic the Alpine mountain system had come into existence. At about this time the Rockies on the western side of North America were also being formed.

The southern continent of Gondwanaland was in existence from the Upper Carboniferous to the Jurassic. The continental sedimentary deposits of the Gondwana system in India which are found in narrow linear basins belong to these periods. The sandstones and shales of the Panchet series of the Gondwana system belong to the Triassic and the sandstones and shales of the Jabalpur series belong to the Jurassic. The Gondwanaland broke up in the Cretaceous period, and connected with this rift large areas were flooded with basaltic lava from fissure eruption in South Africa and western part of Peninsular India. The Himalayan orogeny also started during the upper Cretaceous.

In the Mesozoic rocks are found iron ore in the Jurassic of Europe, coal, petroleum and gas in the Cretaceous of North America and diamond in the volcanic pipes of the Jurassic and Cretaceous of Africa.

Cainozoic Era

This is the latest era in the earth's geological history which began about 65 million years ago. It is normally divided into two periods—Tertiary and Quaternary. But there are some people who prefer to regard the Tertiary and Quaternary as two separate eras. The latter view, however, has not received wide acceptance and the traditional classification still holds good.

The Tertiary period is sub-divided into five epochs. These are, beginning from the oldest, Paleocene, Eocene, Oligocene, Miocene and Pliocene. Tertiary is the period when the old animals were replaced by mammals such as horses, cows, elephants, dogs, pigs, bears, monkeys, etc. There was also considerable development of deciduous trees, flowering plants and grasslands. Tertiary is also the period of mountain building earth movements. The Mesozoic geosynclines were folded and uplifted to form the extensive mountain ranges in the Tertiary. The Tethys sea disappeared from

Europe and Asia by the Pliocene. In India, the folding and uplift of the Himalayas in the eastern part of the Tethys sea began towards the end of the Eocene and reached its peak in the mid-Miocene. Between the upper Miocene of Tertiary and the lower Pleistocene of Quaternary the Siwalik series in the southern part of the Himalayas were formed. In Peninsular India the Deccan trap (lava) was formed on account of fissure eruption in the Creteceous and early Tertiary periods, which is nearly 5,000 ft. thick and covers even at the present time an area of about 2 lakh sq. miles

In this very period the island arcs on the western coast of the Pacific Ocean as also the Rockies in North America and the Andes mountain systems in South America were formed.

The Quaternary period started about two million years ago. It is subdivided into two epochs—Pleistocene and the recent (or Holocene). The Holocene epoch has begun only about ten thousand years ago. Man appeared on this planet only towards the beginning of Pleistocene but the modern man developed in the Holocene epoch. The Pleistocene epoch is also called the Great Ice age or the glacial period, because the northern and middle parts of North America as well as Eurasia were covered with extensive ice-sheet during this period, and large areas were subjected to glaciation. This situation prevailed till about twenty thousand years ago after which the ice started melting on account of increase in temperature, and it is only gradually that the present state has been reached.

The most important mineral of the Cainozoic rocks is petroleum and gas. The lignite of North America also belongs to this era. Metallic minerals are found in igneous intrusions associated with mountain building, the notable among these being tin in Bolivia, copper and silver in the Andes etc.

It is probably not more than 50 thousand years ago that man was born on this earth and the beginning of civilization may be placed at about 20 thousand years ago. But during this relatively short period man by his intelligence, mental ability and power of organization has greatly speeded up the pace of development.

Questions

1. Give a concise account of the Standard Geological Time Scale.
2. Give the probable age of the Earth and the methods by which this has been determined.

Chapter 4

Internal Structure of the Earth

Our knowledge of the internal structure of the earth is inevitably based upon indirect evidences, as it is not possible for man to observe the interior of the earth. The maximum depth of mines and the bore holes dug for petroleum is no more than 5 to 6 kilometers. This depth is obviously negligible in comparsion to the radius of the earth which is 6,371 kms. It is also possible to get some idea about the interior of the earth from the ancient igneous and metamorphic rocks which were initially formed 10 to 15 kms below the surface of the earth but are now exposed on the surface as a result of prolonged subaerial denudation. We also know that hot lava is ejected by volcanoes from the earth's interior but the source of this molten magma is probably not more than 30 to 60 kms deep.

Actually our knowledge about the earth's interior is based largely on the investigations carried out by geo-physicists, mathematicians, geologists and other scientists with the help of highly sophisticated instruments and using extremely complex methods. In this the contribution of geophysics and seismology is probably the greatest. Maximum help has been received from the study of earthquake waves which has provided us with extremely valuable information about the internal structure of the earth.

Thermal and Physical State of the Earth's Interior

Temperature— We all know that as we go down from the surface of the earth towards the earth's interior, the temperature goes on increasing. The average rate of increase of temperature with depth or thermal gradient is 1^0C for every 32 meters of descent, i.e. 16^0 F per 1,000 ft. or broadly speaking 100^0 F per mile. At this rate of increase the temperature at the depth of 48 kms would become so high that rocks and metals cannot remain in solid state, as the temperature should increase to somewhere between $2{,}200^0$ to $3{,}300^0$ F or $1{,}200^0$ to $2{,}000^0$ C. It would appear that the source of magma erupted from volcanoes probably lies at about this depth, as the temperature of the lava inside the craters has been usually found to be 2,200⁰ F which is suggestive of a depth of 30 miles or 48 kms for the source of the magma. If we, however, assume that the temperature goes on increasing at this very rate till the centre of the earth, then the temperature should increase to 400,000⁰ F at the depth of 4,000 miles, which is absurd. On the basis of the findings of several geo-physical studies, it is now generally believed that the rate of increase of temperature found in the earth's interior near the surface undergoes a sharp decrease at greater depth. In the upper layers of the earth an important source of heat is radioactive minerals like uranium and potassium. These minerals are more abundant up to a depth of about 100 kms below which they are found in much lesser quantities. Taking into account the high pressure in the interior, the presence of metals, the melting

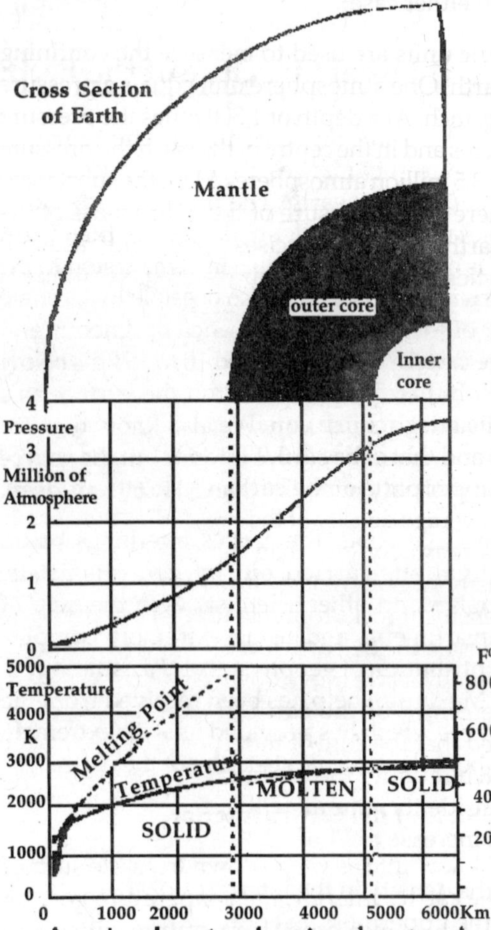

(After Strahler)

Fig 4 — Changes in temperature and pressure with depth in the earth's interior.

temperature of iron under conditions of high pressure and other relevant factors, it has been estimated that the temperature at the bottom of the mantle should be about 3,000° C (5,400°F) and that at the centre of the earth should be about 4,000°C (7,200°F) with variation of ± 1000°C in either case[1].

Pressure— Atmospheric units are used to measure the confining pressure inside the earth. One atmosphere unit equals a pressure of about 14.7 lbs. per sq. inch. At a depth of 1,500 miles the pressure is one million atmospheres and in the centre of the earth the pressure is estimated to be about 3.5 million atmospheres. Here the substances have no weight but there is the pressure of the entire mass of the earth. Pressure in the earth's interior exercises a powerful influence on the physical properties of rocks and their reactions to different forces.

Density— The average density of the earth is 5.5 grams per cubic centimeter or 5.5 gm/cc or just 5.5 . This means that the weight of the earth is 5.5 times the weight of an equal volume of water. The density of granite is 2.6 and that of basalt 3.0 These are the two most common rocks in the upper parts of the earth. In other words, we may say that while the average density of the earth is 5.5, the density of the surface rocks is normally less than 3. From this we may conclude that the density of materials in the core of the earth is much more than the density of rocks found in the uppermost layer, and probably the materials in the earth's core are 10 to 12 times heavier than water. Actually as we move down from the surface of the earth towards the centre of the earth, the density goes on increasing. It is believed that the core of the earth consists of iron. The density of iron is 8 but under the confining pressure of 2 to 3 million atmospheres inside its volume will decrease appreciably and its density should increase to 11 or 12.

It is believed that the density in the intermediate layers, lying between the core and the uppermost layer of the earth, should be some where between 4 and 6. The intermediate layer consists primarily of dunite rocks which are composed mainly of the minerals of olivine group. The density of dunite is 3.3, but under the influence of high pressure this may increase to between 4 to 6.

---

1. *The New Encyclopaedia Britanica, Macropacdia*, Vol 6, 1977, p. 50.

Some people previously believed that the materials in the earth's interior were not very different from the surface rocks and differed from the latter only in being denser and heavier on account of the high pressure above them. But scientific investigations have now proved that there is a limit beyond which the density of rocks and metals cannot be increased by pressure, and the density in the earth's interior is too high to be explained merely by pressure. In other words, the materials of the earth's interior are basically different in their composition from the surface rocks and the earth's core is composed of heavy materials. In fact there are evidences to show that the core of the earth is metallic in composition and there is predominance of heavy metals like iron and nickel. The main cause of the earth's magnetism is the presence of iron and nickel in its core in large quantities, as these two metals have very high magnetism. On this metallic core is a layer of heavy rocks called the mantle. And on the top of the mantle is the layer of ordinary rocks called the crust or the lithosphere.

It is clear and there is unanimity on this point that the earth is composed of several shells or layers the density of which goes on increasing with depth. Different scholars have tried to classify these layers in different ways and a knowledge of these may be useful as these names are still used in different contexts.

Earth's Internal Structure according to Suess

According to the Austrian geologist Suess (1835-1909), below the sedimentary rocks of the continents there are three principal layers— the upper layer, the intermediate layer and the central core which have been called by him sial, sima and nife respectively.

1. *The Upper Layer*— Just below the sedimentary rocks of the continents there is a layer of rocks which resemble granites and gneisses. The land-masses or the continents are composed

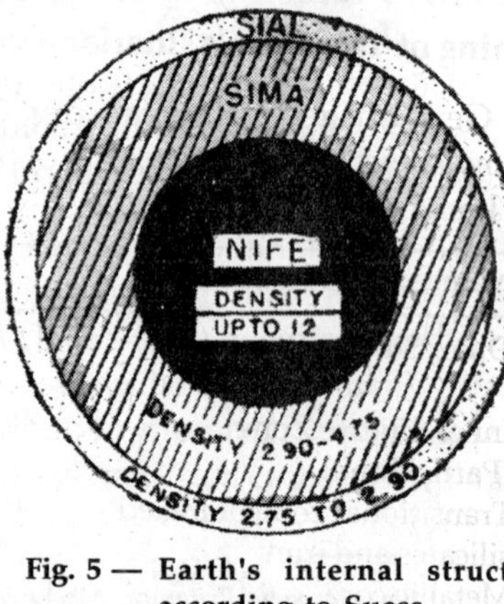

Fig. 5 — Earth's internal structure according to Suess.

of these rocks. Suess calls this layer by the name of Sial or Sali, as the two principal elements of the rocks of this layer are silicon and aluminium. The chemical symbol for silicon is Si and that for aluminium is Al. We get the word Sial by combining these two symbols. The specific gravity of this layer is 2.75 to 2.90. It has a predominance of acid rocks.

2. *The Intermediate Layer*— Below the Sial is a thick layer of heavy rocks. It has a predominance of heavy, basic igneous rocks such as, basalt and gabbro. This layer has been called Sima, as the principal elements in the rocks of this layer are silicon and magnesium. Sima is a combination of Si and Ma. The density of this layer is 2.90 to 4.75. In the opinion of Suess, sima is very much thicker than the sial.

3. *The Internal or the Central Core*— The third and the lowest layer is constituted by the core of the earth. It consists of dense and heavy materials with predominance of heavy metals like nickel and iron. (Fe). Suess therefore calls it Nife (Ni+Fe). Its average density is a little below12.

Fig. 6—Zoning of the earth's interior by Van Der Gracht.

Zoning of The Earth's Interior according to Van Der Gracht

German geologist Van der Gracht presented in 1928 the following scheme of the zoning of the earth's interior on the basis of the knowledge then available :

Name of the layer	Density	Thickness of the Layer
1. Outer Silicate Zone	2.75 to 2.90	60 kms (Below the continents)
2. Inner Silicate Zone (Partly Sima)	3.1 to 4.75	1,140 kms
3. Transitional zone of mixed silicates and iron	4.75 to 5	1,700 kms
4. Metallic core	11 or more	

This can be clearly understood with reference to Fig 6.

The Views of Holmes

Holmes has suggested that instead of Sial and Sima, it is preferable to use the terms crust and substratum for the upper and middle layers. He has given the following classification of these layers.

Sial	—	Upper Layer	Crust
Sima	—	Intermediate Layer	Crust
		Lower Layer	Substratum

The chief merit of Homles' classification is that it helps us in understanding the influence of the internal forces on the formation of surface landforms.

Seismology and the Earth's Interior

During the last few years there has been considerable increase in our knowledge about the constitution of the earth's interior. This has been mainly possible through the detailed studies of earthquake waves. Thus seismology has made a very significant contribution to our understanding of the internal structure of the earth. The different types of earthquake waves, their nature and speed have been given in the chapter on Earthquakes. Here we shall confine ourselves to a discussion of the various conclusions about the internal constitution of the earth derived from the study of earthquake waves.

Based on a detailed analysis of the differential speed of the different types of earthquake waves as they pass through the interior of the earth, it is possible to divide the earth into three concentric shells or layers called the core, the mantle and the crust. The earth's internal structure may be said to consist of concentric shells, each of which transmits seismic waves at a different speed. The boundaries between the different shells are marked by seismic discontinuities at which the speed of transmission of seismic waves suddenly changes. These seismic discontinuities are very distinct between the inner and outer core and between the outer core and the mantle, reflecting major changes in composition.

(i) Since the 1980's our knowledge about the interior of the earth has further improved with the rapid advances in equipment and computer technology. Each of the three shells has district physical

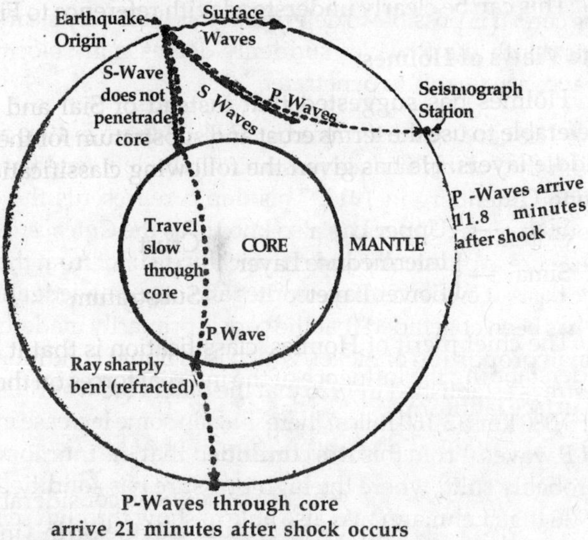

Fig.7 — Differential speed of the earthquake waves in the earth's interior.

and chemical properties. The outermost and the thinnest shell is the crust. Below the crust is the mantle which extends to a depth of 2891 km and is subdivided into the upper and the lower mantle. Finally at the centre of the earth is the core which is also subdivided into the outer core and the inner core. Their depth, volume, mass and density are given in the table below.[1]

	Depth (Km)	Volume % of Earth' total	Mass % of Earth' total	Density
Crust	0 - Moho	0.9	0.5	2.60 - 2.90
Upper mantle	Moho - 670	27.4	17.8	3.35 - 3.99
Lower mantle	670 - 2891	55.4	49.2	4.38 - 5.56
Outer core	2891 - 5150	15.6	30.8	9.90 - 12.16
Inner core	5750 - 6371	0.7	1.7	12.76 - 13.08

The Earth's Core— It has been found that at a depth of about 2,900 kms. (1,800 miles), the *S* earthquake waves which can pass through only solid objects, suddenly disappear. Further, at this depth the velocity of the *P* earthquake waves which can travel through solids, liquids as well as gases, abruptly decreases from 13.7 kms per second to 8 kms per second. This has been identified as the outer limit of

1. Hancock, Paul L and Skinner, Brian J. (eds) (2000). *The Oxford Companion to the Earth*, Oxford Univ. Press, p. 276

the core. It is possible to determine this limit very clearly because at this depth the *P* waves suddenly decrease in velocity and the *S* waves are unable to penetrate.

The core was discovered by R.D. Oldham in 1906 and its depth from the earth's surface was accurately delineated at 2900 km by Beno Gutenberg in 1912. This depth represents the core-mantle boundary or CMB and is also known as the *Gutenberg dicontinuity*. The core is physically and chemically distinct from the mantle. On the basis of evidence of meteorites and our knowledge about density it has been concluded that the core is primarily made of iron with a small proportion of nickel, which is in liquid condition. Sir Harold Jeffreys established in 1926 that the outer core was fluid. At a depth of 5,080 kms (3,160 miles) there is again some increase in the velocity of *P* waves. From this it is concluded that the interior of the core is probably solid where the high pressure has solidified the iron. In 1936 Inge Lehmann was also able to show through seismic studies that the inner core was solid. The outer core-inner core boundary is called the *Lehmann discontinuity* after her. Thus it is possible to subdivide the core into two parts—the inner core and the outer core. The radius of the inner core is approximately 1,250 kms (1,400 miles). In other words, the combined radius of the inner and the outer cores taken together is 3,500 kms or about 2,200 miles which is more than half of the radius of the earth. It is estimated that the maximum temperature of the core is 5,500°C (10,000°F) or slightly less and the specific gravity is 13. The outer core has sometimes been also referred to as the Shell. The inner and the outer cores taken together are equivalent to the nife of Suess. On the basis of seismic velocity and density structure coupled with laboratary experiments at high pressure and temperature, it is believed that the inner core may be almost pure iron, and the outer core an iron alloy with 10% of lighter elements such as oxygen, sulphur, nickel and silicon. That the outer core is liquid and the inner core solid is consistent with all seismological observations as well as studies on tides and the earth's rotation, both of which require a liquid core. The liquid outer core also acts as a giant self-exciting dynamo in which rising convection currents of liquid iron carry electric currents and interact with the earth's magnetic field. The liquid outer core is the source of the earth's magnetic field[1].

1. Flower, CMR-Earth structure in *'The Earth'* edited by P.L. Hancock & B.J. Skinner, 2000, Oxford University Press, p. 280

The Mantle— Above the core and below the crust there is a thick intermediate layer called the mantle. It extends from Moho to a depth of about 2,900 kms (1800 miles). This is solid, as both *P* and *S* waves pass through it quickly. The specific gravity varies from 3 to 3.5 in the upper part of the mantle to about 4.5 or more in the lower part of the mantle. The mantle is composed of dense and rigid rocks having predominance of minerals like magnesium and iron. These rocks are probably very similar to dunites and peridotites. It is possible that the mantle is subdivided into thinner layers each containing a distinctive group of minerals. But broadly speaking, we may subdivide it into two parts — the lower mantle or the *mesosphere* and the upper mantle the upper most part of which is called the *asthenosphere*. The lower part of the asthenosphere like the mesosphere, is solid, but the upper part of the asthenosphere is plastic and is in a partially molten condition. The velocity of the earthquake waves decreases in the asthenosphere and the asthenosphere is, therefore, referred to as the *Low Velocity Zone*. This low-velocity zone for S waves in the upper mantle extends up to about 220 km but the upper mantle extends from the base of the crust to a depth of 670 km. Beneath the asthenospehere the seismic velocity as well as density increase steadily down to 400 km depth and then again sharply at 670 km depth. The lower mantle continues from 670 km down to the core-mantle boundary ie about 2900 km. The velocity then increases

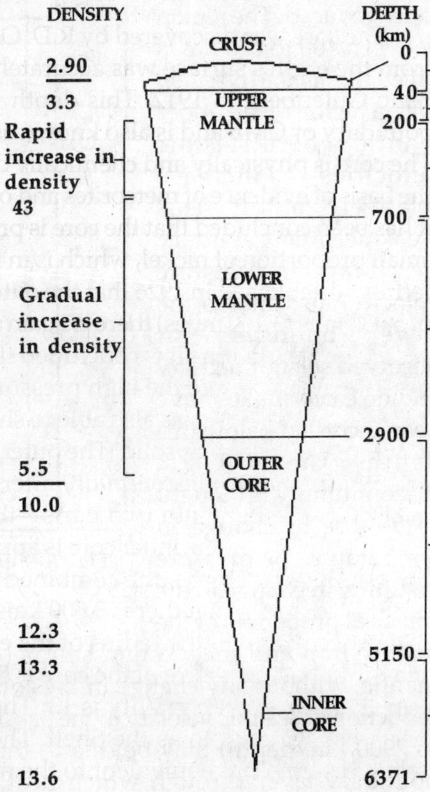

Fig. 8 — Division of the earth's interior into core, mantle and crust.

steadily through the inner mantle but in the basal 200 km of the mantle it is much reduced. The region between 400 to 670 km is called the *mantle transition zone,* a region of the upper mantle where seismic velocities increase relatively rapidly. This weak seismic boundary (ie jump in seismic velocity) also marks the lower boundary of observed seismic activity and no earthquakes have been recorded below this depth. The 670 km discountinuity is thought to be due to change in temperature or pressure resulting in change in the physical properties of the mainly iron and magnesium silicate minrals that make up the mantle, without any change in the composition of minerals. The reduction of seismic velocity in the basal 200 km of the mantle (2700 to 2900 km. depth) may be due to the shell acting as a thermal boundary layer through which heat is conducted rather than convected as it is through the rest of the mantle. It may also be partly caused by vigorous chemical interaction between the silicate mantle and the irone core.[1]

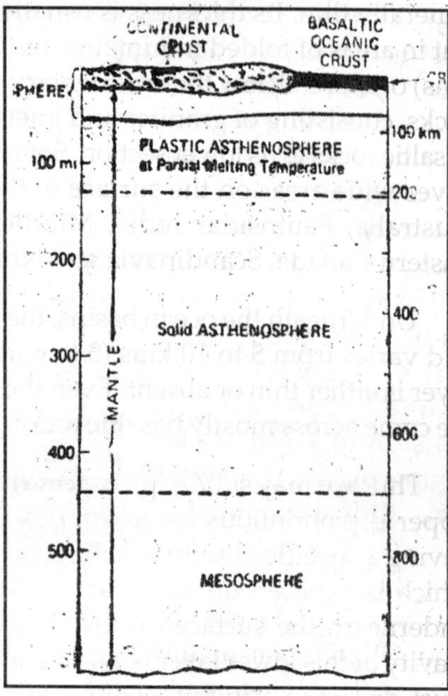

Fig. 9—Sub-division of the mantle and the crust

The Earth's Crust—This is the uppermost and the thinnest layer of the earth. Its thickness varies from 16 to 40 kms (10 to 25 miles). In the continental areas its thickness is about 40 kms (25 miles) and has been found to vary from 30 to 65 kms (18 to 40 miles). It is composed of various kinds of silica-rich rocks of different ages. In its uppermost part we find sedimentary rocks. This sedimentary layer is not continuous over the entire surface of the earth and is

1. Ibid, p.279

generally thin. Its thickness is usually less than 2 miles (3.2 kms), but in areas of folded mountains this may increase to 20 miles (32 kms) or more. Below the sedimentary cover is a layer of crystalline rocks, consisting of granites and gneisses in its upper section and basaltic rocks in its lower section. Some times these crystalline rocks cover wide areas on the surface of the earth, such as in Western Australia, Peninsular India, Middle Africa, Brazilian plateau, Eastern Canada, Scandinavia and North-east Asia.

Underneath the ocean basins, the thickness of the crust is less, and varies from 5 to 10 kms (3 to 6 miles). Here the sedimentary layer is either thin or absent. Even the granitic layer is absent, and we come across mostly basaltic rocks.

Thus we may subdivide the earth's crust into two parts—(i) an upper discontinuous layer composed mostly of the continents, having a specific gravity of 2.65, and (ii) a lower continuous layer which is exposed on the floor of the ocean basins but is found underneath the surface on the continents. The average specific gravity of this lower layer is 3.0 or a little less. Thus the continental crust differs in structure, compositon and origin from the crust beneath the oceans. The continental crust is granitoid in composition whereas the oceanic crust is basaltic. The average thickness of the continental crust is 38 km but beneath the great mountain systems such as the Himalayas, Alps or the Andes it may exceed 50 km. The total volume of continental crust is less than 1% of the earth's total. It has increased through earth history but much of it is ancient dating from the Archaean. In contrast, the oceanic crust is young and thin and has been formed since the jurassic. Its average thickness is 7 kms. It is also uniform in composition.

The contact zone of the crust and the mantle is called *Mohorovicic or Moho discontinuity*. It represents the base of the earth's crust. Here the rocks are different in chemical composition from those below and above but both crust and mantle are made of silicate rocks and the compositional difference between them is slight. Even so it is sufficient to cause the earthquake *P* waves to suddenly increase in velocity from 6 to 7 km per second in the crust to 8.2 kms per second in the upper mantle. Moho is found at the depth of 45 to 60 kms and sometimes 70 km from the surface in the principal fold mountain areas, as crust is thicker in these areas. On the other hand, Moho is found only at the depth of 5 to 7 kms below the floor of the oceanic plains, as the crust is thinner here. Outside the mountainous zones, Moho is usually met with at the depth of 30 to 35 kms below the continental surfaces. It is the shallowest seismic discontinuity and is named after the Croatian geophysicist Andrija Mohorovicic who first discovered it in 1909.

Lithosphere actually includes the crust (continental as well as oceanic) and the uppermost part of the upper mantle i.e. the uppermost comparatively rigid part of the plastic asthenosphere. It is composed of mainly brittle rocks and is 70 to 100 kms (40 to 60 miles) thick under the continents but only 4 km under mid-ocean ridges where new lithosphere is formed. Oceanic lithosphere is over 90% mantle, consists mostly of olivine and has high strength and density. Continental lithosphere may have a maximum of 60% mantle and with predominantaly crustal minerals like feldspar and quartz, is much weaker and relatively light. Below the lithosphere is the plastic and partially melted asthenosphere whose average thickness is 130 to 160 kms (80 to 100 miles) though it extends down to about 220 km. As the thin layer of lithosphere is located on the partially melted layer of the asthenosphere, the lithosphere is unstable and is liable to be easily deformed. This has important implications for the recent theories of plate tectonics, sea-floor spreading as well as mountain building and volcanic action, and these are discussed in detail in the appropriate subsequent chapters.

Because the lithosphere is rigid, heat passes through it mainly by conduction, which is a less effective mechanism than convection which characterizes the deeper ductile mantle. Geothermal gradient in the lithosphere is therefore very much higher than that in the mantle below and the temperature increases rapidly downwards in the lithosphere but the rate of increase slows down further below in the mantle, Though the mantle beneath the lithosphere is hotter. The lithosphere may thus be also conceived as a thermal boundary layer.

Questions

1. Give a reasoned account of the constitution of the earth's interior.
2. How does seismology help in understanding the internal structure of the earth and what are its main findings?
3. Discuss the principal characteristics of the crust, the mantle and the core of the earth.

4. Write short notes on the following :-
   (a) Sial and Sima
   (b) Low Velocity Zone
   (c) Thermal state of the earth's interior
   (d) Moho.

Chapter **5**

The Distribution of Land and Water

Characteristics of the Distribution of Land and Water

71% of the surface of the earth is covered by sea and only the remaining 29% is land. While 39.4% of the northern hemisphere is land, only 18.6% of the southern hemisphere is land. In other words, in both the hemispheres water occupies more space than land. But since 75% of the earth's land area is located north of the equator, the northern hemisphere is referred to as the land hemisphere, while the southern hemisphere is called the water hemisphere. In contradiction to this general arrangement of land and water, however, the north pole lies in the sea, while the south pole is located on land.

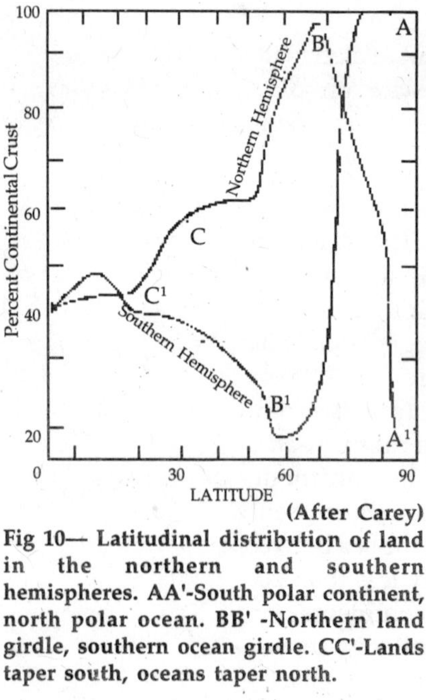

(After Carey)

Fig 10— Latitudinal distribution of land in the northern and southern hemispheres. AA'-South polar continent, north polar ocean. BB' -Northern land girdle, southern ocean girdle. CC'-Lands taper south, oceans taper north.

Another characteristic of the distribution of land and water is that the continents and oceans are roughly triangular in shape, and with the exception of the Antarctica continent all the other continents extend as huge north to south wedges between the oceans. Each continental wedge is composed of two continents- (1) North and South America, (ii) Europe and Africa, and (iii) Asia and Australia (or Oceania). In the western hemisphere both North and South America are distinctly triangular in shape, and in the eastern hemisphere the landmass comprises an union or coalescence of two triangles whose bases are joined together. The base of the triangular landmass lies in the north and their apex lies in the south. On the contrary, the base of the oceans lies in the south and they are narrow towards the north.

The third characteristic of the distribution of land and water on the surface of the earth is their anti-podal position. The landmass of Antarctica is anti-podal to the Arctic Ocean, Australia is antipodal to the North Atlantic Ocean, North America is anti-podal to the Indian Ocean, and Europe and Africa are both anti-podal to the Pacific Ocean. The south Atlantic Ocean lies on the opposite side of East Asia. Altogether nearly 95% of the land is anti-podal to the ocean.

Yet another characteristic but related to those mentioned above is the existence of a continuous circumpolar land ring round the Arctic ocean, while the Antarctica continent in the south stands alone and aloof surrounded by vast stretches of ocean.

Besides the continents there are other landmasses found in the form of big and small islands. The island of Greenland which is included in North America is the largest of them all, with an area of nearly 2,175,000 sq. kms. Most of the islands are, however, much smaller in size. Most of the bigger islands are situated relatively close to the continents, not more than 200 kms away, but many of the smaller islands lie far away from the continents in the heart of the oceans. These are therefore called *ocean islands* as opposed to the bigger islands close to the continents which are called *subcontinental islands*. Most of the subcontinental islands are located on the continental shelves and the sea between the island and the continental landmass is usually less than 200 metres deep. Their structure is also similar to the continents. A third type of islands, different from the two types mentioned above is the *island arcs*. They are found in groups where small as well as big islands are found mixed, and each group constitutes a garland and has distinctive geological structure. Some of these are extensions of continental mountain ranges, but most are located on submarine ridges and are related to volcanic activity. Such islands are found off the coast of eastern Asia in the western Pacific Ocean.

In fact the Pacific Ocean is very different from the other oceans.

It covers nearly one-third of the surface area of the earth. It is surrounded on all sides by extensive ranges of folded mountains, and along its coasts, in association with the island arcs or even otherwise, are found deep trenches characterized by intense seismic and volcanic activity.

The Hypsographic Curve

If we draw a graph where the horizontal scale represents the percentage of the earth's surface and the vertical scale represents the depths and elevations of land above the sea surface, we get a curve which is known as the hypsographic or the hypsometric curve or the cumulative height curve (Fig. 11). This curve gives us a generalized cross profile of the continental surfaces and the ocean floor and indicates their average height and depth and their frequency. It will be seen that very high lands on the continents or

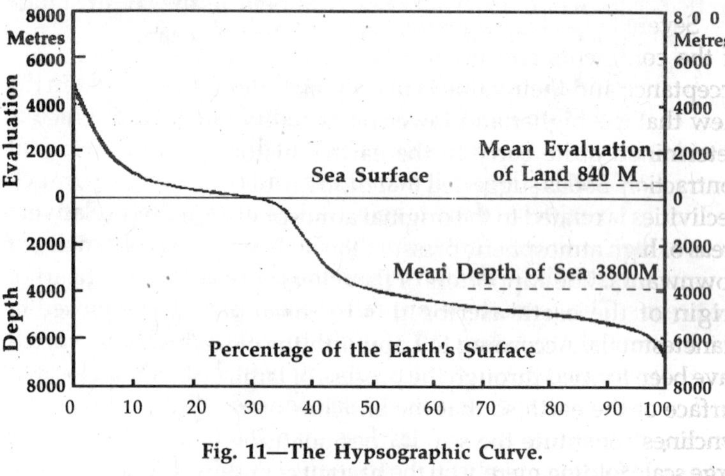

Fig. 11—The Hypsographic Curve.

very deep portions on the oceans occupy a relatively small area of the earth's surface. Most of the land (nearly 21% of the earth's surface) lies between the sea surface and +1,000 metres, and most of the ocean floor (22% of the earth's surface) lies at the depth of -4,000 to 5,000 metres. Between the two prominent and comparatively level surfaces of the continents and ocean basins there is a sharp slope known as the *continental slope.* The continental slope is thus an important earth feature which separates the two fundamental hypsographic levels—one of the continental plateaus and plains and the other of the abyssal oceanic plains. The boundary between the continents and ocean basins is thus extremely sharp, and the continental masses really extend up to the outer edge of the continental shelves. Where deep trenches are found close to the island arcs the boundary between the continents and oceans becomes still sharper.

We have seen that there is important difference in structure between the continental crust and the oceanic crust. The continental blocks are composed of granitic rocks which we call sial. Below the granitic layer there is a layer of relatively heavy basaltic rocks called sima. We find basaltic rocks on the floor of the oceans. Thus the landmasses are composed of relatively lighter materials which are floating on the heavier layer of the sima.

Origin of the Continents and Oceans

Several hypotheses have been put forward regarding the origin of the continents and oceans. None of these have found wide acceptance and their value is mainly historical. Kelvin was of the view that the higher and lower parts of the earth's surface were determined quite early in the earth's history in the process of contraction. Sollas suggested that the distribution of elevations and declivities is related to the original atmospheric pressure so that in areas of high atmospheric pressure the molten material was pushed downwards. The supporters of the planetesimal hypothesis of the origin of the earth ascribe this to the unequal collection of planetesimals. According to Lapworth the continents and oceans have been formed through the process of large scale folding on the surface of the earth so that the anticlines represent the land and synclines constitute the sea. It does not appear logical to invoke large scale folding merely on the basis of contraction and cooling of the earth. In 1907 Love gave a mathematical garb to the hypothesis of Lapworth and tried to prove that since the geometrical centre of the earth and its centre of gravity are not the same, there will be a series of harmonic deformations in the shape of the earth resulting in upwarpings and depressions and the formation of the continents and oceans.

In the beginning of the present century a new hypothesis was proposed by Jeans which was later supported by Sollas. According to this view, after the earth had given birth to the moon the earth became pear-shaped in the process of attaining rotational stability. On this pear-shaped earth the neck of the pear would form the ocean and the edges will constitute higher land. Jeans postulated that these two hemispheres in the process of cooling will press against each other due to mutual attraction, resulting in equatorial buldge and the formation of higher land. Sollas tried to show that it is possible to explain the present distribution of land and water with the help of Jeans' hypothesis, but people were not convinced. In this conne-ction it may be pointed out that Osmond Fisher suggested that the depression which was left on the earth after the separation of the moon forms the Pacific Ocean. This hypothesis had the merit of explaining the principal asymmetry on the earth's surface as also the difference between the Atlantic and Pacific coasts.

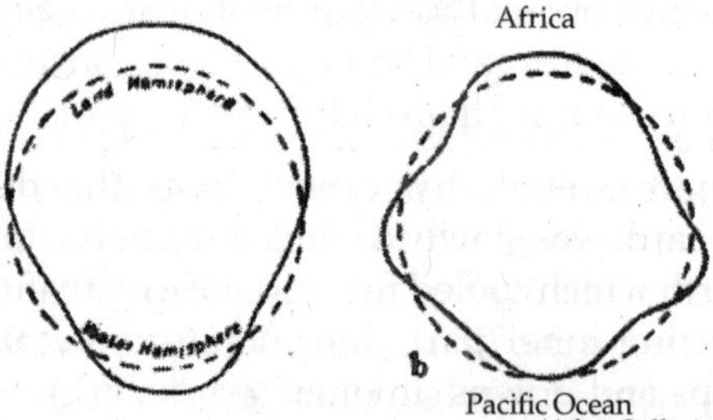

Fig. 12— Hypothesis of Jeans and Sollas

The Tetrahedral Theory—In 1875 Lothian Green proposed an attractive theory in his tetrahedral hypothesis based on a certain organization and symmetry in the distribution of continents and oceans. He pointed out that the distribution of land and water shows a tetrahedral arrangement. A tetrahedron is a three dimensional figure bounded by four equilateral triangular faces, like a three sided pyramid with a triangular base and a pointed top. Lothian Green stated that the earth was like an inverted tetra-hedron which stood on its top. In such a position the four oceans will occupy the four flat faces of the tetrahedron and the upper horizontal face is occupied by the Arctic Ocean. On the edges of the upper horizontal

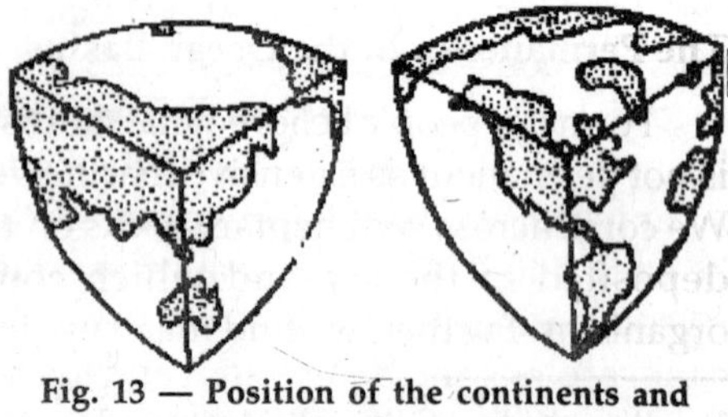

Fig. 13 — Position of the continents and oceans on the tetrahedron.

face are found the landmasses which form a ring round the Arctic Ocean. On the three vertical edges are the continents extending from north to south—North and South America on one edge, Europe and Africa on the second edge, and Asia and Australia on the third edge. At the bottom lies the south pole round which extends the land of Antarctica. He pointed out that the shape of the earth cannot be an exact geometrical tetrahedron, and its edges which constitute the continents will be convex rather than sharp and angular.

The basis of Lothian Green's hypothesis was the then prevailing view that the earth was slowly cooling and contracting. The upper part of the earth which cooled first did not exactly fit in with the slowly contracting inner part, and therefore, became wrinkled resulting in ups and downs. In other words, in such a situation, the surface of the earth is changing into a form in which the area is larger in proportion to its volume. In fact, this is a natural law that the sphere, the body with the smallest surface in proportion to its volume tends to collapse into a tetrahedron, which has the largest surface in proportion to its volume. Green's theory has an apparent basis in the above fact.[1]

In recent years J.W. Gregory has tried to present the above theory in a more detailed and systematic manner with reference to the characteristics of the distribution of land and water. But this hypothesis has now been rejected as according to the laws of mathematics, a big body like the earth cannot maintain its tetrahedral shape in gravitational equilibrium. Even if the earth shows a tetrahedral tendency, the weight of the edges and corners of the tetrahedron would be so much that they would settle down and the earth would return to its spherical shape.[2]

The Permanency of the Ocean Basins

Formerly people believed that the distribution of land and water is not permanent as extensive marine deposits are found on land. We come across sedimentary rocks on the continents which were deposited in the sea and which contain remains of marine organisms. Further, we find many evidences of the invasion of the

---

1. Wooldridge, S.W. and Morgan. R.S. —*An Outline of Geomorphology*, Longmans, 1961, p. 33.
2. "A tetrahedral deformation does not correspond with a figure of equilibrium and were such deformation to start, it would bring into being forces which after a short time would resotre the spherical state" Ibid., p.33

coastal areas of continents by the sea as well as of the uplift of the submerged margins into land. But later on, it came to be realized that the continents and ocean basins are permanent features. In other words, they are the result of an original plan, and they were formed more or less permanently in the initial stages of the earth's history. In reality, the ocean floor has always remained the ocean floor and has never been uplifted to form continents. Similarly the continental blocks are also permanent. Temporary and local changes are not ruled out, but there has not been such a revolutionary change which changed sea into land and vice-versa. The following arguments have been put forward in support of the permanency of the ocean basins:-

1. In the olden ages there were no instruments by which the depth of the oceans could be measured, and the general belief was that the oceans were not so deep. But it was discovered in the latter half of the 19th century that the ocean was quite deep in extensive areas and only the coastal margins were shallow. Under the above circumstances it was considered improbable that the deep ocean floor could ever be raised to form continents.

2. Oozes of a special kind are found on the floor of the deep oceanic plains. Sedimentary rocks formed from oozes are nowhere found on the continental surfaces. On the other hand, the sediments of the stratified rocks found on land resemble the sand and mud usually present on the continental shelves and slopes. The absence of deep sea oozes in the sedimentary rocks of the continents indicates that the deep sea plain has never been raised to form dry land.

3. The detailed study of sedimentary rocks has led scientists to believe that they have been essentially formed in relatively shallow coastal waters.

4. In the oceanic islands, far removed from the influence of continents, the ordinary sedimentary rocks are usually absent.

5. The materials found below the ocean floor are denser and heavier than the continental rocks. In such a situation if the ocean floor is changed into continental landmass, it should result in a marked change in the density of the submarine materials. In the existing state of our scientific knowledge such a change in density is ruled out. The continents are built of the relatively lighter rocks of the sial. The sialic layer is extremely thin on the oceanic floor and

the oceanic floor is primarily built of the heavier rocks of the sima. It looks therefore improbable that the heavier materials of the oceanic floor would change into lighter sialic materials of the continents.

Hypothesis of Continental Drift

According to the concept of sial and sima proposed by Suess it was impossible for the continents to sink to form ocean basins or for the ocean basins to rise to form continents. But the biologists in their study of the geographical distribution of certain organisms had found evidences which suggested that the present continents were probably joined together in the geological past. They assumed in the first instance that the continents were probably joined together by land bridges, but according to the hypothesis of Suess, the sinking of these land bridges and their dismemberment were a physical impossibility. Gradually it came to be realized that while the continents and ocean basins may be essentially permanent features, this does not imply that their relative geographical positions have always remained the same. In 1910 F.B. Taylor put forward the hypothesis of the horizontal displacement of continents. His objective was to explain the origin of the fold mountains of the Tertiary age. He was searching for a clue which might explain the north-south alignment of the fold mountain ranges in North and South America in contradistinction to their west to east alignment in Eurasia. He pointed out that there were two landmasses in the Cretaceous age—Lauratia surrounding the north pole and Gondwanaland surrounding the south pole. He postulated that the moon became the earth's satellite during the Cretaceous and was then much nearer the earth. The resulting tidal forces were supposed to alter the rate of earth's rotation and to draw the continents away from the poles. Jefferys has pointed out that if the tidal force of the newly captured moon were sufficient to displace continents and raise mountains on the scale envisaged, the friction involved would have brought the earth's rotation to a standstill within a year. In the opinion of Holmes neither the tidal force of the moon nor any other external force could bring about changes in the location of the continents. This force had to come from within the earth. The primary merit of Taylor's hypothesis, however, lies in the fact that he set into motion a new line of thinking unravelling the problem of the distribution of continents and oceans.

In 1912 Wegener proposed his famous theory of continental

drift which throws light on a number of problems connected with the distribution of land and water. This theory has been discussed at great length in a subsequent chapter. It may, however, be pointed out that though Wegener's theory was criticized bitterly for sometime, recent studies have brought forth strong evidences in its support. His theory has also found support from the modern concepts of ocean-floor spreading and plate tectonics. These form the subject matter of the chapters that follow and need not detain us here.

Major Structural Units of the Earth

We may divide the entire earth into three major structural units or macro-structures :-

(i) The floor of the ocean basins which is relatively rigid and stable, but which is influenced by faults and is liable to slow spreading. This is composed mainly of basaltic rocks.

(ii) The continental blocks which are also rigid and stable, but which are liable to fracturing. They are composed of granite rocks with a thin and discontinuous covering of sediments. These are sometimes referred to as continental platforms.

(iii) The geosynclinal areas which are unstable and are the seat of most of the earthquakes. This is in fact the most sensitive part of the earth's crust where intense tectonic activity takes place and the mountains are formed. They constitute either oceanic trenches near the continental margins or they are the seats of mighty fold mountain ranges.

Questions

1. Describe the principal characteristics of the distribution of land and water over the globe.
2. Discuss the important hypotheses regarding the origin of continents and ocean basins.
3. In what sense can we regard the ocean basins as permanent ? What are the evidences in favour of their permanence ?
4. Write the short notes on the following :-

   Tetrahedral Hypothesis, Hypsographic Curve, Continental Crust and Oceanic Crust

Chapter 6

Materials of the Earth's Crust - Rocks and Minerals

The Earth's crust is composed of rocks and minerals of various kinds and includes the continents as well as the ocean basins. Rocks are aggregates of minerals and minerals are mixtures of chemical elements. All objection on the earth are basically formed of these elements. The total number of known elements is 99. Man has known the resence of 9 of them since every early times.These include copper, gold, iron, nickel, sulphur, mercury, carbon and tin. More than 98% of the earth's crust is composed of only 8 elements but there are many other elements which are present in very small quantities. The 8 more abundant elements and their percentage by weight of the crust are given below :-

Chemical elements	Symbol	Percentage of the crust by weight
Oxygen	O	46.71%
Silicon	S	27.69%
Aluminium	Al	8.07%
Iron	Fe	5.05%
Calcium	Ca	3.65%
Sodium	Na	2.75%
Potassium	K	2.58%
Magnesium	Mg	2.08%

Oxygen is the most abundant element and is found mixed with a number of other elements. Nearly 47% of the earth's crust is formed by this element alone. The remaining six elements – aluminium, iron, calcium, sodium, potassium and magnesium account for about 24%. The remaining nearly 1.5% of the crust is formed by a number of other elements which occur in very small quantities. These include titanium, phosphorus, carbon, manganese, sulphur, barium, chlorine, radium, platinum, gold, silver and others.

Elements are rarely found in pure form. Gold, sulphur, carbon and a few others are found in pure form. But most elements occur as chemical compound with other elements. Some minerals are composed of only one element such as copper, gold etc, but most minerals are chemical components of two or more elements. In mineral definite elements are found mixed in fixed proporion[1]. Generally every mineral has a definite crysal form. Every mineral has a fixed chemical composition which can be written in chemical symbols and by determining the proportion of elements by chemical analysis the mineral can be identified. For example, quartz is a mineral which has one part of silica and two part of oxygen. Its chemical symbol is therfore sio_2. In all quartz these two elements will be found in the above proportion. Similarly in pyrite there is one part of iron and two parts of sulphur and the chemical symbol is Fes_2. Muscovite is a mixture of potassium and aluminium in a fixed proportion.

The number of minerals is about two thousand, but with the elements, the crust or lithosphere is made mostly of only a few minerals. There are only 20 minerals the mixtures of which are found in most rocks. Quartz is the most common mineral which is a compound of silicon and oxygen. It disintegrates into sand. Another important mineral is feldspar which is a mixture of silica, aluminium and oxygen in which sodium, Calciam and potasium are also found in small quantities. Ferromagnesian minerals are another group of minerals which consist of mostly silica mixed with iron, magnesium and other elements.

Since there is predominance of oxygen, silicon, aluminium, iron and calcium in the crust, these elements also dominate the minerals. Accordingly minerals are classified into the following groups.

1. **Oxides** - In this group of minerals oxygen is the principal element in which other elements are found mixed.
2. **Carbonates** - Examples are calcite and dolomite. Calcite is a mixture of calcium, carbon and oxygen and dolomite is a mixture of calcium, magnesium carbon and oxygen.

(A given mineral always includes the same constituent elements in the same proportion and in most cases it has its own distinctive crystalline form - Finch and Trewartha - Physical elements and geology, 1942, p. 24.)

3. **Sulphides** - these are simple mixture of elements. Otherwise they are not important. Pyrite and galena are examples.
4. **Sulphates** - Gypsum is an example.
5. **Silicates** - This kind of minirals is a mixture of silicon and oxygen with some other elements. Examples are feldspar, muscovite etc.
6. **Chlorides** - Example is halite or rock salt which is a mixture of chlorine and sodium.

Rocks

The materials of the earth's crust are called rocks. They are combination of mineral particles. Most rocks are combination of two or more minerals. For example rock granite is essentially a mixture of two minerals – quartz and feldspar, but other minerals are also found mixed in small quantities. There are rocks which consist essentially of only two minerals. For example, sandstone is essentially a mixture of two minerals – quartz and feldspar, but other minerals are also found mixed in small quantities. There are rocks which consist essentially of only one mineral. For example, sandstone is essentially formed only of quartz and marble is entirely calcite. Other minerals which can also be called rocks are gypsum, dolomite, sulphur etc. There are several rocks which consist of essentially only one mineral but the name of the rock is different from the mineral. For example rock sandstone is made of quartz, limestone is made of calcite, marble is made of either mineral calcite or dolomite, rock salt is made of mineral halite. Some rocks have been formed by chemical disintegration of organic matter like coal or some types of limestone.

A mineral has a fixed chemical composition, but rocks do not have a fixed chemical composition and different minerals are mixed together in varied proportion.

Further in rocks the component minerals exist separately and not in chemical combination. Moreover the component minerals in different speciments of the same type of rocks are not identical in kind or amount[2].

Minerals number 2000 but only 20 are important from the point of view of rock formation, Minerals, however, mix in large variety of ways and proportion so that the number of rocks is also very large.

Types of Rocks

The various kinds of rocks are grouped together into three principal types based on their mode of formation– (a) igneous, (b) sedimentary and (c) metamorphic.

(a) **Igneous rocks:** These rocks have been formed by cooling of molten magma lying inside the earth. The hot magma may cool either inside the earth or on the surface of the earth after ejection or mid-way in the passage through which it is forced to come to the surface. Depending upon the place of cooling it is subdivided into three divisions – Plutonic rocks, Volcanic rock and hypabyssal rocks. (Fig.13 A)

1. **Plutonic rocks:** When the hot magma cools and solidifies deep down inside the earth, it is called plutonic rock. The hot magma inside the earth cools very slowly and takes time to solidify. As it cools very slowly, it solidifies into big crystals and its texture is coarse crystalline or coarse granular. Examples of plutonic rocks are granite, gabbro and pegmatite.

2. **Volcanic rocks:** When the molten lava comes up through some vent or passage and solidifies on the surface, it is called volcanic rock. Here the hot molten material cools down quickly and as a result either no crystals are formed or if formed they are extremely small. Its texture is therefore very fine crystalline or very fine grained. Basalt is the best example of volcanic rocks in which the crystals are very small and the rock is fine grained. The Rajmahal hills in Jharkhand are made of basalt. So is the extensive

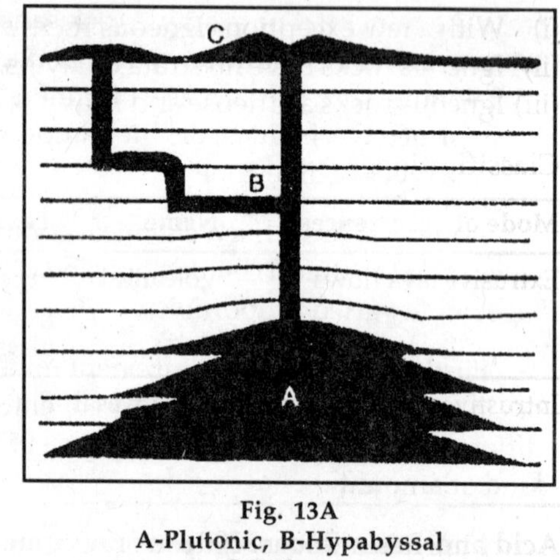

Fig. 13A
A-Plutonic, B-Hypabyssal C-Volcanic

Deccan lava Plateau in Peninsular India. Sometimes the volcanic lava cools down so fast that no crystals are formed and the rock may be as smooth as glass. Obsidian is an example of this type which is called natural glass. Its best example can be seen in Yellow Stone Park of U.S.A..

3. **Hypabyssal rocks:** When the molten magma while coming up through the volcanic vent solidifies mid-way before coming to the surface the rock so formed is called hypabyssal. In this case the cooling is neither too fast nor too slow but midway between the two. Consequently the crystals are neither so big as plutonic rocks nor so small as volcanic rocks. Such rocks are plentiful in Singbhum of Jharkhand. Hypabyssal rocks are found either as vertical walls (dykes) or in horizontal position (sills)

Plutonic and hypabyssal rocks together are called intrusive igneous rocks as they are formed inside the earth. On the other hand volcanic rocks are formed on the surface of the earth. Intrusive igneous rocks can be exposed on the surface only after a long period of denudation.

Characteristics and classification of igneous rocks

We may summarize the characteristics and classification of igneous rocks as below:

Characteristics:

(i) With a few exceptions igneous rocks are mostly crystalline.
(ii) Igneous rocks have no strata or layers.
(iii) Igneous rocks are dense and heavy.

Classification:

Mode of occurrences		Name	Texture	Example
Extrusive lava flows		Volcanic	very fine grained glassy	Basalt, obsidian
Intrustive	Minor intrusive	Hypabyssal	Fine grained	Dolerite
	Major intrusion	Plutonic	Coarse grain	Pegmatite, granite, gabbro

Acid and Basic Rocks: Igneous rocks may be classified into two major divisions on the basis of chemical composition- acid and basic. Acid rocks have a high proportion of quartz (silica) and feldspar and are light coloured and are light in weight. Granite and ryolite are examples of acid rocks. Basic rocks have no quartz and have predominance of coloured minerals such as augite, hornblende and biotitie. These minerals are rich in iron and manganese which make them heavy and dark-coloured. Gabbro, dolerite and basalt are examples of basic rocks.

(b) **Sedimentary Rocks:** Rivers, glaciers and winds carry fragments of different kinds of rocks and deposit them in low lying places especially seacoasts or lakes. With passage of time these fragments of rocks are deposited in layers and are cemented together by some cementing material like silica, clay, rock salt or carbonates. The lower layers are also pressed by the overlying layers. Such rocks are called sedimentary or stratified as they formed in by sediments and are found in layers or strata. Example of such rocks are sandstone, shale and conglomerate. Sandstone has been formed basically from sand, shale from clay and conglomerate from rock fragement and clay. Sedimentary rocks may also form from the decomposition or deposition of dead organism like trees and corals. Coal has been formed from buried forests and limestone from dead corals in the sea. A good example of sedimentary rock is the vast pile of sandstone rocks of Vindhyan mountain in central India which has been uplifted without any folding.

Types of sedimentary rocks: On the basis of composition sedimentary rocks may be divided into four types:

(a) Arenaceous or sandy rocks– These rocks are primarily made of mineral quartz and they are porous. Sandstone and grit are examples.

(b) Argillaceous or clayey rocks such as clay and shale. Such rocks have very small pores and water con not easily pass down through them. They are impervious.

(c) Calcareous rocks– They are primarily made of line. Limestone is an example. Such rocks are formed from remains of marine organisms or animals from deposition of lime present in water. Limestone is an example.

(d) Carbonaceous rocks– Such rocks have preponderance of carbon, such as coal and lignite. They are formed by chemical decomposition of natural vegetation.

Conglomerates which are formed by cementing of various fragments with either sandy or clayey soil do not come under any of the above four categories.

Sedimentary rocks have two important characteristics which distinguish from igneous or metamorphic rocks. Their mode of formation is such that they have strata or layers. Secondly only sedimentary rocks contain vegetable or animal fossils.

(c) **Metamorphic rocks:** When igneous or sedimentary rocks are so changed under pressure and heat that their original character is completely lost, they are called metamorphic rocks. Metamorphism or change takes place under two conditions.

(i) When igneous rock in molten condition tries to come to the surface through vents, it exerts a lot of pressure over the sides of the vents and the hot magma makes the neighbouring rocks so hot that their mineral arrangement is completely changed and they become a different kind to rock. This is called contact metamorphism as the metamorphism or change takes place on account of contact with the hot magma,

(ii) At the time of formation of folded mountains earth movements exert intense lateral pressure on the sides and there is enormous increase in temperature on account of intense pressure. In this way metamorphism takes place over an extensive area. This is called regional metamorphism. An excellent example of regional metamorphism is provided by the extensive marble rocks in the Pre-Cambrian Aravalli mountain of Rajasthan. Metamorphic rocks are crystalline but the different minerals are re-arranged in separate layers and the rocks break (separate) easily on these layers. This quality of being separated on a particular surface is called cleavage or schistosity and the rocks which posses this property are called schists. In metamorphic rocks there is another kind of parallel arrangement of minerals in which mineals are arranged in separate bands and the rocks are separated easily along the junction of the bands. This is called foliation and is found in gneiss, slate and schists. Metamorphic rocks are formed from both sedimentary and igneous rocks.

Examples of metamorphic rocks which formed from sedimentary rocks are marble, quartzite and slate. Marble is metamorphosed limestone, quartzite is metamorphosed sandstone and slate is metamorphosed shale. Example of rocks which have become metamorphosed from igneous rocks is amphibolite which is metamorphosed form of basic igneous rocks like basalt and gabbro and consists mostly of minerals horneblende. Schist and gneiss can be formed by metamorphism of either sedimentary or igneous rocks. When schist or gneiss is formed by metamorphism of sedimentary rocks, it is called para-schist or para-gneiss respectively and when schist or gneiss is formed by metamorphism of igneous rcks, it is called ortho-schist or ortho-gneiss. Mica-schist is formed by metamorphism of shale and other clayey rocks is an example of metamorphism of sedimentary rocks, and horneblende schist which is metamorphosed basalt is an example of metamorphism of igneous rocks. Gneiss generally is formed by metamorphism of granite (igneous) but example of gneiss formed by metamorphism of sedimentary rocks are also found.

Cycle of Rock Formation or Rock Cycle

There is a continuous and unbroken process of the formation of the three main types of rocks which has been changing one type of rock into another type and by this process the minerals of the crust are undergoing new arrangement. This cyclic process of rock

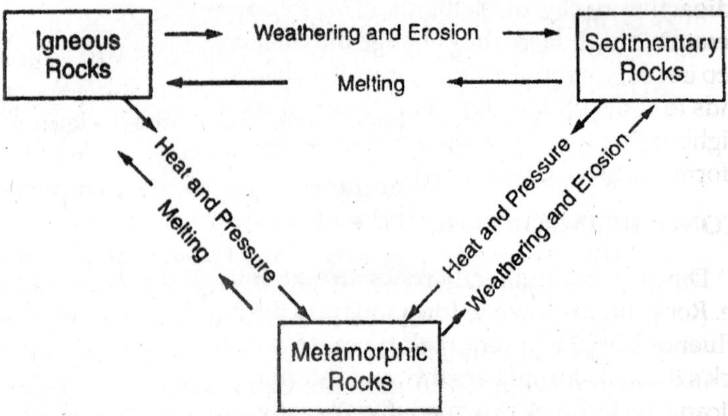

Fig. 13B - Rock Cycle

formation is known as rock cycle. Igneous rocks are the first to be formed by cooling of hot magma. By weathering and erosion the igneous rocks are broken, transported and deposited in layers on the sea coast to form sedimentary rocks. Then during the process of mountain building intense pressure and heat cause enormous physical and chemical changes in the rocks and metamorphic rocks are formed. Both igneous and metamorphic rocks change into sedimentary rocks by weathering and erosion. The sedimentary rocks change into metamorphic rocks under intense pressure and heat. The metamorphic rock change into igneous rocks by melting and cooling. The igneous rocks change into sedimentary rocks by weathering and erosion. The processes under which the different types of rocks are formed constitute a single system in which the materials of the earth's crust are being continuously cycled and recycled.This cycle of rock change has been shown in Fig. 13B. There are two environmental elements in this system – one is the surface environment where heat and pressure both go slow and then the interior environment where there is high pressure and intense heat. In the surface environment there is sudden reduction in pressure and the rocks are broken and the sediments are deposited. In this environment igneous, sedimentary as well as metamorphic rocks are inflended by atmosphere and water and there is chemical change in their minerals and they break to form sediments.

When the sedimentary rocks reach the deep inside environment radioactive decay of elements causes heat and the containing pressure is high.Here they change into metamorphic rocks. In the deep inside environment sometimes it is changed into magma which tends to come up through vents. In the vent it heats and melts the neighbouring rocks. On coming to the surface it cools and solidifies to form extrusive igneous rock.

ECONOMIC IMPORTANCE OF ROCKS

Directly and indirectly rocks are extremely important in our life. Rocks break down to form soils and the type of soils is greatly influenced by the parent rock types. In area of ancient and hard rock's the soils formed are thin and infertile, but sedimentary and volcanic rocks break down relatively easily and the soil so formed is deep and friable. The arrangement and type of rocks determine the amount of run off or the amount that will penetrate and accumulate as ground water and the digging of wells.

Rocks are the source of different kinds of minerals and include such minerals as coal, petroleum and uranium which are sources of power, precious stones as well as variety of metals like iron, copper, aluminium, gold, silver etc. There is a close relationship between the type of rocks and the type of minerals. Metallic minerals are found in metamorphic and igneous rocks. Coal and petroleum are found in the sedimentary rock and country's mineral wealth is determined by the rocks it possesses and the mineral wealth determines the industrial development of a country.

INFLUENCE OF ROCKS ON TOPOGRAPHY

The topography of a country entirely depends on regional geological structure and the processes of erosion. There are special attributes of rocks which are important in this connection. These include permeability, porosity, solubility, grain size, the form of joints, etc. There are rocks which have joints and bedding planes or small holes which allow water to enter into tem. These are called permeable rocks. Sandstone is a highly permeable rock. In areas of sandstone most of the rain water goes underground and rivers and their tributaries are fewer in number. River valleys are shallow. On the other hand shale is an impermeable rock. It has such small pores that water can not enter into it easily and the pores are closed due to surface tension and it does not permit water to enter into it. Clay is porous but impermeable. Chalk is another rock which is porous but impermeable. Water can enter into clay or chalk but can not go down into it. Water which can not go down but accumulates and causes granular disintegration and chemical change in the rocks. Erosion is however reduced as part of the run off is able to go down it.

Broadly speaking it can be said that in permeable rocks erosion is less due to limited run off and in impermeable rocks there is more erosion due to high surface run off. For example in area of hard limestone and soft shale, the permeable limestone makes plateau and ridges while the impervious shale makes valleys.

Permeability of rocks also depends upon joints and bedding planes and in rocks like granite, gneiss and limestone water enters sufficiently downwards. Joints and bedding planes allow water to penetrate sufficiently downwards and erosion becomes relatively easy.

Limestone is soluble in water mixed with carbon dioxide. As a result much of the rain water sinks underground through joints and forms underground rivers by solution which sometimes appear on the surface, caverns with weird stalacites and stalagmites and other distinctive features.

Clay is not soluble but is easily eroded as it is soft and forms areas of low relief. Granite is a plutonic rock which forms deep down within the earth by cooling of magma and is generally found in great masses and forms domes. Domes of granite and gneiss are found in Chota Nagpur Plateau and in the plateaus of Peninsular India. All such massive rocks like gabbro, gneiss etc break rather easily at edges and the relief is dome- shaped. On the other hand basalt is an igneous rock which has been formed by cooling of magma on the surface of the earth. It is found spread over a large area and forms plateaus. It has very fine crystals and therefore it does not brak as easily as granite or gabbro which are coarse grained and have big crystals. A large area of basalt in Gujarat and Maharashtra forms a huge plateau which consists of a series of low stair like plateau and where long period of erosion has resulted in the formation of the highly fertile black cotton soil.

Chapter 7

The Theory of Isostasy

Isostasy is derived from the Greek word *Isostasio* meaning 'in equipoise' or in a state of balance. American geologist Dutton was the first to use the term isostasy in 1889 for the state of gravitational equilibrium that exists between continents and ocean floor and the mountains, plateaus and plains on a rotating earth, in accordance with densities of their underlying rocks. Dutton was of the view that the height of these different features was related to their density. In other words, the higher features were composed of rocks of lower density. For instance, the continental blocks whose average height is about 5 kms above the ocean floor are composed of lighter rocks than the latter, and to compensate for their higher height the continental blocks go deeper into the earth so that the continents and ocean basins may remain in equilibrium. The same argument applies to the mountain ranges. The excess mass of lighter materials which rises in the form of mountain chains on the earth's crust is built of lighter rocks which is floating on the sub-stratum of denser rocks. The hypothesis of the sialic rocks floating over the sima of the sub-stratum had been first put forward by Sir George Airy of Britain in 1855. The use of the word floating, of course, did not imply that the sub-stratum was in a liquid condition.

The Basis of the Doctrine of Isostasy

The first hint that the mountains are not merely masses of rocks resting on an unyielding crust came from the Andes Expedition of 1735 led by Pierre Bouguer. He found to his surprise that the deflection of the plumb line towards the towering mountain was very much less than he had estimated and recorded his suspicion that the gravitational attraction of the Andes is much smaller than that to be expected from the mass represented by these mountains.

Similar discrepancies were observed in 1854 during the trignometrical and geodetic survey of the Indo-Gangetic plain by Sir George Everest, Surveyor General of India. When the difference in latitude between Kalianpur and Kaliana (603 kms due north, about Long 78^0 E) was determined by both astronomical and triangulation methods, it was found that the two results differed by 5.236 seconds of arc. The latitudinal distance measured with the help of the astronomical method was less by 5.236 seconds of arc which is equal to a distance of 108 metres (550 ft.) on the ground. The discrepancy was considered to be due to the attraction exerted by the enormous mass of the Himalayas and the Tibetan Plateau and the deflection of the plumb-bob used for levelling the astronomical instrument at Kaliana which was close to the Himalayas. This error does not arise in the triangulation method.

In 1855 A. Pratt went deeper into this problem. He assumed the density of the Himalayan rocks to be the same as that of the continental rocks (2.75) and made a minimum estimate of the mass of the mountains, and then calculated what the corresponding gravitational effects would be at the two places in the plains to the south. He found that the deflection of the plumb line towards the mountain should be 27.853 seconds at Kaliana and 11.968 seconds at Kalianpur. The difference between the two was 15.885 seconds which was more than three times the observed deflection of 5.236 seconds. In other words, the gravitational attraction of the Himalayan was much less than that expected from the mass represented by these mountains. Sir George Airy, then the Astronomer Royal was the first to offer a solution and put forward his Floatation Theory in 1855 which is now widely accepted. The apparent gravitational anomaly has since been confirmed by the results of innumerable measurements.

Airy's Views

According to Airy the continents which are made of lighter sial are floating over the sub-stratum which is built of the denser sima. He suggested that the lighter sial of the Himalayas is floating over the denser material of the sima lying underneath. He also made it clear that the Himalayas were not just a surface feature but that the light rocks of the mountain extended as roots deep down into the denser rocks below. Just as a large part of the boat remains immersed in water to keep it afloat, so are the Himalayas floating on the denser rocks below and the weight of mountains is balanced by light materials penetrating into considerable depths. This is, of course, the normal principle of floatation and can be most easily understood with the help of icebergs floating in water. When the ice-bergs float in the sea, only one-tenth of the ice-berg is above water and ninetenths remain immersed in water. If we assume the average density of the continental rocks to be 2.67 and that of the sub-stratum as 3, as suggested by Joly, then nearly eight times the height of the rocks above the sub-stratum should penetrate into the denser sub-stratum. Underlying the visible mountain ranges of the Himalayas these rocks of relatively low density must extend to depths nearly eight times as great into the sub-stratum. In other words, the density of the rocks must be relatively low down to considerable depths below these mountains. It is this low density down to considerable depths which reduces the gravitational attraction of the mountains, as this results in a deficiency of mass in the columns representing the visible mountain ranges. It is important to remember that while the earth's major relief features are thus balanced by the underlying differences of density, the individual peaks and valleys are not separately balanced or compensated in this way, as the crustal rocks have enough strength to maintain and hold in position the minor relief features of the surface.

Airy showed that the landmasses could be conceived of as wooden blocks of different heights floating in water. The blocks emerge by amounts which are proportional to their heights, so that the block which rises highest above the water is also the block which goes deepest into the water. These blocks are said to be in a state of

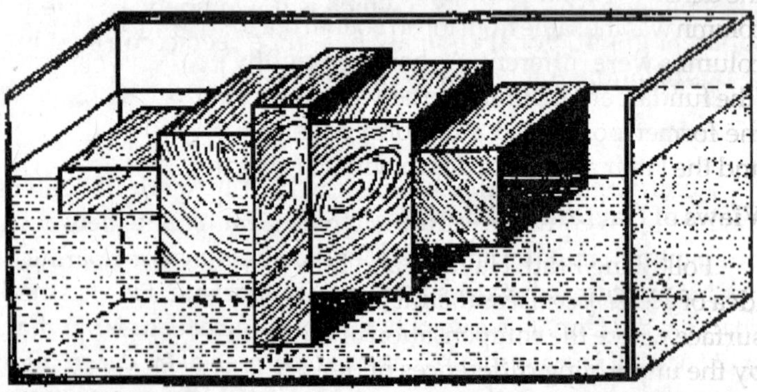

Fig. 14—Wooden blocks of different height floating in water.

hydrostatic balance or equilibrium. Similarly, the landmasses are composed of similar density or rocks, but they penetrate deep into the sub-stratum in proportion to their height. Thus Airy "assumed that the continents and islands are resting, or rather floating, in a denser mass, and that excess of matter above the upper surface of the 'sub-stratum' is balanced by deep projection of the lighter material into the sub stratum. In short, the sial masses are in hydrostatic equilibrium"[1].

Isostasy is therefore 'the corresponding state of balance between extensive blocks of the earth's crust which rise to different levels and appear at the surface as mountain ranges, plateaus, plains or ocean floor[2].

The Views of A. Pratt

It has already been stated that Pratt had calculated the deflection of the plumb line at Kaliana and Kalianpur and had found the difference to be more than the observed deflection. According to him the gravitational attraction of the Himalayas was less than the mass represented by these mountains because the mountains were made of much lighter materials i.e. rocks of much lower density. He stipulated that the rocks of the mountain chains should have a lower density than those of the plateaus, rocks of the plateaus lower than the plains and the plains lower than the rocks of the ocean floor. In other words, in his opinion there was an inverse relationship between height and density. Higher the column lesser the density, and smaller the column greater the density. Pratt believed that there was a level above which these density changes were found and below that level the density was uniform. In other words, the density of a particular column was the same from top to bottom but the density of the different columns were different. As has been rightly pointed out by Boure, "the fundamental difference between Airy's and Pratt's view is that the former postulated a uniform density with varying thickness, and the latter a uniform depth with varying density.[2]

Views of Hayford and Bowie

Following Pratt, Hayford and Bowie of America conceived the idea of a *level of compensation*, a little over 100 kms below the earth's surface where the compensation of the earth's major relief features by the underlying differences of density is said to be complete. In other words, it is the level of uniform pressure. Above this level there

1. Steers, J.A.,—*The Unstable Earth*, 1937, Indian Reprint,p.74
2. Holmes, A.—*Principles of Physical Geology*, Nelson, 1966, p. 27.
3. Quoted in Steers, J.A. *op. cit.*, p. 74

is variation in density and the density is inversely proportional to height for contiguous columns of equal cross-section. Below the level of compensation the density is uniform in that particular layer of the earth's interior. This can be easily understood with reference to Fig. 15. There are four columns of interior plain, plateau, coastal plain and offshore region respectively which go down to the level of compensation. There is difference in their height but they are kept in balance due to their differing density. The pressure of all the columns

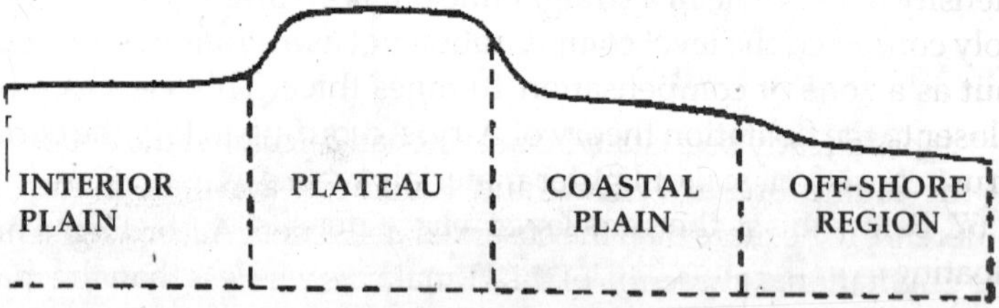

Fig. 15—Level of Compensation

is thus equal at the level of compensation. Bowie tried to explain the concept of the level of compensation and the theory of isostasy with reference to a particular experiment. He took eight pieces of iron, silver, copper, lead, zinc, pyrite, tin and nickel, each of equal width and thickness but of varying heights. When these eight pieces were

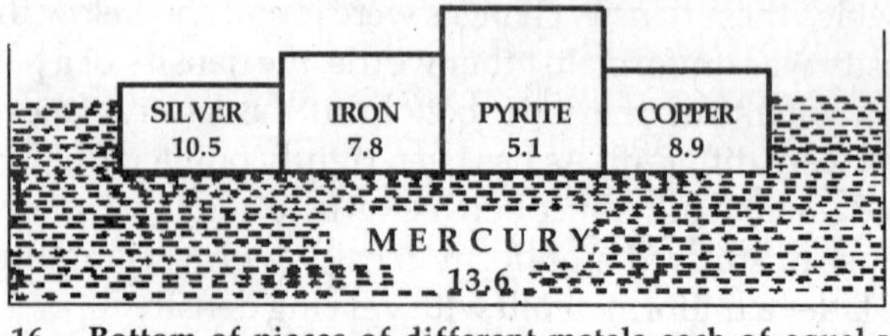

Fig. 16 —Bottom of pieces of different metals each of equal width and thickness but varying height lies at the same level in the basin filed with mercury.

placed in a basin filled with mercury, it was found that the bottom of all eight pieces lay at the same level, but their height above the mercury differed. The pieces with lower density floated higher and those with higher density stood lower.

Views of Joly

Joly has criticized the views of Hayford and Bowie and supported the views of Airy. First of all, Joly has objected to the depth of the level of compensation as suggested by Hayford, as at a depth of 100 kms the temperatures would be so high that all rocks would become liquid. According to Joly there is a ten-mile thick layer below the areas having similar low density. The areas of low density penetrate into this 10-mile thick layer, while where the surface rocks are of higher density, the heavier materials below rise to higher levels. In other words, in the layer containing rocks of different density the bottom of the different columns of varying density does not lie in a straight line but is at different levels. Thus Joly conceived the level compensation not as a straight line surface but as a zone of compensation 10-miles thick. This view is much closer to the floatation theory of Airy. According to Joly the earth's crust consists of sial of lighter materials whose average density is 2.67. Below this is the sima layer whose density is 3.00. The sial is floating over the sima like icebergs.

Views of Arthur Holmes

While supporting the views of Airy, Holmes agrees that the higher relief features of the earth's surface are composed of lighter materials and their major portions lie buried in the heavier and denser rocks below to keep them in a state of hydrostatic balance. He supports the root hypothesis of Airy and states that exploration of the crust by earthquake waves confirms the inference that mountain ranges have roots largely composed of sialic rocks, going down to depths of 50 to 60 kms. In coastal plains the thickness of the sial and other crustal rocks is only 30 kms or less and beneath large parts of the deep ocean floor the sial is either extremely thin or totally absent. Fig. 17 based on Holmes shows a diagrammatic section through the crust and the upper part of the mantle and illustrates the relationship

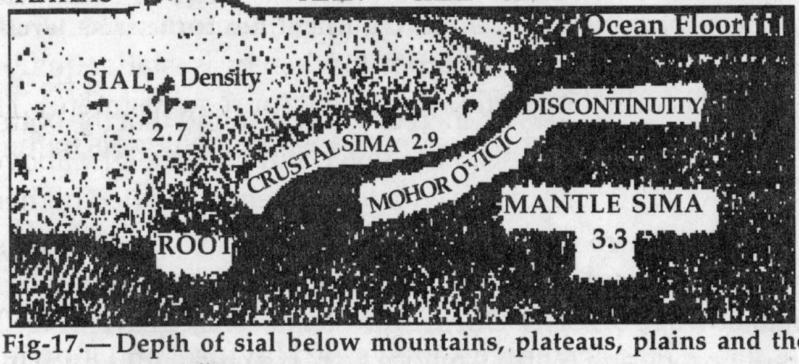

Fig-17.—Depth of sial below mountains, plateaus, plains and the ocean floor.

between surface features and crustal structure. In reality this is but an elucidation and confirmation of Airy's views in the light of more recent knowledge about the distribution of sial, crustal sima and mantle sima as revealed by earthquake waves. The concept of isostasy originated well before the modern ideas of lithosphere and asthenosphere. When the Moho discontinuity, separating the light crust from the denser mantle, was discovered, it was natural to equate the crust with the solid floating layer of isostasy and the mantle with the fluid sub-stratum. It is now known that it is more appropriate to consider the lithosphere (crust and uppermost mantle) as the floating layer, while the asthenosphere provides the fluid sub-stratum.

The theory of isostasy is no longer a theory but is accepted as a fact. It helps us to understand the internal structure of the earth, the relief features of the earth's surface, the drifting of continents, the relative rise and fall of land and sea and a host of other problems. It is thus of major importance in controlling the topography of the earth's surface.

Isostatic Adjustment

When an uplifted landmass or mountain is worn down by the agents of erosion, the load on the underlying column of the crust is reduced by the weight of the material that has been eroded away. At the same time a neighbouring column underlying a region of delta and sea-floor where the denuded material is being deposited, receives a corresponding increase of load. At the base of the crust the pressure exerted by the unloaded column is decreased, while the pressure exerted by the loaded columns is increased. In response to this pressure difference in the mantle there is a sub-crustal slow flow of sima from under the loaded column towards the base of the unloaded column. The loaded column therefore sinks while the unloaded column rises. This process, whereby isostatic equilibrium is restored is called isostatic adjustment (or re-adjustment). This is made possible by a compensating transfer of materials in the sub-crust.

Sometimes certain processes may disturb the pre-existing isostatic balance at a more rapid rate than can be restored by sub-crustal flowage of rocks. During the Pleistocene Great Ice Age large parts of Europe and North America were covered with a thick mantle of ice-sheet and under the load of ice large areas were depressed. When these ice-sheets melted away, these regions were quickly relieved of an immense load of ice and the land started rising and the process is still continuing. The raised beaches of Finland and Scandinavia show that an uplift of about 250 metres has already taken place during the last 8,000 years and the region is still out of isostatic balance and is rising.

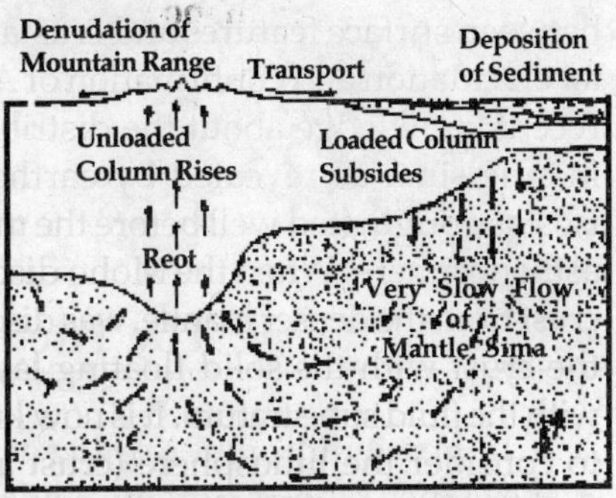

Fig.18— Isostatic re-adjustment.

Questions

1. Discuss the theory of isostasy and its implications.
2. What do you understand by Isostasy ? Discuss in this connection the views of Airy and Pratt.
3. Give an account of the observations on which the theory of isostasy has been founded and show how it helps in the interpretation of the major relief features of the earth.
4. "The outer crust of the earth is floating on a denser sub-stratum". Discuss this statement and bring out its significance.
5. Explain the concept of the level of compensation and its relevance to isostasy.
6. Write short notes on :—

   Level of Compensation; Airy's Root Hypothesis;

   Isostatic Adjustment; Zone of Compensation.

Chapter **8**

The Theory of Continental Drift

The possibility of the drifting of continents was first suggested by the French scholar Antonio Snider in 1858, but this was so much opposed to the scientific view then prevailing that nobody took notice of it. In 1910 F.B. Taylor of America invoked the hypothesis of horizontal displacement of continents or continental drift with a view to explaining the distribution of mountain ranges, but for various reasons the hypothesis of Taylor received scant attention. German professor Alfred Wegener was the first to put forward this idea in the form of a theory in 1912 and published a detailed discussion of it in 1915. But on account of the First World War (1914-18) this could not attract the attention of many scientists. In 1924 an English translation of Wegener's book was published. Since then it has attracted much attention and publicity, and a huge literature has grown around this theory.

Wegener was primarily a climatologist and the main problem he wanted to explain was the change of climates in the geological past. There are several geological evidences to show that there have been important and large scale changes in the climates of the world in the geological past. The occurrence of coal in the cool middle latitudes indicates that the climate of these regions was hot and humid in the carboniferous times and the natural vegetation was dense forest which ultimately resulted in the formation of huge coal deposits. On the other hand, in India, South Australia, Brazil or South Africa where the climate is warm today, there are evidences of cold climate and the presence of glaciers in the carboniferous period. The only conclusion that can be reached from this is that either the climatic zones have moved or if they have not, then there has been movement of the landmasses. There are several difficulties in postulating shifts in the climatic belts of the world as their geographical distribution is governed primarily by the sun. It, therefore, appears to be more probable that the landmasses have changed their position.

Pangaea and its Distribution

Wegener collected evidence in support of his views from a very wide area and on the basis of these evidences he postulated that in the carboniferous period all the continents were joined together and formed one huge landmass. Wegener called this huge landmass *Pangaea*. Australia, Antarctica, Peninsular India, Africa and South America together formed the southern block of this landmass. Geologists had recognized the presence of such a landmass even earlier on the strength of other evidences and they had called it Gondwanaland. North America, Europe and Asia formed the northern block of Pangaea and was called Laurasia. These two blocks were separated by a long shallow inland sea called the Tethys. To this combination of Gondwanaland and Laurasia, making up the whole land area of the earth, Wegener gave the name of Pangaea, a Greek word, meaning all land. Pangaea was surrounded on all sides by an extensive water mass called the *Panthalasa* or the Primeval Pacific Ocean. Wegener also imagined that in the carboniferous period the south pole was near the South African coast and the north pole lay in the Pacific Ocean. Snider had suggested earlier that the Carboniferous coal measures of Europe and North America were tropical and these continents were then near the equator, implying that South Africa lay near the south pole. Because of the distribution of the Permo-carboniferous glaciation, Wegener also inferred that the Carboniferous south pole occupied a position close to the South African coast.

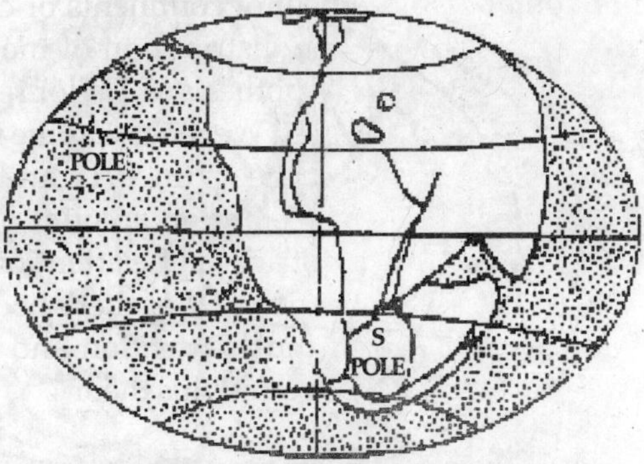

Fig. 19—Pangaea. according to Wegener.

According to Wegener the super-continent of Pangaea started breaking up in the carboniferous period, and the present shape and

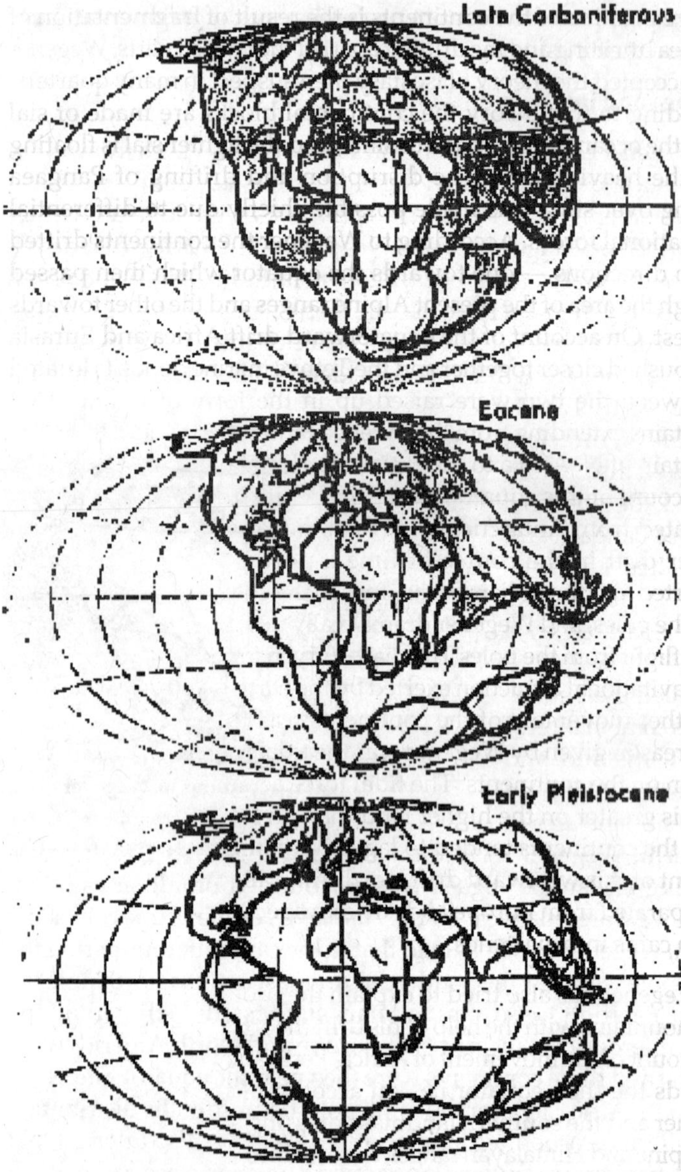

Fig. 20 —Wegener's reconstruction of the distribution of the continents during Carboniferous to Pleistocene. Africa is placed in its present position as a standard of reference. The shaded areas on the continents represent shallow seas.

relative position of the continents is the result of fragmentation of Pangaea by rifting and the drifting apart of the broken parts. Wegener fully accepted the theory of isostasy then suspect in many quarters. According to this theory the continental blocks are made of sial while the ocean floor is made of sima, and the lighter sial is floating over the heavier sima. The disruption and drifting of Pangaea floating over sima was made possible chiefly due to differential gravitational forces. According to Wegener the continents drifted in two directions — one towards the equator which then passed through the area of the present Alpine ranges and the other towards the west. On account of the equatorward drift Africa and Eurasia were pushed closer together and the Tethys marine deposits located in between the two were raised up in the form of mighty fold mountains extending from the Pyrenees and the Alps, and the Atlas mountain of N. Africa to the extensive Himalayan ranges of Asia. On account of this equatorward drift Peninsular India and Africa separated from Antarctica and Australia, and as a result of their further drift in the course of time a portion of Panthalasa got converted into the Indian Ocean. In fact, the continents drifted away from the poles, and Wegener dramatically described this as *Polflucht* or the flight from the poles The reason given for this movement was the gravitational attraction exerted by the earth's equatorial buldge. The other movement of the continents was towards the west. Its main reason given by Wegener was the tidal force of the moon and the sun on the continents. The tidal force acted like a brake; as the effect is greater on the higher land than on the lower levels of the crust, the continents tended to lag behind or drift to the west. On account of this westward drift North America and South America got separated from Europe and Africa respectively and the Atlantic Ocean came into existence.

Wegener has also tried to explain the problem of formation of fold mountains with the help of his drift theory. As indicated above, on account of the movement of Africa, Peninsular India and Eurasia towards the then equator i.e., on account of their coming closer together and the southern landmass pressing against the northern, the Alpine and Himalayan ranges were formed in the Tertiary period in the region occupied by the Tethys sea. Similarly, the Andes and the Rocky mountain ranges were formed as a result of the westward drift of the Americas and the resistance offered by the rocks of the ocean floor to this movement. Between the two immense rafts a trail of fragments was left behind to form the islands of the West Indies.

Wegener prepared a geological chronology of the drift of the continents and the related position of the poles and the equator. The disruption of the southern block of Pangaea took place mostly during the Mesozoic era, while the westward drift of North America took place mostly during the Tertiary. The North Atlantic Ocean was formed during the Pleistocene period. By the time the continents had reached their present positions, Antarctica found itself stranded over the south pole, Africa lay across the equator, India had been attached to Asia, and Australia and New Guinea had drifted far to the south in the Pacific.

Evidences in favour of the Drift Theory

Wegener collected evidence in support of his theory from different sources and these may now be summarized :-

(i) First of all, he was struck by the geographical similarity between the opposing coasts of the Atlantic Ocean. The outlines of the coasts on the two sides of the Atlantic are such that they can be easily joined together and one appears to be a detached portion of the other. The eastern coast of South America can be fitted into the western coast of Africa and similarly the eastern coast of North America can be fitted against the western coast of Europe. This was called the "jig-saw fit" of the opposing coasts of the Atlantic Ocean by Wegener. Thus according to Wegener the Atlantic ocean was an enormously widening rift, with the sides as closely matching as the lines of a torn drawing.

(ii) There is remarkable similarity also in the geological structure and history of the lands located on the two coasts of the Atlantic Ocean. On both the coasts the structural and relief features are found tranverse to the coast. The best example is provided by the Appalachian mountains of North America which come right up to the coast and then continue their trend across the North Atlantic Ocean in the old Hercynian fold mountains of South-West Ireland, Wales and Central Europe. Similarly the trend of the Caledonian mountains of North America is south-west to north-east along the eastern coast and into the sea and then they re-appear on the other side of the Atlantic in Ireland and Scotland in the same direction (Fig. 21). On both the coasts there is great similarity in the structure and direction of the Hercynian mountains and their truncated end can be jointed together. The opposing coasts of the South Atlantic Ocean in Africa and Brazil display even greater resemblance in their structure and rocks.

(iii) The distribution of the Permo-carboniferous glaciation presents a powerful proof of the fact that at one time these landmasses were assembled together, since the evidences of this glaciation are found in Brazil, Falkland island, South Africa, Peninsular India as well as Australia. It is difficult to explain this extensive glaciation on the basis of the existing distribution of land and water. In the opinion of Wegener, all these lands were assembled together to form Pangaea, and the south pole was located at the time some where near Durban off the present coast of South Africa.

(iv) Yet another evidence of the joining together of the continents and their subsequent drifting is provided by the presence of fossil remains of land animals and plants and of fresh-water species in distant lands now separated by oceans as these species could migrate

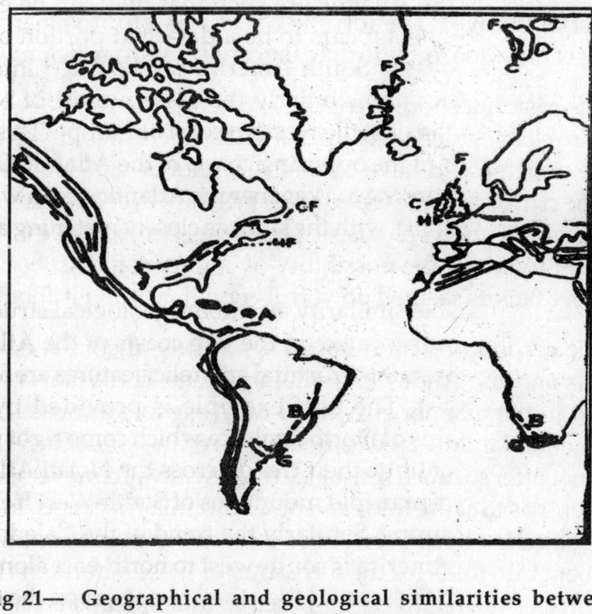

Fig 21— Geographical and geological similarities between opposing coasts of the Atlantic ocean.

CF— Caledonian mts, HF—Hercynian mts, A—Alpine fold mts, B—Brazillides, G—Gondwanides.

freely across united continents but not across an intervening ocean.

Criticism of Wegener's Theory

It is clear that Wegener had amassed an imposing array of evidences in support of his theory and some of this evidence was undeniably convincing, But so much of the theory was based on speculation and inadequate evidence that it provoked a lot of criticism and controversy and people were not inclined to accept his views :—

(i) The greatest criticism of his theory has been that the force which were invoked by Wegener were hopelessly inadequate to cause drift. The westward drift of the continents was ascribed by Wegener to the tidal force of the moon and the sun, but this movement is only possible when the tidal force is ten thousand million times more than at present, and if the tidal force really increased to this dimension then it would have stopped the rotation of the earth within one year. The reason given for the equatorward drift or flight from the poles was the gravitational attraction exerted by the earth's equatorial buldge. "The force is a real one, but it is many many million of times too feeble to drag the continents from their moorings."

The basic question, of course, is whether there has been drift. If there are clear evidences of drift, then the mere inability to point to a force which could have caused it cannot be considered to be a genuine criticism of the theory. The fact, however, remains that the geophysicists and geologists of the time seemed to be extremely reluctant to admit the possibility of continental drift because no operative factors seemed powerful enough to bring it about.

(ii) The second objection raised against the theory was that the Rockies and the Andes mountain systems could not have been formed in the manner envisaged by Wegener, because if the sial (continental landmass) was floating over the sima, then the sima could not offer so much resistance to the westward movement of the continents as to cause folds and build mountain systems. In answer to this objection it has been stated that it is not necessary for the sial to be bent into folds for the formation of these mountain chains; it were, in fact, the relatively weak sedimentary deposits on the edge of the continental landmasses which were folded to form the mountains. Wegener has not clearly stated this, but it has been pointed out by Holmes that it is possible for the coastal sedimentary deposits to be raised into mountains and for the ocean floor to be bent into trenches.

(iii) The third objection was that the jig-saw fit of the opposing coasts of the Atlantic Ocean on which so much stress was laid, was not so complete. But considering our knowledge of the earth's crust, to expect that after the breaking and drift of the continents there would be an exact fit of the opposing coasts of the Atlantic is unnatural.

(iv) A fourth objection closely related to the third was that although there was similarity in the structural and stratigraphical features of the two coasts of the Atlantic, it would not be quite correct to conclude that one was an extension of the other and that they were joined together. It has been pointed out in reply that it would be unnatural to look for exact identity as in the course of the drift parts of the sial would have thinned out and got lost. In fact such broken fragments of sial may be seen in the islands of the West Indies as also the group of islands lying between S. America and the Antarctica. The similarities between the geological features of the two coasts of the Atlantic were confirmed by later workers. E.B. Bailey has found a remarkable and significant resemblance between the Palaeozoic mountain systems of Europe and North America. For example, the Caledonian fold mountain ranges get closer to the Hercynian fold mountain ranges towards the west in northern Britain and start meeting each other in South Wales and South Ireland. These very ranges cut each other across the Atlantic in Pennsylvania and after that the Hercynian mountain ranges begin running north of the Caledonian mountain ranges (Fig. 21)

Alexander du Toit, a famous South African geologist presented several evidences in his book '*Our Wandering Continents*' published in 1937 in support of continental drift. After detailed investigations he showed that striking correspondence existed in the sediments, fossils, earth movements and igneous intrusions of the two sides of the Atlantic. According to him both the regions had essentially the same geological history in Palaeozoic and early Mesozoic times, and it seems highly probable that they were then very much closer together than now. It is possible that continental materials from a width of 300 to 800 kms got lost in the course of breaking up and drifting which is confirmed by the existence of abundant remains of siallic materials on the Atlantic floor. Du Toit, like Taylor however, thought it more logical that a double landmass existed during the Palaeozoic rather than the single landmass of Pangaea. It would appear that before the Tertiary Alpine- Himalayan orogeny, Laurasia and Gondwanaland were separated from one another by a long and perhaps interspersed seaway called the Tethys by Suess.

It is true that Wegener presented his case more like an advocate than like a scientist, emphasizing the evidences in his support and ignoring those which went against his theory. Till about 1950 his theory remained highly controversial, but after that the status of the theory has udergone remarkable change on account of the confirmation it has received from new directions. As new investigations have proceeded and new facts have been brought to light, they have provided further powerful and irrefutable evidence in support of the theory and have led to its wide acceptance. The confirmation of continental drift has come particularly from palaeomagnetic studies as well as from the new theories of sea-floor spreading and plate tectonics. These theories are examined in detail in the next two chapters, but here we may summarize the other new evidences in support of continental drift which have led to its confirmation.

New Evidences in Support of Continental Drift

The new evidences in support of continental drift have come from four sources:—

(a) Computer estimates of the topographic fit of the continents,
(b) New investigations regarding geological similarities between the southern continents,
(c) Palaeomagnetic studies, and
(d) Evidences of sea-floor spreading.

Computer fit of the Continental Coasts—Wegener had emphasized the geographical similarities between the Atlantic coasts. In 1958, S.W. Carey with the help of a relatively simple experiment showed that the fit of South Atlantic coasts was so good that it could not be called accidental. Then in 1965 E.C. Bullard, Everett and Smith demonstrated with the help of computer that the African and South American coasts could be fitted into each other at a depth of 500 fathoms (1,000 metres) in the South Atlantic and similarly ir. the North Atlantic the coasts of Europe and North America could be fitted into each other (Fig. 22). In other words, there is such a great resemblance in the outlines of the continental slope on both the coasts of the Atlantic that the entire Atlantic Ocean can be nearly closed and with the exception of the Niger delta the gap or overlap does not exceed 90 kms anywhere. So far as the lands surrounding the Indian Ocean is concerned, the problem of reconstruction is more difficult. In 1969 W.P. Sproll and R.S. Dietz prepared a computer fit of Australia and Antarctica, and in 1970 A.G. Smith and A. Hallam prepared a computer fit of S. America, Africa, Arabia, Australia, Antarctica along the 500 fathom isobath, using only those fits which they considered to be geologically sound. This map is strikingly similar to the map prepared by du Toit in 1937 though the knowledge of the geological structure of these regions at that time was very incomplete.

Geological Similarities — If Africa and South America were joined together, then we should expect the geological structure of

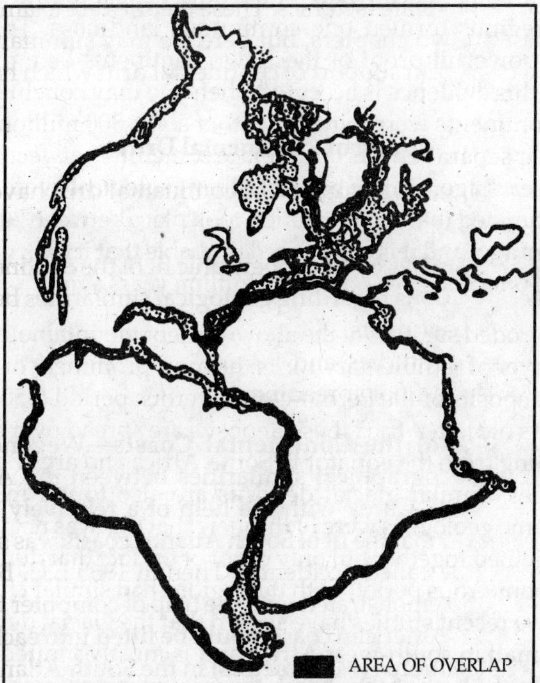

Fig. 22— Computer fit of the Atlantic coasts prepared by Bullard and associates.

the two to be similar in the period prior to their disruption. Wegener had stated this as one of the evidences in support of his theory, and later in 1937 du Toit had confirmed this view by putting forward some striking examples of this similarity. But at the time of du Toit the exact age of the geological structure was not accurately known. It is now possible to determine their exact age with the help of isotopes and radiometric dating. Data relating to such isotope ages of rocks in eastern South America and western regions of Africa have been plotted on maps, and both have been found to have rocks of the same ages and similar structure. P.M. Hurley (1967) and his associates found great similarity in the geological age provinces of western Africa and Eastern Brazil and their geographical distribution. In the western parts of western Africa the age of rocks is about 2,000 million years and in the eastern parts about 550 million years. The age of the rocks in Brazil is also the same, and the line dividing these two age provinces has been found in Brazil exactly at the same place where one would expect to find it, had these two regions formed one continuous landmass. This fact provides a powerful proof of these two continents being joined together. If this evidence is accepted, then one may conclude that these two continents were joined together some 500 million years age, and their separation has taken place sometime between 500 to 50 million years ago. On the basis of other geological evidences it has been estimated that the separation took place between 200 to 50 million years ago, and it is considered probable that it took place in the Lower Cretaceous period i.e. 100 million years ago.

On the eroded surfaces of the above ancient mountainous areas is found a layer of sedimentary rocks the most prominent of which are glacial deposits of the Permo-carboniferous period (350 to 250 million years old). In Africa these deposits are spread over a wide area extending from the equator to South Africa and are known as Dwyka series. Similar glacial deposits are also found in South America. Some geologists were of the view that this was no proof of their being joined together but only proof of the fact that during the permo-carboniferous period both the regions had similar climates. But extensive recent studies have shown that the glacial deposit is relatively small in south-west Africa and is mostly limited to the ancient valleys. On the other hand, there is extensive evidence of glacial erosion, the direction of glacial flow being from east to west. From this it has been concluded that south-west Africa was covered by an actively eroding ice-sheet and the eroded debris were being deposited towards the west. In the Parana basin of eastern Brazil extensive deposits of glacial boulder clay called tillites are found which are sometimes as thick as 600 metres. In other words, these glacial deposits have ten times the thickness of the Pleistocene glacial deposits of northern Europe. The direction of ice flow in the Parana deposits is from south-east to north-west, i.e. the source of the glacier lay some where south-east of the present Brazilian coast from where large quantities of debris were obtained. The scarcity of glacial deposits in south-west Africa coupled with abundant evidences of glacial erosion seem to suggest that the source area of the glaciers was south-west Africa. Further, we find erratic blocks of coloured quartzites and cherts in Brazil. These rocks are not found anywhere in Brazil but they are found in south-west Africa. Similarities between other old sedimentary rocks of eastern Brazil and South west Africa lead one to the same conclusion.

Another striking similarity is the great flood of basalt that inundated the Triassic deserts of both the continents in the early Jurassic times. In the Parana Basin the basalts cover an area of a million sq. kms and reach a thickness of 1.5 kms. In southern Africa a smaller area of basalts remains, but considering the large number of basic dykes and sills in the Karoo Basin and the enormous thickness of basalt preserved in the eastern side, the volume of lava erupted appears to be even greater than in South America. "Such spectacular signs of unusual activity in the depths would seem to be a significant preparation for the major break-up of Gondwanaland and the appearance of seaways between the separating land. Palaeomagnetic evidence based on these basaltic rocks indicates that separation of Africa from South America began while this igneous activity was in progress. But no actual proofs of the existence of the South Atlantic before the Lower Cretaceous have so far been detected."[1]

Both Wegener and du Toit had laid stress on the similarity in the plant and animal fossils in support of the theory of continental drift. A significant one among these is *glossopteris* flora which is found in the sedimentary rocks of the Gondwanaland in South Africa, India, Australia as well as South America. Marsupials such

1. Holmes, Doris L.—*Holmes Principles of Geology*, Third Ed., E.L.B.S. and Nelson, 1978 (Reprinted 1981), p. 632

as kangaroo and opossum, animals which keep their babies in bags on their bodies are found only in Australia and South America, and the fauna of Indonesia is completely different from Australia. It appears that Australia was connected with South America at one time, and has separated from the latter at a subsequent date and has come closer to Indonesia.

Palaeontological evidence of the former union of Africa and South America was provided by the fossil remains of a small reptile called *mesosaurus* in the Permian deltaic deposits of both Brazil and South Africa, as it was not possible for this animal to swim across the ocean. In 1967 fossils of fresh water amphibians were discovered in the Antarctica. These animals are called *Labyrinthodonts*, and they developed in the Devonian period long before the disruption of the Gondwanaland. They could neither live in salt water nor could they swim for long distances. In 1969 Edwin Colbert and a few others found reptile fossils of the Triassic period in the Antarctica. Most of these were *Lystrosaurus*, a reptile which lived in swamps and lakes and is found in several other parts of the world. It could not have crossed the ocean to reach Antarctica. This is considered to be an important discovery in favour of continental drift.

Another important evidence in support of continental drift is that we find the development of distinctive faunas in the different parts of the world in and after the Mesozoic era but before the Mesozoic there is great similarity in the faunas throughout the world.

Palaeomagnetism— According to Wegener there has been change in the position of the poles in the different geological periods which he called polar wandering. Recent palaeomagnetic studies have made it possible to locate the exact position of the poles in the different geological periods. These studies have revealed that the magnetic declination and inclination are different in the rocks of different ages, but the position of the poles in the same geological period has been found to be different in the different continents. Since the pole cannot be at two places in the same period, it has been concluded that the position of the continents has changed after a certain period. It has also been found that if all the contients are assembled together in the position of Pangaea, then the position of the poles becomes the same.

Creer (1965) calculated the position of the palaeomagnetic poles for various geological periods from reliable palaeomagnetic data and showed that it is possible to find whether or not there has been a movement between two continents by comparing their poles wandering curves. If the curves have the same shape and can be superimposed one upon the other, then there has been no relative movement. With the help of these curves he found that South America and Africa were together across the south pole during the Palaeozoic era. More recently Creer (1973) has plotted the positions of the palaeomagnetic poles on a map of Pangaea for the periods Cambrian to Cretaceous. This Pangaea is composed of Laurasia as reconstructed by Bullard and associates and Gondwanaland as reconstructed by Smith and Hallam. Creer found that the poles for the Ordovician period were so widespread that they could not be reconciled with the existence of Pangaea at that time. The coming of the poles closer together in the periods from lower Carboniferous to the Jurassic indicates the existence of the Pangaea in a form similar to that tested. For the Cretaceous period the poles are more dispersed showing the breaking up of Pangaea during that period.

Palaeomagnetism has thus provided the most reliable proof of continental drift and palaeomagnetic evidence does indicate the existence of Pangaea in some form at the end of the Palaeozoic era. But even more significant than the palaeomagnetic studies of the continental rocks have been the paleomagnetic studies of the rocks of the ocean floor which have not only confirmed continental drift but have provided the basis for the new hypothesis of sea-floor spreading (See Chapter 8). As we shall see in the sequel, in addition to the compelling nature of much of the geological evidence for continental drift, the most unequivocal proof of the movements of continets now comes from palaeomagnetism, especially the linear magnetic anomaties produced by sea-floor spreading. These mark the ancient positions of plate boundaries and with their help plate and hence continental motions can be reconstructed. The palaeomagnetic and geological evidences have now made it possible to trace the history of continental drift back to the pre-cambrian times. This shows recurring episodes of continental break-up, drift and collision since earlier times, each cycle lasting for a few hundred million years and have been called the Wilson cycle. after J. Tuzo Wilson, one of the pioneers of plate tectonics[1]. Pangaea which

1. Searle, Roger- Continental Drift in *The Earth*, Op cit, 2000, p. 159

started breaking up in Mesozoic, was formed from the earlier convergence of Lauratia and Gondwanaland in the Silurian and Lauratia itself was formed from a number of earlier continental blocks.

Questions

1. Give a concise account of Wegener's theory of Continental Drift.
2. Discuss the main evidences in support of the theory of Continental Drift ?
3. What are the arguments for and against Wegener's theory of Continental Drift.
4. What is Pangaea ? What is its relationship with the present continental masses ?
5. What are the main criticisms of Wegener's theory of Continental Drift ? What is the present status of the theory ?
6. What is the present status of the theory of continental drift ? What are the new evidences in support of this theory ?
7. "Continental Drift is a fact rather than a theory". Examine the validity of this statement.

Chapter 9

Sea-Floor Spreading

The hyptothesis of sea-floor spreading was first put forward by Harry Hess in 1960. During the Second World War and shortly there after a number of investigations had resulted in considerable increase in our knowledge about the ocean floor. Between 1950 and 1960 the scientists had been able to establish three important facts about the ocean floor. (i) Firstly, the crust below the ocean floor was found to be only 6 to 7 kms thick, whereas below the continental surfaces this thickness was 30 to 40 kms. (2) Secondly, formerly people knew only about the existence of mid-Atlantic ridge, but it was discovered that mid-oceanic ridges were present in all the oceans and were not confined to the Atlantic. Their summits are 2 to 3 kms above the deep sea plain. These ridges contain large scale evidences of faulting and volcanicity and are frequently visited by earthquakes. (3) Thirdly, nowhere the rocks of the ocean floor were found to be older than the Cretaceous period which began about 135 million years ago.

On the basis of the above facts the realization dawned that the ocean floor was not an old and inactive part of the earth's surface but possibly the youngest and the most active. In 1960 Harry Hess proposed that the ocean floor was mobile and hot magma rises up from the earth's mantle by convection currents along the middle of the mid-ocean ridges and moves away on either side and is finally lost in the oceanic trenches situated along the continental coasts. It is through this process of spreading that the ocean floor has been built. In other words, the ocean floor is a relatively young feature of the earth's surface. It is constantly being regenerated at the mid-oceanic ridges and is subject to continuous lateral spreading until it is destroyed in the trench system and becomes re-incorporated into the mantle (Fig. 23). On the other hand, inspite of the fact that the continents are old and permanent features, they passively drift apart and together on the backs of the mantle generated convection currents.[1]

It may be mentioned that Arthur Holmes had earlier suggested that mantle convection currents could be responsible for continental drift but at that time there were no adequate geophysical data to subtantiate his views. Hess thought that near the mid-oceanic ridges

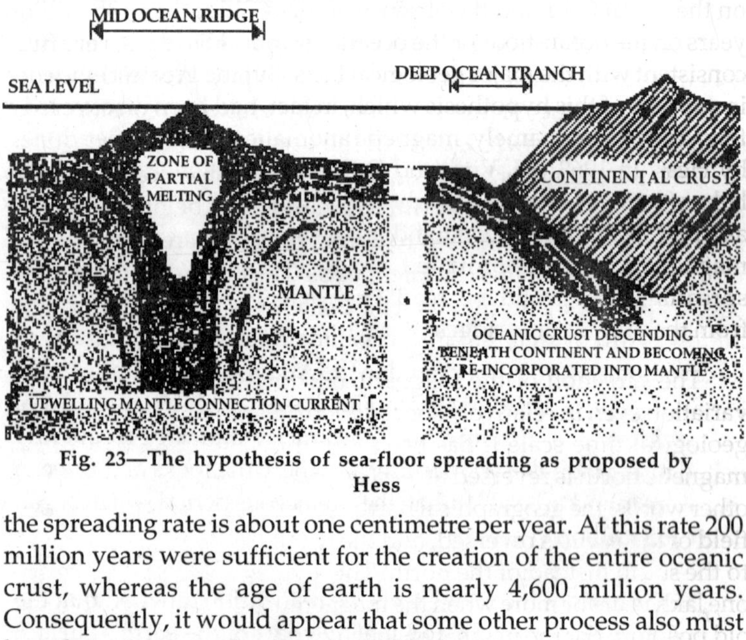

Fig. 23—The hypothesis of sea-floor spreading as proposed by Hess

the spreading rate is about one centimetre per year. At this rate 200 million years were sufficient for the creation of the entire oceanic crust, whereas the age of earth is nearly 4,600 million years. Consequently, it would appear that some other process also must have been in operation by which the newly created crust is being destroyed at a balancing rate. Hess suggested that this destruction process happened in the deep ocean trenches, such as the Chile-Peru trench situated off the western coast of South America, where the relatively cold oceanic crust descends into the mantle and is ultimately incorporated into it, thus resulting in the thickening of

1. In summary, Hess suggested that despite their great age and apparent permanency continents have been and are being, passively drifted apart and together on the backs of mantle-wide convection cells. In contrast the ocean floors are young and ephemeral features of the earth's surface, constantly being regenerated at ridge crests and destroyed in the trench systems. Vine, F.J.—Sea-Floor Spreading, in *Understanding the Earth* edited by Gass, Smith and Wilson, ELBS, 1973, p.234

the continental margins and formation of mountain ranges along the coast.

It was possible to explain a number of things with the help of Hess's hypothesis, such as, continental drift, occurrence of earthquakes near the mid-oceanic ridges and absence of sediments on their summits, association of active volcanic islands with the mid-Atlantic ridge, the existence of only a thin cover of sediments on the ocean floor and the absence of rocks older than 135 million years on the ocean-floor or the oceanic islands. His ideas were fully consistent with the major geological facts. But the greatest evidence in support of this hypothesis which, in fact, had been discovered a few years earlier namely, magnetic anomalies, was not mentioned by Hess. In 1963 F.J. Vine and D.H. Matthews showed with the help of analysis of data already existing as well as those collected as a part of the International Indian Ocean Expedition of 1963, how magnetic anomalies provided a magnetic record of ocean floor spreading that could be read from each ocean floor outwards from the mid-oceanic ridges.

The direction of the magnetic north undergoes some slight variation even in relatively short periods. But with reference to the geological time scale it has been found that the direction of the magnetic north is reversed at least every one lakh years or more. In other words, the geographical north remains fixed but the magnetic field of the earth is reversed, and the needle of the compass points to the south instead of the north. This process goes on for another one lakh years or more when the magnetic north again returns to its old position. For the past seven lakh years the magnetic field of the

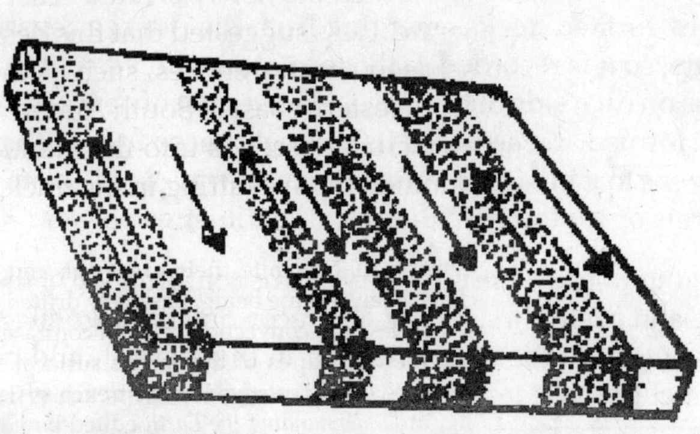

Fig. 24—Strip reversals of magnetism.

earth is normal. This duration is longer than the average of 4 to 5 lakh years. Nobody knows the cause of these reversals but they have been occurring regularly throughout the geological past, and their evidence is preserved in the rocks of the continental and oceanic crusts, especially in the basalt igneous rocks. Before the publication of Hess's hypothesis, some geophysicists had found reversals of magnetism in alternate parallel strips in the bottom rocks of North-East Pacific Ocean which was different from the random pattern of reversals found on the continents. Each strip of rock was magnetized in a particular direction but the rocks were magnetized in opposite directions in alternate parallel strips i.e. there were strip reversals of magnetism as shown in Fig. 24

Although in 1963 reversal of the magnetic field of the earth was a controversial subject, Vine and Mathews considered the strip reversals of magnetism in the oceanic crust to be significant and suggested that as new ocean floor is built on the ridge crest and spreads on both sides of the crest, there is intermittent reversal in the magnetic field of the earth, and, therefore, we should find in the basalt rocks on either side parallel to the ridge axis alternately normally and reversed magnetized material. In other words, whereas Hess in his hypothesis had imagined the ocean crust to be a sort of conveyer belt where new crust on the ridge crest spread in opposite directions on either side on the ridge, Vine and Matthews considered it to be a conveyor belt cum tape recorder which provided compelling and detailed evidence of sea-floor spreading. "As the moving convection cell carries basaltic crust away from the ridge, slowly but continuously, the magnetic reversals taking place every few hundred thousand years or so leave their imprints on the newly formed rocks emerging from the ridge; as every reversal occurs, so it is recorded in the formation of reversely-magnetized strips on each side of the ridge. Once a strip of new ocean floor has been formed and acquired its magnetic recording, nothing further happens to it but it continues to move away slowly from the ridge, at a rate of anything up to several centimetres per year"[1].

Four things were necessary for the confirmation of the views of Vine and Matthews. Firstly, the recording of magnetic anomalies i.e. reversal of the magnetic field in the rocks should be in strips parallel to the ridge axis. Secondly, these strips on either side of

1. Francis, P.—*Volcanoes*, Penguin Books, 1976, p. 33

ridge should be symmetrically dispersed about the ridge i.e. the strips on either side should be arranged in the same sequence. Thirdly, the same sequence of anomalies should be reproduced across all ridge flanks in the different oceans. Fourthly, the rocks situated at a greater distance from the ridge should be older than those situated near it.

Within the course of the next four to five years so much data was available from extensive investigations that the first three facts stated above were readily confirmed. In other words, the strips were found to be parallel and symmetrical to the ridge and the sequence of anomalies was found to be the same in all the oceans. An excellent and oft quoted example at that time was the magnetic pattern found near the Reyknjanes ridge, south of Iceland, which is a part of the Mid-Atlantic ridge. Here the magnetic strip reversals are very clear and impressive (Fig 25). By 1965 magnetic anomaly data were available from all the oceans in adequate quantities and it was possible to analyse them with the help of the computer. These studies established beyond doubt that in all the oceans the same sequence of magnetic anomalies parallel to the mid-oceanic ridges was present. Assuming that the rate of sea-floor spreading has been 2 cms per year on each side of the ridge in the South Atlantic, it has been found that during the last 76 million years there are evidences of 171 magnetic reversals in similar sequence in every ocean.

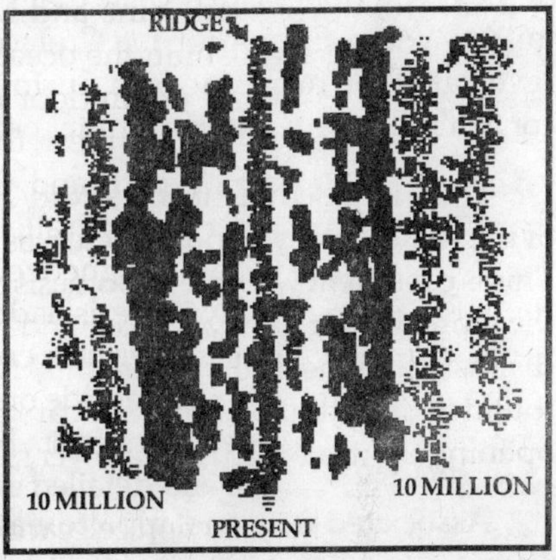

Fig.25—Magnetic strip reversals near the Reykjanes ridge, south of Iceland.

But it remained to be proved that the rocks furthest away from the ridge were the oldest. This was extremely important. This work was started in 1969 under an extensive American programme known as JOIDES (Joint Oceanographic Institutions Deep Earth Sampling). This organization selected 31 spots in the Atlantic and

53 in the Pacific Ocean for the sampling. From these investigations it was found that the age of the marine sediments above the basaltic layer goes on increasing with distance from the ridge. In the Atlantic Ocean the age of the marine sediments near the ridge was found to be less than 7 million years, while the age of the sediments furthest away from the ridge has been calculated to be more than 160 million years. These investigations also confirm the fact that during the entire Cenozoic era the sea-floor spreading has taken place at the rate of about 2 cms per year on each side of the ridge. From this it can also be concluded that nearly one half of the existing deep ocean floor has been built during the Cenozoic era i.e. during the last 65 million years which is but 1.5% of the entire geological period. Further, it appears to be highly probable that the age of the remainder of the ocean floor is Mesozoic, that is, it is not older than 65 to 225 million years. Except for the Cretaceous, there is no evidence of reversal of the magnetic field during the Mesozoic era and this is considered to be a relatively quiet period.

There is no sediment on the Mid-Atlantic ridge and the thickness of the sedimentary deposit increases with distance from the ridge. This is consistent with the hypothesis of sea-floor spreading, because the more distant parts are older and have had a longer time at their disposal to deposit the sediments. No Pre-Jurassic sediments are found in the Atlantic, and it appears that the Atlantic Ocean started opening up nearly 200 million years ago.

Associated with and approximately at right angles to the mid-ocanic ridges are fracture zones called *transform faults* that offset the median axis of the ridge. Only in a relatively short section of the fault are fault movements in opposite directions, and beyond the material on opposite sides of the fault is moving in the same direction. Earthquakes are associated with only that portion of the transform fault where movements are opposed. Such faults were first recognized and named transform fault by Tuzo Wilson in 1965, in distinction from the more familiar transverse faults found on continents, where all the rocks move in opposite directions on opposite sides of the fault.[1] The Atlantic Ocean is split into numerous roughly latitudinal strips that move slowly away from the ridge at

---

1. Wilson, T.J.—"A new class of faults and their bearing on Continental Drift", *Nature*, 1965 Vol, 207, pp. 343-7

slightly different rates, inevitable with sphere, and the strips are separated by great fracture zones along the transform faults and marked by huge submarine escarpments.[2] With great intuition, Wilson recognized that the apparent offsets of the ridges might be original features formed when the continental crust split apart and new oceans were born. The mid-Atlantic ridge, including its offsets, for example, is parallel to shores of the bordering continents.

In the Pacific Ocean, the spreading site ridge is not located in the middle but between Antarctica and Newzealand from where it turns eastwards to the California coast. The southern section of this ridge is called the Pacific-Antarctic ridge and the remaining part is known as the east Pacific Rise. From the evidences of magnetic anomalies, the age of the rocks of the ocean floor and the deposition of sediments it has been inferred that the rate of spreading there is up to 6 centimetres per year. North Pacific is spreading towards the north-west while the movement in the Southern Pacific is towards the south. Evidence of spreading is also provided by the composition of volcanic rocks. In the mid-Pacific the volcanic rocks are basalt which shows that the ocean floor is new, but in the volcanoes near the coasts the rock is andesite. The mid-Pacific volcanoes provide further evidence of sea-floor spreading. Of all the volcanic islands in the mid-Pacific, the island of Hawaii is the most prominent. This is located in the east and contains active volcanoes. In the island located further west we find extinct volcanoes where there has been erosion after the cessation of eruption, and coral reefs have developed and atolls have been formed as a result of sinking. In other words, the oldest islands are to the west and the youngest islands to the east. From this we may conclude that the centre of volcanic activity is moving towards the east or that the crust is moving west with reference to the 'hot-spot' in the earth's interior from which the volcanoes are erupted. The apparent rate of migration of the volcanicity is the same as the rate of sea-floor spreading, which cannot be easily rejected as mere coincidence[1].

The Indian Ocean started developing in the Jurassic period, but there is difference of opinion regarding the arrangement of the continents before the drift. According to one view Australia was

---

1. Ollier, C.D.—*Tectonics and Landforms*, Longman, New York, 1981, p.19.
2. Ibid., p.21

linked to India, there being a good Permian fit between India and Western Australia. According to the other view India lay close to Antarctica and both India and Antarctica were located between Africa and Australia, this view being corraborated by computer fit. This much is, however, clear that once India was separated from the main landmass, it started moving north at the rate of approximately 16 centimetres per year. There is an extensive mid-ocean ridge in the Indian Ocean. It starts from the Red Sea in the north, continues to the Seychallies under the name of Carlsberge ridge and further south is known as the Indian Ocean ridge. In fact, in the eastern Indian Ocean there are several separate spreading sites, the most important being the 90^0 E. Ridge.

A mid-oceanic ridge is found between Africa and Antarctica and there is a ridge surrounding the Antarctica continent. Antarctica separated from Australia in the Eocene period. In the maps of Wegener also this separation is shown to have taken place at a fairly later date.

	Opening of the Atlantic Ocean	*Fragmentation of Gondwanaland*
	Opening of Red Sea, Gulf of Aden, Gulf of California and initiation of Galapagoes Rise	
Tertiary	Opening of the extreme North Atlantic and the Arctic Ocean	Separation of Australia from Antarctica and India from the Seychelle Bank.
Upper Cretaceous	North Atlantic extends northwards to form the Labrador sea and Bay of Biscay.	Initiation of Pacific Antarctic ridge between Newzealand and West Antarctica.
100 M.Y.		
Lower Cretaceous	Opening of the South Atlantic.	
Lower Jurassic	Partial opening of North Atlantic between N. America and Africa	Separation of Australia and Antarctica from Africa and S.America
200 M.Y.		
	Separation of North America from Eurasia.	

On the basis of the above studies as also on the basis of other evidences relating to the age of igneous rocks, marine transgressions etc., a probable time-table of continental drift has been prepared which is given in the table above. The estimated rate of spreading of 2 centimeters per year is adequate for separation of South America from Africa and the formation of the South Atlantic Ocean in about 200 million years. At this rate 40 million years are enough for the separation of Australia from the Antarctica.

We thus find that the hypothesis of sea-floor spreading is based on investigation of diverse kinds which relate to the form of the ocean floor and the distribution of sediments on it, the distribution of faults and earthquakes, the structure of the earth's interior and the magnetic field of the earth and its periodic reversal. This concept enables us to explain many surface features as well as the internal processes of the earth. As a result of the studies relating to sea-floor spreading and palaeomagnetism, increasingly convincing evidence is being made available for continental drift. In fact, the hypothesis of sea-floor spreading has now developed into a more sophisticated, extended and comprehensive theory of Plate Tectonics, in the context of which the terms continental drift and sea-floor spreading appear to be a bit out of date. This new theory is the subject matter of the next chapter.

Questions

1. Explain fully the hypothesis of sea-floor spreading and its significance.
2. "At the present time there is no serious obstacle to the acceptance of sea-floor spreading and continental drift as facts rather than theories." Substantiate.
3. Show how the hypothesis of sea-floor spreading forms the basis of the new theory of Plate Tectonics.
4. Discuss the concept of sea-floor spreading and the evidences supporting it.

Chapter 10

Plate Tectonics

The hypothesis of plate tectonics is an extended and more comprehensive version of the theory of sea-floor spreading. This is a great unifying concept which "draws sea-floor spreading, continental drift, crustal structures and world pattern of seismic and volcanic activity together as aspects of one coherent picture."[1] It is based on the çoncept of sea-floor spreading proposed by Harry Hess and confirmed by Vine and Matthews by their interpretation of linear magnetic anomalies. The term plate was first used by Tuzo Wilson in his definition of transform faults in 1965, but the hypothesis of plate tectonics was first outlined by W.J. Morgan of the Princeton University in 1967, basing himself on the findings of Wilson regarding transform faults and their relationship with the mid-oceanic ridges. More or less concurrently but independently D.P. Mackenzie of Cambridge and Parker of America had arrived at conclusions similar to those of Wilson. Thus Wilson, Mackenize and Parker through their published research findings laid the foundation of this new hypothesis, though it was really proposed by Morgan. It first came to be known by the name of New Global Tectonics but after sometime the term Plate Tectonics gained currency.

Division of Crust into Plates

Morgan established that the mid-oceanic ridges form the boundary between two plates that are moving relative to one another and the transform faults that intersect the mid-oceanic ridges lie on small circles that are concentric about a pole of rotation. "The movement of a plate is defined by the position of its pole of rotation and its angle of rotation about the rotation axis; its rate of movement varying with distance from the pole of rotation, being nil at the pole and reaching a maximum on the

1. Oxburg, E.R.—Plate Tectonics in *Understanding the Earth*, edited by Gass, Smith and Wilson, 1973, ELBS Ed., p.263.

equator relative to the pole of rotation". Morgan divided the earth's surface into 20 plates, and determined their poles of rotation by drawing great circles at right angles to the relevant groups of transform faults. Le Pichon (1968), however, simplified the concept of plate tectonics by dividing the earth into six major plates and a few smaller ones after taking into consideration the spreading rates calculated from magnetic anomalies as well as the strike of the transform faults intersecting the mid-ocean ridges.

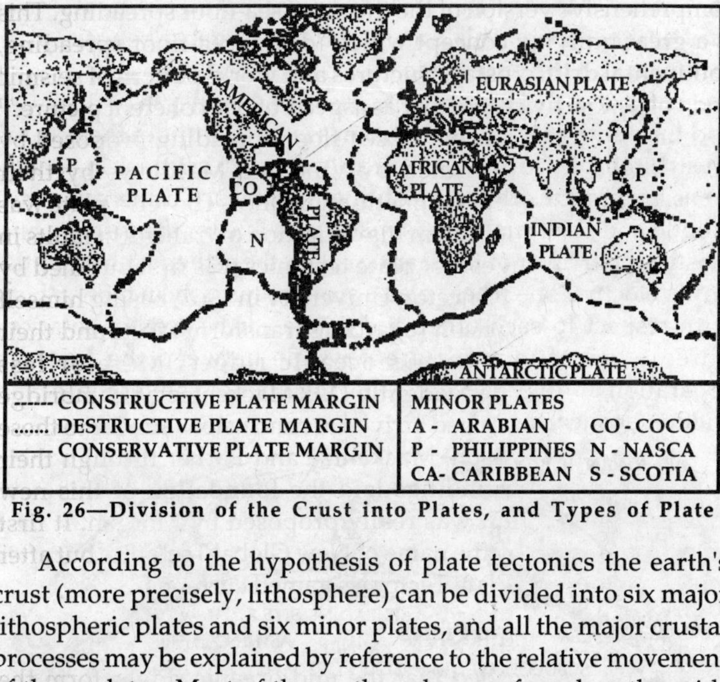

Fig. 26—Division of the Crust into Plates, and Types of Plate

According to the hypothesis of plate tectonics the earth's crust (more precisely, lithosphere) can be divided into six major lithospheric plates and six minor plates, and all the major crustal processes may be explained by reference to the relative movement of these plates. Most of the earthquakes are found on the mid-oceanic ridges and the two coasts of the Pacific ocean. By taking these major linear belts of seismic activity as the base the earth's crust has been divided into six major and six relatively minor plates which are named below (See Fig. 26):

Major Plates : 1. Indian Plate or Indo-Australian plate, 2. Pacific Plate, 3. American Plate (also divided into North American and South American), 4. African Plate, 5. Eurasian Plate, 6. Antarctica Plate.

Minor Plates : 1. Arabian Plate, 2. Philippines Plate, 3. Cocos Plate, 4. Caribbean Plate, 5. Nasca Plate or the East Pacific Plate, 6. Scotia Plate.

The Plates are the inert aseismic regions of the earth bounded by narrow mobile belts which are characterized by seismic and volcanic activity or by orogenic belts (where continents form the leading edges of plates). Each plate is about 70 to 100 kms thick and its configuration is not related to the distribution of land and water. Most of the plates include both continental and oceanic lithosphere. This is an important difference between plate tectonics and the earlier ideas of continental drift. The area of the plates is fairly large in comparison to their depth or thickness and the depth of the plates becomes even less under the oceanic crust. Lithospheric plates are on an average 20 km thick in oceanic areas and 80 kms thick in continental areas. All the plates are mobile and their mobility is made possible by the viscous and plastic upper layer of the mantle called the asthenosphere or the Low Velocity Zone. These plates are continuously in motion both with respect to each other and to the earth's axis of rotation. Three types of motions are possible between the plates—separation or divergence, closing together or convergence, and friction or shearing.

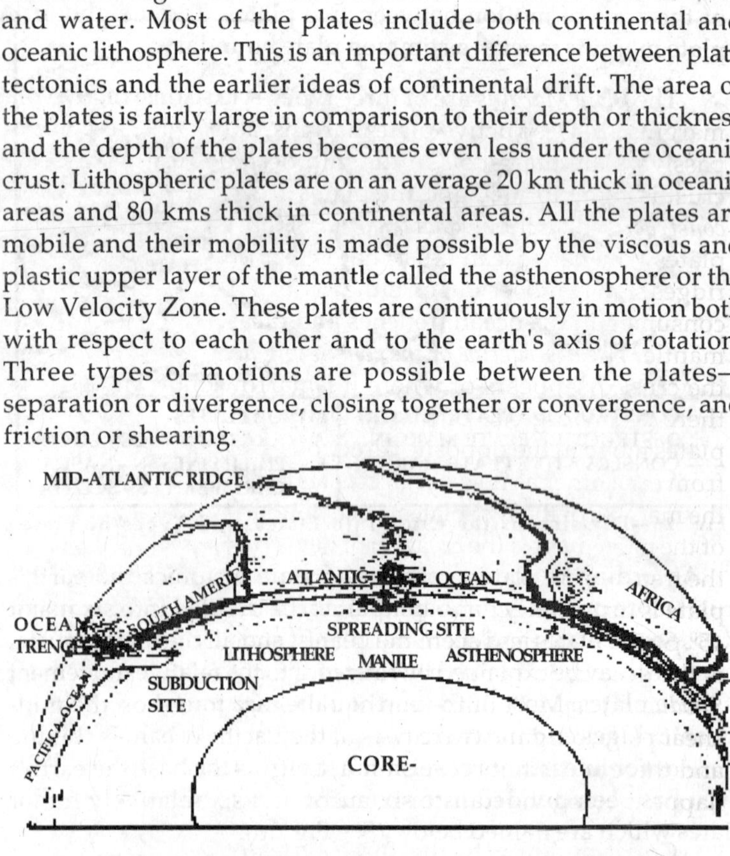

Fig. 27—Molten material rises from the asthenosphere on the mid-ocean ridges and forms new oceanic crust. Here there is divergence of plates. On the west S. America and the western part of S. Atlantic form one plate and on the east Africa and the eastern half of S. Atlantic form another plate. The new ocean crust slowly moves out on either side of the mid-ocean ridge and is finally consumed into the trenches and returns to the asthenosphere.

Plate Boundaries and Plate Margins

Nearly all seismic, volcanic and tectonic activities are confined to the plate margins and are related to the differential motion of the neighbouring plates. The boundaries between the different plates have therefore a special significance. The distinction between plate margins and plate boundaries must be clearly understood. The *Plate margin* is the edge or marginal part of a particular plate, while the *plate boundary* is the surface trace of the zone of motion between two plates. Two neighbouring plate margins meet at a common plate boundary.

The *Plate Margins* are of three types— (a) Constructive plate margins, (b) Destructive plate margins, and (c) Conservative or passive plate margins. On the mid-oceanic ridges new ocean crust is being formed and the plates are growing; this is called *constructive or accreting plate margins*. Here there is divergence of plates. The new ocean crust that is formed on the mid-oceanic ridges slowly moves out on either side of the ridge and is finally consumed in the ocean trenches and disappears down into the mantle. The *destructive or the consuming plate margins* are where the crust disappears or where it is narrowed by folding. Here there is convergence of plates. On the mid-oceanic ridges the plates move in the opposite directions, but they do not separate from each other as new ocean crust is constantly being added on the margins of the two plates. Such boudaries where the net effect of the movement is the creation of new crust are called *Sources*. At the trenches the plates converge or come closer together and one plate is thrust under another plate at an angle of approximately 45^0. Such boudaries where the net effect of the plate movements is the destruction or narrowing of the crust are called *Sinks or Subduction zones*. A third kind of plate margin is conservative or shear plate margins where two plates just glide past one another and there is neither creation nor destruction of the crust. This happens along the transform faults.

Corresponding to the three types of plate margins and relative motion, *three types of plate boundary* are also recognized : divergent, or accretionary plate boundaries, convergent or destructive plate boundaries and transform or conservative plate boundaries. Divergent boundaries are found at mid-ocean ridges where new lithosphere is created and spread on the seafloor on either side of the ridge. Convergent plate boundaries occur at the deep sea trenches and the related subduction zones as also zones of compressional deformation between converging plates in continents (such as the Alpine Himalyan zone) but in the latter case they are much broader than normal plate boundaries and are broad continental convergence zones rather than boundaries. "Transform boundaries are those where plates slide past one another with no change of area; they join one plate boundary to another, and can thus *tranform* one type of tectonic boundary (e.g. extensional) to another (e.g. compressional), hence the name. They are found in both continental and oceanic lithosphere.[1]

Creation of new crust

An important fact to remember is that with the exception of three plates —Nasca, Pacific and Cocos, all the other plates contain both continents and oceans. But only those parts of the plates which contain the oceanic crust essentially take part in the processes of plate creation and destruction. The process of the creation of new crust at the mid-oceanic ridges and its spreading are responsible for the formation of only the thin layered oceanic crust and not of the thick continental crust. The geo-chemical composition of the continental crust is different and it is composed of rocks of relatively low density. The density of the continental crust is so low that it does not sink into the mantle but continues to stay at the surface.

Basic Assumptions of Plate Tectonics

There are three basic assumptions underlying the hypothesis of Plate Tectonics :—

(i) There is spreading of the sea floor, and new oceanic crust is being continually created at the active mid-oceanic ridges and destroyed in the trenches.

(ii) The area of the earth's surface is fixed and during the last 600 million years the radius of the earth does not appear to have increased by more than 5% . In other words, the amount of crust consumed almost equals the amount of new crust created.

(iii) The new crust that is formed becomes a part and parcel of a

1. Searle, Roger- Plate Tectonics in '*The Earth*' op.cit, 2000, p.829

plate which normally includes both continental and oceanic crusts, although there are some plates which are almost wholly composed of oceanic crust. The process whereby one plate is consumed by and disappears under another plate is called *subduction.*

It has been revealed clearly by seismic studies that the continental crust is thick and unlayered or irregularly layered, and on the contrary the oceanic crust is thin, layered and uniform. About 7 kms below the ocean floor is the Moho or M-discontinuity and in these 7 kms of thickness there are three horizontal and virtually undisturbed layers. The uppermost layer consists of sediments whose maximum thickness is about 1 km and which is absent on the ridge tops. Below this are two layers roughly one and four kms thick respectively. Had there been any notable crustal deformation such as we find in the continental orogenic belts, the layers of the oceanic crust would not have remained uniform and horizontal but would have been buckled and distorted, but this is not so. One arrives at the same conclusion also from the simple and regular pattern of magnetic anomalies found on the ocean floor for long distances starting from the mid-oceanic ridges. In other words, deformation tends to occur not within the plates but around their margins.[1] If we extend the lines of the active mid-oceanic ridges and trenches towards the continents, we find that all these lines meet zones of continental tectonic activity. For example, the trench system of Indonesia meets the eastern edge of the Himalayan mountain system. Further, we find that all the zones of continental tectonic activity are located on either the continuation of the above lines or on plate margins, that is, alongside sources or sinks. For instance, the Andes mountains are located on the western margin of the (South) American plate near the Peru trench. If the continental continuation of oceanic sinks and sources are regarded to be continuation of the plate boudaries, it becomes possible to recognize a world-wide system of plates encompassing both continents and oceans.[2]

We have seen that the plates are in continual motion in relation to one another. Simultaneously, they are also joined

1. Oxburg, E.R.-op. cit, p. 264.
2. Ibid., p. 264

together on the surface of the spherical earth. This means that the movements of all the plates are to some extent inter-dependent, and any change in either the direction or rate of movement of one plate is not possible without concomitant changes in the other plates. Morgan has explained that these movements and their rates can be examined with reasonable accuracy with the aid of the laws of spherical geometry. As the plates are spherical shields moving across the surface of the earth, their motion can be defined by a pole of rotation and by an angular rotation velocity.[1] The direction of relative spreading is not perpendicular to either the mid-ocanic ridges or to the linear arrangement of magnetic anomalies but is parallel to the transform faults.

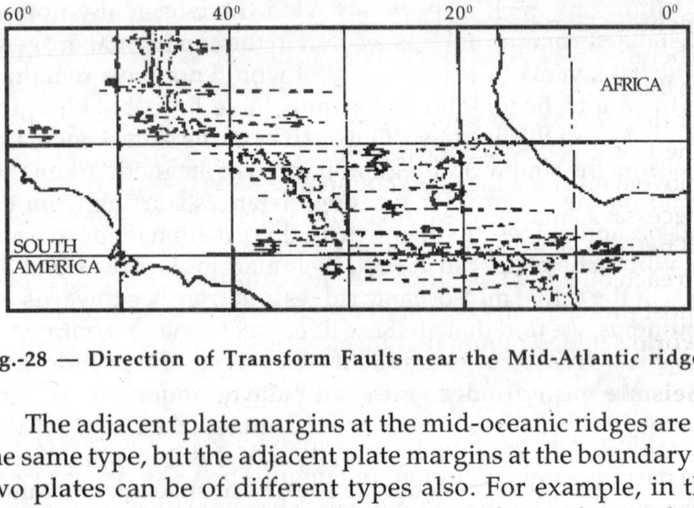

Fig.-28 — Direction of Transform Faults near the Mid-Atlantic ridge.

The adjacent plate margins at the mid-oceanic ridges are of the same type, but the adjacent plate margins at the boundary of two plates can be of different types also. For example, in the trenches we find the destructive margin of one plate and the conservative margin of the other plate. There is no geometrical limitation on the shape of the plate either. It can be of any shape and any combination of margins can be found on its boundaries. But at the level of the entire globe there is one limitation that operates and that is that the pattern of plate boundaries and their relative velocities should be such that there is no change in the surface area of the earth. The western boundary of the African

1. Boillot, G. — *Geology of Continental Margins*, Translated by A. Scarth, Longman, London, 1981, p.12.

plate is the mid-Atlantic ridge and the eastern boundary is the Indian Ocean ridge. Both are active centres of spreading and the direction of transform faults of both is roughly east-west. There is no trench in between these two ridges. Thus the African plate has two opposed constructive margins, and it should, therefore, be spreading in the east-west direction. This means that if there is an increase in the area of some plates, then there should be proportionate decrease in the area of other plates, otherwise the surface area of the earth can not remain constant. Thus in principle it is possible for some plates to be destroyed altogether, and it is also possible for the mid-oceanic ridges and the deep oceanic trenches to change their relative position either in relation to each other or in relation to the conservative features on the earth. But we know that for physical reasons a plate which includes continental crust, can never be destroyed completely. Therefore whenever the continental portion of a plate comes close to the trench or sink, possibly there is change in boundary motion or the relative motion between the two plates stops or ends. In the event of the second possibility, adjustment in the rate of movement along the other boundaries of the two plates would be necessary, and the creation of one plate as a result of the union of two plates becomes possible. Where two continents come closer to each other on two sides of a sink, this possibility seems to be most probable and it is under such conditions that some of the world's folded mountain ranges appear to have been formed.

Seismic and Volcanic zones and Plate tectonics

Most of the earthquakes and volcanoes of the world are found in narrow, semi-continuous but definite belts. These belts are for the most part confined to the plate boundaries. In fact, seismic belts have been used as basis for the delimitation of plates. Seismicity is limited to mid-oceanic ridges, trench-arc systems, major fracture zones and the yound orogenic belts, and all these features are indicative of plate boundaries. On all plate boudaries, and especially on the mid-oceanic ridges shallow-focus earthquakes are found, but intermediate and deep-focus earthquakes whose focus is deeper than 200 kms are mostly met with in the oceanic trench zones. This is the area where the plate descends starting from the surface of the trench to depths of 300 to 400 kms which may sometimes extend to depths of 700 kms.

The inclination of this zone is towards the continental landmass and the angle of inclination is 30⁰ to 80⁰, though in most cases it is 45⁰ to 60⁰. Such inclined seismic focus zones are found near the entire Pacific coast. They are called *Benioff zones*, as the seismologist Hugo Benioff was the first to discover them. As a general principle it may be stated that the seismicity of the Benioff zones is related to the downward motion of the lithosphere.

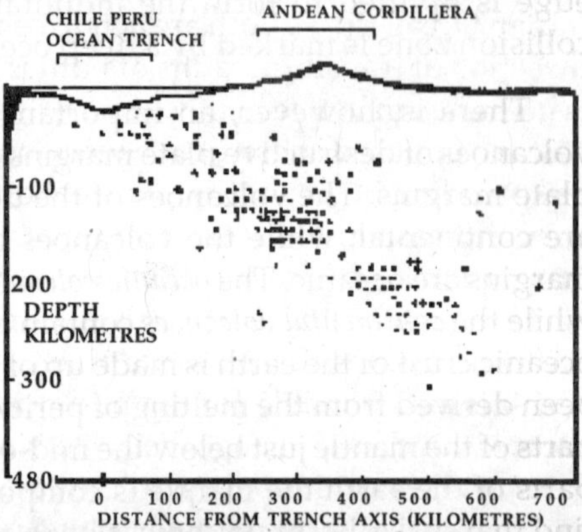

Fig.29 — Distribution of the focii of earthquakes near the Chile-Peru trench.

Fig. 29 shows the Benioff zone near the Chile-Peru trench. About 350 kms east of the trench lies the summit line of the Andean Cordillera which stands nearly 15 kms above the trench. The focii of the earthquakes are limited to a narrow zone. This zone is inclined towards the continental coast at an angle of 60⁰. It extends below the landmass to a depth of about 300 kms and comes to the surface in the trench. According to the principle of sea-floor spreading the Pacific Ocean is moving towards South America from the East Pacific Rise, but South America is moving west from the mid-Atlantic ridge. Thus at the junction of the two plates there is a steady-state continuous collision, and on the site of the collision we find a major belt of active volcanism which is but a part of the Pacific Ring of Fire, which encircles the entire Pacific coast. As a result of the collision of the two plates, the advancing continent overrides the ocean floor, which is thrust down under the continental plate. On account of the generation of heat the downgoing slab melts and produces andesitic magma. Part of this magma is intruded as batholiths and part is erupted as andesitic volcanoes. We thus find a long belt of volcanoes in the Andes mountains. The downward progress of the slab is recorded by a series of earthquakes that show a fairly consistent increase in depth with distance from the plate edge making an inclined earthquake zone called the Benioff zone. The continental edge is uplifted to form the mountains (the Andes) and the collision zone is marked by a deep ocean trench.[1]

There is, however, an important difference between the volcanoes of destructive plate margins and those of constructive plate margins. The volcanoes of the destructive plate margins are continental, while the volcanoes of the constructive plate margins are oceanic. The *oceanic volcanoes* contain basaltic rocks, while the *continental volcanoes* contain andesite rocks. The entire oceanic crust of the earth is made up of basalt and this basalt has been derived from the melting of peridotites found in the upper parts of the mantle just below the mid-oceanic ridges. Since in all parts of the earth the mantle is composed of similar rocks, we find that such distant volcanic islands as the Hawaii and Iceland are composed of the same type of basalt rocks. In contradistinction of this, the magma of the continental volcanoes is derived from three different sources—some from the mantle below the downgoing oceanic plate, some from the partial melting of the oceanic plate and some from the melting of the lower part of the granitic continental crust. Besides, some oceanic sediments on the oceanic plate may also melt. The result is that the continental volcanic rocks are very different from the oceanic volcanic rocks and are called *andesites*. This name has been derived from the Andes where all the volcanoes contain this type of rocks. The properties of andesite (i.e. percentage of silica, density, colour etc.) are in between those of basalt and granites. Its colour is light brown, and it contains 60% silica and small white crystals are found in a matrix of fine particles. These white crystals are called phenocrysts and are a distinguishing feature of andesites.

It is important to remember that we find basalts on the constructive plate margins (mid-oceanic ridges) and andesite rocks on the destructive plate margins.

Rift Valley Volcanoes

There are some volcanoes which are not related to either constructive or destructive plate margins. Plate tectonics helps

1. Ollier, C.D.—op, cit, 1981, p. 42

usto understand their distribution as well. The most notable of such volcanoes are those in East Africa which are associated with the East African Rift Valley system. The Great Rift Valley of East Africa extends from the Zambezi in the south to as far as Ethiopia and the Red sea in the north over a distance of nearly 3,000 kms. Throughout the East African Rift system we find a chain of volcanoes. The rocks erupted by these rift valley volcanoes are a special kind of basalts called alkali basalts which have a relatively high percentage of sodium and potassium.

Till recently it was believed that the Red Sea is a rift valley where the sea has entered the depressed zone. But the floor of the Red Sea is not wholly composed of continental rocks, and on part of its floor are found basalts resembling those of the ocean floor. It is possible that the Red Sea is a proto-ocean where there has been no development of a mid-oceanic ridge but oceanic crust is being created. In other words, the Red Sea is becoming wider and Africa and Arabia are separating from each other. In all probability Red Sea was originally a rift valley in which for hundreds of thousands of years, volcanic activity has continued, and it became a sea about 20 million years ago. Similarly in the East African Rift system where volcanic activity has been going on for the past several million years, it is possible that a sea is being created and Africa is dividing into two parts. According to the latest views on the subject probably this may not happen, as the Atlantic Ocean is becoming wider so rapidly that Africa is pressing against Arabia with the result that the Red Sea may become narrower or close altogether.

Mid- Plate Volcanoes

In addition to the rift valley volcanoes there are some volcanoes which are situated in the middle of the plates. These are called mid-plate volcanoes. Examples are to be found in central Sahara, but probably the best examples are to be found in the Hawaii islands in the Pacific Ocean. Apparently the location of these volcanoes is difficult to explain with reference to the theory of plate tectonics. But these volcanoes are those which were originally formed near the mid-oceanic ridges and have since migrated from their place of origin with the spreading of the ocean floor. They are related to the hot spots below in the mantle which are fixed, but the oceanic plate is moving. The result is that the volcanoes are found stretched in a linear belt with the young active volcanoes towards the eastern end and the extinct and the old volcanoes towards the western end of the belt, as the crust is moving west over a deep-set hot spot from which the volcanoes have erupted.

Continental Collisions

Plate tectonics offers an acceptable, plausible explanation of the process of continental drift. There is a general agreement that before continental drift all the continents together constituted one landmass round the south pole, and India was not attached to the mainland of Asia. India was probably situated between Africa and Antarctic, and there was sea between India and the mainland of Asia which has been called Lauretia. When continental drift began nearly 200 million years ago and the vast parent continent started breaking up, India moved rapidly towards the north and ultimately joined the Asian mainland about 60 million years ago. At about this time Africa also drifted towards the north, and the sea between Europe and Africa was narrowed down to its present width. As a result of this great continental collision the Alpine -Himalayan mountain system was formed, which extends as far as the ocean in the east. In this entire extensive mountain system no volcanoes are found. There are some famous volcanoes located in the Mediterranean Sea (Vesuvius. Stromboli, Etna etc.), but they are found south of the Alpine system and not within it. When Africa drifted towards Europe, then the oceanic crust in between the two was thrust under the continental crust and thus a destructive plate margin was formed. This plate margin is more complex than the Andean plate margin. In the Andes the volcanoes are found on the mountain summits, but in the Mediterranean region they are found in several arcs which cut across the Italian and Greek islands and pass towards the Asia Minor.

The situation in the Himalayas appears to be even more complicated. Here also there existed an oceanic crust between the Asiatic mainland and India, and there are evidences to show that there existed here a destructive plate margin where the oceanic crust has been destroyed. It has been suggested that there was a chain of active volcanoes here as well but they probably became extinct millions of years ago and have since been denuded away. There is no evidence to indicate the presence of any considerable section of oceanic crust going down under the Himalayas into the mantle. Thus the normal Benioff zone which we find in the Andes does not appear to be present here.

Mechanism for Plate creation and movement

In 1948 Beno Gutenburg of California proposed that there is a *Low Velocity Zone* in the upper part of the mantle about 80 kms below the earth's surface where the velocity of the earthquake waves suddenly slows down. This idea was then not accepted, but the researches conducted between 1960-70 have confirmed his view, and the existence of such a low velocity zone is now an accepted fact. This low velocity zone or LVZ is world-wide but its characteristics change with the differences in the characteristics of the crust in the different parts of the world. Broadly, however, it can be said that this is found at a depth of 50 to 250 kms and is commonly about 150 kms thick. As the velocity of the seismic waves decreases with increase of temperature, the temperature in LVZ must increase rapidly. In fact, this layer is extremely hot and the mantle is either in molten state or very near molten state. The rocks are therefore very weak and have very low shear strength. Possibly this is the region where the 100-150 km thick plate gets decoupled from the deeper mantle. This hot and relatively weak layer situated just below the 100 km thick relatively strong lithosphere is called *asthenosphere*.

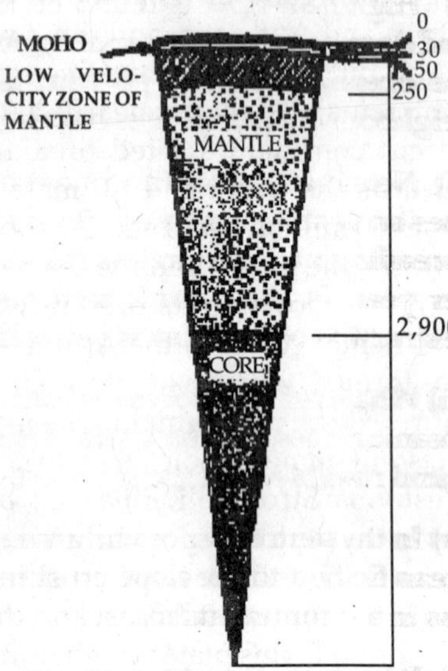

Fig. 30 — The Low Velocity Zone lying between the mantle and the crust.

Our knowledge about the thickness of the plates is very limited. But it is believed that it is at least 5 to 7 kms thick below the ocean floor. To the extent the continental crust is related to the plate movements, the thickness of the plates below the continents is considered to be from 40 to 100 kms.

The creation of plates and the causes of their movement are an extremely complex and controversial subject,, as we know comparatively little about the physical properties, temperature distribution, distribution of radioactivity etc in the earth's interior. In any event, in any explanation of the causes of plate movements, it is necessary to bear in mind the following facts, because any possible mechanism will have to operate within the framework of these limitations and conditions :—

(i) High heat flow is found on the active mid-oceanic ridges, and decreases with distance from the ridge. On the contrary, in the oceanic trenches no heat flow is found, though in the neighbouring island-arcs heat flow is met with at depth.

(ii) Near the mid-oceanic ridges the spreading of the ocean floor goes on symmetrically on either side of the ridge and the rate of spreading varies from less than one centimetre to 6 centimetre per year. In the oceanic trenches the rate of consumption or destruction of the crust is between 5 to 15 centimetres per year.

(iii) Whatever the process of formation of new crust on the mid-oceanic ridges, the thickness of the newly formed crust has been found roughly to be the same irrespective of its rate of spreding.

(iv) In the centre of spreading the ridge rises 2 to 4 kms above the ocean floor, and the slope on either side of the summit becomes less in a symmetrical manner.

(v) It has been found from gravity observations that the ridge is in a state of isostatic equilibrium. On the other hand, the trenches (where deposit of sediments is found) are in a state of isostatic disequilibrium and here we find the maximum negative gravity anomaly.

(vi) There are great differences in the distance between sinks and sources. At some places this is only a few hundred kilometres while at other places it is up to 3,000 kms

(vii) The lithosphere has strength and the big lithospheric plates can also move considerable distance without internal deformation. The length of some plates is twenty times or more than their thickness. In this kind of length to thickness ratio it is not possible for compressional or tensional stress to travel from one edge of the plate to the other unless the frictional resistance at the base of the plates is very poor.

On account of the presence of high heat flow near the ridges several persons have suggested that the movement on the surface is related to some sort of thermal convection in the earth's interior. Recently D.L. Turcotte and E.R. Oxburgh have put forward the concept of thermal boundary layers based on thermal convection. On account of thermal instability in the mantle, heat is conducted upwards through the movement of the molten material. Once

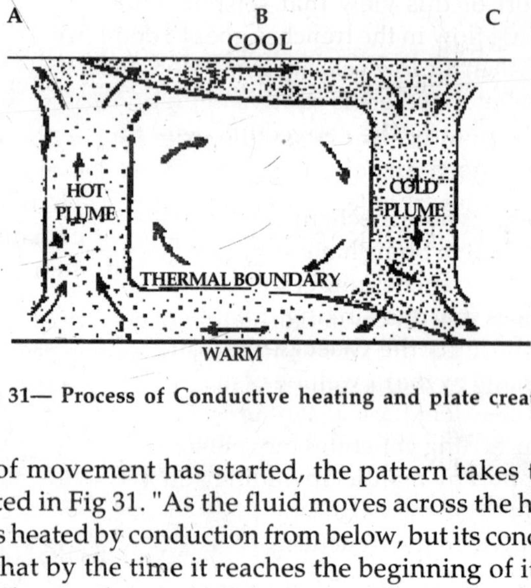

Fig. 31— Process of Conductive heating and plate creation.

this kind of movement has started, the pattern takes the form respresented in Fig 31. "As the fluid moves across the hot lower surface it is heated by conduction from below, but its conductivity is so low that by the time it reaches the beginning of its ascent conductive heating has extended only a small way into the flow—a warm thermal boundary layer has formed. The ascending flow is formed by the coalescene of two warm boundary layers to give a hot 'plume.' As it approaches the upper surface the plume flow diverges once more to give the symmetrical upper horizontal flow. The hot fluid now moves along a cold surface, and as it flows latterly it loses heat by conduction upwards with the generation

1. Oxburg, E.R.—Plate Tectonics in *Understanding the Earth* edited by Glass, Smith and Wilson, 1973, p.277.

of a progressively thickening cold boundary layer. A descending cold plume is produced by the convergence of two such boudary layers. The interior of the cell plays no part in the heat transfer process, and within it there is an adiabatic temperature gradient. The heat is picked up, transported, and lost by a kind of thermal conveyor belt formed by the outside of cell.[1]" According to this hypothesis the plate represents the cold thermal boudary layer which is highly viscous but where the upper part is quite rigid. Asthenosphere or the low velocity zone is the only layer which can generate and support such a convection flow. But this layer is relatively thin. Therefore some people have suggested that the source of the convection flow lies in the mantle. It is pointed out in support of this view that seismic evidence shows that the descending flow in the trenches goes to depths of 700 kms which is deeper then the low velocity zone. Moreover, it is also necessary to remember that there is relative migration of sources and sinks and consequently the convection cells may also change their position.

Other possible mechanisms for plate movements have also been suggested. Some have suggested that plates slide from the ridges to the trenches under the influence of gravity. Others are of the view that the sinking crust beneath the trench systems constantly pulls the crust in that direction and this is being compensated by the upwelling of magma at the ridges. Still others hold the view that the constant upwelling of magma at the ridges and its spreading generates movement in the plates and they are pushed apart by mantle upwelling and in order to maintain the surface area of the earth one plate is being subducted under another plate in the trenches.

Although several possible driving mechanisms for plate movements have been suggested, it seems probable that plate movements are not solely attributable to any one of these but result from a combination of all acting in varying degrees on individual plates.[1] The process of convective movements would appear to be the most probable mechanism, but nothing definite can be stated about the source of the convection flow in our present state of knowledge. Most probably, the source lies in the

---

1. Vine, F.J.—Sea Floor Spreading in *Understanding the Earth,* Ibid. p.249

Low Velocity Zone but one cannot say so with any definiteness, and it is possible that some hot spots in the mantle are the source of the flow. The geometry of convection flow is also extremely complex and is related to the distribution of sinks and sources and their relative activities and movements. Convection appears to be necessary to maintain the volume of the earth in steady state. As the ocean floor is moving towards the continents and is entering the mantle near the oceanic margins in the trench systems, it is necessary for materials to be upwelling by convection from below to keep it in balance.

Objections to Plate Tectonics

Although plate tectonics has been a powerful and all embracing principle during the last 40 years or so, there are several problems to which it has not been able to offer a satisfactory solution. The principal objections pointed out against plate tectonics may be stated as follows:—

(i) The length of the spreading ridge is far greater than the subduction zones. In other words, the rate of creation of new crust appears to be much greater than its rate of destruction. This means that the rate of plate movements becomes accelerated in some places. But we seem to know nothing about this at present.

(ii) Plate tectonics is unable to explain why subduction is limited to the Pacific coasts while spreading is found in all the oceans.

(iii) Sometimes there are evidences to show the movement of the same plate in opposite directions which is impossible.

(iv) The Benioff zone is not present equally in all the probable places. For example, intermediate and deep focus earthquakes are absent in North America.

(v) It cannot be stated with any definiteness that each plate behaves like a unit, and some people have proposed an increase in the number of plates.

(vi) Plate tectonics has failed to provide a (completely) satisfactory explanation for mountain building. There are several mountain ranges, such as the eastern Highlands of Australia, Drakensburg of S.Africa and Sierra Delmar of Brazil which cannot be related to plate tectonics. The relationship between plate tectonics and mountain building is discussed in detail in the chapter on mountain building.

Conclusion

Plate tectonics is a revolutionary and comprehensive principle. During the last 40 years it has been accepted as the most powerful and fundamental theory explaining and correlating so many major features of the earth. It has provided a satisfactory explanation of the distributional pattern of earthquakes and volcanoes in the world. It has confirmed our views about continental drift and sea-floor spreading, and has thrown light on the mode of formation of the oceanic and the continental crusts. The objections which have been voiced against plate tectonics are basically not against this theory, but arise from our inadequate knowledge of the mantle and related matters, and more especially the processes that operate in the earth's interior. There is no doubt that this theory has given a new turn and direction to geological and gemorphological thought, and with its help the composition of the earth's crust and its manifold processes can be viewed and understood with reference to a great unifying principle.

Questions

1. Give a critical account of the theory of Plate Tectonics.
2. What is Plate Tectonics ? How does it help in understanding the geographical distribution of earthquakes and volcanoes in the world ?
3. Into how many plates has the earth's crust been divided ? What is the difference between plate boundary and plate margin ? What are the different types of plate margins and plate boudaries and what is their significance?
4. Write short notes on any three of the following : —

   (a) Low Velocity Zone, (b) Subduction zones, (c) The mechanism of plate tectonics, (d) Constructive plate margins, (e) Benioff zone.

Chapter 11

Earthquakes and the Earth's Interior

When there is sudden disturbance or movement of rocks in the earth's interior, vibrations spread out in all directions from the source of the disturbance. An earthquake is the passage of these vibrations through the earth's crust. Taking the earth as a whole earthquakes of one kind or another take place almost every minute, but the great majority of them are so feeble that their occurrence is not felt by man except through highly sensitive instruments. Severe earthquakes which can cause serious damage to life and property in populated areas, also take place on an average once every two or three weeks, but fortunately most of them originate on the ocean floor below the continental slopes.

Causes and Classification of Earthquakes

Earthquakes are caused either by volcanic explosions or by sudden movement of rocks generally along fault planes. Accordingly earthquakes are distinguished as volcanic earthquakes and tectonic earthquakes. *Volcanic earthquakes* are caused by gas explosions or the updoming and fissuring of volcanic structures and their foundations, and occur either simultaneously with eruption or more commonly in the period preceding an eruption. They are generally of shallow origin and their area of disturbance is relatively small and rarely exceeds a few hundred square kilometres. Their intensity however, may be quite high near the volcano itself. *Tectonic earthquakes,* on the other hand, are felt over a much wider area and are much more common. The great majority of earthquakes are in fact tectonic in origin and are due to sudden earth movements, generally along faults. 'The term tectonic (GK *tekton*, a builder) refers to any structural change in rocks brought about by their deformation or displacement. The cause of tectonic earthquakes is thus the building up of stresses in rocks until they are strained to breaking point when they suddenly rupture and move. The fault movements themselves.......may be vertical, horizontal or oblique[1]. When the rocks are very near their breaking point, an earthquake may be triggered off by some extraneous agent such as a heavy rainfall or flood, a high tide, sudden change in barometric pressure or tremors from another earthquake. The principal shock lasts only a few seconds or at the most a minute or two in exceptional cases. This is generally preceded by *fore-shocks* and followed by a series of *after-shocks* which are comparatively mild. "The foreshocks represent the preliminary loosening and removal of small obstruction along a fault plane or zone. When these have been overcome, the main movement occurs. But complete stability is not restored immediately, and moreover the jolt often disturbs adjoining fault blocks. The after-shocks represent the minor movements as the region gradually settles down again".[2]

Tectonic earthquakes may originate at depths which may vary from only a few kilometres to over 700 kms. Based on the depth of their origin they are classified as :

1. *Shallow or Normal*, when the depth of origin is less than 60 kms.
2. *Intermediate*, when the depth of origin is between 60 and 300 kms.
3. *Deep*, when the depth of origin is between 300 to 700 kms, the maximum depth so far recorded being 720 kms.

Earthquake Intensity and Magnitude

The earthquake originated usually some kilometres below the surface. The place of origin of the earthquake below the ground is called the focus. The *focus* may be either point-like or elongated. The point or line on the surface vertically above the focus is called the *epicentre or epicentral line*. The vibration radiates in all directions from its source or the *seismic focus*. It reaches the epicentre first on the surface because epicentre is the shortest distance between the focus and the surface of the earth. The intensity of the earthquake is also maximum at the epicentre, but here the vibration is felt in the form of up and down movements, and structures and other objects here move up and down but not to and for with the result that the damage is comparatively small at the epicentre. The maximum

---

1. Holmes, Doris L. — *Holmes Priciples of Physical Geology*, ELBS & Nelson 1978, p. 568.
2. Holmes, A — *Principles of Physical Geology*, Neloon, 1945, p 361.

damage is usually done where the earthquake waves reach the surface obliquely and at the same time their intensity is sufficiently great. This happens at some distance away from the epicentre. Here the move-ment is to and fro and this kind of movement results in lot of damage to buildings. But far away from the epicentre the intensity of the earthquake decreases, and normally the damage decreases as we move further away from the epicnentre. The maximum damage, however, is not at the epicentre but in the areas close to the epicentre.

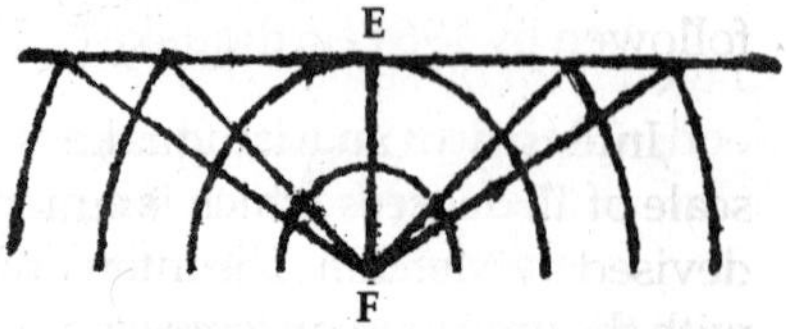

Fig. 32— Seismic focus (F) and the Epicentre (E)

Based on his study of the catastrophic Lisbon earthquake of 1755, John Mitchell was the first to recognize in 1760 that the earthquake spreads out in waves from the region of the initial shock and the destructiveness (or what we call *intensity*) of an earthquake decreases outwards from the source of the disturbance. In 1902 a scale of intensity based on the amount of damage to various types of structures was developed by the Italian seismologist Mercalli. The damage, however does not depend only on intensity but also on the nature of the surface materials, the building standards and design and population density. Consequently, methods were devised to determine the total amount of energy released during an earthquake, this measurement being known as *magnitude*. The Richter scale, devised in 1935 by C.E. Richter and developed since then, is used to describe the magnitude of an earthquake. On this scale magnitude is assigned according to the maximum motion recorded by the seismograph i.e. it corresponds to the highest intensity of the earthquake. Though the intensity of an earthquake varies as it spreads outwards it has only one magnitude, which refers to the total energy released during the quake. The Richter scale is logarithmic, so that an increase of 1 magnitude signifies 10 times the ground motion and the release of about 30 times the energy. Magnitudes of 2.5 can only be detected by seismograph and are not felt by human beings, while the most catastrophic earthquakes have magnitudes of 8.1 or more. Some of the world's major earthquakes of this century-Andes of Colombia/Ecuador 1906, Sinkiang, China 1911, Kansu, China 1920, Japan 1933, North Assam 1950, Chile

1960, Alaska 1964 — all had magnitudes of 8.4 or more, the maximum recorded being 9.5 during the Chile earthquake of 1960, followed by 1964 earthquake (9.2) of Alaska

Intensity of an earthquake is now expressed by reference to a scale of 12 degrees which is a modification of a scale of 10 degrees devised by Mercalli. The intensity scale is shown in the table below with the approximate corresponding magnitudes. The intensity is now determined from seismograph records by the maximum rate of change of movements of the ground as well as the duration and number of jerks and tremors i.e. by the maximum acceleration of the ground. In the same unit the average acceleration of gravity is 9800 mm/s^2. When this value is exceeded, we have the highest intensity (intensity XII) of the earthquake.

Modified Mercalli Intensity Scale with corresponding Magnitudes

Intensity	Characteristics Effects	Maximum acceleration of the ground	Magnitude corresponding to highest intensity reached
I	Detected only by seismographs.	10	3.5
II	Feeble: Felt only by sensitive people	25	
III Slight	Felt by people indoors especially on upper floors.	50	4.2
IV Moderate	Felt by people while walking. Windows, doors distubed. At night some awakened. Rocking of loose objects, including standing vehicles.	100	
V Rather strong	Felt by nearly everyone, most awakened. Unstable objects overturned.	250	
VI Strong	Felt by all; many frightened and run outdoors. Trees sway and suspended objects swing. Damage by overturning and felling of loose objects.	500	4.9 to 5.4
VII Very strong	General alarms; Everybody runs out doors. Plaster falls and walls crack. Damage negligible in buildings of good design and structure but considerable in poorly built structures.	1000	5.5 to 6.1
VIII Destructive	Car drivers seriously disturbed. Fall of chimineys, columns, walls. Considerable damage to ordinary/poorly constructed buildings with partial collapse. Sand and mud		

	ejected in small quantities. Changes in well water.	2500	6.2 to 6.9
IX Ruinous	Ground cracks and buildings shifted off foundations. Considerable damage in substantial buildings with partial collapse. Pipes break open.	5000	6.9
X Disastrous	Ground badly cracks. Many good buildings destroyed and rail lines bent. Landslides on steep slopes. Water splashed over river banks.	7500	7 to 7.3
XI Very Disastrous	Few buildings remain standing. Bridges destroyed. Broad fissures in ground. Pipes, cables and railways thrown out of action. Great landslides & Floods.	9800	7.4 to 8.1
XII Catastrophic	Total destruction. Ground rises and falls in waves. Objects thrown into air.		> 8.1 (maximum recorded 8.9)

The line joining places which experience the earthquake at the same time is called a *homoseismal line*. Homoseismal lines are oval or elliptical in shape and run around the epicentre. A line drawn through places with the same earthquake intensity (i.e. destructiveness) is called an *isoseismal line*. Each line encloses a roughly circular or elliptical area according as the place of origin or the focus is point-like or elongated. From the intensity at the epicentre and at a given point on an isoseismal line both of which are known, it is possible to estimate the depth of the focus.

Effects of Earthquakes

The *geomorphological effects* of earthquakes are not spectacular, but they do result in local and relatively minor but sudden topographic changes. Vertical displacements along faults are common. On one side of the fault the surface rocks are raised while on the other side they may be depressed. Due to the passage of land waves fissures gape open at the crests which may close again as the crests turn into trough but sometimes a broken undulating surface with many linear fissures may be preserved. Broad fissures may also open up on the ocean floor and slides develop on valley sides and hill slopes and avalanches are set into motion on snow-covered mountains. Glaciers are broken and where they terminate in the sea, icebergs suddenly become abundant. Large quantities of loose sediments may slump down the continental slope and be redeposited on the seaward side of the slopes. In alluvial plains, the sandy deposits get compacted with the passage of earthquake vibrations, and as a result of the decrease in volume water is pushed up the water-filled sand and escapes with great force to the surface through fissures with sand and mud. Ground water may be disturbed by earthquakes in other ways. Lakes may be drained off by the opening of cracks, and new lakes may be formed in depressions.

Considered from the *human point of view* the effects of earthquakes are far more important and serious, resulting in considerable loss of life and property. As the energy released by an earthquake travels along the earth's surface it causes the ground to vibrate in a complex manner by moving up and down as well as from side to side. The amount of destruction on account of the vibrations depends upon (a) the intensity and duration of the vibration, (b) the nature of the materials on which the structures are built, and (c) the design of the structure. The maximum destruction is wrought in the densely populated areas where within a few seconds, thousands of buildings may collapse, chimneys are overturned, roads crack, rail lines are bent and bridges are destroyed. In the case of earthquakes of high intensity thousand of people may die, and the entire life of a city or region may be completely disorganised and disrupted within a moment. The appalling loss of human life is mainly related to the secondary events which are triggered by the earthquake such as the collapse of buildings, fires, landslides, floods and seismic sea waves. Gas mains and chimneys may be torn apart and start fires which may spread rapidly. In the San Francisco earthquake of 1906 far more damage was done by fires than by the earthquake itself. In the Japanese earthquake of September 1923 fires that broke out when the people were just cooking their mid-day meals devastated the cities of Tokyo and Yokohama. In some areas landslides or landslips may prove equally disastrous. In the Loess Highlands of China two lakh persons were killed in 1920 and one lakh in 1927 on account of landslips of loess which overwhelmed dwellings, buried villages and towns and blocked river courses, causing floods.

In the 1964 Alaskan earthquake most of the deaths were caused by seismic sea waves which extended along much of the west coast of North America. The giant seawaves which are frequently associated with earthquakes of high magnitude are called *tsunami* in Japan (*tsu, harbour; nami* waves). These destructive sea waves are caused by displacement of water due to sudden change of level of the ocean floor either by the fault movements responsible for the earthquake or by submarine slumping and turbidity currents resulting from the sudden movements. They usually result from topographic changes on the ocean floor brought about by powerful earth movements along deep-water coasts, such as along the sides of the great Pacific trenches. Whether the sea-floor displacement is up or down, the momentum of a vast body of moving water carries it far beyond its position of equilibrium. The resulting ebb and flow takes the form of a series of oscillating waves. In deep water these waves have wave lengths of hundreds of kilometres (i.e. distance between crests) and travel at speeds of 700 to nearly 1,000 kilometres an hour, but their crests are less than a metre high, so that they go unnoticed by ships in the open ocean. The energy transmitted is, however, immense because the whole depth of water is involved. Therefore when these waves reach shallow coastal waters and narrow bays and inlets, they have been known to grow into a wall of water 30 metres or more high and inflict widespread damage on coastal towns and ports along thousands of kilometres of coast. Tsunami generated by a major earthquake (magnitude 8.5) on the western slopes of the Japanese trench in 1933 raised waves to heights of 27 metres along the shores of the Pacific bays and inlets of Japan and drowned thousands of people. Nearly 150 disastrous tsunamis have visited the Japanese coasts since records of earthquakes first began be the kept there following powerful earthquakes originating beneath the deep trenches of the Western Pacific. Similarly the Pacific coast of South America has a long history of destructive earthquakes and tsunami. Outside the Caribbean the Atlantic shores are, however, more or less free from tsunami.

Tsunami waves are rare in the Indian Ocean. But an earthquake of the magnitude of 9 on the Richter scale, the strongest in 40 years and the fourth largest since 1900 hit Indonesia early in the morning on December 26, 2004 with epicentre 250 km southeast of Sumatra coast. It sent powerful tsunami tidal waves roaring across the Indian ocean that hit the densely populated coastal areas in several south Asian countries including Maladives, Sri Lanka, eastern coast of India including the Andaman and Nicobar Islands, Thailand, Burma and Malaysia. It caused unimaginable damage to life and property as the high tidal waves 10 m or more carried away whole willages and even larger settlements. Over 3 lakh people were killed (over 2 lakhs in Indonesia alone) and many more rendered disabled and homeless. The largest number of people died in Indonesia, Sri Lanka, India and Thailand and it is feared that probably one- third of the population of Andaman and Nicobar Islands has been wiped away. It could be history's coastliest disaster according to the U.N.

The damage and destruction caused by earthquakes can be best understood by reference to specific earthquakes which have occurred in the different parts of the world. On June 16, 1819 the Kutch area in India was visited by an earthquake of high magnitude which resulted in a 150 km long (over 90 miles) east to west fault in the northern part of the Rann of Kutch. The ground rose north of the fault and subsided below sea-level in the region south of the fault. In the town of Bhuj alone nearly 2,000 people lost their lives. On January 26, 2001 the area was again visited by an earthquake of over 7 magnitude with epicentre at Bhuj, near Jamnagar which killed over 30,000 people and rendered 5 lakhs homeless. Luckily the industrial town of Jamnagar was not affected. The North Bihar earthquake of January 15, 1934 is also well-known for its devastations. Fissures opened up at several places through which sand and water were ejected and fertile fields were covered with sand. Roads cracked and bridges were damaged. There was large scale loss of life and property and the towns of Muzaffarpur, Darbhanga and Munger were almost completely destroyed. Nearly ten thousand people died, although the earthquake occurred during day time when most people were outside their homes. Including Nepal, this earthquake was felt over an area of nearly 19 lakh sq. miles.

On August, 15, 1950 North Assam was visited by an earthquake which had a magnitude of 8.6 Thousands of houses collapsed in Dibrugarh, Lakhimpur, Sadiya and other towns. Gaping cracks and fissures opened up at several places which swallowed up trees, animals and men, and left the roads unserviceable. Railway lines were buckled and twisted. The sand and water which came out of the cracks flooded large areas. In the hilly areas there were large scale landslides on valley sides and boulders of rocks blocked the flow of streams at some places, and when after a few days these broke open, there was such a heavy rush of water that large areas were suddenly flooded and many people were drowned.

On September 1, 1923 a major earthquake occurred in Japan

with a magnitude exceeding 8.2. This is known as the Sagami Bay earthquake of 1923. Gaping fissures opened on country roads and the railway lines were buckled. Dry land was submerged under the sea and in other areas land under the sea was uplifted above sea-level. In the Tokyo Bay the sea-floor was depressed at places by nearly 1,500 metres. In the Sagami Bay the landward parts of the continental slope were uncovered by 300 to 450 metres by the slipping down of vast masses of sediments, and the seaward parts were correspondingly built up where this sediment was redeposited. Large scale fires broke out in Tokyo and Yokohama and added to the toll of death and destruction. Altogether 2.5 lakh people were killed and 5 lakh houses destroyed.

Distribution of Earthquakes : Earthquake Zones and Plate Boundaries

Earthquakes are not randomly distributed over the globe but they tend to occur in narrow continuous belts. These earthquake belts encircle large seismically quiet regions which constitute the plates of lithosphere (See Chapter 9). The plates are in continuous motion with respect to one another, and this relative movement of plates or plate motion is the fundamental cause of the earthquakes. Most earthquakes occur on the boundaries between lithospheric plates and arise directly from the motions between the plates, though there are some that cannot be so simply related to the movements of the plates. We may identify three well-defined belts or zones of seismic activity in the world where most earthquakes originate. These earthquake zones are— (i) a Circum-Pacific zone, (ii) a Mediterranean and Trans-Asiatic zone, and (iii) a zone following the mid-oceanic ridges with an extension along the East African rift valley system.

(i) **The Circum-Pacific Zone**—The Circum-Pacific zone follows the oceanic trenches and the associated island areas where plates converge and the oceanic lithosphere is thrust down into the asthenosphere and re-melted, and this melting supplies the magma for the volcanic arcs which occur behind the trenches. On the western side of the Pacific, this zone, starting from Alaska runs towards the south parallel to the Kurile, Marianas and Philippine trenches, beyond which it divides into two branches, one going west parallel to the Indonesia trench and the other towards the Keramac-Tonga trenches to the north-west of Newzealand. On the

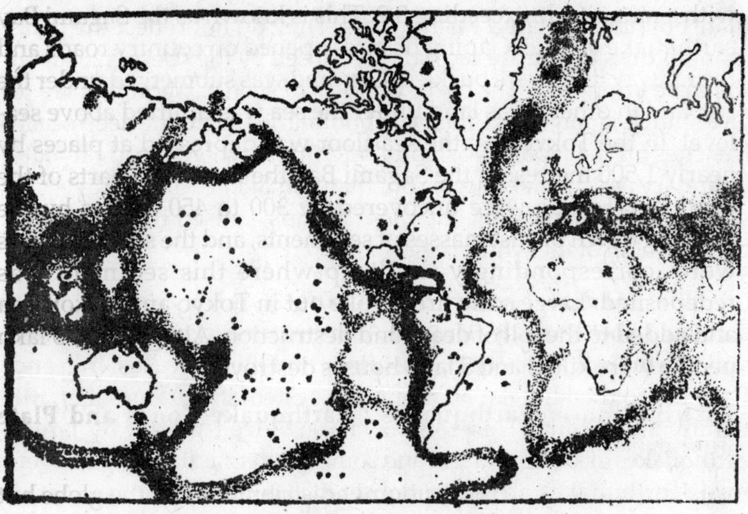

Fig. 33—Geographical distribution of earthquake belts.
(after Doris Holmes)

eastern side of the Pacific, the earthquake zone follows the west coast of North America, being particularly important in California, although there is no ocean trench associated with it there. It continues southwards parallel to the middle American trench and further south along the Peru and Chile trench on the west coast of South America. Shallow, intermediate as well as deep earthquakes are found to occur along the Circum-Pacific belt, but deep earthquakes are practically restricted to this zone.

We have already referred to the existence of inclined earthquake zone known as Benioff zones along the Pacific coasts where the focii of earthquake deepen from shallow through intermediate to deep in a landward direction from the trenches. This inclined earthquake or Benioff zone depicts the route along which the oceanic lithosphere descends into the mantle along the trenches, and its downward progress is recorded by a series of earthquakes which have their focii at various depths and many are as deep as 300 to 700 kms below the sea level.

Further, the earthquakes are not restricted to the plate boundary itself, but occur over a broad zone, several hundred kilometres wide, adjacent to the plate boundary. Such earthquakes may be called plate boundary related earthquakes. They do not reflect the plate motions directly but are secondarily caused by the stresses set up at the plate boundary. The best examples of such earthquakes are to be found in Japan where the plate boundaries are in the deep ocean trenches off the Japanese islands, and that is where the great plate boundary earthquakes occur. But many smaller earthquakes occur scattered throughout the Japanese islands, caused by the overall compression of the whole region.

(ii) **The Mediterranean and Trans-Asiatic zone** — This earthquake belt extends along the Alpine mountain system of Europe and North Africa, through Asia minor and the Caucasus, Iran and Pakistan to the Himalayan mountain system including Tibet, the Pamirs and China. This zone is characterized mostly by larger earthquakes of shallow origin and some of intermediate origin. Deep focus earthquakes are almost absent. This belt is not associated with oceanic trenches but with the Tertiary and Recent orogenic belts where continental plates collide and the lithosphere buckles under the force of the collision forming the great mountain ranges.

(iii) **The mid-oceanic ridges and the African rift system zone**— This zone lies mostly along the mid-oceanic ridges and the transform faults and contains mostly earthquakes of the shallow variety. These constitute major fracture zones where the plates diverge and new oceanic crust is being formed by the upwelling of magma on the mid-oceanic ridges. An extension of this belt is to be found along the Red Sea and the rift valleys of East Africa.

Stick Slip Friction and Elastic Rebound

Where the relative motion of the plate is parallel to their common boundary, slip occurs along great faults which form the boundary.

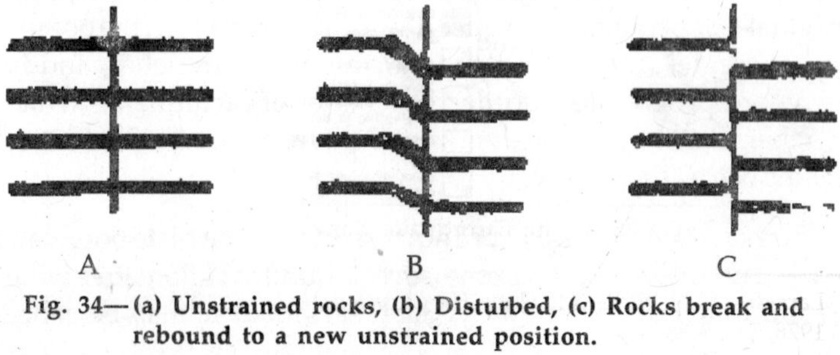

Fig. 34—(a) Unstrained rocks, (b) Disturbed, (c) Rocks break and rebound to a new unstrained position.

As the plates move past each other, the motion at their boundaries does not take place by continuous slippage but in a series of rapid jerks, and each jerk produces an earthquake. Under the pressure and temperature conditions of the shallow lithosphere the frictional sliding of rocks occurs in a series of jerky movements, interspersed with fairly long periods of no motion or sticking. This property has therefore been called stick-slip. In between the period between major earthquakes strain slowly builds up near the plate boundary in response to the continuous movement of the plates. The strain is ultimately released by an earthquake when the frictional strength of the plate boundary is exceeded. This pattern of strain build-up and release was discovered by H.F. Reid. From his studies of the 1906 San Francisco earthquake he deduced his theory of elastic reboud. This is illustrated in Fig. 34. In 34 (a) the rocks are unstrained. They are first (b) disturbed by the slow relative movement between two plates causing strains within the fault zone that finally becomes so great that (c) rocks break and rebound to a new unstrained position.[1]

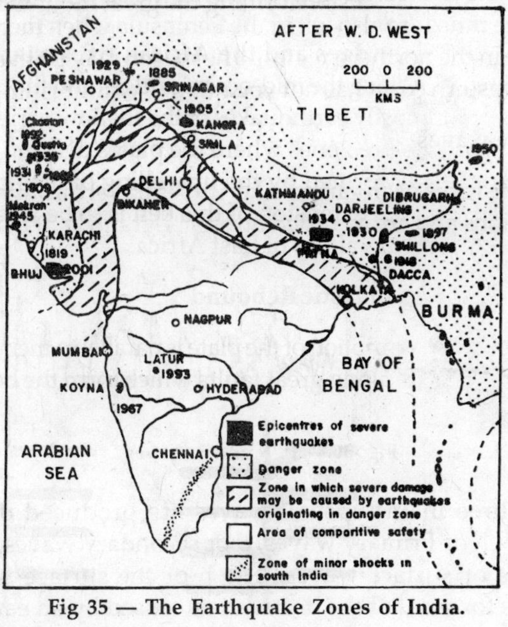

Fig 35 — The Earthquake Zones of India.

---

1. Lapedes, D.N. (Ed)—*Mc Graw Hil Encyclopedia of the Geological Sciences*, 1978, p. 196.

The Earthquake zones of India

The earth-quake zones of India have been shown in Fig. 35. It will be observed that the epicentres of most earth quakes in the Indian region lie in the Himalayan mountain belt. This is a region of marked instability and is characterized by several thrust planes, and though the mountain building movements have ceased, some adjustments are still taking place resulting in movement along fault planes and finding expression as earthquakes. The north-western (including Baluchistan and the Pamirs) and the north-eastern corners of the Himalayas are particularly vulnerable as there are sharp changes in strike and the rocks are under great strain and stress. The Indo-Gangetic alluvial trough is mostly affected by earthquakes originating in the first zone, but changes appear to be still taking place at the bottom of this structural trough, giving rise to occasional earthquakes. The Sind earthquake of 1819 and the Bihar earthquake of 1934 had their focii in this trough.

The Peninsular part of India is a stable block. Occasionally, however, some feeble shocks are felt along the margins. Sympathetic shocks some times are also felt in the Peninsula when there is major earthquake in the northern region. The reason may be that the fault along the western coast has not yet attained equilibrium.

Earthquake waves

From the focus of an earthquake, waves travel through the earth in all directions and are recorded on seismograph kept in recording

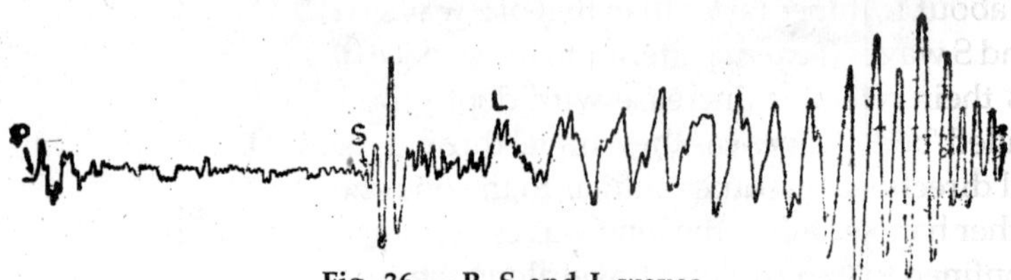

Fig. 36.— P, S and L waves

stations. Three main types of waves are produced during an earthquake: *P* or Primary waves, *S* or secondary waves, and *L* or long waves or surface waves. The L or the surface waves are responsible for most of the destruction caused by an earthquake, while the main significance of P and S waves is in the study of the earth's interior. *P* waves travel fastest; therefore they are the first to reach the seismic station and be recorded on the seismograph. After some irregular oscillations, the *S* wave arrive also followed by a train of oscillations, after which the L waves which are the slowest arrive. The *P* or primary or preliminary waves are "push and pull" waves like those of sound in which each particle vibrates to and fro in the direction of propagation. These are also known as compression-dilation or longitudinal waves. Like the sound waves, the *P* waves pass through solids, liquids and gases alike. The *S* waves are tranverse or shear waves, in which the motion of each particle is at right angle to the direction of propagation. These are called secondary waves, because they appear after the *P* waves. The *S* waves can travel only through solids and not through liquids.

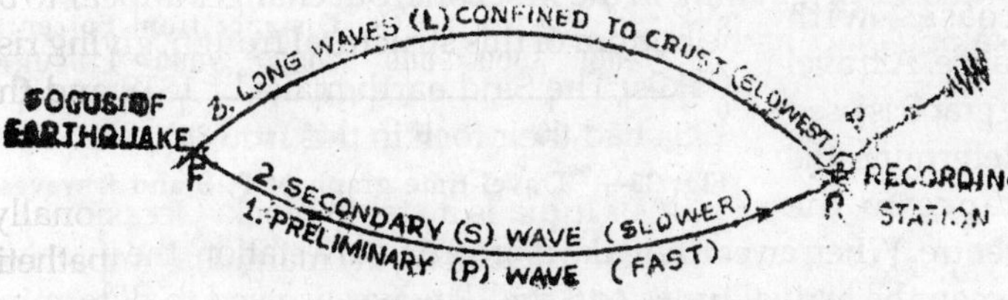

Fig.37— The paths of the P, S and L waves to the earth's surface.

The *P* and *S* waves travel at different speeds through the same kind of rock, because while the speed of *P* waves depends on the density and compressibility i.e. resistance to compression of the rocks, that of the S waves depends on the density and rigidity i.e. resistance to distortion or shearing of the rocks. Ordinarily the speed of *P* waves is about 1.7 times faster than that of S waves. The path of both the P and S waves from the interior to the surface of the earth is concave, as their velocities increase with depth. Both P and S waves are called 'body waves', as they travel through the body of the earth in all directions by routes similar to the one shown in Fig 37. On the other hand, the L or the long waves are 'surface waves' as they are confined to the crust and travel along the surface of the earth along a convex path after reaching the epicentre from the seismic focus. They are recorded on the seismograph after the *P* and *S* waves because of the greater complexity of their paths through the crustal layers, but their oscillations are more prominent than the earlier waves. The L waves fade out below certain depths, but they can pass through both land and water. They move through the land in a circular path similar to the motion of water in ocean swells. These waves are connected with the main shocks of the earthquake and are responsible for most of the upheaval and destruction.

When the times taken by *P* and *S* waves to travel to seismological stations in various parts of the world are plotted against the respective distances from the epicentres, it is found that the difference in the arrival times of the first *P* and *S* waves steadily increases with distance. A travel time graph is used to determine the distance to the epicentre. When an earthquake is recorded at a station, the inter-val between the arrival times of *S* and *P* waves is used to determine both the distance from the epicentre and the time of origin of the earthquake. At least three such distance determinations from well spaced stations are necessary for locating the position of the epicentre fairly accurately, and for greater accuracy several determinations are necessary. There is effective international cooperation between recording stations in different parts of the world for exchange of information relating to position of epicentre and the time of origin.

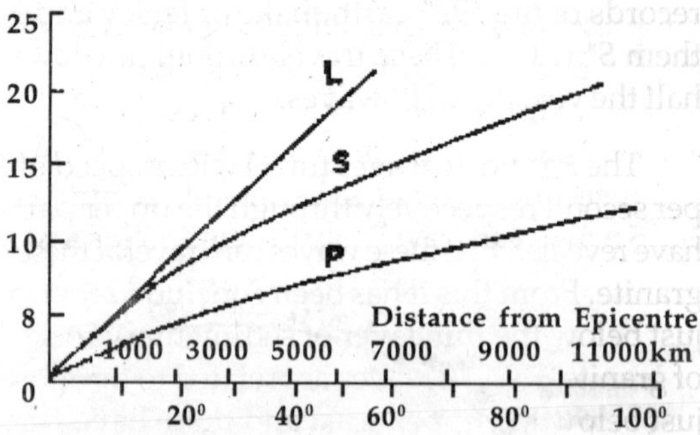

Fig. 38— Travel-time graph of P, S and L waves

The Earth's Interior as Revealed by Earthquake Waves

In has already been briefly indicated in chapter 4 that *P* and *S* waves have been extensively used in finding out the nature and thickness of the different layers that constitute the earth's interior. Recent studies have revealed some special categories of *P* and *S* waves which have helped us in estimating the nature of the earth's crust and its dominant layers. In the course of a study of earthquake that occurred in the Kalpa Valley of Yugoslavia, A. Mohorovicic found two waves very similar to *P* and *S* waves but with lower velocities. These have been called *Pg* and *Sg* waves. While the P and S waves are refracted into the mantle before coming up to the surface again, the Pg and Sg waves travel directly to the surface through the crust. V. Conard in his study of the 1923 earthquake in the Austrian Alps found another wave called P*—a kind of *P* wave intermediate in velocity between *P* and *Pg*. In his opinion this was related to a crustal layer lying between the upper part of the crust and the mantle. Then Jeffreys identified the corresponding *S* pulses in the records of the 1926 earthquake of Jersey and Hareford and called them *S** waves. These travel through the lower crustal layer with half the velocity of *P** waves.

The *Pg* and *Sg* waves travel with a speed of 5.4 kms and 3.3 kms per second respectively through the upper part of the crust. Studies have revealed that these waves can travel at these speeds only through granite. From this it has been concluded that the upper crust lying just below the thin layer of sedimentary rocks consists essentially of granite. P and S waves are related to the intermediate layer lying just below the upper crust and these have velocities intermediate between the *Pg-Sg* and *P,S* waves. This indicates the existence of an intermediate crustal layer which consists of plutonic and metamorphic rocks with a predominance of basalt. The higher speed of the P and S waves suggested that there is a lower layer below the intermediate layer which is composed of denser rocks like dunite and peridotite, as the velocity of P and S waves increases with increasing density. While the seismic evidence for division of the crust into two or three layers is not universally present, there is general increase in the velocity of seismic waves with depth through the continental crust, and the values for the upper part suggest rocks of granitic composition and those for the lower part of the crust are indicative of more basic rocks.

A great deal of seismic data indicates that the average thickness of the continental crust is about 35 kms, though it varies between 20 and 50 kms. The maximum of about 50 kms occurs beneath the great mountain ranges, while all the old shield area have a crustal thickness of around 35 kms.

So far as the oceanic crust is concerned, the Moho is found at a depth of 10 to 11 kms. The Moho is overlain by three crustal layers. The uppermost layer is sedimentary with a thickness varying from zero in mid-oceanic ridges to several kms close to the foot of the continental slopes. Below this is a layer about 1 to 1.5 kms thick with *P* velocities which would fit some sedimentary as well as basaltic rocks. This has, however, been proved to be basaltic by dredging. Below this is the lowest or the main oceanic crustal layer about 5 kms thick, with P velocities indicative of basic rocks. It is supposed to consist of either peridotite or metamorphosed basaltic rocks. The granite layer below the sedimentaries which is clearly marked in the continental crust, is absent in the Pacific ocean and is found only in thin patches in the Atlantic and Indian oceans. By and large the oceanic crust resembles the basaltic layer of the continental crust.

So far as the main P and S waves are concerned, we have seen that they descend through the Moho into the mantle before coming up to the surface. The velocity of P waves is 7.8 to 8.1 kms per second near the top of the mantle and increases to about 13.6 kms per second at the bottom of the mantle i.e. at a depth of about 2,900 kms from the surface. In the case of *S* waves the velocity increase is from 4.35 kms per second in the upper part to 7.25 kms per second in the lower part of the mantle. These waves that go up to such great depths, emerge on the surface at a distance of about 11,500 kms (105^0) from the epicentre of the earthquake concerned. At stations beyond this distance within the next 4,500 kms no *P* and *S* waves are recorded, although the *L* waves appear at the right time. At distances greater than 16,000 kms the P wave starts appearing again and continues to be recorded right up to the antipode of the epicentre (i.e. up to the anti-centre). For every earthquake of sufficient magnitude which can be recorded to such great distances, there is a ring-like *shadow zone* which is free from the *P* and *S* waves. The shadow zone was first discovered by Oldham in 1906, and he suggested that this was due to the fact that the earth has a core which acts like a spherical lens by refracting the deeper waves inwards and concentrating them in the antipodal region, so that they become absent in the surrounding shadow zone. Because of strong refraction the velocity of *P* waves is greatly reduced as they pass from the mantle into the core and the S waves are not transmitted at all. This discontinuity

Fig. 39— Shadow Zone cast by earth's core in the case of a Japanese earthquake. (After Doris Holmes)

shows that while the mantle is solid, the core is liquid. It was at first thought that the entire core was liquid. The change from mantle to the core is sharply defined and there is no transition zone. But in 1936, a Danish seismologist Inge Lehmann indicated that the inner portion of the core was probably solid. Till 1962 it was believed that the core was divided into an inner solid layer and an outer liquid layer separated by a transition zone about 150-300 kms thick. But later studies seem to suggest that the transition zone is probably an independent zone with fairly sharp boundaries on both sides.

Our present state of knowledge about the internal zones of the earth from seismic data computed by Harold Jeffreys, K.E. Bullen, E.C. Bullard and B.A. Bolt and summarized by Doris Holmes is presented in the table below:

Zones	Depth (and Radius) of boundaries in km.	Velocity of P waves in km./s	Density gm/cm³	Pressure approx. in atmospheres
Crust (Continental)[1] A	Sea-level (6371)		2.7	
			2.8	
Moho Duscontinuity	33 (6338)		2.9	9,000
		7.9-8.1	3.32	
	50 Low velocity zone	7.8		
Upper Mantle B	250	8.1		
C	413 (5958)	8.97	3.64	140,000
	720 (deepest earthquakes)	10.70	4.29	270,000
....	984 (5387)	11.42	4.64	382,000
Lower Mantle D				
Oldham or Gutenberg Discontinuity.	2898 (3473)	13.64	5.66	1,368,000
		8.10	9.71	
Outer core E	4703 (1667)	10.31	11.76	3,180,000
			10.31	
Transition F	5154 (1216)	....	c 14	c3,300,000
		11.23		
Inner core G	6371 (Centre)		c16	c3,600,000

1. In oceans zones Zone A extends from the ocean floor at a mean depth of 4.8 kms to the Moho at a depth of about 10-11 kms below sea level.

Questions

1. How do earthquakes originate? Discuss the world distribution of earthquake zones and its relationship with plate boundaries.
2. Describe the causes and effects of earthquakes with examples.
3. Describe the types of earthquake waves. How have they helped in understanding the nature of the earth's interior ?
4. Write short notes on :—

   Seismic focus, Epicentre and Anti-centre, Tsunami, Intensity and magnitude of earthquakes, Volcanic and tectonic earthquakes.

Chapter 12

Volcanic Landforms and Intrusive Features

The Geomorphic Significance of Volcanism

Volcanism is a special case of igneous activity in which the molten rock or magma is transferred from below or within the crust to the outer surface. The magma is conveyed to the surface essentially along tube-like conduits and the extrusion of lava builds distinctive conical or dome-shaped landforms which owe their origin entirely to endogenetic forces and are in no way related to gradational processes. Volcanic activity produces dramatic and strikingly beautiful landforms in a relatively short period of time. Localized volcanic activity may impose these features upon a landscape with dramatic suddenness, completely burying the existing landscape and presenting an entirely new type of initial surface to the agents of erosion.

The volcanic landforms may form in any latitude or climate but have an unmistakably similar, distinctive and symmetrical appearance. In course of time they are, of course, modified by the climatically controlled processes and thus afford an opportunity to study varying erosional rates and forms under different climatic conditions on the initial volcanic landscape.

Volcanoes have gained renewed geomorphic interest in recent years because of new techniques which have made possible the exact dating of eruptive rocks. By measuring the amount of rock eroded from a dated volcanic landscape, it is now possible to measure the absolute rate of landscape modification and to reconstruct the original volcanic features.

It is important to remember that the products of volcanism form most of the oceanic crust and part of the continental crust. Oceanic volcanism which mainly takes the form of non-explosive fissure eruption from the active mid-oceanic ridges is responsible for building almost all the crust beneath the ocean floor. Much of the ocean floor has constructional topography of lava flows and in addition to a large number of volcanic islands, thousands of eroded volcanoes present as seamounts lie hidden below sea-level. Essentially similar basaltic lavas form some of the largest plateaus on the continental surfaces, such as the Deccan Plateau of Peninsular India and the plateaus of Iceland and the north western United States. About 2 million sq. kms of land surface have been covered to depths of 500-3,000 metres by lava flows in geologically recent times, and probably the entire earth has received a layer of volcanic ash and dust of varying thickness.

Nor must we forget that the volcanic processes are really responsible for all the water on the earth's surface, as water brought from the earth's interior by volcanic activity has allowed the creation of the hydrosphere as well as the atmosphere.

The Nature of Volcanoes

A volcano is commonly defined as a deep vent in the earth's crust through which molten rocks or hot lava, ash and hot gases are ejected from the earth's interior to the surface of the earth. The process of ejection of hot materials from a volcano is known as *eruption*.

When eruption takes place, there is usually a violent explosion preceded by an earthquake. At the site of the explosion the broken fragments of rocks are first thrown to heights, clouds of water vapour and other gases appear, the atmosphere is covered with smoke, clouds and dust, and finally the extremely hot and molten lava comes to the surface.

Fig 40— The Volcano and its Crater

The magma which comes to the surface through the volcanoes from the interior of the earth does not consist only of molten rocks but includes hot liquids, gases and solid materials in mixtures of different proportions. When the magma is encased under pressure, it contains large quantities of gas and water vapour. As the gas-filled magma rises towards the surface and the pressure over it is reduced, the gases tend to separate and erupt to the surface by forcing through it, and there is an explosive eruption. It is also possible that the water entering the crust comes in contact with the hot molten rocks inside and is suddenly converted into vapour thereby compelling the lava to move upwards and resulting in volcanic eruption. It is pointed out in support of this possibility that most of the volcanoes are found near the seacoasts and generally a large amount of water vapour is released with the volcanic eruptions.

Lava and other materials erupted by the volcano collect round the vent. The accumulation is maximum close to the vent and decreases with distance from the vent, with the result that it leads to the formation of a conical hill. Thus normally a volcano is associated with a conical hill from the top of which there is periodic eruption. But the hill is a product of the eruption, and the volcano is the vent in the centre through which the lava is ejected. In most volcanoes, there is a funnel-shaped depression called the *crater* on the top of the cone. The base of the crater is connected with the vent through which the lava rises to the top. When the flow of lava stops, the vent is usually sealed with the solidified lava. The crater is usually the result of outward explosion or of inward subsidence, but sometimes it may be formed due to accumulation of the ejected material round the vent.

Frequency of Eruption

In some volcanoes the eruption continues uninterruptedly for a long time, but in most volcanoes it is discontinuous. In some volcanoes the period of rest or quiet is fairly long and these are extremely dangerous. Based on the frequency of eruption it is customary to divide volcanoes as *extinct, dormant and active,* though it must be remembered that many that are considered dormant may well become active any day. Volcanoes is which the eruption has completely stopped and is not likely to recur are called *extinct*. Examples are the Pope mountains in Burma or Mt. Kilimanjari in Africa. Examples of extinct volcanoes are also found in Mauritius, Malagassy and several other islands in the Indian Ocean. Volcanoes which have been quiescent for a long time but in which there is a possibility of eruption are called *dormant* volcanoes. Fujiyama of Japan and Krakatoa of Indonesia are examples of dormant volcanoes. Fujiyama has been quite for a fairly long time, but it is believed that it may erupt any time again. Krakatoa is located between Java and Sumatra. It was quiet for nearly 200 years and then suddenly erupted in 1883 which continued for several months. The eruption was so violent that nearly two-thirds of the island of Krakatoa disappeared. Though the island was uninhabited, the explosion raised 40 m high waves along the coasts of Java and Sumatra which resulted in the death of about 36,000 persons and the sound of the explosion was heard 5,000 kms away in Australia. Dense volcanic clouds rose to a height of 27 kms while the burning cloud of ash and rock fragments rose as high as 80 kms.

The volcanoes which continue to erupt periodically are called *active* volcanoes. Mona Loa in the Hawaii Islands, Etna in Sicily and Vesuvius in Italy are examples of active volcanoes. Prior to the eruption of 79 A.D. Vesuvius had remained dormant for so many centuries that people had taken it to be an extinct volcano, but in 79 A.D. it erupted so violently that the city of Pompei was buried under its burning ashes and the town of Herculanium was submerged under the flood of hot mud. During the last 1900 years the Vesuvius has erupted 16 times, and the last three eruptions have taken place in 1872, 1906 and 1944. Cotopaxi in Ecuador is the world's highest active volcano and is 19,600 ft high.

Products of Volcanic Eruption

The products of volcanic eruption may be either solid, liquid or gaseous. The solid rock fragments of different sizes which are ejected with the eruption are together called *pyroclasts*. When the volcanic eruption is explosive, a lot of rock fragments are thrown into the air which fall down near the vent or the adjacent areas. First to come out are the rock fragments which are present in the vent. In dormant volcano the vent is plugged with solidified lava, and this is thrown out in fragments before the lava is ejected. Together with these are also the rocks which constitute the wall of the vent. Sometimes the top of the hill may be blown off due to the explosion. In most eruptions, considerable solid fragmental materials are ejected, and the adjacent areas experience virtually a rainfall of ashes and stones. Besides the broken rocks fragments, part of the lava which is thrown up with the rapidly rising gases also solidify before falling down on the surface of the earth in the form of bomb-rains.

The pyroclastic materials are sometimes collectively called *tephra*. It consists of fragments of different sizes and shapes and varies from fine dust to big angular fragments. They have been given different names according to their size. The large angular rock fragments are called *blocks or brecia*. When the molten lava is thrown into the atmosphere and its bigger drops solidify before falling down,, they assume round, oval or pear-shaped forms. These solid rounded materials are known as *volcanic bombs*. When size resembles peas or they are as small as glass marbles, they are called *lapilli*. When they have a vesicular structure, they are referred to as *pumice, scoriae or cinders*. When the volcanic material consists of sand size and finer tephra, it is called *volcanic ash*, and when the materials are still thinner, they are called *volcanic dust*. On account of the heavy rainfall which sometimes accompanies the eruption, the volcanic dust and ash start flowing as mud and may cause almost as much damage as the hot lava.

The bigger fragments like lapilli, cinders, bombs and blocks fall down close to the volcanoes, but the volcanic ash and dust may be airborne for many kilometres before settling down. The rock which is composed of a mixture of ash, dust, lapilli and cinder is referred to as *tuffs*. Sometimes this may also contain admixture of bombs and blocks along the slopes of the volcanic hill.

The most important product of the volcanic eruption is, of course, the liquid lava. Its temperature is usually between 900^0 to 1200^{0}C The lava containing a higher percentage of silica is called acid lava, while lava with lower percentage of silica is called basic lava The acid lava has a high melting point and is usually viscous. It therefore flows slowly and cannot travel very far. On the other hand, the basic lava has a low melting point. It is more liquid and flows to longer distances before it solidifies. Further, its speed of movement is also faster. The cooling of basic lava results in the formation of basalts. The composition of the molten lava is a major factor in landform development because of its influence on viscosity. Sillicic or acid lava builds high steep sided cones while basaltic or basic lava produces a flatter cone of great diameter.

Sometimes the surface of the solidified lava may be smooth but more usually it is quite rugged. In the case of both acid and basic lavas the escaping gases make the surface of the lava vesicular and full of small holes. Besides, the upper surface cools earlier and forms a crust while lava underneath the surface continues to flow. This results in cracks in the crust and the parts thus broken are carried down. The vesicular fragments of the lava crust, spongelike in texture are called *scoriae*. The surface formed by the lava deposit is thus normally rough and irregular and is full of cracks and holes into which the rainwater can easily penetrate. Two contrasting surface textures may develop on the lava flows which are commonly known by their Hawaiian names; *aa* for the angular vesicular scoriaceous surface, and *Pahoehoe* (or ropy) for smoothly twisted, convolute surfaces which develop on hotter and more fluid lava. An extreme case of *Pahoehoe* lava occurs when the fluid lava either erupts under water or quietly flows into water. Blobs or lobes of lava upto a metre in diameter form tough flexible skins and pile up like sandbags or pillows while their interiors are still molten. This is called *pillow lava*. It is an evidence of subaqueous eruption and is commonly found on the ocean floor.

Of the volcanic gases by far the most important is steam. Lot of steam is ejected with the eruption and its condensation usually results in heavy rainfall. It should, however, be remembered that in addition to the volcanic steam, water-vapour present in the atmosphere may also be responsible for the rainfall. Besides steam (60 to 90%), other volcanic gases include carbon dioxide, nitrogen and sulphur dioxide, and there are very small quantities of hydrogen, carbon monoxide, sulphur and chlorine as well. After being ejected, these gases mix in the atmosphere and they do not have any influence on the volcanoes as such. But these gases are quite active and full of energy and have high temperature, and it is possible that upward rise of these powerful gases from the mantle is an important cause of the volcanic eruption. In any case, these gases assist the lava in coming up and in creating a passage for its ejection. Sometimes inflammable gases like hydrogen and its compounds may be present in large quantities, but in most volcanoes the flames associated with volcanic eruption are due to the red-hot rock fragments and the reflection of the red lava in the rising clouds.

Types of Volcanic Eruption

Volcanic eruptions are divided into two principal classes, based on the mode of eruption—(a) Central eruption, and (b) Fissure eruption.

(a) **Central Eruption**— Here the eruption takes place through a central vent or mouth. The rock fragments, ash, lapilli as well as liquid lava are ejected with more or less explosion and collect around the mouth. In this type of eruption, the volcanic eruption is confined to a pipe-like vent and after the eruption, cone and crater structure is developed. A good example of this type is the Cotopaxi volcano in Ecuador.

Sometimes in addition to the main hole, there may be satellite holes on the slope of the hill through which lava may escape at lower elevations. Etna volcano is example of this type of central eruption.

In central eruptions, the nature and intensity of the eruption show great variation according to the amount and pressure of gases and the viscosity of the lava. It is said that when the lava is basic (basalt) and therefore less viscous, the eruption is peaceful and there is absence of explosion. On the contrary, when the lava is acid and therefore more viscous, the eruption is explosive. This may be true to a certain extent but, it cannot be accepted in the form of a law, because the amount of gases in the magma, their pressure and action appear to be no less important.

(b) **Fissure Eruption**— Some eruptions instead of being restricted to central vents take place along a fissure or series of fissures or cracks without any explosive activity. On account of the absence of explosion, pyroclastic materials are absent. Fissure eruptions do not build large volcanic cones but more commonly they build lava plateaus and plains. In some places small cones and craters may be formed after the eruption has slowed down and the cracks have closed and volcanic action is possible only at a few points. Layers of ash are sometimes found between the various lava flows, but the total amount of ash in comparision to the huge volume of the lava ejected is very small.

Sometime the lava rises up along the cracks but reaches the surface only at isolated points so that the eruption is linear but discontinuous. The volcanoes of the Hawaii islands are an example. In these there is absence of explosion and the lava is usually thin basalt. Lava is not ejected from a crater but comes out from the lateral cracks along the hillslope and results in the formation of small-hillocks and cinder cones.

Fig. 41—The Deccan Lava Plateau of Peninsular India

In fissure eruption there is no pipe leading directly to the magma in the earth's interior but there is a long fissure and the eruption may be either along the entire fissure or at selected points along the fissure. In 1783 the Laki eruption in Iceland was of this type and the length of the fissure or crack was 32 kms. In 1886 the Travera eruption in Newzealand was from a fissure nearly 15 kms long. The Travera eruption was, however, explosive and a large quantity of ash was released. But ordinarily the fissure eruption is quiet and not explosive.

Examples of fissure eruption are rare in the present day world, but excellent examples are available from the geological past resulting in the formation of several extensive lava plateaus, built mostly of basaltic lavas. An excellent example is the Deccan lava plateau of India covering an area of over two lakh sq. miles. The eruption took place towards the close of the Cretaceous age. The basalt lava which was poured out of the fissures inundated an area of nearly five lakh sq. miles, as the remnants of the old lava plateau are to be found as far as Ranchi district in Jharkhand in the east to Rajamundry in the south. The Deccan lava plateau is built up of a succession of lava flows which from different layers one above the other. The region is commonly referred to as the Deccan Trap as the successive lava flows with their different thicknesses have resulted in step-like structure. Over this entire area the average thickness of the lava is 2,000 to 5,000 ft, but varies from about 200 ft. near Belgaun to over 7,000 ft. near the Mumbai coast. Another good example of lava plateau built by fissure eruption is the Columbia plateau in the U.S.A. spread over the States of Washington, Oregon, Nevada and Idaho. Here the basalt lava covers an area exceeding one lakh sq. miles and has an average thickness of 3,000 to 4,000 ft. This eruption took place in Miocene times and the plateau contains thin layers of lava flows. The Snake river has cut deep canyons in it going at places to depths of 5,500 ft. below the surface and has exposed the different basalt layers.

Types of Central Eruptions

Volcanoes of the central eruption class may be sub-divided into five types based on the nature and intensity of eruption together with the differences in the ejected materials.

(a) *Hawaiian Type*—Volcanoes of this type show less explosive activity and the eruption is peaceful. In other words, Hawaiian activity is essentially effusive. The gases are not so powerful and lava is of the thin basalt variety. Whatever gas is there is ejected relatively quietly and separately from the lava and the effusion of basalt lava is dominant which erupts quietly from the pit-like crater or the lateral cracks and either spreads widely over open slopes or pours down valleys as lava rivers. There is little or no tephra. Sometimes fountains of basaltic spray rise up with the gases and

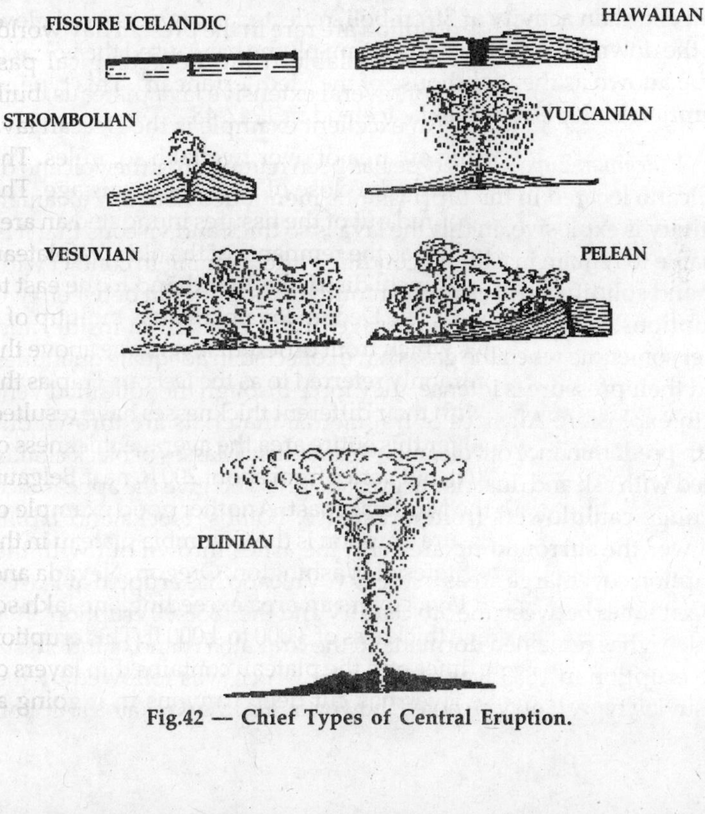

Fig.42 — Chief Types of Central Eruption.

when the wind is strong, these lava pieces are stretched into long shiny threads known as 'Pele's hair' in the Hawaii islands after Pele, the goddess of fire in those islands. This type of eruption is especially characteristic of the Hawaiian islands and hence its name. The basalt plateaus of the Columbia Plateau and Iceland provide other examples of this type of eruption.

(b) *Strombolian Type*—In this type the basaltic lava is not quite as thin as in the Hawaiian type, and the gases come out either continuously or interruptedly with moderate explosive action. The explosion is relatively mild, and besides liquid lava fragmental materials are also ejected and scoriae and bombs are also formed. Normally eruption is intermittent and fountains of lava are ejected at regular, rythmic intervals varying from a few minutes to about an hour, and flows are unusual. Neither in the Hawaiian nor in the Strombolian type of volcanoes do we find clouds of black smoke. Stromboli is an island in the Lipri group of islands, north of Sicily in the Mediterranean sea, where this type of eruption is found. The lava-fountain activity of Stromboli, reflected at night as a red glow on the down side of a towering steam plume has caused the volcano to be known as the "lighthouse of the Mediterranean". This type of eruption took place in Hekla, Iceland in 1947-48.

(c) *Vulcanian Type*—This type has been named after the volcano of Vulcano located in the Lipri islands mentioned above. Vulcanian activity is explosive. In this the lava is so thick and viscous that it is unable to remain in a liquid condition after coming in contact with air and solidifies and seals the mouth of the crater in between two eruptions. This blocks the passage of the gases from inside, and after sometime when the gases have collected in adequate quantities and their pressure is intense, they force through the solidified vent with explosion. All sizes of fragmental materials are thrown out with predominance of volcanic ash and dust. Masses of black clouds filled with ash and dust rise to great heights and give the appearance of huge cauliflowers from a distance. Bombs, blocks and lapilli shower the surrounding area and the ashes thrown out with the eruption cover large areas in the sky. Vulcano has erupted at least a dozen times between the 5th century and the 1888-90 eruption after which it has remained dormant. In the Krakatoa island of Indonesia the eruption in 1883 was of this type which was followed by the vesuvian type. It may be noted, however, that the Krakatoa eruption is considered to be a special variety of the Vulcanian type called *Phreatic eruption*, because it was caused by ground water or sea water entering the magma chamber and flashing into steam. The major eruptions of many volcanoes begin with a vulcanian phase where an obstructed vent has to be first cleared.

(d) *Vesuvian Type*—In this type of eruption there is violent explosion due to the intensity of the gases and the lava comes out with great force. Lava comes out first from the lateral cracks, but the gases keep on accumulating in the main vent. When the pressure has eased a bit on the top on account of the ejection of lava from the lateral cracks, then under the influence of the intense gases the lava comes out rapidly with the explosion, the gases forming cauliflower-like clouds. These clouds rise to great heights and look very shining and bright, and cause downpours of ash over considerable distances. In the Krakatoa eruption of 1883 such clouds had risen to heights of nearly 80 kilometres in the atmosphere.

Sometimes the amount of ash present may be small and for several kilometres around the rounded clouds may consist principally of gases and water-vapour. An eruption of this type was witnessed in Vesuvius in 79 A.D. which was first recorded by Pliny. This type of explosive and violent Vesuvian eruption is therefore sometimes distinguished as the *Plinian type*. Some consider the Vulacanian and Vesuvian types to be essentially similar, and account of the Vesuvius eruption of 79 A.D. are often quoted as example of Vulcanian activity.

(e) *Pelean Type*— This type of volcanic eruption is violently explosive and the lava is highly viscous. At the time of the eruption the dense lava solidifies and closes the mouth of the crater and a dome is formed there. After sometime, the powerful gases trapped inside either break through this obstacle or come out along the hill-slope, and then an exceedingly dense mass of hot, highly gas charged lava mixed with fragmental materials and ash flows down the slope like an avalanche. These have been called '*nuees ardentes*' or glowing clouds. An explosion of the Pelean type differs from the Vulcanian in that the very hot gas and lava mixture is not thrown upwards but spreads downslope as *nuee ardente*. The nuee ardènte is extremely dense but highly mobile on account of the gases and moves rapidly down the hill slope almost without friction. *Nuee ardente* is most terrifying form of explosive volcanic activity. It is, however, soundless, as inspite of the fact that it has the velocity of winds in a hurricane, the glowing fragments in the nuee ardente are cushioned by hot gas and are not abraded by the turbulent motion. In fact most of the volume of nuee ardente is hot gas and super-heated steam and the quantity of tephra is small. In 1902 in Martinque island of the West Indies there was terrific explosion of Mount Pelee and the nuee ardente from this explosion completely destroyed the town of St. Pierre located at the bottom of the hill and all its inhabitants were killed[1]. The temperature in the nuee ardente that devastated St. Prerre is believed to be between 650^0 and 1060^0C.

It may be pointed out that whereas in the Strombolian and Vesuvian types the gases come out of the crater through the central vent, in the Pelean type mostly thick lava comes out of the crater which cogulates and takes the form of an obstructive dome. It is only after this that the gas-filled magma starts coming out from the subsidiary vents and cracks along the slope, and the amount of gases is so much that they rise into enormous clouds into the air.

Volcanic Landforms

Volcanic landscape consists of assemblages of very distinctive landforms built largely by accumulations of solidified lava and of fragmentary volcanic products. Though in some cases they may be partly modified by erosion, they contrast strongly with most other landscapes which are largely shaped by the processes of erosion. The constructional volcanic landforms may be said to fall into two broad categories — (a) cones and fields of cones, domes and related features, (b) volcanic plateaus and plains. There are, however, complexes consisting partly of extruded domes of lava and partly of volcanic plateaus. The volcanic landscape may be further diversified by negative features — explosively blown out hollows, the largest of which are calderas. Lakes may occupy some of these

---

1. "These downward-rolling explosive blasts, one of which wiped out St. Pierre in 1902, are commonly referred to as *nuees ardentes*. Perret who has observed them at close quarters defines *nuee ardente* as an avalanche of an exceedingly dense mass of hot, highly gas-charged and constantly gas-emiting fragmental lava, much of if finely divided, extraordinarily mobile, and practically frictionless, because each particle is separated from its neighbours by a cushion of compressed air." Holmes A—*Principles of Physical Geology*, Nelson, London, 1966, p.309

explosion-made depressions but they also occupy valleys dammed by solidified lava or warped by differential subsidence caused by vulcanicity.

The viscosity of the magma is a major factor determining the type of volcanic features produced, as viscosity determines both the type of eruptive activity and the products that are erupted. We may begin with the forms built by the least viscous, effusive lava eruptions i.e. the forms built by basaltic lava, and then consider the forms built by more viscous (medium silica content) andesitic lava characterized by explosive eruption, and finally the landscape built by the extremely acid and extremely viscous ignimbrite (high silica content).

Basaltic Lava Forms :Domes, Shield Volcanoes and Lava Plateaus

Basaltic Domes or Shield Volcanoes

Lava of basic composition is less viscous and hence less explosive than andesite. Consequently the proportion of tephra or pyroclastic materials is very small. There is voluminous lava flow which spreads out over large area like a thin sheet because of its low viscosity, covering the pre-existing topography. Such outflows of basaltic lava build large lava domes, shield volcanoes and also lava plains and plateaus. Basaltic domes and shield volcanoes are essentially the same, the difference being one of size only. The basaltic domes grade upward in size to become shield volcanoes. Both the basaltic domes and shield volcanoes are built as a result of repeated accumulation of lava of successive flows piled around localized vents, the resulting domes being relatively flat, having gentle surface slopes and composed almost entirely of lava. Both these characteristics are related to the mobility (lower viscosity) of basaltic lava due to its low content of silica and the associated property of releasing gas rapidly and quietly, so that explosive activity is minimal and the lava travels a considerable distance before solidifying, thus covering a large area like a thin sheet.

Basalt domes and shield volcanoes would thus appear to originate in a manner somewhat similar to andesite cones i.e. as a result of emission from localized vents, but in other respects they have more in common with lava plains and basalt plateaus. Basalt emerges from fissures rather than pipe-like vents, though the distinction between these is somewhat arbitrary, as lava rising up along fissures may reach the surface only at isolated points or pipes may develop where feeding fissures intersect.

Fig. 43—Shield Volcano.

The basaltic domes built of successive flows piled around a central vent are called *exogenous domes,* to distinguish them from *endogenous domes* or cumulo-domes of very viscous acid lava which may rise upward into a crater but be too stiff to flow and too degassed to explode, resulting in a skinned over or bulbous protrusion of lava from the vent (fig 46).

Exogenous domes when sufficiently large and extensive become *shield volcanoes,* though there is no definite size criteria. Shield volcanoes typically developed in Iceland are between 100 m and 1,000 m high and their base diameter is about twenty times their height. They are very flat cones of moderate size with small summit craters. Basaltic domes (or shield volcanoes) of the Hawaiian type, found in the Hawaii islands, Samoa and other volcanic islands in the Pacific and Indian Oceans have a more domical form and are generally elongated because of a concentration of feeding fissures beneath the volcano along rift zones or faults that encircle the growing dome. The crest of the dome becomes a broad, convex-upward plateau around a rimless crater or a summit caldera. The basalt domes or shield volcanoes of the Pacific and Indian Ocean volcanic islands are very large. The size of the Hawaiian shield volcanoes is particularly large — two large and three smaller domes merge to form the large island of Hawaii. Mauna Kea and Mauna Loa, the two major shields have summit elevations of 4,206 m and 4,175 m. above sea-level. The large size of such domes can be comprehended when it is realized that much of their bulk is below sea-level, and the two largest (Mauna Kea and Mauna Loa) rise to heights of 8 kms and more from the ocean floor. Including its submerged portion, Mauna Kea is the highest mountain on the earth. The dome form is not perfect in the shield volcanoes of Hawaii and the angle of slope is very small, only 3^0 to 11^0 around Mauna Loa. A large central area on Mauna Loa is fairly level because of its tendency to subside with the formation of fracture sinks which become filled up with lava flows that issue from fissures within the sink. Outflows from radial fissures that open on the mountain flanks also continually build the gentle slopes of the flanks.

Among islands of the Pacific Ocean, there are basalt domes in all stages of dissection by erosion. Two maturely dissected domes form the island of Tahiti, and two overlapping domes, dissected sub-maturely and partly submerged form the Banks Peninsula, Newzealand.

Basalt Plateaus and Plains

Related closely in origin to basalt domes and especially to dome such as that of the Hawaii island, are *basalt plateaus.* As has been pointed out earlier (*ref. fissure eruption*) they have been built by lava "which has risen through fissures, not grouped in rift zones but sporadically distributed, so that flows from a multitude of sources have spread out and interfingered to make a pile of sheet thousands of feet thick, thus forming a plateau."[1] Ancient basalt plateaus are found in various parts of the world. The most extensive is the Deccan plateau of India spread over an area of over 5,00,000 sq. km, an account of which has been given earlier. The other well-known basalt plateau is the Columbia plateau of western U.S.A. covering as area of over 2,50,000 sq. km. These are now in various stages of erosion and the landscapes so produced are characterized by stepped escarpments, mesas and buttes. A recently built and comparatively little dissected example is the Snake river basalt "plain" 52,000 sq. km in extent, in Idhao, U.S.A.

"*Basalt plains are,* as it were, embryonic forms of basalt plateaus. Consisting of much thinner and less extensive piles of lava sheets, they are common throughout the world; they are, indeed, with associated accumulations of coarse scoariaceous material forming mounds and cones, the most prevalent of volcanic landscapes".[2] They may consist of a single layer or at the most a few superposed layers of lava which solidified on gentle slopes or after running down on to the valley floors. They form nearly level surfaces hundreds of sq metres in extent. Some valleys may be blocked by

1. Cotton, C.A.—"Volcanic Landscapes", in *The Encyclopaedia of Geomorphology* edited by Rhode W. Fairbridge, Reinhold. N.Y. 1968. p.1197
2. Ibid, p. 1198

lava and may form lakes. When the plains forming lava sheets are partly destroyed by erosion, distinctive landforms develop. Basalt is commonly more resistant than the underlying rocks. Consequently as the margins are under cut, they become escarpments. Gorges are formed and where the impounded rivers overflow across lava dams waterfalls result. Of these, there are good examples in the North Auckland district of Newzealand. Because of the hardness of lava that buries a plain or valley floor, inversion of relief commonly develops as the land surface is lowered by erosion, and the lava-filled valley becomes in its turn a ridge.

Andesitic Volcanic Forms

While basaltic lava builds basalt plateaus and plains with associated domes and mounds consisting almost entirely of lava with little pyroclasstic materials, andesitic landscapes are characterised by great cones and cone clusters with surrounding plains built up by deposits of shower, lahar and nuee ardente origin. Andesite is lava of intermediate composition with medium silica content. Most such lavas are explosive and 50 to 90 percent of their product consists of fragmental materials, much of which is built into cones. Cones may consist either entirely of pyroclastic materials heaped around the crater (*ash or cinder cones*) or of alternating layers of pyroclastics and of lava (*strato volcanoes*). With an increasing proportion of lava layers, the outer layer may consist entirely of lava producing a lava cone. Ash cone and strato volcano forms are, however, more common. Many of the world's largest volcanic mountains are cones consisting of andesitic materials.

Ash or cinder cones are built where eruptions are of the explosive type and there is predominance of pyroclastic materials, such as rock fragments, lapilli and ash. The pyroclastic materials collect round the crater and their deposit becomes less and less as we move away from the crater. Successive eruptions result in the accumulation of different layers of pyroclastics which slope outwards. But in the crater the deposits slope inwards as the ash tends to fall towards the crater. In general the ash of cinder cones are perfectly conical

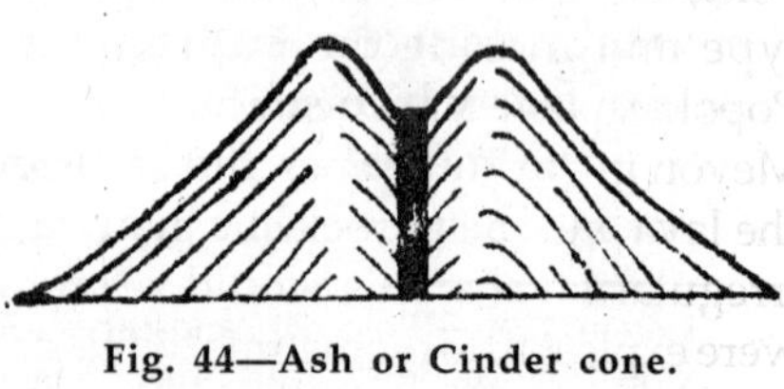

Fig. 44—Ash or Cinder cone.

and symmetrical, for on their slopes the material has come to rest at the angle of repose. As a result of explosive ejection that carries pyroclastic materials to a distance beyond the upper slopes, and partly as a result of the washing down of ash and its deposition as alluvial fans and of the avalanching of newly deposited hot and mobile lava, the base of the cone is long and stretched and the lower slopes of the cone are gentler than its upper slopes (Fig. 44). The slope profile as a whole is thus concave. Near the crater there is predominance of coarser materials and the slope of the cone is 30^0 to 40^0 which is the maximum slope of repose for such loose materials. Some of the fragments that fall on the top and sides slide downhill but this goes on decreasing with distance from the vent, and towards the lower slopes the hill is made up mostly of ejected materials which have fallen directly from the air. If at the time of the eruption, there is a strong wind blowing from some direction, the ash and rock fragments are deposited largely on the leeward side of the crater and the cone formed under such condition is unbalanced

The feeding pipe that leads up through the crater to the summit where the cone- building eruptive ejection takes place, does not change appreciably during the growth of the cone. Thus large cones are sharp and steep at the apex, while the small cones look bluntly truncated so that the small cones are better described as mounds than as cones.

Seen at close quarters the ash slopes of growing cones are not smooth but are generally fluted with radial furrows due to sliding of hot ash and the slopes may also be dissected to some extent by ravines cut by erosion between eruptions.

Composite or Strato-Volcanoes

When the cones have a layered structure, with alternating ash and lava, they are described as starato-volcanoes or composite volcanoes. Most of the volcanic hills of the world are of the composite type and include the Fujiyama in Japan, Vesuvius in Italy, Popokatepatal in Mexico, Shasta in the U.S.A., Cotopaxi in Ecuador, Meyon in the Phillipines and several others. In such volcanic hills the lava and the pyroclastic materials are found in successive but irregular layers. This would seem to suggest that the volcanoes were explosive for sometime, while at other times the lava overflow was quiet. In the initial stages of their formation, they are cone-

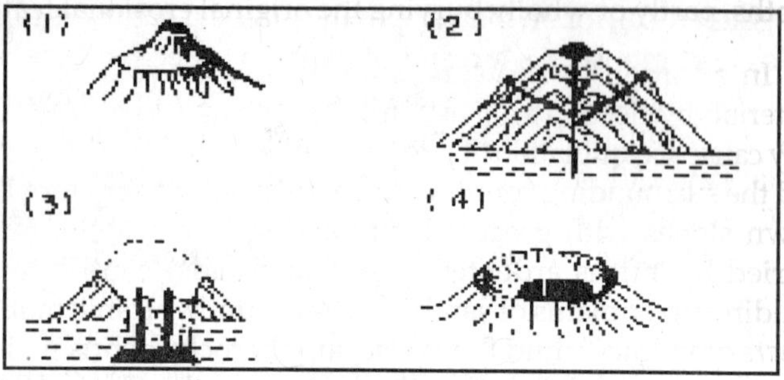

Fig. 45— Some Volcanic Forms
1. Secondary cone within crater, 2. Composite Volcano, 3. Caldera, 4. Crater Lake

shaped like an ordinary ash volcano. But as the volcano rises in height, the lava encounters greater difficulty in reaching to the top of the vent, and it finds it easier to escape to the surface from the lateral vents and cracks. Therefore in the bigger volcanoes the lava comes out from the lateral vents and cracks in the later stages, resulting in the formation of *subsidiary or satellitic or parasitic cones*. The symmetry of even the most perfectly conical volcanoes is commonly marred to some extent by the upgrowth of one or more parasitic cones on the flank, as in the case of the Fujiyama in Japan and the Egmont in Newzealand. Etna, the highest volcano in Europe (10,870 ft) has more than 200 such subsidiary vents and cones.

Large cones like the Fuji and Egmont which from a distance seem perfectly symmetrical are venerated as objects of great beauty,, by they are not common, and multiple cones and volcanic mountains of irregular form are the rule rather than the exception.

Andesitic Volcanic Fields

Around and among the clustres of andesitic volcanic cones are fields of widespread ash and coarse debris which may sometimes be of regional dimensions. The fragments of exploded lava yield large quantities of dust, sand, pumiceous blocks and blocks. Apart from the layers of ash which are deposited near the crater to build the cone, much fine materials as well as coarse fragments are ejected with such force that they fall beyond the cone slopes. The finest materials are carried by the wind and deposited over a wide area. The surrounding country is thus covered and built by debris that fall as shower and accumulate over the nearer parts to considerable depths, partly or wholly burying the original erosional features.

In addition to showers, many andesitic volcanoes supply materials by ejection as "*nuees ardentes*" and as '*lahars*' both of which may cause accumulation of large quantities of debris on lower slopes and the surrounding country. '*Nuees ardentes*' are very mobile, flow down slopes with great velocity and spread widely. Materials carried by 'nuees ardentes' (glowing clouds) contribute to the building up of plains of sandy debris around volcanic mountains. *Lahars* or volcanic mud flows also supply rocky debris which may form extensive plains, the plains west of the Central volcanic mountains of the North Island of Newzealand being a good example. A mixture of fine debris and water, derived generally from a lake in the crater of a dormant volcano, bursts out when eruption is resumed, forming mud that rushes down the mountain side. The ratio of debris to water is commonly so high that the density of the mixture is great enough to buoy up and carry along a large quantity of boulders. During the slow growth of some andesitic volcanoes, repeated lahars have distributed over the surrounding country both fine debris and lava blocks and boulders derived from near the mountain tops.

Acid Domes

Sometimes even when the eruption is not explosive and the lava wells up and overflows quietly, the lava may be acid and not basic. Where the lava is acid with a high percentage of silica, it is viscous and pasty and cannot flow far from the vent. The deposits consist entirely of lavas with no rock fragments or ash and are dome-shaped. The dome formed is quite convex with steep sides and usually there is no sign of the crater. It resembles an inverted bowl in form and is called *Cumulo-domes*. Examples of such domes are found in the Reunion Island, north of Madagascar in the Indian ocean as well as the Auvergne region of France.

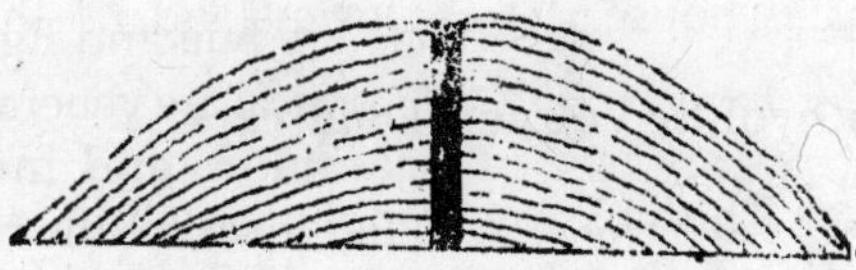

Fig. 46 — Acid Cumulo Dome

Sometimes the lava solidifies as soon as it comes out of the vent and then a vertical dome may be formed in the crater itself. Such domes are called *plug domes or tholoids or volcanic necks*. A good example of this is provided by the crater dome of Mont Pelee,

Martinique, in the West Indies. This was formed at the time of the 1902 eruption and stands like a backbone on the top of Mont Pelee. Volcanic necks that occupy former conduits today form geomorphic features as a result of the differential erosion of the various materials and appear quite striking in mature landscapes of low relief. In most cases they stand up like a stump, crag or tower. A famous volcanic neck is that of Arthur's Seat, the site of Edinburgh castle, Scotland, consisting of compact igenous rock of massive basalt called plugs. Also well known are the impressive 'Puys' of the Auvergne in Central France. The term *Necks* is more general and includes volcanic pipes from which the original rock fragments were not completely expelled.

The volcanic neck may sometimes crown a shield-shaped volcano as can be commonly seen in the Cascade Mts. of the Pacific North-west (Washington, Oregon). Some of the most striking necks are those found in the southern part of the Colorado Plateau (i.e. the Hopi Buttes in North-Eastern Arizona) where they are scattered across a broad plateau landscape and are extremely well displayed under the semi-arid climate.

Radial dikes may sometimes be laid bare by deep erosion of a volcanic neck, a classic example being that of Shiprock in New Mexico where a volcanic neck with prominent radiating dikes are seen in N.E. New Mexico. Radiating dikes are also found in the Glasshouse Mts. of Queensland, north of Brisbane, Australia.

Rhyolitic Volcanic Landscape

This volcanic complex constitutes yet another distinctive type of landscape. This is exemplified in the North Island of Newzealand by the features of the Taupo-Rotorua volcanic field.[1] The most striking landforms of this region are great non-basaltic volcanic plateaus of ignimbrite. The ignimbrite is an acid high silica rock composed of welded particles derived from magma with gas content and emitted in great volumes from fissures. The magma boils over the rim of the fissure as a highly fluidized system of pumice, glass pieces and crystals, all surrounded by very hot and highly compressed gases which reduce the internal friction. The bulk of this material spreads sidewards from the eruption fissure as a

1. This section is based on C.A. Cotton in the *Encyclopaedia of Geomorphology* edited by R.W. Fairbridge, 1968, pp. 1199-1200

vigorous overflow of pyroclastic ash (glowing clouds) of very low viscosity, which spreads out extremely quickly owing to its high mobility and covers enormous surfaces with the hot deposits which weld together with their own weight into thick, solid and glassy sheets. One of the several sheets now recognized in the Taupo-Rotorua region had volume of at least 50 cubic miles and flowed for more than 40 miles from its source.

Some of the features of basalt plateaus are duplicated in the ignimbrite plateaus, but because of the thickness of the sheets, stepped structural escarpments are not usually developed. Where, however, the plateau margins are steepened by erosion, the ignimbrite forms vertical faced escarpments, generally fluted by large scale columnar jointing. The plateau surface may not be smooth because of the incomplete welding (agglutination) of particles towards the top of the sheet and gaseous emanations which may indurate the sandy layer by introduction of new mineral matter.

Another feature of acid volcanicity in some localities is a grouping of steep-sided domes and short flows of rhyolite lava. As a rule extrusion of rhyolite follows ignimbrite eruption. Acid lava is extremely viscous and when it disintegrates explosively, the resulting pyroclastic materials are usually ejected so violently that they are unable to build cones around the vents. Thus, a rhyolite volcano usually grows as a protruded mass of lava that assumes a roughly domical shape (sometimes spinous). These domes have an irregular form and may not at first sight look like volcanic accumulations. Sheets of ejected rhyolitic pumiceous ash and pumice brecia may also mantle the surrounding surface.

In regions of large scale acid volcanocity, the emission of large volumes of ignimbrite lava, the protrusion of rhyolite domes and the ejection of much airborne pumice results in subsidence of the roof over the underlying magma. The surface is let down and within the area of subsidence there is much normal faulting and warping. Numerous lakes of volcanic origin are associated with such volcanic fields.

Volcanic Craters and Calderas

We come across primarily two kinds of topographic depressions in the volcanic regions. These are craters and calderas. In fact, the difference between them is one of size only. Craters are smaller depressions, while the caldera covers a wider area. Just above the volcanic vent there is usually a rounded bowl or funnel-shaped depression which is called the *crater*. Its diameter is normally a mile or less. The crater is formed as a result of either the blowing up of the rocks from the violent explosion or due to the slumping or collapse of the rocks near the vent mouth after the extrusion of the lava or both. Sometimes the materials which come out of the vent are deposited around its mouth and form a circular ring with a depression in the middle. Such craters are called *Ring craters or Crater rings*. In such craters the walls are made up of the erupted ash, lapilli and rock fragments and there is very little lava. Examples of such craters can be seen in the Eiffel volcanic region of Germany and in the western parts of the Rift Valley area of Africa. In these crater depressions small lakes are found which are called *Maars or Maare* in Germany. In these areas there has been formation of craters or crater rings but there has been no formation of volcanic cones.

The *caldera* is an extensive rounded volcanic depression whose diameter is many times greater than that of a crater. It is formed either due to the blowing off of the top of the volcanic cone through violent explosion or the collapse of the overlying rocks due to lack of support after a considerable amount of lava has been ejected from the lava reservoir. It was formerly believed that explosion was the prime reason for the formation of the caldera, but now subsidence or collapse of the overlying rocks is considered to be the more plausible cause, the reason being that the broken pyroclastic materials found there are much too small in proportion to the size of the hollow (depression). It is possible that both the processes have played a part in the formation of many calderas, and the composite explosion- collapse theory appears to be more probable. A huge caldera with a diameter of 4 miles was formed at the time of the Krakatoa explosion in 1883 in which both explosion and collapse played a part. The caldera of Mt. Katmai in Alaska which was formed at the time of 1912 explosion is a result of both explosion and collapse. The Crater Lake of Oregon is located in a caldera. This caldera is roughly circular and its diameter is about 6 miles. The area of the Crater Lake is about 20 sq. miles and its depth nearly 200 ft. It is surrounded by steep hills which rise from 500 to 2,000 ft. above the lake level. Formerly there was a huge composite volcanic cone here, and the caldera was formed in the Pleistocene period by explosion and collapse.

On the top of the Barisan Highlands in north-west Sumatra there is a huge caldera, having an area of nearly 1,000 sq. miles in a section of which Lake Toba is situated. It is believed that this caldera came into existence in the Pleistocene period primarily by violent explosion, because tuffs and other materials found in large part of Sumatra and Malaya are almost equal in amount to the materials which could have been blown off by the explosion.

The Lonar Lake in Berar, Buldhana District, Maharshtra is located in a circular caldera which is surrounded on all sides by 500 ft. high steeply sloping hills. Its diameter is about 1.25 miles. The solid rocky and level bottom of the lake as well as the absence of pyrocalstic materials in the neighbourhood both seem to suggest that the caldera possibly owes its origin to collapse rather than explosion.

Solftaras

Sometimes steam and other kind of gases continue to come out of the volcano even after the eruption of lava and ashes has ceased. The stage of the volcano when it emits only steam and gases is called solftara stage. This name has been derived from the Solftara volcano near Naples which continued to give out steam and gases for a long period after the eruption of 1198 had ceased. In almost all volcanic regions of the world examples of volcanoes in the solftara stage can be found.

Geysers and Hot Springs

In some volcanic regions fountains of boiling hot water are found in which the hot water is thrown off into the air at regular intervals. Such intermittent fountains in which hot water and gases are released at intervals are called *geysers*. Examples of geysers can be seen in Newzealand, Iceland and in the Yellowstone park in the U.S.A The mouth of the geyser is connected with the hot interior by a pipe. In the interior at the bottom of the pipe the water is heated and is changed into steam after reaching a certain temperature. The steam forces the hot water to ascend along the pipe which when forced up comes out on the surface in the form of a fountain. After some steam and water have come out on the surface, the pressure is reduced in the pipe and consequently the upward rise of water also stops. After some interval when the water has again heated to form enough steam, the water again rises up with the steam. It is this which is responsible for the intermittent burst of water and gases from the pipe in the geysers. In the Great Geyser of Iceland there is a saucer-shaped depression above the pipe. When the geyser is not in action, this depression is filled with water having a temperature of 170° F to 190° F., while about 100 ft. down below in the pipe the temperature is 260°F. A famous geyser in the Yellowstone park is the Old faithful. It has earned this name because of the regularity in the interval between two consecutive throws of water. This interval is 66 minutes and the geyser water rises to a height of 110 to 160 ft.

Sometime the water does not rise into a geyser but hot water flows continuously from a spring. Such hot springs are quite common in the volcanic regions. Iceland, Yellowstone Park (U.S.A.) and the North Iseland of New Zealand are particularly well-known for this. It is important to remember that hot springs may occur also in regions which are not volcanic, and the source of heat may be radio-activity. The hot springs of Rajgir (Rajgir hills) and Munger (Kharagpur hills) in Bihar belong to this category.

World Distribution of Volcanoes

At the present time the total number of active volcanoes in the world including those that have been in eruption during historical times, would not exceed 800, but in addition there are innumerable extinct volcanoes which must also be taken into account in considering the world distribution of volcanoes. After the appearance of man on the earth's surface, most of the globe has been relatively free from large scale volcanic activity.

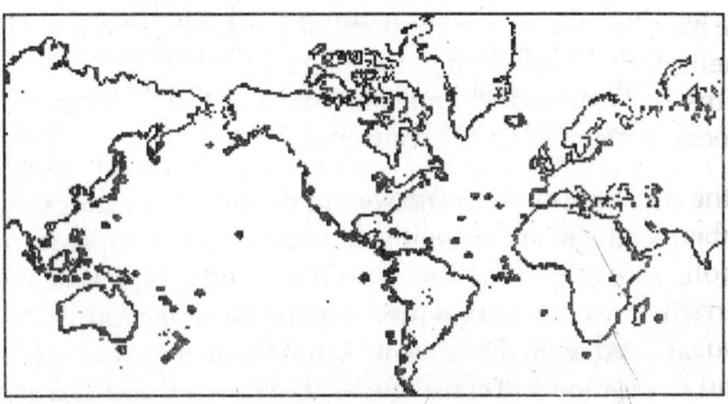

Fig. 47— Distribution of Volcanoes over the globe.

The volcanoes are not distributed over the world in a haphazard fashion. In fact the volcanoes which are active today and those that have been active in the past show a definite pattern of distribution and are found in definite belts. As we have seen in the Chapter on Plate Tectonics these belts show a close association with plate boundaries. Broadly speaking, the distribution of volcanoes is found in the following three belts :

(a) Volcanoes associated with the Circum-Pacific and Alpine young orogenic belts,

(b) Volcanoes situated on or near the mid-oceanic ridges,

(c) Volcanoes associated with rift valleys.

(a) **Volcanoes associated with the Circum-Pacific and Alpine young orogenic belts** — The most prominent zone of volcanic activity is to be found along the Pacific coast which encircles it on all sides in a semi-continuous but definite and sprawling belt. This is called 'the Pacific girdle or ring of fire". Starting from the Andean region of South America it extends northward through Central America (and the West Indies), Mexico, Western U.S.A. and Canada to Alaska, From Alaska it extends through the Aleutian islands towards the islands off the eastern coast of Asia and passes through Kamchatka, the Kurile islands, Japan, the Philippines and further south to the Celebes, New Guinea, Solomon islands, New Celedonia, New Zealand and Antarctica. The volcanic belt of the Indian Ocean which passes through Andamans, Sumatra, Java, Bali, Sunda and Burma meets the Pacific belt near the Malacca islands. This major belt of active volcanism surrounding the Pacific is associated with the destructive plate margins where the oceanic crust is subducted into the trenches under the continental margins. The downgoing slab melts to produce andesitic magma which is intruded as batholiths and erupted as andesite volcanoes, rich in silica. The volcanoes are situated on the continental side of the trenches.

In the Alpine-Himalayan orogenic belt the volcanoes are fewer in number, but it is in this belt that we find the Vesuvius, Etna, Stromboli, Vulcano and other famous active volcanoes of the Mediterranean region. East of this region there are volcanoes in the Caucasus, Elburz mts, and in Iran and Baluchistan. But the main Alps and the Himalayas are remarkably free from volcanoes. It is believed that this may be due to the intense lateral pressure on the rocks of this system and the absence of vertical cracks. A few sporadic groups, however, occur at some distance from the ranges of the orogenic belt, and the world's highest plateau Tibet has several regions of hot springs.

(b) **Volcanoes on or near the mid-oceanic ridges**— These are oceanic volcanoes associated with the constructive plate margins. Here the magma rises from the mantle through the rifts in the mid-oceanic ridges to form new oceanic crust, and the lava is basaltic and poor in potash. Through the process of sea-floor spreading some of these have migrated away from the mid-ocean ridges and have become extinct.

(c) **Volcanoes associated with rift valleys**— There are some volcanoes which are associated not with plate margins but with rift valleys, the most notable of these being the volcanoes associated with the East African rift system. Here the lava is very rich in alkalis and is referred to as alkali basalts, rich in potash and sodium.

The boundary between the basalts of the oceanic crust and islands and the andesite volcanic rocks of the Circum-Pacific belt is called the *Andesite Line*. It is essentially the boundary between the continental crust with sial and the oceanic crust without sial, and follows the oceanic trenches. The Andesite Line along the Pacific and its continuation around the Indonesian arc is one of the fundamental geological boundaries of the earth, and is the line where oceanic lithosphere is subducted along Benioff zones.

Intrusive Features or Plutons

We may now briefly consider some of the intrusive igneous features which form when the magma solidifies within the earth's crust instead of erupting on the earth's surface. These features appear on the landscape only after erosion has removed the covering rock. It is, however, important to remember that intrusive and extrusive igneous activity frequently occurs together.

Rock bodies which form by the consolidation of magma below the surface of the earth and are now exposed due to erosion are collectively called *plutons*. An elaborate terminology based on the shape of the plutons and their relationship with the adjacent rocks has developed.

Dykes and Sills

The intrusive features are grouped according to their shape (i.e. tabular or massive) and their relationship to the parting planes of the rocks into which they have intruded. When the intrusions cut across the pre-existing structures, they are said to be *discordant*, but where the magma intrudes and consolidates along a bedding plane, the intrusion is said to be *concordant*. Tabular masses have a sheet-like appearance and when discordant they are known as *dykes or dikes*, and when concordant are called *sills*.

In the case of *dykes* the magma ascends through a vertical fissure, which on cooling becomes a vertical sheet of rock with roughly parallel sides cutting across the bedding planes of the pre-

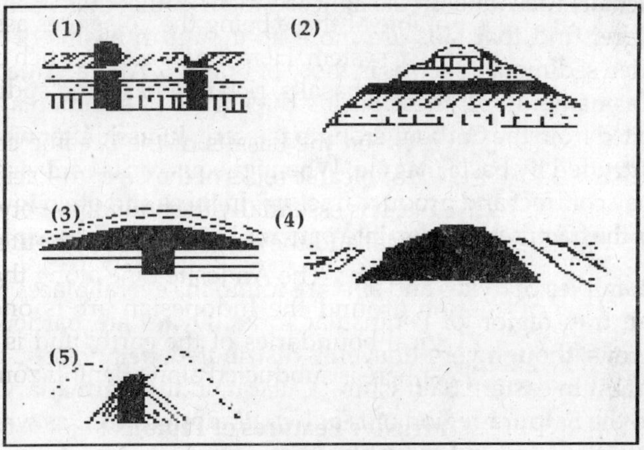

Fig. 48— Intrusive Features.
1. Dyke, 2. Sill, 3. Loccolith, 4. Batholith, 5. Volcanic Neck.

existing rocks. Dykes are much longer than they are wide and are often more resistant than the rock into which they intrude. Frequently, therefore, they stand as ridges or project as cliffs along the coast when exposed at the surface by erosion. Though their width varies greatly, commonly they are 2 to 7 metres wide, while their length varies from a few metres to many kilometres. Dykes are quite numerous in some areas of igneous activity. Good examples of swarms of dykes are found along the western coast of Scotland. Most are dolerites but along the margins the rocks are of finer grain and resemble basalts due to rapid cooling by the walls or loss of gases. The wall rocks show the effects of thermal metamorphism. Many dykes die out upwards indicating that the internal pressure of the magma became insufficient to raise it further. These dykes fail to reach the surface. Other dykes may, however, represent feeders of lava flow of plateau basalts and are under sufficient hydrostatic pressure not only to reach the surface but build great thickness of lava flows. The high pressure magma in such cases obviously comes from great depths in comparision to the dykes that fail to reach the surface.

Tabular masses which are concordant i.e. intrude and solidify along the bedding planes of pre-existing rocks are called *sills*. Basaltic magma which ascends under great hydraulic pressure finds it much easier to penetrate laterally along the bedding planes of thick sedimentaly rocks and lift the light sediments rather than itself. We, therefore, find that sills are more abundant in basins of thick unfolded sediments, such as in those of Brazil, S. Africa, Antarctica and Tasmania. In all these basins thick piles of sediments were deposited from the Carboniferous to the early Jurassic after which it was intruded by basic magma. When igneous rocks cool within a sill, they contract and produce fracture. In these and other igneous bodies the fractures develop into patterns known as columnar joints.

Examples of dykes and sills are found in several places in the Deccan trap region of Peninsular India. Dykes are particularly numerous, though very unevenly distributed, being particularly important in eastern Saurashtra, Gujarat, in the Narmada Valley and in the Satpura region of Madhya Pradesh as well as western Andhra Pradesh. In Saurashtra they show a radial disposition around Amreli and Jasdan and also around the plutonic mass of Girnar hills.

The Deccan dykes vary considerably in their dimension. Some dykes in the Satpura mountain are 60 m. thick. Many are several hundred metres long. While a few have been traced over a length of 30 to 50 kms. Most of the Gondwana coalfields are bounded by dykes and sills especially in the Satpura region of M.P., more especially in Upper Gondwana formations and in the Gondwana basins of Rewa, the Damodar and Assam. A few sills also penetrate the Jurassic strata in Kutch. Thick sills are sometimes met with in Rewa and Satpura regions. The intrusives of Satpura and Rewa are of Deccan trap age and those of the Damodar and Assam probably of Rajmahal age.

Laccoliths

Intrusive structures which resemble sills except for their lens shape are called laccoliths. Like sills they are concordant intrusions. When the magma is more viscous than the basaltic magma, it may find it easier to arch up the overlying strata into a dome-like shape instead of spreading widely as a relatively thin sheet, provided its hydraulic pressure is sufficient. Such lens-shaped relatively shallow igneous intrusions arch overlying sedimentary strata to produce topographic domes. The name laccolith was given to such inflated intrusive forms by the famous geologist G.K. Gilbert in 1877 when he first discovered these in the Henry Mountains of Utah in the Colorado Plateau of U.S.A., where a number of them are now known to exist. Until recently it was thought that the laccoliths were fed by magma rising from below and spreading out along bedding planes until its movement was obstructed by marginal cooling and consolidation. In the process, it also uplifted the strata near the feeding channel by vertical thickening and produced a dome-shaped roof. But recent studies have shown that each group of laccoliths in the Henry Mts. and, indeed, in the entire Colorado Plateau occurs within a large structural dome. " At the middle of each dome is a stock around which the laccoliths are clustered like the petals of a lop-sided flower. "[1] initially these domes themselves were mistaken for laccoliths, but "the real laccoliths are tongue-like buldges which appear to have been fed sideways from the stocks". The laccoliths of the Colorado plateau were intruded in late Eocene or Miocene times and surface arching must have resulted. The drainage pattern suggests that the rivers flowing on a surface a thousand or more metres above the present landscapes, were diverted by the rising laccoliths and migrated laterally off the domes.

Lopoliths

Lopoliths are large concordant intrusions which, unlike laccoliths, have a spoon-like or saucer-like form (*Gk lopas* a shallow basin). The sagging is probably related to the transfer of a large amount of magma from the depth of the upper part of the crust. But lopoliths are not merely very thick sills that have subsided into a spoon-like shape. Their structure is extremely complex and is little

1. Holmes, Doris L—Ibid, p. 167.

understood as yet. The type example is the Duluth gabbro, which extends from the south-west corner of Lake Superior for about 193 kms to the north-east. The Bushveld Complex in S. Africa includes a cluster of several lopoliths which together cover an area almost as large as Scotland.

Radial Dykes

The fractures that sometimes develop when the roof rocks above the intrusion are upheaved and stretches into a broad domal form, are likely to be radial or concentric. These fractures are filled up by the rising magma and form radial dykes. The roof of a volcano is often found to arch up before a great eruption. This may explain the existence of the radial pattern of dykes around certain old volcanic centres where the superstructure of lava and tuffs has been removed by denudation Radial dykes are laid bare by deep erosion of a volcanic neck. A classic example is of that of Shiprock in N.E. New Mexico where a volcanic neck with prominent radiating dykes are found. Radiating dykes are also seen in the Glasshouse mts of Queensland, north of Brisbane, Australia. In India radial dykes are found in Saurashta around the Girnar hills.

Volcanic Necks and Plugs

When the rocks of an extinct volcanic cone have been removed by erosion, the more resistant pipe-like intrusion which represents the conduit may be exposed and form a conspicuous feature of the landscape. The term *volcanic neck* is generally used for such features including those volcanic pipes which contain considerable quantities of the broken fragments of the original rocks or agglomerates which were not completely expelled. The impressive 'puys' of the Auvergne in Central France are a well known example. A striking example of volcanic neck consisting mainly of agglomerates is to be seen in Le Puy, Velay, Haute Loire, France. The volcanic necks that consist of compact igneous rock such as the famous basalt of the Castle rock of Edinburgh, Scotland are more commonly called *plugs*. In most cases the volcanic neck or plug stands up like a stump, crag or tower, reflecting the differential erosion of the various materials, The neck may sometimes crown a shield-shaped volcano, as in the Cascade Mts. (Washington, Oregon, Pacific NW). Some of the most striking necks are to be found in southern part of the Colorado plateau e.g. the Hopi Buttes in north-eastern Arizona. Here they are scattered across a broad plateau landscape and are extremely well-displayed under the semi-arid climate.

Batholiths

Batholiths are the largest intrusive bodies, massive in shape, and when they occupy the core of mountainous areas (such as the Sierra Nevada batholith), they extend for hundreds of kilometres. The term batholith was first used by Suess in 1888 to describe the major granitic intrusions emplaced within the cores of orogenic belts. He envisaged batholiths as being shield-shaped, with steep walls extending down to unknown depths. Daly (1914, 1933) defined batholiths as granitic intrusions of a large size (having cross-section of 40mi^2 or over 100 km^2) and having no visible or inferable floor of older rock. Intrusive plutons similar to batholiths but covering less than 40 mi^2 are called *stocks*. The batholiths are commonly discordant i.e. cut across the enclosing rocks, but on a grand scale most of the batholiths are elongated parallel to the dominant structure or the fold-axes of the orogenic belts within which they are emplaced. The emplacement may take place in various ways, such as underground cauldron subsidence, whereby granite magma fills a cavity developing above a large subsiding cylindrical block several kilometres in diameter (earlier view of Suess); fusion of older formations by forceful injection of hot magma as if a red-hot poker had pierced the crust, with the evolution of batholiths *in situ* (later view of Suess); stoping, a process whereby magma eats its way upward by shattering roof rocks and breaking off small blocks that sink back into the magma to be assimilated at depth (view of Daly); or perhaps metasomatic replacement, i.e. atom-by-atom replacement of the older rock by moving solutions; or by rheid flowage (as in salt diapers or domes) due to the long duration of the deforming stress.

Batholiths when eventually uncovered by erosion, form huge granitic domes. They are a major ingredient in the mosaic of ancient rocks comprising the continental shields. The Idaho batholith is a granite mass exposed over an area of 40,000 sq. kms. The Sierra Nevada batholith of California constitutes most of the high central part of the great mountain block. Dendritic drainage pattern is often well developed on an eroded batholith.

Questions

1. What are volcanoes and how are they formed? Describe the products of volcanic eruption.
2. Describe the different types of volcanic eruptions and give examples.
3. Write a reasoned account of the different forms and types of volcanic cones.
4. Give a reasoned account of the landforms resulting from volcanic activity.
5. Write short notes on :—

   Fissure Eruption, Calderas, Composite Volcanoes, Solftaras, Geysers and Hot Springs, Shield Volcanoes, Cinder Cones, Dykes and Sills, Laccoliths and Batholiths.
6. Explain the distribution of volcanoes in the world and its relation to plate tectonics.

Chapter 13

Landforms of Tectonic Activity

The landforms found on the earth's surface are the result of complex interaction between the endogenetic or the internal earth forces acting through volcanic and tectonic processes and the exogenetic or external forces, acting through the agents of denudation. It is useful in this context to think in terms of two basic categories of landforms— initial landforms and sequential landforms. The *initial landforms* are produced directly by volcanism and tectonic activity and include volcanoes and lower flows, downdropped rift valleys and mountain ranges elevated in zones of recent crustal deformation. The energy for lifting molten rock and the crustal masses to produce the initial landforms is derived from within the earth itself and manifests itself in the form of heat flow, volcanism, earthquakes and crustal movements. This internal heat source is largely a product of radio-activity in rocks of the earth's crust and mantle, and is the fundamental energy source for the motion of the lithospheric plates.

The *sequential landforms* are those which follow in sequence after the initial landforms are created. The initial landforms are modified and shaped by the processes and agents of denudation which carve them into a number of sequential landforms. Every landscape is thus merely the existing stage in the eternal conflict between the internal and external forces. "As lithospheric plates collide or pull apart, the internal earth forces spasmodically elevate parts of the crust to create initial landforms. The external agents persistently keep wearing these masses down and carving them into vast numbers of smaller sequential landforms"[1]. As the internal earth forces act repeatedly, new initial landforms keep coming into existence as the old ones are modified or destroyed.

---

1. Strahler, Arthur N. and Strahler, Alan H.— *Elements of Physical Geography* 3rd Ed. John Wiley & Sons, New York, 1984, p. 242.

Tectonic Landforms

It is customary to describe the landforms that result from crustal movements as tectonic. It is, however, convenient to restrict the term to those landforms that are sufficiently undissected by erosion so that the shape of the deformed surface can be discerned. In nature, of course, all degrees of transition are found between purely tectonic and wholly erosional landscapes. The distinction is nonetheless of value, because if a landform is classed as tectonic, the assumption is that the deformation has been recent or rapid or both in comparision to the rate of erosion. Tectonic landforms are largely the result of Quaternary or of late Cenozoic earth movements and in most the movement continues till today. They are of great value in working out the geological history of a region.

Types of Earth Movements

The crustal or earth movements may be slow or sudden. The sudden movements though more spectacular and causing earthquakes, are less significant geomorphologically than the slow movements which are essentially responsible for the major tectonic features. The slow movements may be either (a) vertical or radial, or (b) horizontal or tangential. The vertical earth movements are called *epeirogenetic or epeirogenic movements* in which large sections of crust are uplifted (or subside) without disturbing or crumpling the strata and form continental masses and large plateaus (Fig 49 and 50). Examples of horizontally bedded uplifted blocks are provided by the Russian Platform or the Vindhyan mountains in India which are essentially a plateau with escarpments.

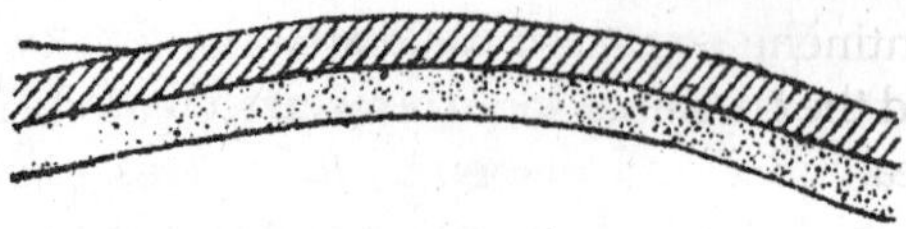

Fig. 49— Epeirogenetic or vertical movement.

The horizontal or tangential earth movements are called *orogenetic or orogenic movements*. These are characteristic of the relatively weaker or flexible zones of the earth's crust and are responsible for mountain building. They may be either (i) movements of contraction due to forces of compression, or (ii) movements of extension due to forces of tension. Compression in one part of the crust, of course, implies tension in another part, and the two are, therefore, inter-related. Compressive forces, however, dominate in the areas of mountain building resulting in the large scale crumpling and folding of rocks. Tension results in the stretching and extension of rocks and in faults and fractures, thought these may sometimes also result from excessive compression and contraction.[1]

We may thus study the effects of the orogenic movements under two principal heads :— (a) Crustal bending or folding, and (b) Crustal fracture or faulting. In the Chapter on Plate Tectonics, we have seen the distinction between these two different expressions of tectonic activity. Along converging lithospheric plate boundaries the tectonic activity is that of compression. In the subduction zones, sedimentary layers of the ocean floor are subjected to compression as the descending oceanic plate forces them against the fixed plate, and in continental collisions the inervening sedimentary rocks of the continental margins are subjected to even severer compression, resulting in the formation of massive folded mountains. On the other hand, in zones of rifting of continental plates; the brittle continental crust is pulled apart by tension and yields by faulting and the crustal block in between moves down to form a depressed area.

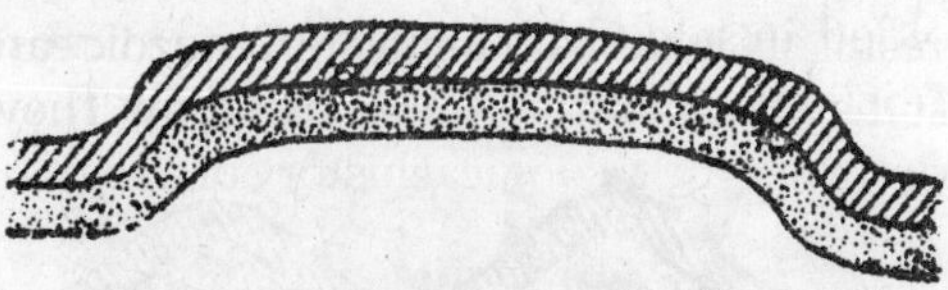

Fig. 50— Epeirogenetic movement with vertical bends.

Types of Folding

When there is continental collision, the intervening sedimentary rocks of the continental margins come under strong forces of compression, and begin to contract, and the horizontal strata begin to fold into wave-like undulation, consisting of alternating arch-like upfolds called *anticlines*, and trough-like downfolds called *synclines*. Thus the initial landform associated with an anticline is

1. "Whether a given rock will fold or fault depends upon its physical properties and upon the magnitude and relations of the forces acting upon it. In general when the maximum degree of contraction by folding has taken place, a continuance of the force will lead to faulting". Wooldridge, S.W. and Morgan R.S.—*An Outline of Geomorphology*, Longmans, 1961, p.59.

a rounded mountain ridge, while that associated with a syncline is an an elongate, open valley (Fig.51).

The shape and size of the folds depends upon the intensity and direction of the compressive forces and upon the nature of the rocks which compose the different strata. Where the compression is relatively mild or more or less equal from both sides, the folds are simple or symmetrical and are called open folds. Simple or symmetrical folds are rather rare but a very well-known example of open folds is provided by the Jura mountains of France and Switzerland. The strata consist of Jurassic limestones and were folded in late Cenozoic (Miocene) times. Though some of the anticlinal arches have been partially removed by erosion, each

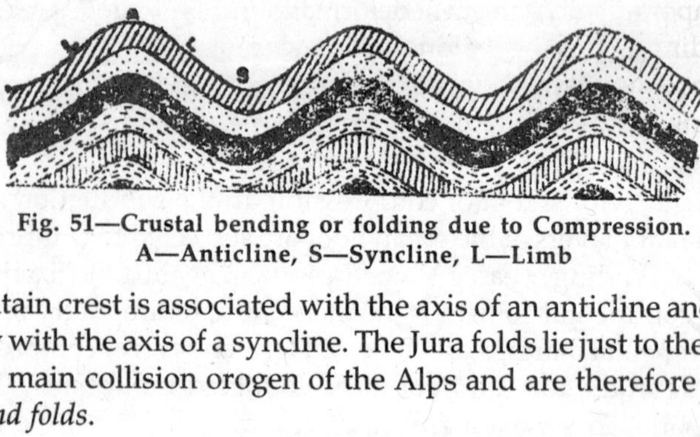

Fig. 51—Crustal bending or folding due to Compression. A—Anticline, S—Syncline, L—Limb

mountain crest is associated with the axis of an anticline and each valley with the axis of a syncline. The Jura folds lie just to the north of the main collision orogen of the Alps and are therefore called *foreland folds*.

The two sides or flanks of a fold are called its *limbs*. The plane that bisects the angle between the two limbs is called the *axial plane*, and the line of intersection of this plane with a bedding plane (the *hinge -line*) gives the direction of the *axis of the fold* at that place. If the axial plane is vertical and the axis horizontal, the fold is upright or symmetrical. If the axial plane is inclined, the fold is also said to be *inclined*. Depending on their orientation, the folds may be *symmetrical, asymmetrical, one-limb vertical or overturned* (Fig. 52). With further intensity in compression both the limbs of the fold may be almost parallel. Folding of this type is called *isoclinal*. Deep within the intensely deformed core of an orogen, folds not only become overturned or isoclinal but they are doubled over in reclining or recumbent forms, both the limbs lying one above the other. These are called *recumbent folds*. With further increase in pressure the recumbent folds may be sheared or sliced by fault planes lying almost horizontal; these are called *overthrust faults.* Further development of the structure results in forward movement of the rocks of the upper limb along the plane of shearing. The plane of shearing is a thrust plane and the structure an *overthrust fold.* The sheet of rocks that has moved forward along the thrust plane and may be highly contorted in the process, is called a *nappe,* this being the French word for 'cover sheet' or 'table-cloth'. In the Swiss Alps examples of complex nappe structure are revealed along steep mountain side, though glacial action has removed much of the nappe structure.

Thrust sheets are known to have moved horizontally for several tens of kilometres. It has been suggested that although plate collision is an important mechanism of deformation, gravity sliding or downhill sliding of the sheet along gently sloping thrust planes under the force of gravity also appears to have taken place. Gravity sliding or gliding can take place on very gentle slope which develops during that stage in mountain building when the igneous intrusions are

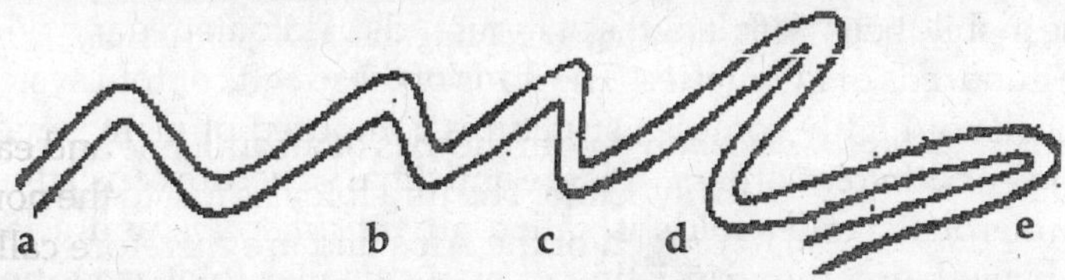

Fig. 52— Types of folding
(a) Symmetrical, (b) Asymmetrical, (c) One-limb vertical, (d) Isoclinal, (e) Recumbent.

emplaced. These igneous bodies thicken the crust and cause some isostatic uplifting, and the strata nearest the intrusive bodies are the most highly deformed and metamorphosed. Some of the folding in the Appalachian mts is believed to have occurred in this manner. The presence of ground water under great pressure beneath a thrust sheet is also thought to have reduced friction and helped the gliding of rocks layers for a long distance.

The Himalayas are not only marked by complex folds but contain overthrust and nappes of great dimensions. The Garhwal and Krol nappes are exposed in the Sub-Himalayas of U.P. The Garhwal nappe contains crystalline gneisses and schists which are intruded by granite. The Krol nappe consists of quartzite. In

Kashmir beyond the Murree thrust is a zone of nappes in which Palaeozoic and Mesozoic rocks have travelled some distance from their original sites of deposition. Granitic intrusions are found in the nappe zones of the central Himalayan range in Simla-Garhwal where highly metamorphosed rocks are found to rest on less metamorphosed and unfossiliferous Palaeozoic and Mesozoic rocks.

Multiple Folding

In some regions, layered rocks have a long history of deformation including several episodes of folding. As the younger deformations do not completely obliterate older deformations, a complex array of folds is found. The relative timing of the various folds can be ascertained by their mutual relationship. The axial surfaces of the older folds are folded by the younger folds, and in some areas folds of different ages may be superimposed upon one another.

Fold Systems

Folds are not generally found singly but occur in festoons in the mobile belts with lengths of thousands of kilometres and width of hundreds of kilometres. The folds of these belts or fold systems consist of highly complex structures composed of many smaller folds of different orders. A large anticlinal structure consisting of numerous smaller folds is called an *anticlinorium*, and a large synclinal structure consisting of many smaller folds is termed a *synclinorium*. In general, the trends of these smaller folds are more or less parallel but in some regions folds belonging to several

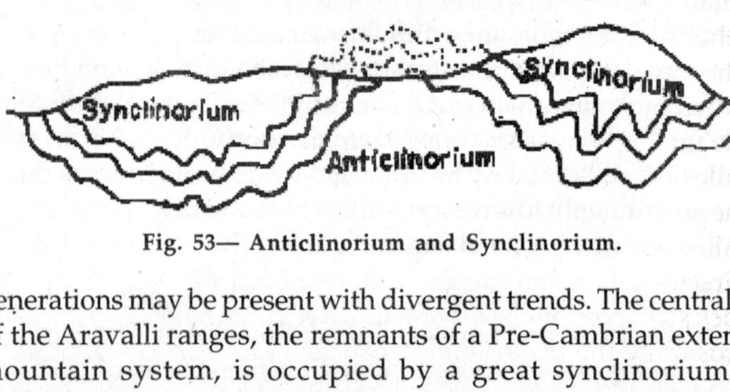

Fig. 53— Anticlinorium and Synclinorium.

generations may be present with divergent trends. The central part of the Aravalli ranges, the remnants of a Pre-Cambrian extensive mountain system, is occupied by a great synclinorium. An anticlinorium may be bordered on either side by a synclinorium, with the smaller folds radiating outwards in a fan-like fashion from the anticlinorium. Such structures resemble a fan and are known as *fan folding or fan structure.* (Fig. 54) This is particularly seen in areas of shale characterized by intense folding where shale becomes slate by recrystalization in closely packed folds.

Most fold systems are asymmetrical with respect to a zone of structural discontinuity. The fold system is more extensively developed on one side of this zone than on the other. On the less developed side of the zone of structural discontinuity, the folds are recumbent or asymmetrical, with steeply dipping thrust faults. On the more extensively developed side, and adjacent to the zone of structural discontinuity, is a region of complex recumbent folds and domal massifs. On the outer side of these complex structures, recumbent folds that have moved towards the margin under the influence of gravity are extensively developed, and in the outer margin the folds are asymmetrical and inclined and are associated with thrust faults having gentle dips.[1]

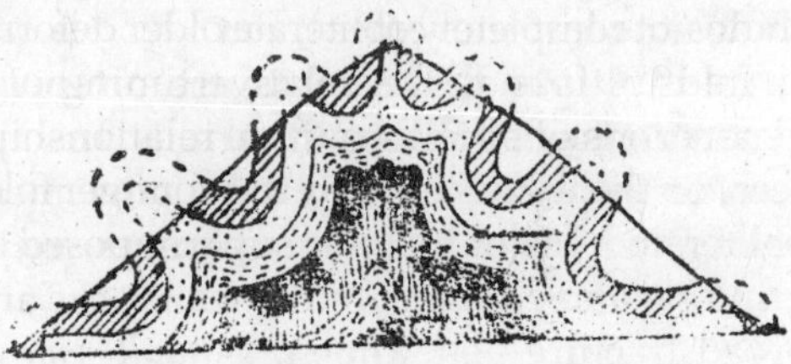

Fig. 54— Fan folding or fan structure

Crustal Fracture or Faulting

Faults are fractures in rock along which the adjacent rock surfaces have been relatively displaced. Some faults may be quite small and their displacements may be measured in inches while others may be miles long with displacements in thousands of feet. The trace of a fault on the earth's surface is known as a *fault line.* When the faults are of great horizontal extent, the surface trace or the fault line can sometimes be followed on the ground for many miles. Its position may be indicated by an escarpment or scarp. If the escarpment is the direct result of movement along the fault, it is called a *fault scarp.* It is a tectonic scarp with a steep slope and is the direct result of faulting and differential movement of the adjacent rock surfaces. On the other hand, a scarp produced by differential erosion of the rocks along an ancient fault is known as *fault-line*

1. Dewey, J.F. and Bird, J.M.—Mountain Belts and the Global Tectonics, *Jour. Geophy. Res.* 75, 2639, 1970.

scarp. In other words, a fault scarp is of tectonic origin while a fault-line scarp is a structurally controlled erosional feature. A fault scarp is thus a relatively fresh feature and can be best seen in regions of recent earthquakes and active faulting. The walls of the fault may be polished by friction between the moving blocks and contain grooves and striations. Such polished and striated surfaces are called *slickensides* (from slick, meaning smooth). The direction of the slicken-sides striations shows the last movement, but since multiple movements are often involved, overlapping generations of striations in various directions may sometimes be seen or the evidences of earlier movements may be completely obliterated. Major fualts are, however, frequently accompanied by grinding and crushing of the wall rocks and there may be a fairly broad fracture belt, known as *fault zone* consisting of sheets of shattered rock.

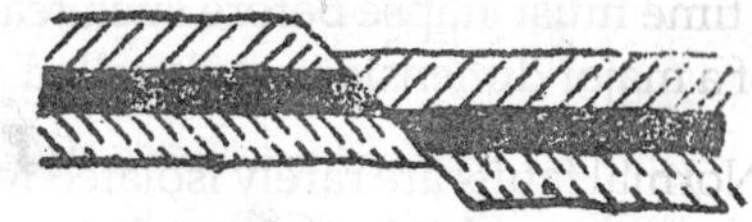

Fig. 55— Fault.

Normal Faults

Several types of faults are recognized depending upon the relative movement of the rocks on both sides of the fault plane. The movement may be vertical, horizontal or both. Faults having primarily vertical movement are called *normal faults*. The plane of slippage or the fault plane is steep (above 45^0) or nearly vertical. The rocks abutting against the fault on its upper face or 'hanging wall' are displaced downwards relative to those against the lower face or 'footwall'. One side is thus raised or upthrown relative to the other, which is downthrown. The terms 'downthrow' and 'upthrow' are purely relative. Sometimes both sides may be uplifted, one more than the other, while in many cases the block on one side is found to have subsided. Normal faults are due to tension and involve an extension or lenghtening of the faulted beds of the crust, and the lateral movement is in the opposite direction (Fig. 56 a).

Regions that have been divided by faults into relatively elevated or depressed blocks are said to possess *block faulted topography*. Topographic features which result from vertical movements involving normal faulting are bold and are preserved in several parts of the earth's surface. Such movements have been particularly active during Tertiary times and are still continuing in some areas.

We have referred earlier to fault scarps. These are steep, straight and cliff-like features produced by normal faults. They range in height from a few metres to a few thousand metres. They may be several kilometres long, and often attain lengths of 300 kms. Fault scarps are gradually worn back by the processes of erosion, but long time must elapse before such features, especially when they are of a major dimension, are levelled out.

Normal faults are rarely isolated features and more commonly they occur as multiple and parallel series of faults, giving rise to a

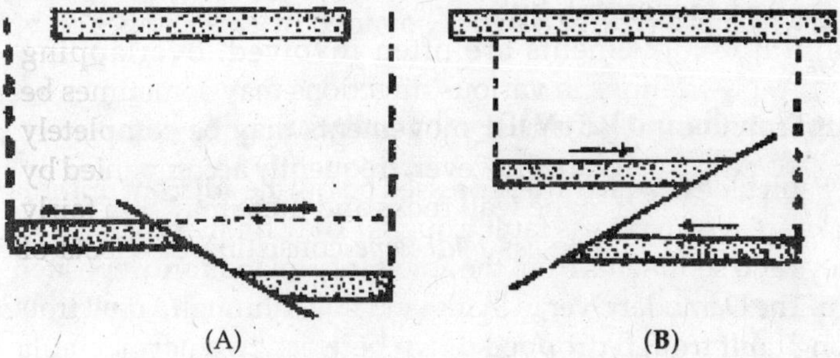

Fig. 56— (a) Normal Fault, (b) Reverse Fault.

distinctive pattern of rock structure and topography. Two important topographic features produced by normal multiple faulting are—(a) Horst and Block mountain, and (b) Graben or Rift Valley.

Horst and Block Mountain

An upstanding fault block between two normal faults is called a *horst*. It may be either uplifted between two normal faults or left upstanding while the blocks on either side have subsided. Horsts make block-like plateaus or mountains with flat top and steep and straight sides. The Shillong Plateau is an excellent example of a horst which was uplifted towards the end of the Tertiary era

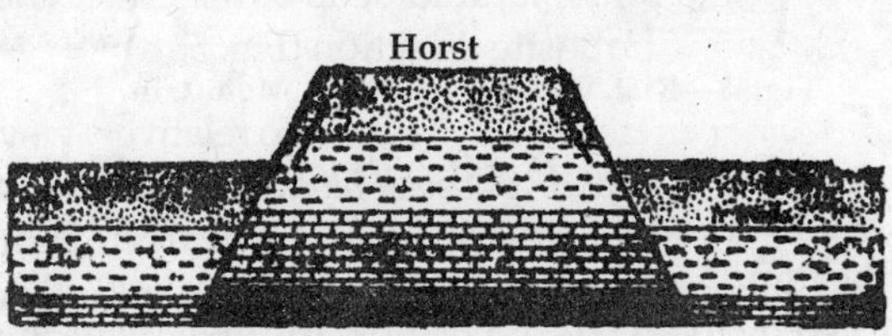

Fig. 57—Formation of Horst.

(Miocene-Pliocene) and is bordered on all sides by faults. The rise is particularly steep or nearly vertical in the west and south where the Dauki fault has a fault zone about 5 kms wide.

When the horst is long and ridge-like, it is called a *block-mountain* The Hercynian massifs of Europe, namely Vosges and the Black Forest and the Harz Mountains, are examples of horst which form block mountains. In the rifted zone of the continents where normal faulting is on a grand scale, block mountains are formed. These faulted mountain blocks may be either tilted or lifted. A tilted block has one steep face, the fault-scarp and one gently sloping side. A lifted block which is a type of horst, is bounded by steep slopes on both sides.

Fault Troughs and Rift Valleys or Grabens

Fault blocks which are depressed below the adjacent upthrown blocks are referred to as fault troughs. They frequently form river valleys and sediments from the adjoining uplands are deposited in them. The Damodar river in Jharkhand flows through a fault trough. A long fault trough dropped down between two normal faults is called a rift valley or graben. A graben is a tectonic valley or trench-like structure bordered by steep and parallel fault scarp, and has relatively flat surface. It frequently occurs in close association with horsts which bound them. Between the horsts of the Black Forest and the Vosges in west Germany the Rhine river flows through a

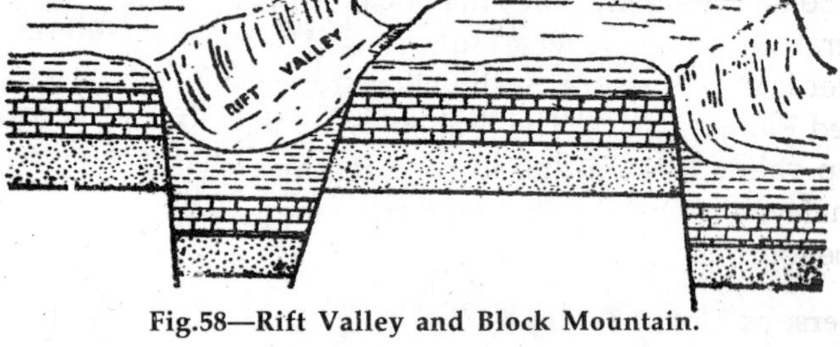

Fig.58—Rift Valley and Block Mountain.

300 km long and over 30 km wide rift valley (the Rheingraben). Similarily, the valley of the Jordan river in Israel is a rift valley. In such cases the important thing to remember is that the valleys have not been carved by the rivers which flow through them, and in fact the valleys are older than the rivers. The most notable example of

rift valley is found in East Africa where the Great Rift Valley is made up of several grabens, above which tilted horsts produce a linear mountainous topography. The Great Rift Valley extends for over 3,000 kms from the Zambezi river to the Red Sea. Here between parallel faults, the fault troughs are occupied by several lakes—Nyasa, Tanganyika, Albert, Edward and so on. The term rift valley was in fact first used by Gregory for the Great Rift Valley of East Africa which he was the first to recognize as a tectonic feature due to faulting. He defined the term to mean a long strip of country let down between normal faults or a parallel series of step faults. There is enough evidence to show that the dominant type of faulting in

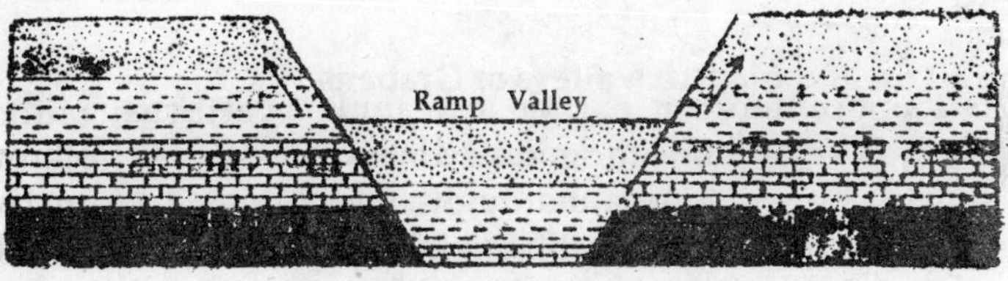

Fig. 59—Ramp Valley.

the East African rift is normal. There is a remarkable uniformity in the widths of most of the rift valleys, about 30-50 kms. The Rhine graben too is of similar width. According to the plate tectonics theory the East African Red Sea rifts may represent an early stage in the splitting apart of the continental crust which is the prelude to the formation of a new ocean.

Sometimes a rift valley may result from the forcing down of a central block along reverse faults caused by fracturing of the strata under lateral pressure. Such valleys formed by compression are called Ramp Valleys. The Brahamputra Valley is an example of ramp valley formed due to compression between the uplifted Himalayas in the north pressing against the horst of Shillong plateau in the south.

Reverse or Thrust Faults

Reverse or thrust faults have shallow inclination, the majority having dips between 20^0 and 30^0. Those with dips less than 10^0 are called *overhtrusts or overthrust faults*. Reverse or thrust faults are caused by compression (Fig. 56b). On account of extreme compression the rocks snap and one strata is pushed over the underlying strata, i.e.

the upper side is displaced above the fault plane relative to the side below. There is thus shortening of the crust in reverse or thrust faults as a result of compressional forces. Conditions under which thrust faults may be formed have been shown in Fig. 60 to 63. In Fig 60 the limb of an anticline has broken, and the broken limb has moved towards the synclinal beds. In fig. 61 overthrust has developed over the middle limb of a recumbent fold, and after snapping the upper fold has over-riden and travelled forward. Some times even when the strata remain horizontal the overlying strata is pushed over the underlying strata after the development of a compressional fault. This is shown in Fig. 62. In fig 63 we find the development of a compressional fault which cuts across the folded layers.

As has pointed out earlier low-angle overthrust faults commonly develop along the middle limbs of recumbent folds. In the Himalayas and the Alps and the other great folded mountain ranges of the world overthrust faults are found on a grand scale,

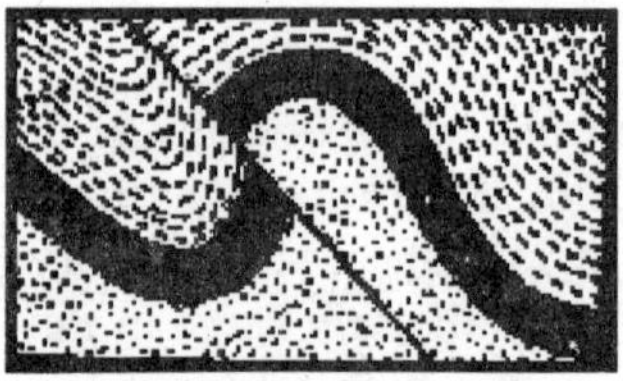

Fig—60

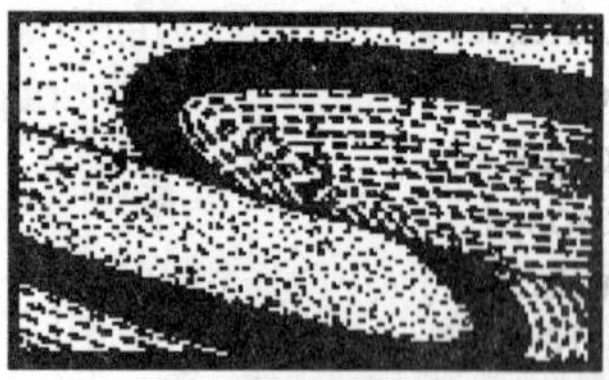

Fig—61

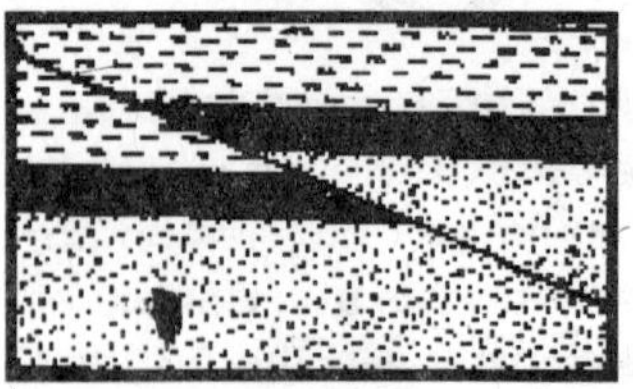

Fig—62

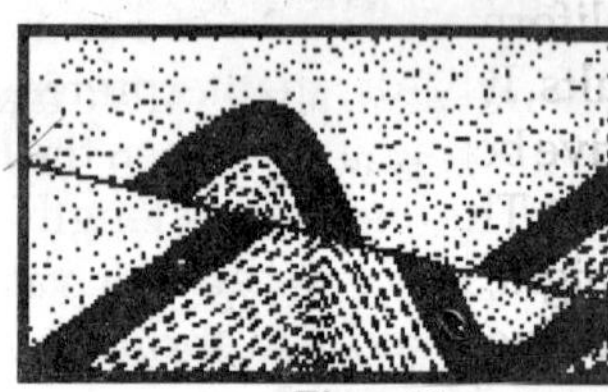

Fig—63

together with the overthrust sheets called *nappes* which have travelled forward many kilometres from their original location. In the Himalayan ıegion, at least three major thrust zones are recognized in the rocks of the Siwalik and Lesser Hiamalayas, in addition to many other less important ones. The southernmost of these thrust zones is called the Main boundary Fault which separates the Siwaliks from the earlier Tertiary and older rocks. The

Siwaliks were involved in the later phases of Himalayas orogeny and they are often found folded, faulted, overthrust and lying at steep angles against other formations. When overthrust, there is often an inversion of the normal order of superposition. There has been so much of movement that even within the Siwaliks the older strata are found thrust over younger strata. Four superimposed thrust sheets have been recognized in the Lesser Himalayas of Garwhal over the Nummulitic rocks. Further north lie other zones of thrust in the central Himalayas and beyond.

Strike-Slip Faults

In strike-slip faults the movements is essentially parallel to the strike of the fault plane. The resulting movement is predominantly in a horizontal direction and no scarp results. Only a thin fault line is traceable across the surface, and in some cases a narrow rift or trench may mark the fault. Strike-slip faults have been known by various names, such as tear or shear-faults, wrench faults and transcurrent faults. In 1967 a new variey of strike-slip faults was identified by Tuzo Wilson which he called *transform faults*. These were first recognized as connecting offset portions of the mid-oceanic ridges as the new oceanic crust moved away from them. The lithospheric plates slide past one another along major transform faults and they constitute one type of plate boundary. The best known of the strike-slip faults is the great San Andreas fault of California, nearly 1200 kms long, which can be followed from the Gulf of California to north of San Francisco with parallel or interlacing faults. It has been interpreted as a transform fault which forms the active boundary between the Pacific plate and the North American plate. Throughout its length the San Andreas Fault appears as a straight, narrow scar which at places becomes a trench-like feature and rivers make an abrupt fall, showing that movement must have occurred in fairly recent times. In fact, surveys have revealed that slow movement along the fault zone is continuously in progress. The growing strain of deformation in progress will ultimately reach the basaltic limit and the sudden jerk will produce another earthquake.

Significance of Faults

It should be noted that faults do not produce only earthquakes. Along the fault planes the rocks are usually pulverized and fractured. This permits ore-forming solutions to rise along the fault planes. Many important ore bodies are therefore found in fault planes. Petroleum too finds its way along fault planes or it may become trapped in porous beds that have been faulted against impervious shale beds. Areas of faulted sedimentary strata are favourite areas of oil exploration and have produced much oil in the past. Another related phenomenon is the rise of ground water along fault planes, so that springs, both hot and cold, are commonly situated along fault lines.

Fault scarps can form topographic barriers across which it is difficult to build roads or railways. They may also give rise to waterfalls.

Questions

1. Describe with diagrams the features resulting from compressional and tensional forces.
2. Discuss the different kinds of faults and their topographic expression. What is the economic significance of faults ?
3. Write short notes on :—

   Rift Valley and Block Mountain; Thrust faults; Transform faults; Types of folding.
4. Give a reasoned account of the formation of tectonic landforms.

Chapter 14

Mountain Building

Major Geological Periods of Mountain Building

In the long geological history of the earth, mountain building is found to be localized both in time and space. There are some specific periods during which mountain building has taken place, and these periods of orogeny are separated by long periods of quiet with little or no mountain building activity. There are four distinct periods when mountain building has been particularly active. The oldest among these is the pre-Cambrian orogeny followed by the Caledonian or mid-Palaeozoic orogeny, the Hercynian or Appalachian or late Palaeozoic orogeny, and finally the Tertiary or Alpine orogeny. "Localized thus in time, mountain building is not less localized in space, since it has affected only comparatively limited tracts separated by wide areas of little disturbed horizontal or gently folded strata."[1]

It is believed that the process of mountain building in the early periods of earth's history must have been different from the later periods. In the initial stages probably the buckling or creasing of the crust might have been more pronounced, but no definite evidence is available for this assumption. It can be stated, however, with some amount of certainty that in the later periods when sedimentary rocks began to be deposited, the process of mountain building has been to a great extent cataclysmic and limited in areal extent.

We may divide the mountains of the world on the basis of their age into three major groups—ancient, old and young.

1. **The Ancient or the Pre-Cambrian Mountain Systems**—These mountain systems are more than 500 million years old and belong principally to the Pre-Cambrian era. They are composed of complex crystalline and highly metamorphosed rocks. They have been

1. Wooldrige, S.W. and Morgan, R.S.—op. cit., p. 69

subjected to repeated sub-aerial denudation for such a long period of time that most of them have been degraded to areas of comparatively low relief. Since they have remained largely unaffected by the later orogenic movements, they are referred to as Stable Blocks or Shields. These remnants of the Pre-Cambrian mountain systems occupy comparatively large areas in comparison to the Caledonian, Hercynian or Alpine mountains which occupy relatively narrow belts. It would thus appear that in earlier times folding was probably a more widespread phenomenon and has become progressively more localized.[1] Remnants of pre-Cambrian orogenies are to be found in the Canadian Shield, the Baltic Shield, the Peninsular Plateau of India, Plateau of West Australia, most of Africa, the Brazilian Highlands and the continent of Antarctica.

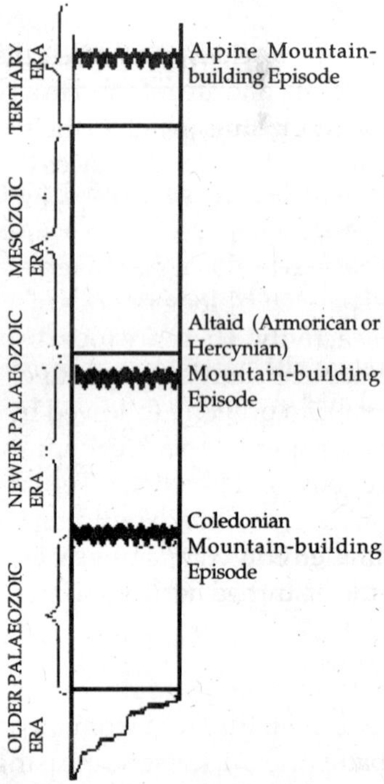

Fig. 64 — Three major periods of mountain building after the Pre-Cambrian orogeny

2. **Old Mountain Belts**— The old mountains vary in age from 100 to 500 million years. They were formed during the Palaeozoic era, principally in the Caledonian and Hercynian periods. They have been base levelled by erosion several times, and their present form is a result of uplift, deformation and volcanicity associated with the Tertiary orogeny in the neighbouring areas. Examples of such mountains are the Highlands of Northern Ireland and Scotland and the Scandinavian Highlands (both Caledonian), and further south the remnants of the Altaid mountains, Armorican and Central massif of France, the Vosges, Black forest and the Bohemian massif (Armorican or Hercynian). The Appalachian

1. Steers, J.A.—*The Unstable Earth*, Methuen, 1937, Indian Reprint, p. 30.

mountains of eastern North America are also Hercynian in age. The maximum stretch of old mountains is to be found in the region north of the Himalayas, in central Asia.

3. **Young Mountain Belts** — These mountain belts are less than 100 million years old. They include the mountains formed during the Tertiary orogeny during the last 75 million years. The principal mountain systems of the world — the Rockies, the Andes, the Alps and the Himalayas belong to this group and are referred to as young folded mountains. These are true mountains which in their height, extent, diversity and majesty know no parallel. At places they still show evidences of uplift, earthquakes and volcanicity, but at the same time they are subject to vigorous erosion by the sub-aerial forces. They are characterized by marked relief, steep slopes and longitudinal valleys.

The major fault block ranges, rift valleys and the volcanic mountain complexes and island arcs are also included in this very group.

Characteristics of Young Folded Mountains

As we have seen the principal mountain systems of the world belong to this category. They are called young, as they have been formed essentially during the last 75 million years. Most of the mountain ranges lie along and run parallel to continental coasts. A long belt of young folded mountains stretches along the western coasts of South and North America (the Andes and the Rockies) and continues along the northern and western coasts of the Pacific in the form of festoons of island arcs. The other belt of mountains consists of the Atlas-Alps and the Himalayan ranges which extend as far east as Burma, Malaya and Indonesia. In other words, the young folded mountains are found distributed in two directions. The first group of mountains stretches from north to south and the second group extends from west to east. These mountains show great diversity in their structure and lithology, but they have some significant common characteristics which are worth noting:-

(1) All the young folded mountains have been formed from sedimentary rocks whose original horizontal stratification has been sufficiently disturbed and deformed as a result of earth movements and there are evidences of folding, fracture and recumbent folding on a large scale.

(2) Marine fossils are frequently found in these rocks, indicating that the sedimentary rocks were deposited in the sea.

(3) The sedimentary rocks are several thousands of feet thick, but the marine organism whose remains are found in them belong to shallow seas. From this it is inferred that in the process of sedimentation the sea-floor seems to have gone on subsiding under the weight of sediments so that such considerable thicknesses could be attained.

(4) The liner extent of the folded mountain ranges is very great, but in comparison to their length their width is relatively small.

(5) Running along the linear extent of the mountains massive granitic intrusions are found.

(6) Their structural features show evidences of intense pressure, and recumbent folds, thrust faults and nappe are met with on a large scale.

The above noted characteristics suggest that the young folded mountains were possibly formed in shallow and narrow but elongated seas by the deposition of sedimentary strata and their subsequent folding and uplift. Such long subsiding seas from which mountains have originated have been called *geosynclines*.

Geosynclines

The term geosyncline is used for such long and relatively narrow ocean basins in which deposition of sediments has continued for a long time and which during the period of deposition has suffered subsidence. It is these sediments which have been uplifted and folded to form folded mountain ranges. It is therefore said that the mountains have been born out of geosynclines. This work has, of course, been accomplished over a very long period of time. The sedimentary basin of geosynclines thus remains in existence through a long geological period. There are landmasses on both sides of the geosyncline whose denudation provides the necessary sediments. The Tethys sea in Asia, the Appalachian geosyncline in N. America and the Caledonian geosyncline in Great Britain provide good examples of geosynclines in different geological times.

Phases of Mountain Building

The conversion of geosynclines into folded mountain systems

requires geologically a long time, and we may identify three definite stages or phases of the mountain building process.

(i) **The Period of Lithogenesis**— This is the initial phase of mountain formation. The rivers bring sediments from the landmass situated on either side of the geosyncline and deposit them in the geosynclinal sea. In other words, this is the stage when these sedimentary rocks are formed which later assume the form of mountains. The geosynclinal basin gets slowly filled up with sedimentary deposits and simultaneously undergoes subsidence under the weight of the accumulating sediments. This process is called *sedimentation-subsidence*. But there is a limit beyond which the floor of the geosyncline cannot subside by this process, and the entire subsidence cannot be explained in this manner. It is surmised that as a result of sedimentation-subsidence there is compression from both sides which further induces subsidence as the two sides of the geosyncline tend to come closer together. This process is called *compression-subsidence*. In this manner the geosyncline becomes deeper and narrower, but the rate of deposition is higher than the rate of sagging or subsidence. Thus it is that the geosyncline tends to be filled up by the deposition of sediments and the sediments become thousands of feet thick.

(ii) **The Period of Orogensis**—The process of actual formation of the mountains begins with further increase in pressure and crumpling of the sediments. Intense compression produces folds in the sedimentary strata, and at places as a result of fracture and the riding of rocks over one another thrust faults may be formed. The process of thrust-faulting and recumbent folding further increases the thickness of the sediments and possibly there is some amount of folding or faulting subsidence as well. In this way rocks of comparatively low density enter the sima to long depths, and the mountains after their uplift have long roots which are much deeper than their height above the surface.

After folding and faulting, the uplift of the region begins. In other words, vertical uplift starts after horizontal compression has completed its task. If we press downwards logs of wood floating in water, we can keep them below water level only by putting adequate weight on them or by pressing the blocks inwards from both the sides. As soon as the lateral pressure is released, the block of wood begins to rise. Somewhat in a similar manner when there is decrease in the lateral compression of the geosynclinal folded deposits, they tend to be uplifted. Further, the lower portion of the deposits expands due to higher temperature below, and in the opinion of some, the thermal expansion of the roots of mountains is an important cause of the vertical uplift of the mountains. Whatever the cause of the uplift, horizontal compression and vertical uplift are two distinct processes in mountain building and the vertical uplift of mountains begins only after horizontal compression has eased or ceased.

(iii) **The Period of Glyptogenesis**— The real effect of uplift is to bring the folded sediments into isostatic equilibrium. The first uplift is sharp and sudden and pushes the mountain upwards, but slow vertical uplift continues thereafter for a long time. In fact, slow uplift may continue concurrently with sub-aerial erosion, and the form of the mountains is the resultant of these two opposing forces.

The three phases of mountain building outlined above in a general way may be also called *the three phases of the evolution of geosynclines*. In the first phase there is sedimentation and sinking of the geosynclines. In the second phase there is folding followed by uplift of the sediments, and in the final phase after the mountains have been formed there is the end or disappearance of the geosynclines.

The Concept of Geosynclines and its Development

The concept of geosynclines is due to Hall and Dana. In 1859 J. Hall showed that the Appalachian mountains of eastern U.S.A. were formed by the uplift of sedimentary deposits in a subsiding trough. In 1873 J.D. Dana called such troughs geosynclinal troughs and subsequently they came to be known by the name of geosynclines. Considering the time it was propounded, it was a revolutionary concept, because it assumed inversion of relief on a large scale so that what was once a trough had been converted into high mountains.

The view of Hall and Dana were based on their studies of the Appalachian geosyncline. They had found there evidences of very thick deposits of shallow water sediments. In their opinion, therefore, the geosyncline was an area of shallow sea situated on continental margins. The floor of the shallow sea went on subsiding with the accumulation of the sediments and through this process it was possible for thick piles of sediments to be deposited in relatively shallow water (Figs. 65 and 66).

Based on his studies of the Alpine mountains, E. Hang put forward a different view in 1900, and presented a detailed analysis of geosynclines. According to him the geosyncline was several times longer than its width and it was an area of relatively deep water, because deep-water deposits are also found in it and there is no correspondence between the rates of subsidence and sedimentation. The phases of evolution of the geosynclines have not been simple

Fig. 65— The Appalachian geosyncline.

Fig. 66— Deposition, folding and faulting of the sediments in the Appalachian geosyncline.

and straight forward. In other words, sedimentation has not been a continuous and uninterrupted process but periods of sedimentation have been interspersed with periods of uplift, and the denudation of the uplifted mass has resulted in the deposition of fresh sediments and subsidence. Further, in his view it was more probable that the geosyncline lay in between two continental masses rather than on the coastal margin of one continent.

The study of the orogenic belts shows that the geosyncline was not an uniformly subsiding depression of similar depth, but it was quite deep in some sections while in other sections it was shallow. We do not find deep water deposits of sediments along the entire geosyncline, but along with deep water deposits contemporary shallow water deposits are also found in some parts. Normally, two or more parallel depressions are met with separated by higher parts which may be ridges formed of basement rocks or newly formed ranges of folded mountains. The denudation of these mountains has also supplied debris to the sinking troughs. These barriers of rock which ascend from within a geosyncline system are referred to as *geanticlines*. Another kind of a barrier may be built up by volcanic cones and deposits from volcanoes which may erupt either from the bottom or the sides of the geosyncline.

During 1935-50 H. Stille presented a significant classification of geosynclines. He divides the earth's crust into two major divisions called *cratons* and *ortho-geosynclines*. Cratons are sub-divided into *hochkraton* (i.e. stable continental crust) and *fiefkraton* (i.e. stable oceanic crust). Ortho- geosynclines are also subdivided into two parts—*eugeosynclines* or the main geosynclines and *miogeosynclines* or the smaller geosynclines. Both the types of geosynclines are found parallel to each other and *geanticline* or a land barrier separates the two. *Eugeosynclines* or the main geosynclines are located in the interior of the geosynclinal system further away from land. In the folded mountain belts, these are characterized by more intense earth movements, complex thrusting and folding and large scale metamorphism and granitization or lava flow. Such areas have been called *internides*. They receive their sediments from the rapid denudation of either volcanic islands or the rapidly rising borderland. In these sedimentary rocks there is predominance of mixture of sand and silt or of clay shales which have been deposited in deep water through turbidity currents. Thick deposits of such sedimentary rocks are found in the Alps mountains where they are collectively referred to as *flysh*.

Miogeosynclines are found in the external zone of the geosynclinal system i.e. near the cratons or the foreland. This zone is almost free from igneous activity. The sedimentary deposits include sandstone (found mostly as quartzites), limestone and thin

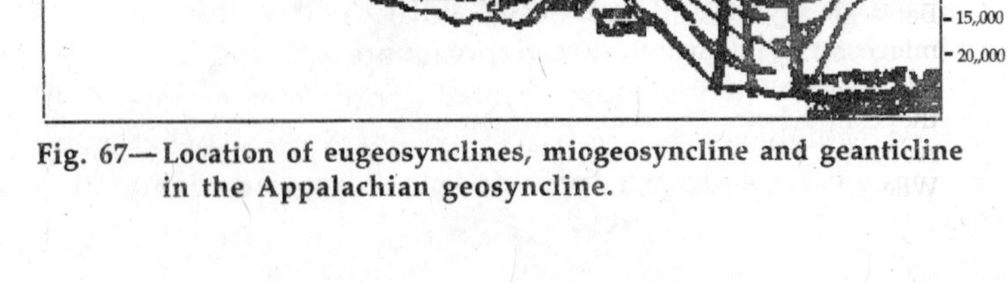

Fig. 67—Location of eugeosynclines, miogeosyncline and geanticline in the Appalachian geosyncline.

layered shales and they contain fossils of shallow water animals and also some coal.

These different elements in the Appalachian geosynclinal system have been shown in Fig. 67. Basically we may distinguish three main elements :—

(a) A relatively shallow coastal zone which contains mostly continental sediments which are only mildly deformed.

(b) A relatively high area in the middle consisting of volcanic rocks (opheolites).

(c) A thick deposit of sediments away from the coast which is highly deformed and contains abundant igneous intrusions.

These elements are found in the same sequence in the Alps as well, but the deposits in the eugeosynclines are quite old and their deformation and the formation of mountains took place much earlier, and their subsequent erosion has supplied the debris for the miogeosyncline. In the Himalayan system too the situation is more or less identical.

In 1965 J. Aubouin in a detailed study of the Mediterranean mountain ranges of the Alps presented an authoritative analysis of geosynclines. He has emphasized the existence of double geosynclines (mio-geosynclines and eugosynclines) and has tried to establish their relationship with the tectonic development of the mountains.[1]

Types of Geosynclines

With the development of plate tectonics there have been attempts to correlate geosynclines with plate tectonics. On this basis, three types of geosynclines have ben identified.

(a) *Pacific type geosynclines* which are found on converging plate boundaries. This type is subdivided into (i) *Andes type* when the

---

1. According to Aubouin, the eugeosynclinal domain (internides) comprises a eugeosynclinal furrow and a eugeosybclinal ridge bordering the ocean. Between the internides and the continenatl crayton (foreland) is the miogeosycline domain (externides) comprising a miogeosyncline furrow and a miogeanticlinal ridge. The two couples...can reach different stages at the same time.... These fundamental units may be paired, according to Aubouin, forming a divergnt bicouple. Wyllie, P.J.—*The Dynamic Earth*, Wiley, New York, 1971. p. 217

oceanic trench lies close to the continent, and (ii) *Island-arc* type where the trenches run parallel to the arc but are away from it, and there is a shallow sea behind the arc.

(b) *Atlantic type of geosynclines* where deposition of sediments takes place on the trailing edge of the continents.

(c) *Mediterranean type of geosynclines* where sedimentation occurs in the sea located between two continents and there is convergence of continental plate boundaries.

Hinterland and Foreland : the Views of Suess and Argand

We know that during the entire Mesozoic era there was an extensive geosyncline called the Tethys sea on the present site of the Alpine-Himalayan mountain system. The Tethys was nearly 3,000 kms wide from north to south and extended as a long linear stretch of water from Gibralter on the west to the East Indies in the east.

North of the Tethys lay a continental landmass called Laurasia which later came to form Eurasia, and south of the Tethys sea there was another continental landmass known as the Gondwanaland which later divided to form the present landmasses of the southern hemisphere as well as India. In this huge geosynclinal sea a thick pile of sediments was deposited starting from the Carboniferous period (Upper Palaeozoic era). A study of these sedimentary deposits shows that on the northern and southern margins of this sea there were shallow continental shelves, while the central portion was more deep. In mid-Mesozoic era (Cretaceous period) the above two stable continental blocks started coming closer to each other as a result of earth movements. According to Suess, both the edges of the geosyncline were not moving towards each other, but the southern continent (especially Africa) was pushing towards the north. The continental landmass which was moving towards the north has been called *Backland* by Suess, and the stationary European landmass against which the southern continent pressed has been called the *Foreland*. Thus the Alpine mountain system was formed as a result of the compression and folding of the geosynclinal sediments on account of the northward advance of backland of Africa. As a rule the alignment of folding is towards the north in Europe and towards the south in Asia. Possibly therefore, the compression was towards the north in Europe and towards the south in Asia.

Argand too has supported the concept of hinterland (backland) and foreland, but in his view, taking into account the continental drift theory of Wegener, both Africa and Europe were moving towards the north, but Africa was moving at a faster rate than Europe. This explains the northward bend of the folds and the nappes in the Alps. According to Argand, on account of the

Fig. 68— Location of Laurasia and Gondwanaland in relation to the present continents.

northward pressure exerted by the hinterland, two parallel *E* to *W* geanticlines were formed in the early stages which divided the Tethys sea into three parts. As the intensity of pressure increased, these geanticlines were raised further and appeared as landmasses or islands. In the subsequent stage or secondary times both the geanticlines were also pushed towards the north and the southern geanticline came closer to the northern geanticline. Towards the close of the Secondary and beginning of the Tertiary era the northern edge of the African landmass forged ahead towards the north and the principal upheaval of the earth movement came during mid-Tertiary times when the hinterland rode above the sedimentary pile and

(After Argand.)

Fig 69— Origin of the Tertiary mountain systems of Eurasia according to Argand.

moved rapidly towards the north, converting in the process the sedimentary deposits into recumbent folds. By the end of the main phase of orogenesis the northern part of hinterland not only covered the nappe but reached the foreland, and after that the middle section was uplifted. In the Eastern Alps we find sedimentaries of the hinterland and the southern continental shelves, which are known as Austrides. In the areas east of the Rhine river we find a covering of these rocks, but west of the river, barring a few exceptions, this cover has been removed by erosion and the structures of the Helvetic and Penine nappes are visible on the surface. This is why the western Alps have provided the key to the understanding of the formation of these mountains. In the eastern Alps, the lower structures lie buried underground, but the structure of the Western Alps can be extended and identified in the Eastern Alps.

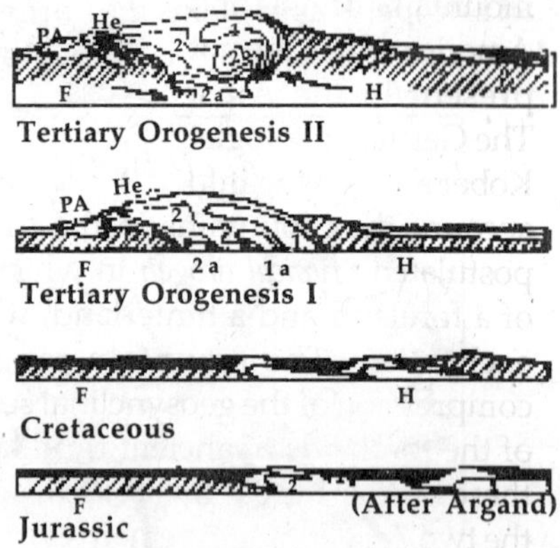

Fig. 70— The different stages in the formation of the Alps according to Argand. F, Foreland; H, Hinterland; PA, Pre-Alps; I, Dent Banche nappe; 1a, Monte Rose nappe; 2, Great St. Bernard nappe; 2a Simplon nappe.

Later on, as we shall see, people have conceived the idea of two forelands moving towards each other in place of hinterland and a foreland. This means that the conversion of the geosynclinal sediments into mountains has been brought about by the coming together of the two forelands, and in some places one foreland has pressed more and in other places the other foreland has pressed harder. The principal propounder of this view is Kober, and his theory is discussed below.

The Geosynclinal -Orogen Theory of Kober

The views of Argand as described above are based mostly on his study of the Western Alps and he has emphasized the movement of the African hinterland towards the European foreland resulting in intense compression of the areas situated in between and their

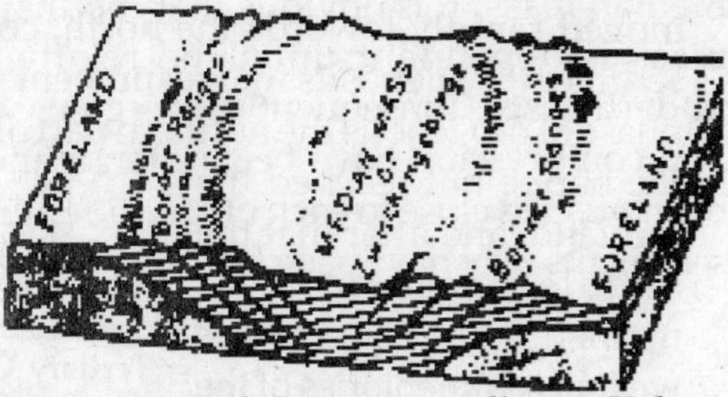

Fig. 71—Typical Orogen according to Kober.

north-ward translation. But there are certain characteristics of the young folded mountains of Asia and America which are not present in the Alps. The German geologist Kober after taking into account the mountains in the different parts of the world, has postulated a *typical orogen* in which there are two forelands, instead of a foreland and a hinterland, which are over thrusting against each other. The mountain ranges are formed as a result of compression of the geosynclinal sediments situated on the margins of the forelands of ancient rigid landmasses or *Kratogen*. He calls these border ranges *Randkatten*, and the zone situated in between the two border ranges remains virtually unaffected by folding. This unaffected middle part between the mountains has been called *Zwischengebirge* (betwixt-mountains) or the *Median Mass* by Kober. There are some evidences in the Alps to show that the thrusting has been from both the sides. In the Dinaric ranges the movement is towards the south and on the southern margins of Europe the movement is towards the north. But there is no clear-cut area in the Alps which may be identified as the median mass. Even so the hypothesis appears to be more comprehensive than that of Argand. In the opinion of Kober, the median mass is present in the Alps between the Alpid in the north and the Dinarid in the south. One median mass has been submerged in the western Mediterranean Sea and the island of Corsica and Sardinia are its remnants. The Hungarian plain located between the Carpathian and the Dinaric ranges is, according to him, another example of the median mass.

In the other mountain ranges of the world, however, evidence of median mass can be seen clearly. The Anatolian plateaus in Turkey, the relatively high land situated between the Elburz and the Iranian ranges in Iran present examples of median mass. In central Asia there are several high basins and plateaus bordered by mountain ranges (Fig. 72). The Tibet plateau lying between Himalayas and the Kunlun is a median mass (Fig. 73). In North America the Basin range region of Nevada situated between the Sierra Nevada and Wasatch range is another good example of median mass. It is obvious that the shape and size of the median mass is dependant upon the relative magnitude of the two underthrusts. Consequently, in regions where the earth movement and compression have been intense and the two forelands have been pushed close to each other, the entire bulk of the geosynclinal sediments may have been folded to form a complex *randketten*. Such

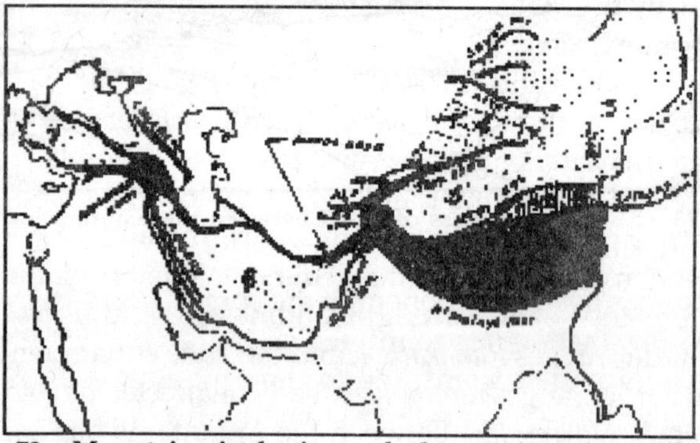

Fig. 72—Mountain-girt basins and plateaus in Asia 1. Tibetan Plateau, 2. Saidam Basin, 3. Tarim Basin, 4. Gungarian Basin, 5. Gobi Plateau, 6. Iranian Plateau, 7. Anatolian Plateau.

a mountain system has been called *Narbe,* and in such cases the median mass is completely absent. In the Swiss Alps we come across such mountains of complex structure. But in areas where the compressive forces have been normal, a median mass, big or small, can be expected between the randkettens.[1]

In his geosynclinal-orogen theory, Kober has tried to explain not only the origin of mountains but has also thrown light on the evolution of continents and ocean basins. While considering the relationship between the ancient rigid masses or kratogens and the relatively mobile geosynclines or orogens, he points out that these ancient rigid masses form the foundation of the present continents,

1. The scheme will be seen to be inherently different from the views of mountain structure suggested by Suess, who was an exponent of the "one-sided' formation of mountain ranges: he spoke of a foreland on to which the folds were forced, and a backland from which the folding force chiefly came. Kober thinks of the squeezing out of the geosynclinal sediments by the mutual approach of two masses, both of which are regarded as forelands. His randkettens are "one-sided", his orogens "two-sided". These mountain building movements are characterized by deep-seated tectonics and a high degree of metamorphism. Steers, J.A.-op. cit., p. 146-7

and the area of these rigid masses has increased by the accretion of the geosynclinal sediments and their formation into mountains, and in this way the present outline of the continents has evolved.

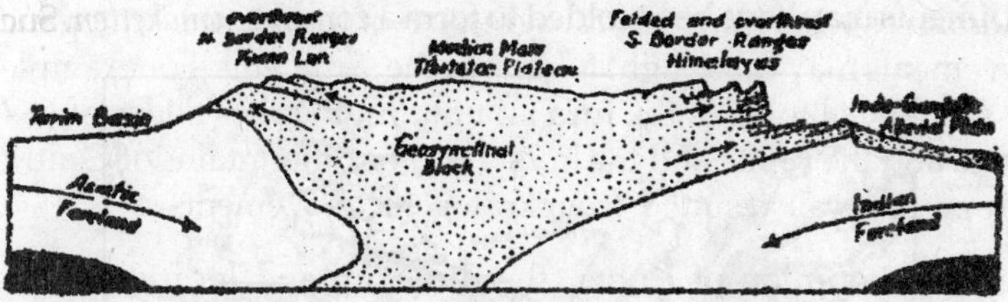

Fig. 73—The median mass of Tibet between the Kunlun and the Himalaya mts.

Kober has been an exponent of the contraction hypothesis. In his view, contraction has continued since the origin of the earth, which has provided from time to time the necessary force for mountain building. According to him, there have been six major periods of mountain building in the geological history of the earth. Out of these three have occurred in the Pre-Cambrian era about which our knowledge is extremely meagre, and three have occurred in the post-Cambrian periods and we know more about them. (see Fig. 64). In his view, during each period of mountain building we find more or less the same sequence of events which may be identified as the general process of mountain formation. First of all, there is formation of the geosyncline in which there is deposition of sediments and contemporaneous subsidence of the geosyncline resulting in the folding of the sediments and their upliftment into mountains. This latter period of orogenesis is also characterized by volcanism and intense metamorphism. After the formation of the mountains, there is a long period of sub-aerial erosion in which the mountains are denuded and ultimately reduced to areas of subdued relief.

The views of Kober are essentially based on the geosynclinal hypothesis of Hall and Dana as expounded later by Haug, and into this he has tried to fit his own views about mountain building. But while the geosyncline of Haug is a narrow depression, that of Kober is a long and wide sea.

In addition to the orogenic movements described above, Kober mentions another kind of movement which is found in the landmasses or kratogen. He calls such movements *Kratogenic movements.* While the orogenic movements cause folding and metamorphism of the geosynclinal sediments up to great depths, the kratogenic movements are responsible for the formation of rifts and fractures in the landmasses. But along the margins of the landmasses where there is transgression of the sea, the kratogenic movements may cause light folding of the sediments. For example, the superficial folds of the Jura mountains, the rift valleys of East Africa and the Rhine and the horsts or block mountains of Central Europe have all resulted from kratogenic movements.

In the opinion of Kober, the direction and location of the mountains have been controlled to a great extent by the location of the rigid landmasses. The basic foundation of Europe is the rigid landmass of Russia and Fenno-Scandia which is surrounded by the folded mountains of different ages— the Caledonian mountains of Scandinavia, the Urals of the Hercynian period and the Carpathian mountains of the Alpine period. In other words, the area of Europe has expanded by the addition of folded zones in different geological ages to the rigid Russian landmass, and the Angara land and Gondwanaland have been joined together by the development of mountain belts in the geosyncline located between the two resulting in the formation of the extensive landmass of Eurasia. Similarly in the other continents as well there are ancient rigid masses which are surrounded by relatively new folded mountains. In reality, these views were originally propounded by Suess according to whom the area of continents has increased by the addition of new orogenetic belts. This view has also been supported by H. Stille.

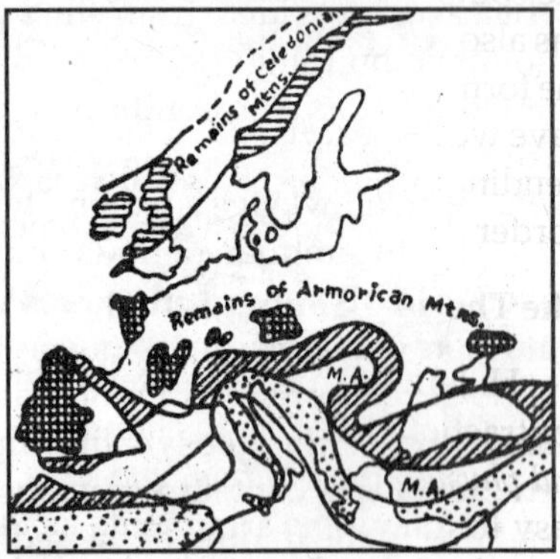

Fig. 74—Location of the old and new mountains in Europe. Main Alps-vertical shading, Dinaric Alps—dotted, M.A.—Median areas.

Kober believes in contraction, and it is contraction which provides the motive force for the compressive stresses. But he fully accepts the theory of isostasy, so that the high mountain ranges are able to stand up on the earth's crust as they are underlain by rocks of low density up to great depths. As a result of the horizontal movement generated by the contraction of the earth, there is accumulation and subsidence of huge quantities of sediments in the orogenic zone. When there is folding in the orogenetic zone, a larger quantity of sediments sinks to great depths than is lifted above. There is a constant conflict between the horizontal and vertical forces, and uplift and sinking are alternately possible in the same zone in the different geological periods.

The orogen theory of Kober has been criticized on the ground that the compressive force produced by the earth's contraction can never be so powerful as to result in the formation of huge mountain chains like the Alps and the Himalayas. Further, Kober's assumption that the two forelands move towards each other and the compressive stresses so produced are responsible for folding, has also remained controversial. While his theory is able to explain the formation of mountains like the Alps and the Himalayas which have west to east trend, it fails to explain the origin of north-south trending mountain chains such as the Rockies and the Andes which border the Pacific coast of the American continents.

The Thermal Contraction Theory of Jeffreys

H. Jeffreys has presented a detailed discussion of his thermal contraction theory in his famous book, *The Earth: Its origin, history and physical constitution, 2nd ed., Cambridge*, 1929. It is not always easy to follow his arguments as they are based on mathematical formulas and calculations. A simpler statement of his views is to be found in his other book entitled *Earthquakes and Mountains* (London 1935). Jeffreys is a contractionist and seeks an explanation for mountain building within the framework of contraction produced by loss of heat in the earth's crust. Contraction also results from decrease in the speed or rotation of the earth but it may be stated with certainty that the contraction so produced has been so negligible in the geological period that it cannot be said to have any significant influence on mountain building.

According to Jeffrrey's calculations, there has been no change in temperature in the interior of the earth, i.e. in the zone from the centre of the earth to about 700 kms below the earth's surface. But the uppermost 700 km-thick layer has experienced temperature decrease. In this upper part, every successive layer cools more rapidly than the layer below it, and, therefore, the upper layers have suffered more contraction than the lower layers. In fact, the hot and less contracting lower layer obstructs and retards the contraction of the upper layer, with the result that upper layer can contract only by spreading out and thinning itself. The solidified uppermost layer can cool only up to a certain point through this process and therefore its contraction also stops. But the cooling and contraction of the underlying layer continue. The result is that upper layer becomes larger in comparison to the lower contracting layer, and does not fit in with the underlying layer. In its effort to adjust itself to the contracting lower layer, the upper layer undergoes compression which produces folding and faulting. In other words, the upper layer undergoes crustal shortening. On the other hand, the contracting lower crustal layer becomes smaller than the hot interior and does not fit in with the latter and the former has to spread or stretch itself to conform to the internal layer. This generates tension, resulting in the formation of fissures and cracks. These fissures and cracks get filled up with the hot and molten material from below. In between the upper zone of compression and the lower zone of tension, there should be an intermediate zone where the contraction is such that it is able to adjust itself to the lower contracting crustal layer. This intermediate zone is a level of no strain. As the earth has gone on cooling, the level of no strain has been sliding downwards from the earth's surface. Above this level, on account of the horizontal compressive stress there is buckling and folding and the formation of mountains.

The thermal contraction theory provides only a partial explanation of the periodicity of mountain building. The compressive stress resulting from contraction goes on accumulating, and the earth movement begins only when this stress exceeds the strength of the rocks. Mountain building continues till such time as the stress and pressure are not relieved. After release or decrease in pressure the period of quiet begins and the compression stress again starts building up. Thus the period of mountain building is followed by a relatively long period of quiet. Assuming that the rigidity of the outer crust is the same as the rigidity of granite, Jeffreys calculated that there have been five occasions when the accumulated compressive stresses have exceeded the strength of rocks and the mountains have been formed. Broadly speaking, this calculation is in correspondence with the major mountain-building episodes, and as we know, other evidences also point to the existence of three Pre-Cambrian and three Post-Cambrian mountain-building periods. According to Jeffrey's theory the interval between the mountain building episodes should go on increasing with the passage of time as the earth goes on cooling, but in reality we do not find that this is so. Jeffreys has also pointed out that the possibility of mountain building is greater in the areas where rocks are soft and flexible, as the hard and less flexible rocks are liable to snap and break.

According to Jeffreys the depth of cooling was more under the oceans than under the continental masses. Further, the rocks underlying the ocean floor are heavy basic rocks which are stronger than the continental rocks. For both these reasons, the direction of the horizontal compressive forces generated in the oceans will be from the oceans towards the continental landmasses, and not from the continental masses towards the ocean as has been assumed in other hypotheses. Thus Jeffreys has propounded here an entirely different process. Through this process the mountain chains were formed along the coasts and parallel to them, and it is possible to explain the existence of the Circum-Pacific girdle of mountains. But the process fails to explain the formation of mountains like the Alps and the Himalayas which extend in a different direction.

For the building of mountains, contraction and compression are not enough, there has to be also uplift. Thus according to Jeffreys' theory, it is essential to assume thinning of the crust below the level of no strain in order to explain the formation of mountains. The upper layer, under the influence of gravity, will collapse towards this thinning lower layer. Further, when as a result of horizontal pressure of compression there is folding and downbending, the pressure on the lower layer increases further resulting in the melting of the lower layer which lifts up the mountains.

Jeffreys is opposed to the idea of continental drift and in his view the oceans and the continents have remained more or less fixed. In his opinion continental drift of the scale imagined by Wegener is impossible, though minor movements in the upper crust are possible which may account for the migration of the poles in relation to the earth's surface, but the migration of poles has not exceeded 5^0 during the entire geological period.

Several objections have been raised against the Jeffreys' theory. We have already referred to some of them in the above discussion.

The contraction of the earth is not only due to loss of heat but some contraction also results from the crystallization of rocks as well as the release of gases and water. According to Jeffreys' calculation there has been a decrease in temperature by 500° C on an average in the upper 400 km-thick layer. This will reduce the circumference of the earth by 130 kms. If crystallization- contraction and loss of gases are also included, the reduction in the earth's circumference will not exceed 200 kms. Jeffreys tried to show that only one-half of the above contraction was adequate for mountain building. In the opinion of other people, however, this was a serious underestimate, and crustal shortening several times more than this is really needed for mountain-building.[1] In reply Jeffreys has pointed out that there is difference between the folding of surface rocks and true shortening of the crust, because in folding there is simultaneously aggregation or compacting of the thick layers of sedimentary rocks. Nonetheless, it is true to say that there are serious difficulties in accepting contraction as the primary cause of mountain building.

There is greater possibility that minor folds and minute puckers will result from thermal contraction and not extensive mountain systems. When we look at the massive Tertiary Alpine orogeny, it appears impossible to believe that only about 200 million years ago there was so much contraction in the earth that mighty mountain systems like the Alps and the Himalayas could come into existence. Several other objections have been raised against the thermal contraction theory and it has now very few supporters. Instead the drift theories in various forms are now being propounded with greater vigour.

Radio-activity Theory of Joly

J.Joly put forward his theory of thermal cycles or the radio-activity theory in 1925 in his book '*The Surface History of the Earth*'. His main purpose is to present a historical account of the earth's surface, but he also tries to explain the formation of mountains. His

1. Holmes has concluded that the calculated reduction of area is seriously in deficit of the amount required to explain mounatin building. Wooldridge, S.W. and Morgan, R.S. : op. cit., p.121

theory is extremely simple, and is based on the latest scientific facts available at that time. Joly accepts the concept of isostasy as well as the fact that the continents are composed of relatively lighter sial whose average density is 2.67, while the ocean floor is composed of heavier materials of the sima with an average density of 3.0. The continents composed of sial (granite) rest on the heavier sima (basalt). If the density of sial and sima is respectively 2.67 and 3.0 then according to the concept of isostasy the amount of sial submerged in the sima should be about eight times the amount which projects above the sima. Taking into account the average height of the continents above the sima or sub-stratum, Joly has calculated the thickness of the sial to be about 30 kms. But wherever mountains have been formed on the surface, the depth of the sial projecting into the sima below increases proportionately.

An important basis of this theory is the radio-activity of rocks. To a certain extent all rocks are radio-active, but in comparision to sima the rocks of the sial are richer in radio-active minerals. Joly is of the view that the amount of heat lost by the earth's surface by radiation is more than counter-balanced by the heat received from the radio-activity of the sialic rocks. In such a situation there is no need for transfer of heat to the sial from the sub-stratum (sima). Consequently, the heat derived from radio-activity is preserved in the sima, and ultimately the accumulated heat becomes so great that the basalt rocks begin to melt. The position below the ocean floor is somewhat different. There is no sialic layer here and in the upper layer of the sima, the heat derived from radio-activity is lost in the oceanic waters through conductivity. But there is no loss of heat in the lower layers of the sima and the accumulated heat is sufficient to melt the basalts.

Joly has shown that assuming the thickness of the sial to be 30 kms, the temperature below the sial should be 1,050°C. The melting point of basalt is 1,150°C. In other words, in order that the sima or sub-stratum may melt, the temperature should rise by another 100°C. Joly has tried to show by mathematical calculation that it may take 33 to 56 million years for the accumulated heat derived from radio-activity to melt the basaltic rocks.

When the sub-stratum becomes wholly or partially liquid, then its density decreases and the buoyancy of the continents is reduced and they sink further down into the sima. As a result the oceanic waters cover the low continental margins and there is transgression of the sea. In such a situation there is deposition of sediments in the shallow coastal waters. In this manner the geosynclines come into existence, and in course of time it is these sediments which are folded and uplifted to form the mountains.

When the sub-stratum melts and is in a liquid condition and the continents are floating over it, the influence of the tidal force is increased. On account of the tidal effect, the sialic blocks start moving towards the west. Through this process the heat is released from under the continents, and the sub-stratum again begins to solidify as a result of loss of heat. The solidification of the sub-stratum increases its density and the continental blocks are pushed up, and there is a fall in the sea-level. In other words, the transgressional seas disappear, and on account of the regression of the sea the coastal sediments appear on the surface.

We have seen that the melting of the sub-stratum results in expansion of the sima, and the expansion is maximum in the oceanic parts. It is obvious that when the sub-stratum again starts solidifying, the maximum contraction will also take place below the ocean floor. The contracting ocean floor exerts pressure on the continents, and the margins of the continents situated between two contracting oceans are subjected to lateral compression with the result that the relatively soft sediments deposited along the coast are folded. This is the first stage of mountain formation. The cooling and re-solidification of the sima take time and the pressure of the ocean floor starts much before the sima has completely solidified. When the entire sima has solidified, then the sialic blocks (continents) located above them rise up. In this way, it is possible to explain the second stage of mountain formation, namely the uplift of folded coastal sediments. In other words, first of all the folds and nappes are formed as a result of lateral pressure and later on after the solidification of the sima, the folded zone is uplifted through the process of isostatic recovery. The theory thus emphasizes the two specific processes of horizontal compression and vertical uplift in the formation of mountains. According to this theory the formation of mountain takes place essentially along the continental margins. Further, the bigger the ocean, the greater will be the pressure exerted by the contraction of its floor, and more widespread will be the formation of mountains. Thus, it may be stated as a general rule that the biggest mountains will face the biggest ocean. This is, of course, true up to a certain point as we find an extensive and mighty chain of mountains surrounding the Pacific Ocean. But this theory fails to explain the absence of mountains parallel to the Atlantic coast or to offer a satisfactory explanation for the location of the great Alpine-Himalayan mountain system.

In his theory Joly has also tried to explain the periods of orogeny and the periods of quiet intervals between them. The entire period of two solid states of the sima i.e. its melting from a solid state and then its re-solidification has been called a revolution. The melting of the sima takes place in 33 to 60 million years. From this it may be concluded that mountain building is cyclic in character, and after a fixed interval of time there is recurrence of mountain building activity. It is true that there have been several distinct periods of orogeny in the geological history of the earth, but the interval between two orogenies has not always been the same. Thus the assertion of this theory that mountain building follows a cyclic order and is repeated at regular intervals of time, is contrary to facts, and this, in fact, is one of its main drawbacks.[1]

Jeffreys has been the greatest critic of the theory of Joly and is in complete disagreement with the latter. We have already discussed the views of Jeffreys earlier in this chapter and it is not necessary to repeat them here. According to Jeffreys the depth of 30 km for the sial proposed by Joly is excessive, and should be around 16 kms on the basis of seismological evidences. If the latter figure is taken as the depth of the sial, then the heat below the sial would be much less than that assumed by Joly. In the absence of adequate heat it will be impossible to proceed further with the theory.

According to Joly during the liquifaction of the sima, the continental masses are pushed towards the west as a result of the tidal force. Jeffreys has tried to show mathematically that there is no known adequate force which may cause westward displacement of the continents. It is not possible to accept this statement of Jeffreys in its entirety, and keeping in mind the views of the supporters of

1. "In short the very essence of the theory, the approximately equally spaced recurrence of similar conditions, seems to be one of its main drawbacks." Steers, J.A. op. cit., p. 185.

continental drift, one cannot state dogmatically that such a force is altogether absent.

With the help of mathematical laws Jeffreys has shown that once the sima has melted under the influence of radio-activity, it cannot re-solidify. If the distribution of the radio-active substances is such that the temperature of the basaltic layer of the sima is above the melting point, then this layer will permanently remain in a liquid state and there is no question of its melting and re-solidification at regular intervals.

Apart from the criticism stated above, other objections have also been raised against this theory. The theory is based on radio-activity about which, especially in the interior of the earth, our knowledge is inadequate. Furhter, the theory presents rather an erroneous account of geosyncline. According to the theory geosynclines develop along the continental margins by marine transgression. The geosynclines are supposed to be long and narrow areas of relatively shallow water in which there is accumulation and subsidence of sediments. The deposition of sediments along the coastal margins of continents is unlikely to yield sufficient sediments for the formation of high mountain ranges. Further, the mountains should be formed all round the continents. For all these reasons, it is difficult to accept the process of mountain formation put forward by Joly.

Daly's Hypothesis of Sliding Continents

Daly propounded his hypothesis of sliding continents in the book '*Our Mobile Earth*' in 1926. His hypothesis is based on the view that there has been down-hill sliding movement of the continental masses, the principal cause of which has been gravity. For continental displacement, he does not invoke any tidal or other force besides gravity. His hypothesis is therefore simple and clear. He is not dogmatic in his views, and accepts that it is not possible to pronounce any definite views on this difficult problem.

Daly assumes that the distribution of land and water was predetermined in the primeval days. The landmasses lay near the equator and the poles. In between these three rigid landmasses lay lowlying areas or the seas which he calls the *mid-latitude* furrows. In the northern hemisphere the Tethys sea was the mid-latitude furrow but in the southern hemisphere there is no definite speculation about a similar depression. In addition, there was the extensive primeval Pacific Ocean. These rigid landmasses as well as the ocean floor were composed of the primitive crust. Thus, in the primeval days the surface of the earth had been divided into land and water. The landmasses were higher than the seas and therefore the equatorial and the polar landmasses sloped towards the Pacific Ocean as well as the two mid-latitude furrows. In the Pacific Ocean as well as the mid-latitude furrows sediments began to be deposited by denudation of the continental domes, and they may therefore be treated as the first geosynclines. The pressure on the ocean floor started rising for two reasons—the pressure of oceanic water and the weight of the geosynclinal sediments. In this manner the sea floor went on subsiding.

In course of time, the downward pressure in the constantly subsiding sea floor will naturally cause some lateral pressure on the higher continental masses. The high continental masses got support from this lateral pressure, and their extent and height increased. The increase in the size of continental domes will lead to increase of pressure on the geosynclinal sediments around the domes, and a time will come when the crust under the geosyncline can no longer stand the strain and will rupture. When this happens a great part of the support of the dome disappears and strong tensional movements will be set up in it, so that large continental blocks may be caused slowly to slide towards the geosynclines. The sliding results in compression, squeezing and folding of the geosynclinal sediments, and in this way the first stage in the evolution of a mountain range is achieved (Fig. 75).

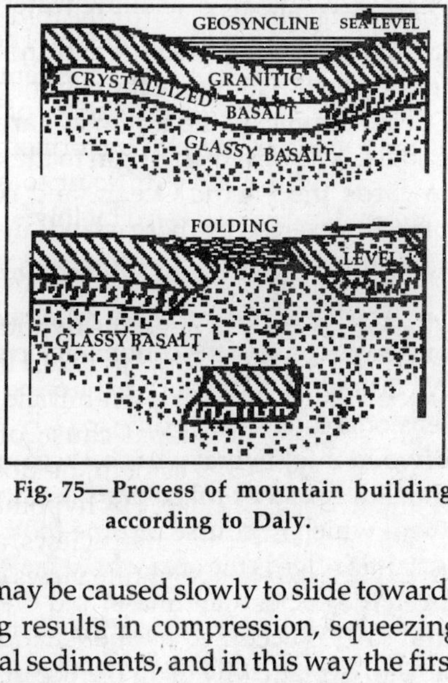

Fig. 75—Process of mountain building according to Daly.

Daly considers the sub-stratum to consist of hot glassy basalt which is slippery on which the crust can easily move. But since in his view the density of the sub-stratum is lower than the crust, the broken part of the crust sinks into the sub-stratum. Further, on account of crustal rupture, the slippery and less dense substance of the sub-stratum rises up and comes directly below the geosynclinal sediments. Even after rupture of the basement of the geosynclines, the geosynclinal sediments do not sink into the sub-stratum, as they are less dense than the latter. But on account of the increasing lateral compression from the sliding continental blocks, the geosynclinal sediments are not only folded but are also pushed downwards towards the sub-stratum. Thus the lower part of the sediments is heated, and melts and expands. The folded sediments are thus raised up, and in this manner the second phase of mountain building, namely vertical uplift is accomplished.

Thus according to Daly, mountains have been formed by the folding of geosynclinal sediments as a result of lateral pressure exerted by the sliding of the continental crust towards the geosynclinal seas. On this basis the west-east trending Alps-Himalayan mountain system has resulted from the sliding of the continental blocks towards the mid-latitude furrow, and the north to south trending Rockies and the Andes mountains have been formed by the sliding of the continental masses towards the Pacific Ocean. Similarly, the mountain arcs and islands off the coast of eastern Asia owe their origin to the sliding of the Asiatic landmass towards the Pacific Ocean, and on account of the downward pressure exerted by these islands deep foredeeps or trenches have come into existence.

The Pacific Ocean was a primeval ocean and there was pressure or compression of the continental masses towards it. But the Indian, Arctic and the Atlantic Oceans came to be formed later as a result of tensional forces. According to Daly, the original continental domes divided into several parts on account of tension and as a result of the sliding of the broken parts, tension rifts were formed in between them which in course of time took the form of oceans. The Mid-Atlantic ridge is the area where the American continent rifted apart from the ancient landmass, and the depression caused by the rift and drift is occupied by the Atlantic Ocean. Daly's hypothesis is able to explain why of all the oceans, the Pacific Ocean is the only one to be surrounded by mountains.

Several objections have been raised against Daly's hypothesis. His hypothesis, it is said, is based on a number of self-proved assumptions which are no better than guesses. Daly assumes an uneven earth surface with separate distribution of land and water from the beginning. In other words, there was the original landmass of Pangaea and the original ocean of Panthalassa. The dome of Pangaea divided into three parts on account of sliding—North Polar landmass, South Polar landmass and the Equatorial landmass and two depressions were formed in between them. He does not go into the question of how all this became possible. It may be assumed that the original crust had two layers—an upper layer of granite and a lower layer of denser rocks. The granitic blocks will tend to aggregate and float on the denser layer below, and in this way one may account for the formation of land and oceans. But one has to remember that even the rigid landmasses are composed of very ancient folded rocks. In other words, the process of sliding must have worked in their case as well. It has therefore been said that his hypothesis contains a lot of guesswork and assumes a distribution of land and water which may help him to prove his hypothesis. But at the same time it is suggestive, and an attempt is made to understand the distribution of land and water. The existence of Pangaea and Panthalassa is not impossible. If the original continent or continents were dome-shaped then sliding under gravitation is much more probable than the force invoked by Wegener for continental drift. But the crustal structure as suggested by Daly is different from the structure given by Jeffreys as deduced from seismic evidences, and more especially it is difficult to accept Daly's assertion that the substratum is less dense than the upper crust.

Daly also presents a rather confusing concept of geosynclines. Ordinarily, geosynclines are long, narrow and relatively shallow seas but he accepts both the mid-latitude furrows and the Pacific Ocean as geosynclines. Furhter, his hypothesis anticipates the formation of mountains from every ocean, irrespective of the extent and depth of the oceans and the amount of sediments that may be deposited in them.

Daly's hypothesis tries to explain the problem of mountain building in a simple manner with the help of gravitation but fails to present a well argued and connected account of the problem. "It is fair comment to say that the cause of the primary 'buldges' which start the slipping has in no sense been satisfactorily indicated"[1].

1. Wooldrige, S.W. and Morgan, R.S.—op. cit., p. 124

The Convection Current Theory of Arthur Holmes

Arthur Holmes put forward his convection current theory in 1928-29. The main objective of this theory is to present an explanation for the processes of mountain formation, but it also throws light on volcanicity associated with mountain building as well as on continental drift. The importance of this theory lies in the fact that it tries to integrate conflicting views regarding mountain building and continental drift.

According to Holmes, it is extremely important to bear in mind the distinction between the solid crust and the liquid sub-stratum. The earth's crust includes the upper layer (sial), the intermediate layer (i.e. upper part of sima) and the lower layer consisting of crystalline rocks. Below this is the sub-stratum which is liquid but is a continuation of the lower crustal layer. Holmes' theory is based on the possibility of convection currents in the sub-stratum. The cause of the origin of convection currents in the sub-stratum is the presence of ratio-active substances there. Disintegration of these substances produces enough heat to keep the sub-stratum in a liquid condition and to generate convection currents. Radio-active substances are more abundant in the upper layers of the crust, but heat from these layers is lost by radiation and conduction, and the temperature therefore does not rise high. The annual loss of heat from the earth's crust by radiation and conduction is about 60 calories per sq. centimetre. This is roughly equal to heat derived from the disintegration of radio-active substances contained in 14 km thick layer of granite, 16.5 km thick layer of granodiorite, 52 km thick layer of plateau basalt or gabbro, and 60 km thick layer of periodotite. Since radio-activity is particularly concentrated in the layers adjacent to the earth's surface it can be stated that the average heat loss from the earth's surface is compensated by 60 km thick layer below it. There is thus no accumulation of heat in the upper and middle layers of the crust. Although radio-active substances are less in the sub-stratum, the heat released by them plus the original heat of the sub-stratum are so great that they are able to generate convection currents, as there is no loss of heat from the sub-stratum by radiation or conduction.

Under normal conditions the average temperature gradient inside the crust is 3^0 C per km, but this increases as a result of the heat derived from radio-active substances. Consequently, the rate of convective circulation rises and convection currents are set up from the sub-stratum upwards. This circulation is primarily influenced by two factors:—

(a) Near the equator the thickness of the crust is more than in the polar regions. As a result the temperature gradient in the equatorial regions is more than the polar regions on account of the larger amount of heat generated by the radio-active substances in the former. Therefore we find ascending convection currents near the equator and descending convection currents near the poles. Near the equator these currents, on rising up, flow horizontally through the crust and start descending down near the poles. As a result of these currents, there is greater movement in the lower layers of the crust than in its upper layers resulting in tension in the upper layers. Through a somewhat similar process the original equatorial landmass was broken up by the convection currents, and the broken parts drifted to the north and the south, and in between the two northern landmasses the Tethys sea came into existence. Holmes has conceived of planetary convection currents very similar to the planetary winds.[1]

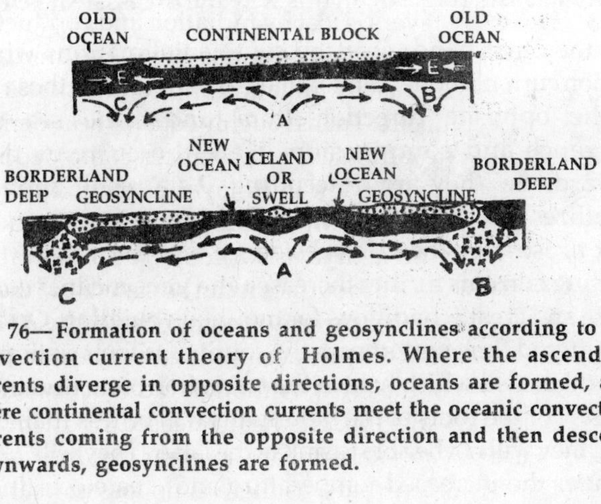

Fig. 76—Formation of oceans and geosynclines according to the convection current theory of Holmes. Where the ascending currents diverge in opposite directions, oceans are formed, and where continental convection currents meet the oceanic convection currents coming from the opposite direction and then descend downwards, geosynclines are formed.

1. Thus the Tethys is regarded as being due to a force tending to pull the original equatorial continents apart, and that force may be the "planetary" convective currents. The subsequent closing of the Tethys is due to later sub-continental convective currents. Strees, J.A.-op. cit., p. 194

(b) The continental crust is richer in radio-active substances than the oceanic crust. Therefore at the same depth the temperature is higher under the continents and the convection currents are more active and powerful. The convection currents originating in the sub-stratum below the continents carry the molten materials upwards as they ascend and then turn towards the oceans. Below the oceans also these convection currents rise up but they are weaker. These weaker currents meet the convection currents from under the continents on the continental margins in the area of the continental shelves and then go downwards. It may be stated that the convection currents do not originate at one place but at several places and their point of origin keeps on changing. The convective process is not a continuous or stable one but is periodic which may wax and wane at one point and re-start from another place.

Where the ascending currents turn outwards in opposite directions under the continental masses, there is tension on account of divergence and stretching of the continental crust. This may lead to rupture and drifting apart of the two portions of the original continental mass. This outward movement, together with the thinning of the crustal material will cause a subsidence and the formation of a new ocean basin (Fig. 76). In this way the excess heat is released.

On the other hand, near the continental margins where the convection currents under the continental masses meet those coming from the opposite direction from under the sea, there is convergence and compression. Here the currents descend downwards i.e. they are descending. As a result amphibolite recrystallizes into heavier eclogite. For this reason as well as on account of isostasy there is subsidence and the velocity of the descending currents further increases and geosynclines come into existence and create conditions for mountain building. On account of the outward flow underneath the continents their sialic margins also get thickened which help in the formation of mountains. Since the density of the roots of the mountains will be less than the sub-stratum, they will not be able to sink in the latter. They will, however, melt under the increased temperature and create conditions for igneous activity.

In the opinion of Holmes, a major difficulty in trying to explain

1. Holmes, A. : Trans. Geol. Soc. Glasgow, 18 pt. 3. 1928-29, p. 581 (Quoted by J.A. Steers).

mountain building by continental drift is that if the sub-stratum is weaker than the sialic blocks floating on it, there will be no folding, and if the sub-stratum is assumed to be stronger, then drift is unlikely. Thus in his opinion "mountain building occurs as a result of rock flowage set in operation by underlying current; it will generally occur provided that the horizontal component of flowage is greater from behind than in front"[1]. In the region of descending currents there is formation of geosynclines on account of subsidence. The sediments derived from the erosion of the continents will be deposited in them leading to further subsidence. Since in the geosynclinal area the continental and the oceanic currents meet, there is lateral pressure from both sides. In other words, folding of the sediments proceeds together with their deposition and the subsidence of the geosyncline.

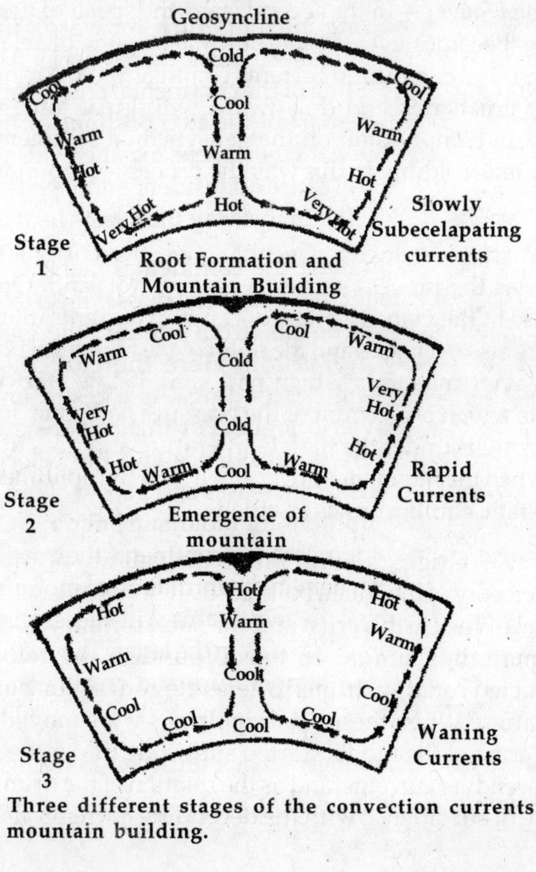

Fig.77— Three different stages of the convection currents and mountain building.

The convectional activity is cyclic, and three stages may be identified during which there is formation of the geosyncline and the folding and uplift of the mountains (Fig. 77):—

(a) *First Stage* — The first stage is of a very long duration. The convection currents are quite strong, and the currents coming from two sources converge and descend near the continental shelves, resulting in the formation of the geosyncline. Sediments are deposited in the geosyncline and there is subsidence of the geosyncline. The sediments, when they reach great depths, become hot and metamorphosed and their density increases. In other words, in the zone of the descending currents the roots of the mountain are formed by geosynclinal subsidence. This is the preparatory stage of mountain building.

(ii) *Second Stage* :— In the second stage the speed of the convection currents becomes even faster, but this stage is of relatively short duration. The convection currents coming from the continental and oceanic crusts descend downward with great force resulting in maximum compression on the geosynclinal sediments and their consequent folding. In this way the process of mountain building starts.

(iii) *Third stage* : — In this stage the convection currents start getting weaker as the process seems to approach its end. On account of decrease in the velocity of the decending currents, the downward pressure also decreases and the uplift of the folded sediments begins. The heavier materials which had sunk below start moving up. Eclogite which on account of the pressure had gone down to great depths, melts under the high temperature and rises up. This is the stage when the mountains are uplifted, and the upliftment continues till isostatic equilibrium is attained.

In 1939 Griggs tried to prove the validity of the convection current theory of mountain building by his experiments. He prepared a model of the earth's crust and the sub-stratum which was made from materials similar to those found in the two layers. He constructed rotating drums to generate convection currents in the substratum. When these rotating drums were moved slowly, the upper crust of the model started subsiding under the influence of the descending currents, and as the rotation of the drums increased, the rate of subsidence with the decending currents also increased.

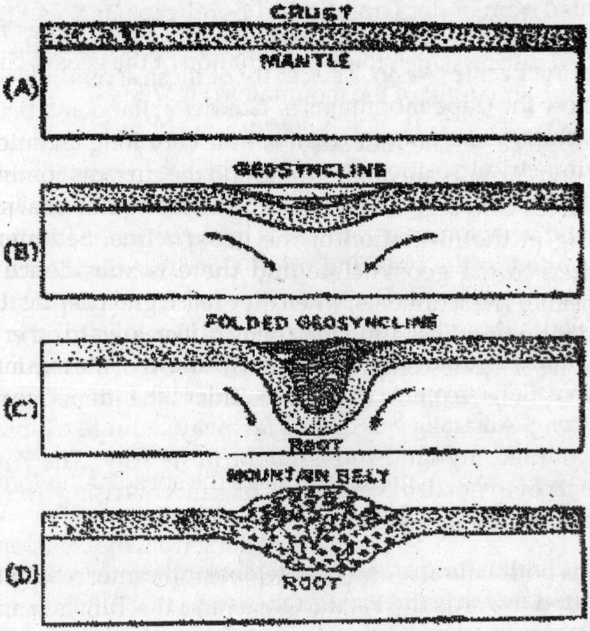

Fig. 78— Tectogene or Root theory of mountains. A-Lithosphere over mantle, B—Convection leads to geosyncline formation, C—Faster convection leads to major downwarp, folding and development of the mountain root, D—Melting of the lower part of the tectogene forms granites. Convection ends and there is buoyant rise.

When the rotation of the drums was decreased and the decending currents, slowed down, the subsided part began to rise up. In this way Griggs confirmed the cyclic arrangement of the convection current. Based on this he presented his *Tactogene or Root Theory of mountains* which has been depicted in Fig. 78 above.

Through his theory Holmes has also tried to explain the distribution of continents and oceans. In his opinion the distribution of carboniferous glaciation can only be understood with reference to continental drift. Towards the close of the Palaeozoic era there were two continental landmasses—the Lauratia and Gondwanaland separated by the Tethys sea, and in addition there was the extensive watermass of the Pacific Ocean. South pole lay near the present position of Natal and the equator near Great Britain. According to Holmes, independent convection currents originated from under Lauratia and Gondwanaland, as a result of which the continental masses started drifting towards the Tethys sea and the Pacific Ocean. The centre of these ascending currents was below the Cape mountains in Gondwanaland and below the Appalachian-Caledonian mountains in Lauratia. On account of the asymmetrical position of the ascending currents with reference to the continental masses, there will be movement towards the north in the land hemisphere and towards the south in the Pacific ocean. This resulted in the breaking up of Gondwanaland, and Africa started drifting towards Europe, and India started moving towards north. South America and Australia drifted towards the Pacific Ocean, and the Indian and Atlantic Oceans came into existence. As the convection currents rising from under the Indian Ocean were quite strong, Australia was drifted far away, as, unlike India, it did not encounter any land impediment in its way. The Antarctic continent could not drift very far, as the currents rising from under the Pacific Ocean impeded its movement.

Thus both Lauratia and Gondwanaland came to be extended and drifted towards the Pacific Ocean and the Tethys sea, and in this process extensive mountain chains came into existence along the continental margins. On the margins of Lauratia, the North American Cordillera, the mountain island groups of East Asia and the northern section of the Alps-Himalayan system came into existence. On the margins of Gondwanaland, the Andes mountains, the mountains of Newzealand and Guinea and the southern section of the Alps-Himalayan system were formed. Holmes also tried to explain the function of geosynclines, median mass, rift valleys and volcanicity with the help of convection currents though he admits that some of his suggestions are highly imaginative.

Although the theory of Holmes is quite interesting and some of his assumptions are valid, our knowledge about the existence of convection currents and their behaviour is extremely limited, and we do not know whether they are powerful enough to cause breaking up and drifting of the continents. Though this theory opened a new direction for understanding the problem of mountain formation, the horizontal movements of convection currents below the crust, their rise and descent, their process being periodic instead of continuous, the re-starting of the process from new locations are all assumptions based on speculation, and for which we have no proof.

At the same time, it is noteworthy that the theory has recently received support and confirmation from studies relating to plate tectonics where convection currents provide the driving mechanism for the movement of lithospheric plates.

Plate Tectonics and Mountain Building

The theory of Plate Tectonics has already been discussed in Chapter 9. We have seen that at one margin of a crustal plate there is found the mid-oceanic ridge where new crust is being formed by the rise of basalt through rifts. Here there is divergence of plates, and the newly formed oceanic crust diverges and moves towards its other margin to be subducted in the oceanic trenches and the mantle. Here near the oceanic trenches there is convergence of two plates, and the oceanic plate under-rides the continental plate and re-enters the mantle. On the remaining two margins of the plates, the plates merely pass across each other and there is only frictional movement. Thus we have three types of plate boundaries :—(a) *Divergent boundaries* or junctures at the mid-oceanic ridges of rifts, (b) *Shear boundaries* or junctures where two plates pass across each other, and (c) *Convergent boundaries* or junctures where two plates collide against each other, and one of the two plates is lost by subduction into the trenches. The divergent and the convergent boundaries are particularly important, because oceanic ridges and rift valleys are formed on the divergent boundaries, and the folded mountain ranges are built on the convergent boundaries.

The young folded mountains of the world—the Alps-Himalayan mountain system as well as the circum-Pacific belt of mountains—are located on convergent plate boundaries where there is a state of collision between two plates. It may therefore be stated as a general rule that where there is convergence or collision of two plates, mountains are formed as a result of compression in the earth's crust.

The convergence of plates is possible under three different conditions :— (i) collision between a continental and an oceanic plate or continent-ocean collision, (ii) collision between two continental plates or continent-continent collision, and (iii) collision between two oceanic plates or ocean - ocean collision.

(i) *Continent-Ocean Collision*— This is the most common type of

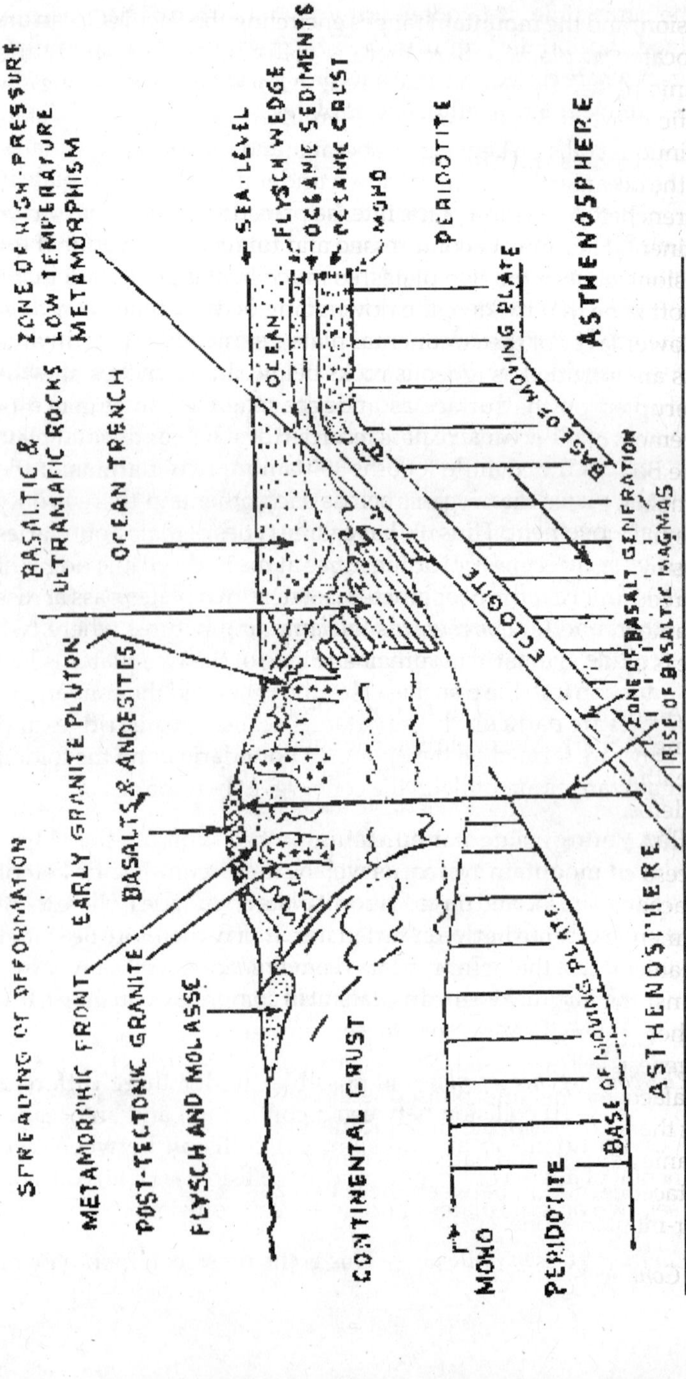

Fig. 79 - Subduction of the oceanic plate below the continental plate, resulting in the formation of mountain and related structures.

collision, and the mountain ranges encircling the Pacific Ocean are all located at places where there is collision of continental and oceanic plates. Its simplest and best example is found along the Pacific coast of South America. Here there exists a steady-state continuous collision between the oceanic and the continental plates, and the oceanic plate is thrust down under the continental plate in the trenches. As a result of the intense pressure the deposits on the continental margin are compressed and folded. On account of the collision between the two plates, so much heat is generated in the Benioff zone that rocks of the downgoing oceanic plate as well as the lower layer of the continental plate get melted. These melted rocks are intruded as igneous rocks below the surface, and some are erupted on the surface as andesite magma. The downward movement of the oceanic plate is recorded by a series of earthquakes in the Benioff zone whose focus is 300 to 400 km or more deep. As the mobile core of the orogenic belt develops, the deformation of the plate margins increases with increasing temperature and intense pressure. There is gravity sliding on account of uplift and thrusting on account of compression. The mobile core pushes the metamorphosed rocks towards the continent and the continental edge is uplifted to form mountains. (Fig. 79)

According to the plate tectonics theory the Andes were formed during the early Mesozoic era. The subduction of the oceanic plate beneath the S. American plate started about this time resulting in the deformation of the Palaeozoic marine sediments. After this the South American plate started moving west which intensified the process of mountain formation in the mid-mesozoic and early cretaceous times, and in addition to the folding of the marine sedimentary deposits there started intrusion of igenous rocks and volcanicity. As the oceanic plate descended further the pressure against the South American plate also increased and there was further intensification of these activities and the area of orogenesis also increased. The eastern part of the Andes is composed of Palaeozoic metamorphosed sediments, in the western Cenozoic part there are Mesozoic and volcanic materials and shallow water oceanic deposits and batholiths ranging in age from Jurassic to the Cretaceous, and in between these two zones is a relatively lower inter-montane zone.

(ii) *Continent-continent Collision*— The Alpine-Himalayan mountain system provides the best example of mountains formed as a result of this type of collision. In Mesozoic times India together with the southern continents formed part of Gondwanaland, and there existed the Tethys sea between the main Asiatic landmass (Lauratia) and the Gondwanaland. After the Mesozoic, Gondwaland started breaking up, and India started moving to the north at the rate of 16 cm per year and joined the Asiatic landmass about 30 to 60 million years ago. As a result the Tethys sea became narrower and ultimately closed. At about the same time Africa also started moving north and the portion of the Tethys sea between Europe and Africa became narrower. As a result of the collision of the Indian and the Asiatic landmasses the marine sediments and the crust located in between the two were folded and thrusted, and the Himalayan mountain chain came into existence from 2 to 30 million years ago. On account of the relative buoynancy of the continents, the upper rocks suffered more folding and thrusting. On the uplifted folded sediments are found deposits of *flysh* derived from the erosion of the former and of *molasse* in the coastal regions. Similarly, as a result of the collision of the African and European plates, the sediments along the continental margins have been folded and thrusted to form the Alps in Southern Europe and the Atlas mountain in North west Africa.

(iii) *Ocean-Ocean Collision*—Where oceanic plates exist on both sides of a convergent plate boundary, the oceanic crust of one plate is subducted under the other plate in the trenches, and the resultant compression leads to the formation of *island festoons and island -arcs.* This type of mountain is especially found off the western coast of the Pacific Ocean and the north-eastern coast of the Indian ocean. Suess was the first to point out that these arc-like island groups are tops of drowned young folded mountain ranges and are extension of the mountain systems found on the continents. Between the continents and the island arcs are shallow seas which are called *back-arc basins*. The sea of Japan is a good example of a back-arc basin. Towards the oceanic margin of each island arc is a deep oceanic trench. It appears as if the trenches owe their origin to the descent of the plate. Here the oceanic plate together with the deposited sediments descends under the adjacent oceanic plate and on account of compression on the continental margin of the trenches, there is formation of metamorphic rocks. When the descending plate reaches a depth of more than 100 kms it starts melting on account of intense heat, and the molten magma rises up as volcanic rocks. The volcanic action takes place at a horizontal distance of 200 kms from the trench towards the continent. The resultant increase in heat and pressure causes further metamorphism of the rocks. With the downward movement of the plate, there is further increase in volcanicity, the type of lava changes and the distance of eruption from the trench also increases. From lesser depths basalt is ejected, and from greater depths the ejected lava is andesite which becomes increasingly alkaline.

The Japanese island-arc provides a good example. Here in the main island of Honshu there are 3,000-4,000 metre high mountains and they are all volcanic. They do not come in the category of fold mountains like the Himalayas or the Alps. But they contain several characteristics of fold mountains. The island of Honshu is situated between the sea of Japan in the west and the Japanese trench in the east, and is mostly composed of basalt and andesite rocks. The subducted oceanic crust melts at a depth of about 120 kms. This is the reason why there is more volcanic activity in the eastern parts of Honshu than in its western part. The existence of belts of metamorphic rocks belonging to two different ages in the eastern and western parts of Japan suggests that Japan has been formed by the union of two island-arcs of different ages. In fact, at present, there are two island arcs—Philippines arc and Mariana arc, immediately south of Japan.

It may be pointed out that sometimes mountains may be formed where there is collision of continent and island-arc. This kind of situation exists in New Guinea where the mountains of New Guinea have come into existence about 20 million years ago as a result of the convergence of the island arc lying to the north and the northern edge of Australia.

It is clear from the above account that in ocean-ocean collision (island-arc situation) and in continent-ocean collision (Andes Cordillera situation), there is found an abundance of igenous rocks (ophiolites) which have been derived mainly from the melting of the down-going oceanic crust and the thermal energy generated by it. But in continent-continent collision (or continent-island arc collision) ophiolites formed under high temperature and metamorphic rocks are less important and the mechanical energy produced by collision is more important. Consequently here the sedimentary and crystalline rocks of the continental crust are primarily deformed.

If it is accepted that the mountain are formed on account of compression of the crust where plates converge, then we may identify four major geological effects of such converging movements. (i) The first possibility is that such movements may result in down-warping of the crust, and geosynclines or linear basins may be formed in which the sediments derived from the adjacent lands will be deposited. (ii) The other effect will be the deformation of the crust resulting from compression. The evidence of this is to be found not only in folds and thrusts of the mountain chains but also in the rocks like slate which have been flattened and have developed a wholly new structure as a result of the pressure. (iii) The third effect is heat produced by metamorphic and magmatic activity. In many mountain ranges, the occurrence of granite is limited to the zones of orogeny. An excellent example is to found in the Andes where the Mesozoic granite is found in a narrow belt of the mountain. (iv) The fourth effect is uplift whereby the mountains really emerge on the surface. The evidences of uplift may come from the sudden change in the nature of new deposits, the cessation of all plutonic activity and other factors. These evidences help in establishing the time of the uplift.

The significance of the above four effects lies in the fact that they help us in establishing the geological history of a mountain belt.

Plate Tectonics and Geosynclines

The hypothesis that plate tectonics controls the geosynclinal cycle and mountain building, provides new answers to some old questions. According to plate tectonics, the shape or size of the ocean basins is not fixed, and they open up or close. At the present time the Atlantic Ocean is opening up, and the Pacific Ocean is closing. Simultaneously the continents also drift. In other words, each continent has a leading edge that moves forward and a trailing edge that follows. For the last 200 million years the Pacific coast of North America has been the leading edge and its Atlantic coast has been the trailing edge. The trailing edge is characterized by tectonic stability, and since the highest parts of the continent are located along the Pacific coast, most of the continental sediments are deposited in the Atlantic Ocean. Thus there is deposition of sediments along the trailing edge, and there is formation of miogeosycline or miogeocline on the continental shelf and of eugeosyncline or eugeocline below the continental slope. On continental shelf the deposition is more thick towards the continental slope, and below the continental slope are found nearly 10 km thick turbidity sediments of sand and silt. This zone is about 250 km wide and nearly 2,000 km long, and is parallel to the coast. In size and its characteristics of rocks and sediments, it resembles an ancient eugeosyncline. On account of the changing movements of the plates, if ever there is formation of a trench and subduction zone here, the Atlantic Ocean will start closing and there will be collision of the oceanic crust desending into the trenches with the continental coast, and the eugeocline will change into a folded mountain zone.

When the sediments of eugeocline are folded, thrust faults usually develop and the sediments of the eugeocline are thrust toward the adjacent miogeocline. Also low density magma rises up on account of the heating of the descending plate into the mantle and invades the eugeocline. Abundant intrusions of magma produced by the intense heat are found in the eugeocline towards its continental margin, but towards the ocean the changes in the sedimentary rocks are mostly brought about by high pressure. The sediments derived from the erosion of the newly upraised mountains in the eugeocline begin to be deposited towards the continent in the miogeocline. Such continental deposits of shale and conglomerate are collectively called *molasse*.

According to the Plate tectonics theory the miogeoclines are ensialic, i.e. their base lies on the continental crust or sial. On the other hand, the eugeoclines are ensimatic i.e. here the deposition takes place on the oceanic crust or sima. This is a departure from the old view which holds that all geosynclines are ensialic. Earlier people had found that the basement of miogeosyncline is formed by granite and it was therefore concluded that sialic rocks must constitute the basement of eugeoclines as well. But in eugeoclines, the thickness of sediments is as great as 35 kms and it can be definitely stated that at this depth the base can only be sima. This is also confirmed by the presence of some rocks of the oceanic crust in the mountains.

Sima being the base of eugeocline, its sediments undergo intense folding, faulting and metamorphism. Here the ocean floor is mobile and presses the sediments against the continental slope. On the contrary, the sediments in the miogeoclines are protected by the continental crust, and only light and open folds are found here.

The floor of the geosyncline is constantly sinking. This can be particularly noticed in the miogeosyncline where the shallow water sediments extend to depths of about 5 kms. When the ocean opens up as a result of sea-floor spreading, the trailing edges of the driting continents push downwards. Then on account of the pile of sediments also there is subsidence of the geosyncline by way of isostatic compensation. For each three meter thickness of sediments the crust subsides by two metres. In the long geological period in which there is rise and fall of sea level a thick layer is deposited on the continental shelf. The excess debris goes down the continental slope in the form of turbidity currents and is deposited below the slope. The two geosyclines form a pair and there is constant interplay between them. In the eugeosyncline the deposition of sediments goes on increasing till it reaches the height of the continental slope. Simultaneously, according to the law of isostasy, the oceanic crust goes on subsiding under the load of sediments and a fully developed sedimentary prism may be as thick as 15 kms. When folding starts, its thickness is further increased. There is also increase in thickness by metamorphism and granitic intrusions and in this way a massive structure is built which goes down to depths of 35 kms. Thus when there is uplift, the building of huge and high mountain chains becomes possible. In this way the area of the continents increases by the peripheral accretion of the sediments deposited below the continental slopes and the "detritus lost to the ocean is eventually carried back to continents and collapsed into accretionary belts that also incorporate new igneous rock".[1]

The mountains are formed after a long and protracted process of convergence of the mobile plates and the uplift of mountains is indicative of the cessation of the converging movements and the

1. Dietz, R.S.: *Geosynclines, Mountains and Continent-Building*, Scientific American, March, 1972, p. 38

welding together of the two moving plates. From this time onwards they may remain in this state for millions of years. In this manner the extent of the rigid crust gradually increases. At some later stage under the influence of the oceanic crust rifts may develop in the continental blocks which are fragmented and disrupted. During this disruption mountain chains continue to evolve on the leading edges of the separating continents, but the older mountains within the blocks may be also destroyed during this period. Since the earth is a sphere, these dispersed continental blocks are brought together again in new combinations to form either one or two super-continents as a result of compression by the growth of new oceanic crust in the surrounding oceans. In time after stabilization, rifting will break up these continental masses yet again and start a further period of continental dispersion, and the process of formation of new mountain will start on the leading edges of the separating continents. The cycle of mountain formation and destruction thus continues, impelled by the mobile plates[1]. There is thus no conflict between the theory of plate tectonics and the idea of geosyncline. On the other hand, the theory of plate tectonics throws a new light on the history of sedimentation in the geosynclines and provides a more satisfactory explanation of the formation of fold mountains, their tectonic history and structural characteristics than has hitherto been possible.

Questions

1. Discuss the current views on the origin of fold mountain chains.
2. Explain the origin of fold mountains with reference to *any one of the* current theories
3. Discuss the mode and phases of formation of young folded mountains and give suitable examples.
4. Explain the validity of the concept of geosyncline in the context of mountain building, and give a concrete example.
5. "Out of the geosynclines have come the mountains." Elaborate this statement with reference of *either* the Himalayas *or* the Alps.
6. Discuss critically the views of *either* Kober *or* Holmes regarding the formation of fold mountains. Give suitable diagrams.

---

1. Sulton, J. : Orogeny in *Understanding the Earth,* op. cit., p. 299

7. Explain critically *any one* of the following theories regarding the origin of mountains :—

   (a) Thermal Contraction theory of Jeffreys,
   (b) Radio-activity theory of Joly,
   (c) Daly's hypothesis of sliding continents,
   (d) Convection current theory of Holmes.

8. Explain the mode of formation of mountains with reference to the Plate Tectonics theory.

9. Write short notes on *any three* of the following : —
   (a) The concept of Geosynclines,
   (b) The concept of Hinterland and Foreland,
   (c) Continent-Ocean Collision and Mountain Building,
   (d) Island-arcs and their formation.
   (e) Continent -Continent Collision.

Chapter **15**

Weathering, Mass wasting and Erosion

Geomorphic Processes

We have seen that the geomorphic processes which shape the configuration of the earth's surface may be divided into two broad categories—the endogenetic or internal processes (hypogene) and the exogenetic or external processes (epigene). The *endogenetic processes* originate within the earth's crust and include volcanism and distrophism. They are essentially processes of land renewal and land uplift. The *exogenetic processes* derive their energy primarily from the sun and have their origin in the earth's atmosphere and are guided by the force of gravity. These are essentially processes of land destruction. They remove materials from one part of the earth's surface and transport and deposit them elsewhere. There is a continuous struggle for supremacy between the external sun-born forces of land destruction and the internal earth-born forces of land renewal. The configuration of the earth's surface is, therefore, never static but is ever changing.

Rocks disintegrate and decompose and move to lower elevations under gravity, and are carried away by running water, glaciers, wind and waves and tides. All these processes through which there is rock wastage and land destruction or land sculpture are collectively described by the term *denduation* (Latin *denudare*, to strip bare). Denudation thus consists of two kinds of processes—one static and the other mobile. The disintegration and decomposition of rocks is a *static process* and is called *weathering*. Weathering makes the rocks more transportworthy by breaking them into smaller pieces. The movement of the rock fragments, big and small, is achieved through *mobile processes*. The mobile processes include mass wasting and erosion. In mass wasting there is transfer of rock material downslope under the influence of gravity. The removal of the weathered material also takes place under the influence of other mobile agents like running water, ground water, glaciers, wind or waves which may carry them to great distances and also break them into smaller pieces in the process. The activities of the mobile processes are together called *erosion*. Erosion includes transport, corrosion and deposition. When rock fragments are moved from one place to another by the agents of erosion, this is called *transportation*. As these rock fragments move, they erode the rocks over which they move and are themselves broken into still smaller fragments in the process. This is called *corrasian*. The debris carried by the transporting agents is sooner or later deposited. This is called *deposition*. There are some authors who do not include deposition under the term erosion and consider it to be a separate process. Following Chamberlin and Salisbury (1904), Thornbury (1969), for example, prefers to use the term *gradation* in place of denudation to include "all those processes which tend to bring the surface of the lithosphere to a common level." The *gradational processes* are subdivided into two categories—those which level down i.e. *degradation* and those which level up i.e. *aggradation*. According to him, degradation includes the processes of weathering, mass wasting and erosion by the mobile agents, while deposition by the various agencies is considered under aggradation.

There is, in fact, a certain amount of confusion in the use of the above terms, and the kind of scientific distinction which has been outlined above between the meaning of these terms, is unfortunately, not observed in actual usage. To a certain extent, this is due to difference of opinion as to what is to be included under certain processes, but it is also due to carelessness and lack of precision in thinking and writing. Erosion is sometimes used in an extremely narrow sense to imply merely cutting or degradation i.e. corrasion and transport, while on other occasions it is used in such a wide sense that it becomes synonymous with denudation or gradation. In fact, of these terms, denudation (or gradation) is the most comprehensive and is the collective term for all the processes of land sculpture whether they are degradational or gradational. Denudation has three aspects—weathering, mass wasting, and erosion (including deposition), it being recognised that sometimes erosion may be used in a more limited sense which may exclude deposition. The three processes of weathering, mass wasting and erosion are so inter-related and inter-dependent that they cannot be separated easily from one another. Weathering by the breaking of rocks encourages mass wasting and erosion, and the transport of materials by mass wasting and erosion in turn encourages the disintegration and decomposition of rocks. Weathering is a preparatory process and makes erosion easier. But it should be remembered that weathering can take place without subsequent erosion and erosion is possible without previous weathering.

WEATHERING

Weathering is the disintegration and decomposition of rock in place. It is a general term which is applied to a group of processes which act at or near the earth's surface and reduce solid rock masses by physical disintegration or by chemical decomposition to the clastic state. It is a process of the breakdown of rocks. Weathering by itself is a static process and does not involve removal of the material. The products of weathering tend to accumulate in soft-surface layer, called *regolith* which covers the bed rock. The rock debris are, of course, subject to gravity and tend to fall or slip down the slopes especially when aided by the lubricating action of water, though water is not the transporting agency in this case. This process of gravitational transfer of mass of rock debris down slopes is called *mass wasting* which we shall consider a little later in this chapter.

Types of Weathering and Controlling Factors

Weathering is of two kinds—(a) physical or mechanical weathering in which the rocks are disintegrated by temperature changes, frost action and organism, and (b) Chemical weathering, in which minerals are decomposed, dissolved and loosened by water, oxygen and carbon dioxide of the atmosphere and soil water and by organisms and the products of their decay. The physical, chemical and biological agents co-operate with one another actively and both mechanical breaking up or disintegration and chemical decomposition of rocks proceed simultaneously in nature.

Broadly speaking, four *factors* may be said to influence the type and rate of weathering—rock structure, climate, topography and vegetation. The term *rock structure* is used to include the many physical and chemical characteristics of rocks. It includes mineral composition and hardness as well as physical properties like joints, bedding planes, faults, and fractures and minute cracks which permit air, water and roots to enter to considerable depths. The minerals determine whether a rock is more susceptible to physical or chemical weathering.

The *hardness of rock minerals* varies from 1 (very soft) to 10 (extremely hard) according to Mohs' Scale of Hardness. Most igneous rocks are hard, as their constituent minerals include felspar and quartz which have a hardness of 6 and 7 respectively and these minerals are very tightly bonded together while crystallizing. Sedimentary rocks are generally much softer, even when they consist of hard minerals. For example, sandstone is largely composed of quartz grains but is weak if the cementing material is soft, such as calcium carbonate or iron oxide. But when the cementing material is hard such as in quartzite which comprises quartz particles cemented by silica, the rock is quite hard. It is a common misconception that hardness is an extremely important factor determining the resistance of rocks to weathering. But the truth is that hardness is one of the less important determinants of resistance of rocks to weathering (though not erosion). This is more so in the case of chemical weathering. But nearly all hard rocks contain weaknesses, either structural or of chemical composition which can be taken advantage of by the chemical processes of weathering.

The chemical composition of rocks is, of course, the most important factor influencing the resistance of rocks to chemical weathering, though jointing and porosity which allow entry of water are also important. Chemical composition also influences mechanical weathering of rocks to some extent. Light coloured rocks like limestone, chalk or marble reflect most of the sun's rays and heat less rapidly than dark coloured rocks like basalt or gabbro, resulting in more marked block and granular disintegration of the latter in the hot deserts.

Rock jointing is an extremely important factor in both chemical and mechanical weathering in a variety of ways. Joints make available a much larger surface area of rocks for attack by chemical processes. They allow the entry of water and oxygen into the rocks, and provide lines of weakness for use by mechanical agents of disintegration such as, frost and ice. The joints or cracks are usually arranged at right angles to the bedding planes in sedimentary rocks, and the bedding planes perform practically the same role as joints in terms of weathering. In igneous rocks the joints develop due to stresses set up by contraction while cooling and are found at right angles to the margins of the intrusion or extrusion. In crystalline rocks such as, granite or gneiss, 'dilation' occurs after the overlying burden of rocks is removed by erosion. As the pressure of the overlying rocks is removed, the underlying rock tends to recoil upwards, and joints parallel to the surface, called sheet jointing are formed. They resemble the stratification of sedimentary rocks and are called 'pseudo bedding planes. ' Joints may be relatively minor and discontinuous in their occurrence or they may occur over long distances in well-defined and persistent patterns. Chemical weathering by rain water is concentrated along joints and bedding planes. Various limestone features caused by solution are closely related to the joint pattern. "In granitic areas, the rock may be divided into cuboidal blocks as a result of chemical attack along vertical contraction joints and horizontal pseudo-bedding planes. In areas of close jointing the granite may be totally decayed because of the comparative instability of the felspar and ferro-magnesium minerals and a kaolin residue containing particles of quartz will be formed. Where the jointing is wider and the decomposition less advanced, rounded subsurface boulders may be produced by the reduction of the joint-bounded rectangular blocks of granite".[1] In periglacial and glacial climates water entering the joints freezes at night. When the temperature of water decreases from 4^0 C to 0^0 C, there is 10% increase in its volume and the rock on either side of the joint is subjected to enormous pressure and breaks into angular blocks after some time. Then the process of exfoliation is dependent on the existence of sheet jointing.

The *climatic factors* include temperature changes and humidity which to a considerable extent decide the rate of weathering as well as whether physical or chemical processes will dominate. Broadly speaking, it may be stated that cold climates and deserts favour disintegration especially where slopes are steep, while chemical weathering is more important in warm and humid climates and areas of low relief. Chemical weathering is most effective in very warm climates, as the intensity of chemical reactions becomes double for every 10^0 C rise in temperature. Chemical weathering is also most active in wet climates, as water is essential for the various chemical processes. Mechanical weathering requires considerable fluctuation in temperature between day and night such as we find

1. Small, R.J.—The *Study of Land forms*, 2nd ed, 1985 Cambridge Univ. Press London, pp. 23-25

in the hot deserts or it may be caused by frost and ice-wedging in peri-glacial climates where there are freeze and thaw cycles. We may make a broad division of the earth's climates into four main types from the point of view of weathering:

(i) *The tropical humid climates* : Here the high temperatures coupled with high rainfall provide very favourable conditions for rapid decomposition of rocks. Even such resistant rocks as granite, gneiss and schist are decomposed to considerable depths in areas of gentle slope, and a thick layer of weathered materials may be found covering the underlying rocks. *Duricrusts or weathering crusts* are widely developed in warm humid and monsoonal climates. They comprise strongly weathered materials cemented by iron, silica or calcium minerals into a very hard layer, several metres thick. The best example of duricrusts is laterite, a regolith rock found widely in the Indian Peninsula and parts of tropical Afrcia, such as, Nigeria and Uganda. Laterite is a kind of vesicular clayey rock composed essentially of a mixture of the hydrated oxides of aluminium and iron, present in variable ratios. The iron oxide generally predominates and gives the rock its characteristic red colour. Laterites form resistance caps to underlying weak material produced by prolonged chemical decay, resulting in the formation of isolated mesas.

(ii) *Arid and Semi-arid climates* : These climates are characterized by marked diurnal changes of temperature. As a result there is alternate expansion and contraction of bare rock surfaces resulting in block and granular disintegration or the physical breakdown of rocks. It is, however, now believed that chemical weathering induced by the presence of small quantities of moisture also play an important part, especially in such processes as exfoliation and 'basal weathering' of steep slopes.

(iii) *Temperate climates* : Most of the weathering processes are found in the temperate climates but none of them are very active. In temperate climates chemical weathering, however, is more active than mechanical weathering and is generally aided by the widespread soil and vegetation cover.

(iv) *Arctic Climates* : In these cold climates mechanical weathering is quite active. The dominant process is frost action, resulting in disintegration of rocks into blocks, boulders and finer materials. Chemical weathering is, however not totally absent. Poor drainage in lowlying areas above the permafrost causes the formation of peats and organic acids.

Topography influences the amount of rock exposure and even more importantly, works through its influence on precipitation, temperature, exposure to the sun and the kind and amount of natural vegetation. In areas of gentle relief there is usually a thick layer of soils which protects the rock from mechanical weathering and the water present in it helps chemical decomposition. On the other hand, in areas of high relief and steep slopes, mass transportation favours landslides, soil-creep and solifluction, and rock surfaces are constantly exposed to mechanical weathering by frost action and insolation changes, while chemical weathering is reduced by rapid run-off, especially if they have no soil or vegetation cover.

The abundance and type of *natural vegetation* determine the extent of rock exposure and the amount of decaying organic matter from which carbon dioxide and organic acids may be derived. The growing rootlets of shrubs and trees exerts pressure as they enter cracks and crevices and break the rocks apart. The rootlets also bind the soil into a mat-like structure so that it remains porous and absorbs water. Grass roots are particularly effective in this. The vegetation, therefore, has also a protective role. Forests break the impact of rain water, prevent soil erosion and regulate surface flow. The natural vegetation thus influences the rate and type of weathering in various ways. Together with the natural vegetation mention may be made also of the influence of earthworms and other burrowing animals such as rodents and termites which play a significant role in preparing materials for removal by rainwash or wind.

Physical or Mechanical Weathering

When a rock undergoes mechanical weathering, it is broken into smaller pieces. One of the significant effects of physical weathering is to increase the amount of surface area (interface surface) available for chemical attack. Physical weathering or mechanical disintegration of rocks takes place in four different ways—frost action and crystal growth, thermal expansion and contraction by temperature changes, organic activity, and expansion by unloading.

(a) *Frost Action and Crystal Growth*—Frost action is one of the most important processes of physical weathering in cold climates. When water fills the pores, cracks and crevices in rocks and then freezes, it expands by more than 9% of the volume and exerts a bursting pressure of about 150 kg to the square centimetre. The rocks are ruptured and fragmented and wedged apart. Frost action by its tremendous force is believed to rupture even the hardest of rocks. It is most effective where there is repeated freeze and thaw and in rocks with fractures and bedding planes. Freezing also penetrates to a greater depth when the ground is bare rather than forest covered. Frost shattering of hard rocks on steep slopes above the timber line is particularly conspicuous especially where joints are well developed. The angular fragments and boulders so formed slide to lower levels and accumulate to form *talus slopes or screes*. The talus slopes formed by these loose aprons of angular debris at the base of cliffs are highly unstable which is well brought out by the term *scree*, which is a Norse word meaning rubble which slides away when trodden on. Frost action produces conspicuous effects in all climates with a cold winter but is particularly significant in the tundra climates of the arctic coastal fringes and the high mountains where it may give rise to several interesting landforms (see Chapter on peri-glacial landforms).

Closely related to the formation of ice crystals in rocks is the disintegration of rocks by the growth of salt crystals. The salt crystals form in dry climates as a result of the capillary action of water containing salts. During the long dry period as the water rises to the surface and evaporates, tiny crystal of salts are left behind in the porous outer zone of the sandstone. The force generated by these crystals lead to grain-by-grain breaking of the sandstone which disintegrates into sand. The zone of rocks lying close to the base of a cliff is the most liable to be affected, because here the underground water seeps outward to reach the rock surface. This process may be responsible for the formation of niches, pits, shallow caves and rock arches sometimes found in sandstone formations.

(b) *Thermal Expansion and Contraction*— Most rock- forming minerals expand when heated and contract when cooled. When rock surfaces are exposed to marked diurnal changes of temperature as in the hot deserts, the alternate heating and cooling results in alternate expansion and contraction which exerts a powerful disruptive force on the rocks. In the hot dry regions the difference between the day and night temperatures often exceeds 30^0 C (50^0F). The intense heat of the day causes the thin surface layer of the rock to expand and to pull it away from the cooler layer within. This process is called *mass exfoliation*. This process was once thought to be of major importance in rock disintegration and formation of spheroidal boulders. But experiments have shown this process to be inadequate under dry conditions to fracture massive rocks. But in the presence of water, repeated alternations of heating and cooling lead to rupture. Further, the various minerals composing a rock have different coefficents of expansion and as a result of differential expansion and contraction of the individual minerals, the rocks undergo grangular exfoliation or disintegration. Thus though experimental evidence of mass exfoliation resulting from repeated heating and cooling is lacking, the diurnal changes are nonetheless significant in producing granular exfoliation or disintegration. Under normal conditions massive rocks must first be weakened by chemical weathering before shells and flakes of rocks can be pulled away and then broken into fragments. Even in the deserts chemical action is present. The heavy desert rains are sometimes near the freezing point and even hailstorms occur. This leads to sudden chilling, and contraction further opens up any cracks that may have developed and water is given an effective entry. Individual minerals expand and shrink and gradually break apart, especially in coarse-grained rocks like granite.

Mass exfoliation or the scaling or peeling of flakes and curved shells of rock just as in onion, appears to be the result of a complex group of processes which include and co-operate with the stresses caused by diurnal expansion and contraction, so that they (latter) become more effective than they appear under laboratory conditions. The final peeling off seems to result from 'fatigue' set up by millions of repetitions of heating and cooling that a shell of rock undergoes before it becomes loose. The stresses are strongly concentrated, and chemical activity is greatly increased at the inner edges of even the smallest cracks.[1]

(c) *Organic Activity*—We have referred earlier to the role of plants and animals in the promotion of both physical and chemical weathering. As the plant roots grow, they wedge the rocks apart and cause the widening of joints and other fractures. Organic

1. Holmes, Doris L. —op. cit., p. 250

activity, however, is of greater importance in chemical than in physical weathering. Dead organisms produce acids as they decay and thus promote chemical weathering. Earthworms, ants, termites and other burrowing animals move material to or near the surface where they are more readily subjected to chemical weathering processes.

(d) *Expansion by unloading* — Another important process of rock disruption related to physical weathering results from unloading i.e. the relief of confining pressure of overlying rock. Unloading occurs when large igneous bodies are exposed through the erosional removal of overlying rock and the resultant reduction in pressure. Igneous and metamorphic rocks formed at great depths are in a compressed state because of the continuing pressure of the overlying rock. On being exposed to the surface they expand slightly in volume. Expansion or dilation accompanying the unloading causes thick shells of rock to break free from the parent mass below. The process which produces thin onion-like layers or concentric large scale fractures is called *exfoliation*. These new surfaces of fractures which develop parallel to the land surface are called *sheeting joints or sheeting*. The individual sheets usually become progessively more closely spaced as the earth's surface is approached. Sheeting is a near surface phenomenon and is best developed in massive rocks like granite, marble, quartzites and thickly bedded sandstone. Where the sheet structure has formed over the top of a single large body of massive rock, an *exfoliation dome* is produced. Exfoliation domes are among the largest of the landforms produced primarily by weathering. The Yosemite valley in California is well noted for such domes. Large scale exfoliation on sandstones in the Colorado Plateau has also been ascribed to splitting off of rock sheets accompanying expansion resulting from relief of rock pressure as erosion has exposed the rocks which were once deeply buried.

Chemical Weathering

In general, chemical weathering is probably more important than physical weathering, This is particularly so in the warm and humid climates of the equatorial, tropical and sub-tropical zones where heat and moisture are abundant. Water is the main agent of chemical weathering. Although water in a pure form is almost inactive, but when mixed with oxygen or carbon dioxide it becomes an active chemical agent. Oxygen, carbon dioxide as well as water-vapour are present in the atmosphere in abundant quantities near the earth's surface. Under their influence the rocks decompose, decay and break into smaller particle size and new secondary minerals are formed. The surface environment of the earth is poorly suited to the preservation of the intrusive rocks which were formed under conditions of high pressure and high temperature. The minerals in igneous and metamorphic rocks were at equilibrium under the high temperature and pressure condition at which they formed, but under the temperature and pressure conditions at the earth's surface some of the minerals are not stable and do not last long in the low temperature and pressure of atmospheric exposure, combined with the chemical action of free oxygen, carbon dioxide and water. In general, chemical weathering results in the retention of those minerals which are at equilibrium at the earth's surface, and the minerals which weather more easily and are least stable at the earth's surface are those which crystallized first at great depths. Mineral stability is therefore in the reverse order of their crystallization. Olivine weathers more rapidly than augite or hornblende. Of the minerals which are common in igneous and metamorphic rocks, quartz and muscovite are the most stable and are retained in their original form as the most common constituent of weathered rocks, as these minerals are the most nearly in equilibrium with the temperature and pressure conditions at the earth's surface and are very resistant to chemical weathering. By extensive studies especially on the silicate minerals which form most of the earth's crust, a mineral stability series or *weathering series* i.e. a list of common silicate minerals arranged in order of relative susceptibility to chemical weathering has been prepared.

Common minerals of igneous rocks have been arranged as a weathering series, with the least resistant mineral at the top and the more resistant minerals progressively lower in the series. This arrangement is essentially that of Bowen's reaction series for the sequence of formation of minerals from a silicate melt.

Olivine—*least resistant*
Augite
Hornblende
Biotite
Albite
Orthoclase
Muscovite
Quartz—*most resistant*

The silicate minerals that crystallize at the highest temperatures and that have the lowest silican-oxygen ratios weather most rapidly.

Chemical Weathering Processes

The chief chemical weathering processes are solution, oxidation, hydration including hydrolysis and carbonation. It should be borne in mind that these processes are inter-related and act in combination.

(a) *Solution*—Solution is not considered to be a distinctive chemical weathering process, because solutions in water are always a part of chemical weathering. Pure water, however, is not a good solvent, though substances like rock salt and gypsum do get dissolved in pure water. Limestone is not soluble in pure water but gets easily dissolved in rain-water which contains carbon dioxide. In fact atmospheric carbon dioxide is dissolved in all surface waters of the land, including rain water, soil water and river water. Where limestone solution is active, we find a distinctive landscape with many unique landforms.

(b) *Oxidation*— The presence of dissolved oxygen in water in contact with mineral surfaces leads to oxidation. Since some dissolved oxygen is always present in rainwater, surface water as well as sub-surface water oxidation is a universal phenomenon. The effects of oxidation are most appearent in rocks containing iron. Iron rusts or oxidizes under the influence of moisture, as the dissolved oxygen changes the ferrous iron in mineral compounds to the more oxidized ferric state. When iron combines with oxygen, the original mineral structure is destroyed and the mineral components are free to participate in other chemical reactions. Iron oxides have yellow or red colour, and a brown surface layer is the first indication of weathering of most rocks, as most rocks contain iron in small quantities. Discoloration by the oxides is readily noticeable and it is generally believed that oxidation is the first to take place, though actually hydrolysis preceds it.

(c) *Hydration and Hydrolysis*— The process of hydration involves the absorption of water. Most rock-forming minerals absorb rain water. This not only increases their volume but also produces chemical changes resulting in the formation of new minerals which are relatively softer and more voluminous. By the process of hydration anhydrite is converted to gypsum and haematite to limonite. Both the above reactions are however reversible upon application of heat, which indicates that there has been no fundamental chemical change. In hydrolysis, on the other hand, there is a chemical change, and both the mineral and water molecules decompose and react to form new compounds. Hydrolysis takes place in an aqueous solution and is a very important chemical weathering reaction of silicate minerals. Carbon dioxide gas dissolves readily in water to form a weak acid. Rainwater absorbs carbon dioxide from the atmosphere and is usually slightly acidic. Carbon dioxide dissolving in soil water is the most common and important way in which water is acidified for hydrolysis, as the soil air may contain up to a thousand times more carbon-dioxide than the free atmosphere (up to 30% as against 0.03% of dry normal atmosphere weight) since the soil air is greatly enriched in carbon dioxide by the decay of humus.

The significant result of hydration and hydrolysis (as also of oxidation) is that the new minerals formed are more easily attacked by the chemical and physical weathering processes. The hydrolysis of exposed granite surfaces results in grain-by- grain break-up of the rock, creating many interesting boulder and pinnacle forms by rounding of angular joint blocks. There is ample moisture even in the deserts for hydrolysis and the grain-by-grain break-up results in coarse desert gravel, consisting largely of quartz and partially decomposed feldspars.

(d) *Carbonation*— Carbonation and hydrolysis are closely linked. We have seen that carbon dioxide (CO_2) dissolved in water (H_2O) forms a weak acid called carbonic acid (H_2CO_2). Rain dissolves some carbon dioxide as it falls through the atmosphere, and additional amounts released by decaying organic matter are acquired as the water percolates through the soil. Carbonic acid particularly attacks minerals which contain iron, calcium, magnesium, sodium or potassium. These elements are soluble in carbonic acid, and the minerals and rocks containing them start decaying under its influence. Calcium carbonate, the chief element of limestone, is insoluble in pure water, but becomes soluble in rainwater and gets dissolved. The impurities in limestone, such as quartz and clay generally remain more or less undissolved and accumulate. Similarly, the rain-water creates limestone caverns by solution and carbonation. In sandstones where limestone forms the cementing material carbonation results in the removal of calcium carbonate (limestone) and with the removal of the binding medium the sandstone starts readily disintegrating.

The effects of carbonation can be seen on other minerals and rocks as well. Carbonic acid disrupts the crystalline structure of feldspar, and the most common by-product of the chemical breakdown of feldspar is clay minerals. The granite which consists mainly of quartz and feldspar, changes into clay by the decomposition of feldspar, but quartz which is resistant to chemical weathering remains substantially unaltered. Some of the quartz particles remain in the soil, but much is washed to the sea where it is the main constituent of the sandy beaches and in time is converted into the sedimentary rock sandstone.

Silicate minerals make up most of the earth's crust and eight elements are the principal constituents of the silicate minerals. Of these, sodium, calcium, potassium and magnesium produce soluble products upon chemical weathering which can be removed by ground water. The element iron combines with oxygen, producing insoluble iron oxides, chiefly limonite and hematite which remain in the soil giving it a reddish brown or yellowish colour. The three remaining elements-aluminium, silicon and oxygen, join with water to normally produce stable clay minerals.

As chemical weathering changes the internal structure of minerals, physical changes also occur in the rocks. Chemical weathering assists in the disintegration of rocks in three different ways—(a) by weakening the coherence between minerals, so that the rock more easily succumbs to the action of the physical agents, (b) by the formation of solutions which are washed down by rain and the rock becomes porous and ready to crumble, and (c) by the formation of altered products which have greater volume than the original material, so that the outer shell swells and pulls away from the unaffected rock underneath.

The process whereby the increase in bulk of the outer layer exerts an outward force strong enough to cause concentric layers of rock to break loose and fall off is known as *spheroidal weathering*. It is best developed in rocks with regular joint systems, like basalt and dolerite, which are readily decomposed. Water penetrates the intersecting joints and attacks each separate block from all sides. As the decay is more at the corners and edges, they tend to become rounded. As the upper shell breaks loose, a new surface is presented to the weathering solutions and the process is repeated, helped by temperature changes. Each successive wrapping surrounding the core becomes more spheroidal than its outer shell, and the angular block is transformed into an onion-like structure of concentric shells of residual material. Cores of more coherent rock may stand out like boulders when their outer wrappings have been washed away. "Residual boulders of granite, sometime of great size, probably result mainly from exfoliation, but with the aid of spheroidal weathering. "[1]

Zone of Weathering

The upper portion of the lithosphere which has been subjected to weathering processes is known as the zone of weathering. Its depth is variable, but in tropical and subtropical regions, especially in area of granitic rocks, it may extend down to 200 to 400 ft. The interface i.e. the surface between the weatherd and the unweathered rocks has been termed the "*weathering front*" by Mabbutt (1961). This term emphasies the fact that weathering comprises a set of dynamic processes which move downwards through the bedrock but at different rates at different places.

MASS WASTING

Mass wasting or mass movement is the term used for the movement of material down a slope under the influence of gravity. It comprises a wide range of gravity-induced movements of slope material, especially that resulting from weathering, but it excludes those in which such material is carried directly by a transporting medium, such as running water, wind or moving ice. The term, however, includes sheet-wash. Importance of weathering, mass wasting and sheet-wash in the general lowering of the land as well as in the shaping of the details of topography in the interstream areas has been only recently recognized. Stream erosion directly affects only a small proportion of the total land surface, while in the interstream areas there is slow but continuous etching of the earth's surface by differential weathering, followed by removal of the weathered materials by mass wasting and sheet-wash. In terms of effective volume, mass wasting is perhaps the most important among the agencies that sculpture and reduce the surface.

---

1. Holmes, Doris: op. cit., p.255.

Mass wasting is viewed as a transitional phenomenon between weathering which is defined as occurring in a place and erosion, which requires as one element transport by some agent or medium. Mass wasting combines elements of both weathering and erosion but occupies an intermediate position separating these two categories of proceses.

Factors favouring mass wasting

Weathering reduces the shearing strength of materials by physical and chemical changes through granular disintegration, hydration and expansion, saturation and loss of compactness, drying and cracking etc. Original rock materials may be subject to mass wasting because of their composition. For example, materials like schist, sand, gravel, shale, soil etc. are easily sheared permitting fall, slide, flow, creep or subsidence. The texture and structure of materials may also be conductive to mass wasting because of their roundness or compaction. Movement may take place along planes of structural weakness, such as bedding planes, foliation, cleavage, fractures and joints, and faults. Changes in environmental conditions including climatic factors and vegetational cover may result in mass wasting. Slope gradient is, of course, always an important factor. Thus a complex of factors is usually involved in mass wasting, resulting in a variety of mass movements.

The role of water in the reduction of shearing strength is particularly important. This is often referred to as 'lubrication' of rock or weathered rock debris. This is really achieved through the introduction of intergranular forces by pore water pressure and the buoynant effect of water. Water does not directly affect the frictional properties of rocks. However, rapid mass wasting frequently takes place after heavy rains or rapid melting of snow cover and does emphasize the effect of water on gravity movements. The role of slope gradient on mass wasting is equally important.

Types of mass wasting

Sharpe (1938) made one of the first attempts at classification of the various types of mass wasting and though several others have been proposed since then, his classification is still widely used. He based his classification on (a) the nature of water (or ice) present in the moving mass, (b) the nature of movement itself, and (c) the speed of the movement. He recognized *four major types of mass wasting* which he called slow flowage, rapid flowage, landslides and subsidence. Hutchinson (1968) proposed a classification of mass movement on slopes based on mechanism of movement, morphology of the mass and the rate of movement, and has identified four major types—creep, freeze-thaw movements, landslides, and mass movements involving sinking. The various types of mass movements are discussed in greater detail in the chapter on slope development.

EROSION

We have already referred earlier in this chapter to the confusion that prevails in the use of the term erosion. One thing is, however, clear. Weathering is not a part of erosion. Weathering is a separate process and can occur without erosion. Similarly, erosion can take place without weathering. Erosion is essentially concerned with the various ways in which the mobile agencies acquire and remove rock debris. The acquisition of materials by the mobile agencies and their transport—i.e. corrasion and transportation are considered to be integral parts of erosion. There is, however, difference of opinion as to whether or not deposition should be included under the term erosion, or should be considered as a separate process.

Some authors do not include deposition under erosion. Thornbury, for example, includes transportation but excludes deposition from the term erosion. There may be some justification for this in certain situations, as for example, when we speak of erosional landforms and depositional landforms. Deposition, however, is an inevitable concomitant of erosion or degradation and sometimes proceeds simultaneously with the process of cutting. It is, therefore, convenient to include the three processes of acquisition and cutting, transportation and deposition under the comprehensive term erosion. It is in this broad sense that we speak of the cycle of erosion. Transportation is crucial to both the cutting down or wearing away of land, as well as to deposition of debris. Erosion is, however, frequently used in a narrower sense meaning thereby just wearing away of land, and in this limited sense erosion, transportation and deposition become three distinct processes. In practice there is considerable confusion in terminology and frequently the terms are used rather loosely.

The principal erosional agents are running water (both concentrated and unconcentrated), ground water, wind, glaciers, waves and currents. They are also, of course, agents of transportation and deposition. Each of the agents does erosion by distinctive processes and gives rise to distinctive landforms. Basically there are five common aspects to erosion by these agencies: (a) the acquisition of rock fragments, (b) wearing away of the surface through impact of rock materials in transit or in some cases by solution, (c) the breaking down of the rock particles by mutual wear while in transit, (d) transportation of the acquired rock debris by the moving medium, and (e) ultimately its deposition somewhere either in transit or at the end. Of all the agents of erosion by far the most important geomorphologically and the most universal is running water, specially because rivers act as the major conduit pipe for all the materials removed from land and ultimately transported to the ocean basins.

Questions

1. Explain the different kinds of weathering and their geomorphic significance.
2. What is mass wasting? What are the factors which are conducive to mass wasting ? What are the major types of mass wasting ?
3. Distinguish between denudation, weathering, Mass wasting and erosion.
4. Write short notes on :— Chemical weathering, Exfoliation, mass wasting, Factors influencing weathering.
5. Give a broad analysis of the geomorphic processes and agents involved in land sculpture.

Chapter 16

Fluvial Processes and Landforms

The adjective 'fluvial' (from Latin *fluvius*, river) refers to the work of rivers but in the context of landscape development it includes the work of both overland flow and stream flow. Thus landforms shaped by running water (i.e. by the fluvial processes of overland flow and stream flow) are called fluvial landforms. Fluvial landforms and processes dominate the continental land surfaces. Landforms made by glacial and wind action occupy relatively small areas of the continental surfaces, and the landforms made by the waves and currents are confined to a very narrow contact zone between the oceans and the continents. The fluvial landforms constitute by far the largest proportion of the environment of terrestrial life and the major areas of food production, as almost all the land under crop cultivation as well as grazing has been shaped by fluvial processes.

The Hydrologic Cycle

There is a continuous movement of water from the oceans to the atmosphere, from the atmosphere to the land, and from the land back again to the sea. This unending circulation of the earth's water is known as the *hydrologic or the water cycle*. It is a giagantic system, powered by energy from the sun. The water that falls on the continents as rain must ultimately make its way back to the ocean. It is this water which is responsible for the wearing down of the earth's land surface. Water, flowing down to the sea over the face of the land, is the dominant agent of landscape evolution. In fact, the fluvial system is so widespread and effective, and operating even in dry regions, that it has been commonly regarded as the 'normal' process of landscape evolution.

MECHANICS OF FLUVIAL EROSION

Overland flow and Channel flow

Since the middle of the twentieth century quantitative methods have been increasingly used to study geomorphic concepts in fluvial systems. Quantification has resulted in greater precision and exactness in previously enunciated but vaguely defined principles. These include hydraulic geometry of overland flow and surface lowering, precise measurements of channel form and patterns, sediment transport and meandering, linear and areal properties of river basins and many other related features. The emphasis in recent studies, starting with Horton (1945) has been to treat river and its tributaries as an open system, resulting in a concept of dynamic equilibrium of energy balance and adjustment in a stream which is time independent, distinct from the Davisian concept of time dependent grade. As water flows downgradient it uses its energy to transport sediment and carve its channel. As the process continues, the channel geometry and longitudinal river-bed profile are constantly readjusted.

Overland Flow— Rain water falling on the surface is removed by infiltration into the soil or rock, by sheet flow over the surface and/or by flow through a system of rills, gullies and finally stream channels. Overland flow is the term used for the movement of water over the surface of the land when it is not confined to distinctive channels. Overland flow is responsible for much erosion before streams attain identity lower down the slope. Soil particles are loosened by rain-drop impact. This is called *splash erosion*. Splash erosion is especially effective in arid and semi-arid regions where the rainfall is sporadic and torrential, the surface is loose and friable and there is no vegetation cover. The loosened particles are entrained in the run-off flow covering a considerable area, resulting in *sheet erosion*. Surface run-off will not flow as a sheet for long. As the water flows further down there will be concentration of flow in a number of tiny superficial and ill-defined channels called *rills*. During high intensity storms rills may be widened by absorption to be called a *gully*.

Streamflow— The rill or gully system develops into a full-fledged drainage pattern as it undergoes systematic changes. There is rapid increase in tributaries by bifurcation at the head and by growth of rills down the side slopes of widening gullies. The total length of stream channels rapidly increases by elongation at the heads where erosion is carried on by rain itself. It enables headward erosion to cut back into the slope above the point at which the headstream starts its career.

The hydraulics of streamflow are extremely complex. Basically, the movement of water in the channel is laminar flow ie streamlined transference of its mass in a downhill direction. Ideally, it is so if the channel were straight and smooth. Actually the flow in stream channel is varied and non-uniform and there is gradual change in velocity and depth to compensate for bends, obstructions, contractions and expansions. Any obstacle in its bed or projecting from the banks causes eddies and turbulence ie the superposition of secondary movements on, and usually at some angle to, the predominant downstream flow of water. Roughness and turbulence both reduce the overall speed of flow, but turbulence is a major eroding and transporting agent.

The *energy of a river* depends primarily upon its volume and velocity, which together are expressed by the term *discharge*. The discharge increases downstream with the addition of water from tributaries. Discharge is expressed in cusecs ie the number of cubic feet per second passing a particular section of a river or in metric measure, cubic metres per second (*m3/sec*). The number of *m3/sec* is obtained by multiplying the rate of flow in metres per second by the cross-section of the river at the particular point in square metres. Mean velocity is the total discharge in m3 / sec divided by the cross-sectional channel area at the stage of average discharge when the river is lower than bankfull. At bankfull or overbank stages the velocity is constant along the whole length of the river. The mean velocity of a river increases in a downstream direction even though the gradient of its bed may steadily decrease. The cause of this apparent anamoly is that as the stream and its channel enlarge down valley, the loss of velocity through friction of the running water on the smoother bed and banks is reduced proportionately. This more than compensates foı the gentler gradient. Further, increased velocity is necessary to cope with the larger discharge in its lower section.

The energy of a river is expended partly in friction with its sides and bed, partly in friction with moving water due to turbulence, and partly in transporting its load.

River regimes or their seasonal variation in volume has received much attention in recent years in connection with schemes of flood-control and hydro-electric development. Further, there has been growing awareness of the erosional ability of catastrophic flood, as a result of sudden and remarkable increase in its discharge and velocity. Rivers have three types of regimes—simple, double and complex. Rivers having simple regimes have one period of high water and one period of low water, closely following the seasonal rainfall regime. The larger rivers of monsoon Asia, such as the Ganga, the Brahamputra, the Irrawady, the Yangtze or the rivers of Peninsular India have floods in summer and low water in winter. Rivers having a double regime have two distinct periods of high water, the result of early summer snow melt and autumn-winter rain, such as the Garonne in western Europe. The equatorial rivers like the Amazon and the Zaire maintain considerable volume throughout the year but have two maxima following two rainfall maxima during the equinoxes. A complex regime is the feature of world's largest rivers with extensive basins covering various climatic regions and receiving numerous tributaries having different regimes. The flow in such rivers may be remarkably consistent but even so there may be periods of flood. Examples are the Mississippi, the Danube, the Nile. The variations in river regimes tend to make the erosional processes in fluvial system rhythmic and pulsatory and extremely complex varying with the seasons and the climate.

Fluvial Processes—The Work of Rivers

The river does three kinds of work—erosion, transportation and deposition. It cuts its sides and bed, this is erosion. It moves the materials cut by its own erosive power as well as those received from rain wash; this is transportation. Then where the flow of the river water slackens, sand, silt and rock fragments get deposited; this is deposition. These different kinds of activity are not quite independent of one another but are inter-dependent. Through the joint action of these activities the river degrades the surface of the land, carves its own valley, creates flood-plains and deltas and leaves the areas of relatively harder rocks as residual hills and plateaus. From the point of view of landscape evolution, the erosion performed by a river is the most important, but there is such a close relationship between erosion and transportation that these two activities cannot be understood independently of each other.

Erosion

The erosion performed by a river is of two kinds—chemical and mechanical. Some of the rocks over which the river flows, such as limestone, are soluble and the river dissolves the rocks along the joints and forms cavities in them. This is chemical erosion. Chemical erosion is performed through solution. But the mechanical erosion or corrosion performed by a river is far more important than chemical erosion, as it is largely through mechanical erosion that the river derives it debris or load. Apart from solution, the erosive work of a river consists of three interacting processes—hydraulic action, corrasion and attrition.

The river not only erodes its sides but also its bed. In other words, cutting is both vertical and lateral. Through the mere force of its flow, the river can cut its bed and its sides. This is erosion by *hydraulic action*. In this the river erodes without the assistance of pebbles and rock fragments and without chemical action, merely through the force of movement of its water and mechanically breaks and removes the rocks in its course. But the rock fragments present in the river water greatly increase its erosive power. These rock fragments strike against the rocks along the bed and the sides and abrade and break them as they move down with the flowing water. In this manner the river erodes its channel by *abrasion* or *corrasion* and makes it deeper and wider. The rock fragments also collide against one another and in the process get disintegrated into small, smooth and rounded pieces of rock and ultimately into sand and silt. This process by which there is a progressive downstream reduction in the size of the materials constituting the load of a river by constant collision, is called *attrition*.

Thus the corrasion done by a river is of two kinds—lateral and vertical. By vertical corrasion the river deepens its channel, and by lateral corrasion the channel of the river is widened.

The erosive action of the river is more evident in the mountainous regions where on account of the steep slope the river flows with great speed. The erosive power of a river depends on the volume of water, its speed, the size and number of rock fragments in the river water and the hardness of the surface rocks over which the river flows. Any increase in the volume of the river water greatly increases its corrasive power. Similarly the power of corrasion of a river varies approximately as the square of its velocity.[1] The size and number of rock fragments flowing with the river water exert

1. Wooldrige, S.W. and Morgan, R.S.—*An Outline of Geomorphology*, Longman, 1961, p. 139

even a more powerful influence on the corrasive power of a river, as it is largely through their help that moving river water corrades and cuts. Without the presence of rock fragments, the flowing water can perform very limited erosion. It is not merely the amount of the load (rock fragments present in the river water) that determines the amount of erosion that a river can perform but the size and the nature of the load are also important. If the load materials are fine and of very small size, they move either in suspension or in solution with the water. Under both these conditions, there is little erosion. On the other hand, if the load materials are relatively bigger in size or are angular, they are dragged along the river bed with the moving water and are able to perform more erosion. In fact there is a positive relationship between the load of a river and its power of corrasion. With no load, there is little corrasion. When the river has a full load, corrasion is equalled by deposition. There is thus some intermediate load when the effective corrasion is at its maximum.[1] Before this, corrasion increases with increase in load.

Transportation

The river carries long distances the rock fragments, sand and silt with its flow. There are several rivers which are over 1,000 kms long and in this way these materials are transported over a thousand kms and more by the river. The river transports these materials principally in two ways. (1) The minute particles of sand and silt float in the river water and flow down along with it. It is because of the presence of these minute particles that the river water is muddy, especially in the rainy season. If the water is allowed to stay stagnant for some time, these particles start settling down at the bottom. (2) The small and big fragments of rock roll down along the bed of the river with the moving water. In the process they also strike against one another and become smaller, polished and rounded.

The transporting power of a river depends on the volume and velocity of its water, its gradient and the transportability of its load. The more the volume of water in a river, the more the materials it can carry. The transporting power of a river increases very rapidly as the velocity increases.[2] A fast flowing river can carry not only

1. Lake. P.—*Physical Geography* (Indian Edition), 1959, p. 324.
2. It is generally regarded as proportional to the sixth power of the velocity. Ibid

more materials than a slow moving river but can also carry materials of bigger size. The velocity of a river determines the size of the rock fragments that it can transport. The transportability of the rock fragments is dependent on their shape, size and number. As has been indicated earlier, sand and smaller particles generally remain suspended in the water. Relatively bigger fragments roll and are dragged along the bed of the river.

The solid materials which a river carries with itself is called its *load*. This sometimes includes also the materials held in solution (*solution load*). In addition, some material of intermediate size too heavy to be held in suspension and lighter than the bed load may bounce or jump along the stream bed (*saltation load*). Thus we may say that the load of a river comprises of three different kinds of materials—the materials in solution, the suspended load (sand and silt consisting of particles finer than the bed load), and the *traction* or *bed load* comprising the larger fragments, coarse sand and pebbles which are carried along the bed. The largest *amount* of given size debris that a river can move in traction as bed load is called its *capacity*. The largest *diameter* (*and weight*) of particles that it can carry as bed load measures its *competence*. Both capacity and competence of a river increase with increase in volume and velocity. The load of a river depends not only on the volume and velocity of water but also on the size of the rock fragments. If the fragments are small, then a river even with less speed can carry them in large quantities. If the load of river is increased, its velocity decreases, and with decrease in velocity it is unable to carry the larger rock fragments which tend to get deposited in its bed.

The more the velocity of a river the more the load it can carry, and more the load the greater is the corrasive power of the river. But after a certain point corrasion decreases as the load increases, as the increase in load reduces the velocity of the stream. As increase in the velocity of a river results in an increase in its corrasive power as well as transporting capacity, the river is able to perform several times its normal work of erosion and transportation during the short periods of flood in the rainy season.

Deposition

As the velocity of a river decreases, it begins depositing its load of rock fragments, sand and silt, since its capacity to transport these materials is dependent upon its velocity. When a fast flowing hill stream emerges from the mountains and reaches a relatively level surface, there is deposition of a huge amount of stony debris, sand and silt, and alluvial cones and fans are formed here. When the river reaches the more level surface of its valley, its flow slackens further, and the river starts depositing sand and silt in its bed resulting in the formation of 'diaras'. Then in times of flood when the river water overtops its banks and inundates the neighbouring areas, there is deposition of sand and silt outside the confines of its channel. Maximum deposition takes place where the river enters the sea. Here the flow becomes very slow, and there is large scale deposition of sand and silt near the river's mouth. Because of the presence of salt in the sea water, the sediments brought by the river do not easily mix in the sea water and get precipitated and deposited near the river's mouth. This results in the formation of deltas.

Course of a River

The point of origin of a river is called its source, and the point where it ends is called its mouth. The course of a river from the source to its mouth may be divided into three parts—upper, middle and lower. For example, the Ganga has its upper course above Hardwar, its middle course from Hardwar to Rajmahal, and its lower course below Rajmahal. In its upper course the river cuts and deepens its channel, and sometimes this section is called the valley tract. In its middle and lower courses the river transports and deposits its load, and in this way forms alluvial plains. Both these sections are, therefore, referred to together as the plain tract.

Landforms formed in the Upper Course

Usually a river has its source in a lake or spring or a glacier. Some of the world's very large rivers originate from lakes, namely, the Sutlej from Lake Mansarowar, the Lena and Yeneisi from Lake Baikal, the St. Lawrence from the Great Lakes and so on. Several rivers in South India have their source in springs. In the Chota Nagpur plateau there are several examples of such streams, a notable one being the Subarnarekha. The Ganga and the Yamuna provide examples of rivers which originate from galciers. The Indus and the Brahamputra also have their source in glaciers close to lake Mansarowar.

In its upper course the river flows through mountainous or

hilly country. Here the slope is steep and the velocity of river water is great. Because of the high velocity and the predominance of angular rocky fragments, the corrasive/erosive power of the river is great. The cutting is mostly vertically downwards and through vertical downcutting the river carves out its valley. The valley is consequently deep and narrow and the river flows in a constricted channel. When the river valley is very deep and narrow, it is called a *gorge*. In the Himalayas there are several gorges which are thousands of feet deep. Probably the most famous is the gorge on the Indus river near Gilgit where the river flows through an extremely narrow but 17,000 feet deep valley.

Where the river flows through a mountainous or plateau country with comparatively small rainfall, the sides of the river valley are less subject to erosion by rain wash, and the river flows for long distances through a deep valley with sides which are steep and vertical. Such long and deep river valleys with vertical sides are known as *canyons*. On the Colorado river in the U.S.A. is the world famous canyon called the Grand Canyon which is nearly one mile deep, 5 to 12 miles wide with narrow inner gorge, and 300 miles long. Canyons are also found on the Brahamputra in Tibet and the Zambezi river in South Africa.

Although in its mountainous course the river continues to cut its valley vertically downwards, normally there is also lateral erosion of the sides from rain water, frost and other agencies resulting in widening of the valley at the top, and thus the river valley gradually becomes V-shaped. Where the rocks are hard and more resistant, the valley sides are steep and the V is narrow. But where the rocks are soft, the river valley becomes wider and the V is more open. The amount of rainfall also influences the width of the valley. Where the rainfall is small, despite the presence of soft rocks, the valley sides are not easily eroded, and the valley continues to be deep and narrow. On the other had, where the rainfall is heavy, barring areas of exceptionally resistant rocks, the valley sides are gentler and the valley is more open. Thus the most favourable conditions for the formation of gorges and canyons are to be found in areas of comparatively low rainfall. The deep gorges and canyons, so characteristic of the Salt Range and Baluchistan are chiefly due to the arid climate, while in the Eastern Himalayas of Sikkim and Nepal where the rainfall is heavy, the valleys are broader with gently sloping sides.

In its mountainous course the river bed is usually littered with rock fragments. With their help the river deepens its valley by vertical corrasion. Where the river flows over soft rocks or rocks which have joints and cracks, the rock fragments flowing with the river water are able to cut hollows on the rocky river bed by friction. These hollows become deeper and rounded on account of the rotary action of the rock fragments under the pressure of moving water. Such round depressions or hollows found on the rocky river bed and caused by localized drilling action of rock fragments are called *pot-holes*. They can be seen in the dry season when the river bed is exposed and is only partially filled with water. Pot-holes are of common occurrence in the rocky river beds of the Chota Nagpur Plateau.

In the river bed at some places the rocks are hard while at other places they may be soft. Where an outcrop of resistant rocks dips

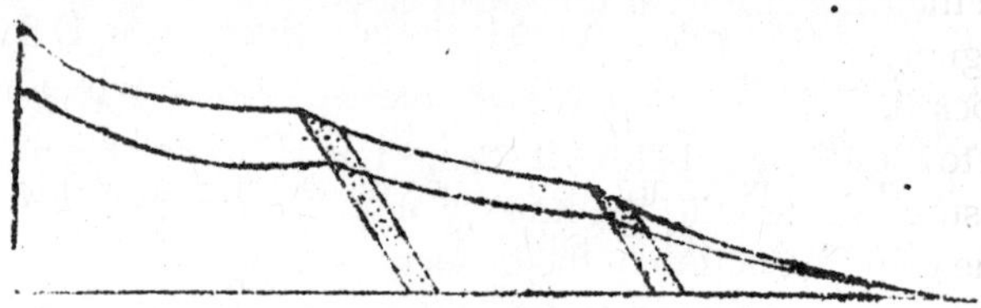

Fig.80—Suitable conditions for rapids.

downstream i.e. in the direction in which the river is flowing, the hard rock being more resistant stands above the bed surface and impedes the smooth flow of water which is first thrown up and is then thrown rapidly downwards resulting in the formation of what are known as *rapids*. Rapids are a stretch of swift-flowing, turbulent and relatively shallow water where the river bed suddenly becomes steep and this is preceded by a resistant outcrop which acts as an impediment. Waterfalls may degenerate into rapids as the rivers approach grade. Rapids may also result from rejuvenation downstream. Rapids and waterfalls are sites of discordance in the longitudinal profile of a river and in the normal cycle of erosion correspond to knick-points and are considered to be transitory phenomena of the early stage i.e. youth.

Waterfalls

When the water of a river suddenly falls down a steep or vertical slope, waterfalls are formed. A waterfall is thus a very steep (commonly almost vertical) fall of some magnitude in the course of a river. The existence of a steep or vertical slope in the course of a river is absolutely essential for the formation of a waterfall. Such a slope may be formed due to differences in structure and relief, differential erosion over hard and soft rocks in the river's bed, or tectonic uplift or subsidence in the river bed. Generally speaking, waterfalls are formed due to differences in rocks and geological structure in the river bed and are indicative of the youthful stage in the life-history of river. Some of the more important conditions under which waterfalls may be formed giving rise to *different types of waterfalls* are described below:

(i) *Cap-rock Falls* : Where a layer of hard rock overlies horizontally or dips upstream over a layer of soft rock, the soft rock is eroded more easily, leaving the hard rock areas upstanding over which the river water falls from a height. For instance, this is likely to happen where the upper layer or the cap-rock is dolomite or some other hard igneous or metamorphic rock and is underlain by a layer of soft rock like shale. When water falls down from the hard rock above to the soft rock below, it scoops out a hollow which goes on increasing in size with the impact of continuously falling water and the grinding action of the broken rock fragments. Where the river has a large volume and a considerable descent, the plunge pool grinding at the base of the fall is quite effective and causes rapid under-cutting of the softer rock below the cap-rock. The cap-rock is thus deprived of support at the bottom, and tends to break back due to its own weight and the weight of falling water. Through the repetition of this process the waterfall tends to recede towards the source of the stream, and a gorge is formed in front of the waterfall on account of its rapid recession. This process is known as recession or retreat of water falls. A well-known example of this variety of cap-rock falls is the Niagara Falls on the Niagara river in the St. Lawrence Valley of N. America. The Niagara Falls are located on the border between the U.S.A. and Canada and lie in both the countries. The cap-rock here is hard and consolidated dolomitic limestone while the underlying layer consists of relatively

Fig. 81— A Waterfall.

softer shale. As a result of the under-cutting of the weaker shale through the process described above, the Niagara Falls have been retreating backwards and the retreat is still continuing. It has been estimated that it has retreated 7 miles upstream since the time of its formation, and its Canadian part is currently moving backward at the rate of 3 to 4 ft. every year. The retreat is comparatively rapid, being dependent upon the rapidity with which the weak formation is under-cut, and there is no time for weathering action of any

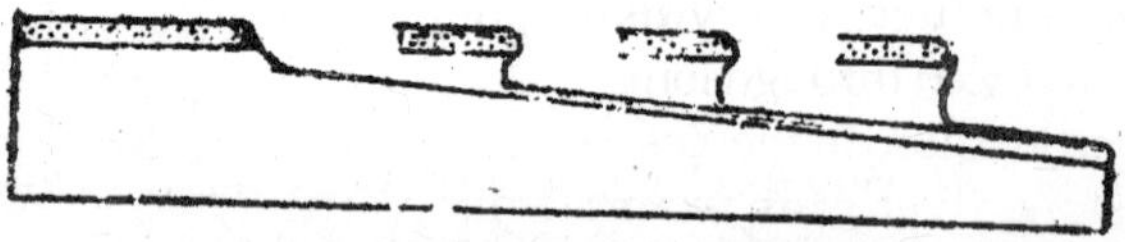

Fig 82 — Recession of Waterfall.

consequence on the valley walls in front of the retreating fall so that the valley in front is narrow and deep and has steep gorge-like walls. Below the Niagara Falls the river flows through a 7-mile long and 360 ft. deep gorge. The famous Dhurandhar falls near Jabalpur on the Narmada formed on crystalline dolomite limestone or Dharwar marbles has a nearly 2-mile long narrow gorge below it.

It Guyana, situated on the northern coast of South America, Kaieteaur Falls on the Potaro river is another example of a cap-rock fall. Here the river descends down a vertical fall of 740 ft. and the cap-rock is hard and flat conglomerate and massive sandstone which are underlain by the relatively softer shales. Examples of cap-rocks falls abound in the world but their nature varies with difference in rock and structure and inclination of the rock layers as well as in the volume of water. *Step falls* develop in strata of more uniform erodibility where the under-cutting occurs along individual bedding planes. The Dasam falls on the Kanchi river, the Hundru falls on the Subarnarekha and the Johna falls on the Garga provide examples of step falls in the Chota Nagpur plateau.[1]

(ii) *Vertical Barrier Falls*— The most favourable condition for the formation of this type of fall is to be found in areas of sedimentary rocks which are intruded by hard igneous rocks in the form of vertical dykes. While the softer sedimentary rocks lower down in the river

1. Singh, R.P. *Geomorphological Evolution of Chotanagpur Highlands, India*, National Geographical Society of India, Varanasi, 1969, p. 69

bed are easily cut, the vertical dykes endure on account of their hardness, and the stream descends over the dyke wall in the form of waterfall (Fig. 83) In such cases, the resistant dykewall stands in the way of the recession of the waterfall which consequently is more stationary, though slowly its height is reduced through long continued erosion. In the Yellowstone park of U.S.A. the fall on the Yellowstone river is an example of a vertical barrier fall. In the upper course of the river are hard volcanic rocks while down below is found greatly decomposed lava, and the dividing line between the two is quite sharp and vertically differentiated.

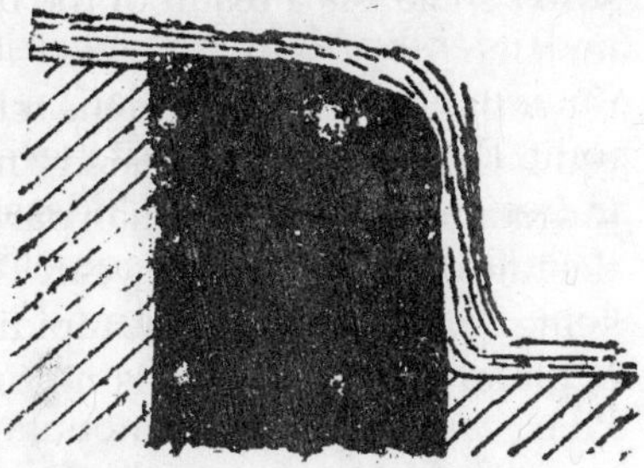

Fig. 83—Vertical Barrier Fall.

(iii) *Plateau Falls*—Plateau falls are found in areas where the rivers descend suddenly from the uplifted plateaus of igneous and metamorphic rocks to the lowlying and softer rocks of the coastal plains. Falls of this kind also have the tendency to retreat. The Congo river while descending down the African plateau to the coastal plain gives rise to the Livingstone Falls. In South Africa the Orange river descends down the plateau through the 460ft. high Aughrabies Falls. In India a good example is the Gersoppa or Jog Falls in the Western Ghats on the Sharavati river in Karnataka which is one of the most spectacular falls in the world and is 850 ft. high.

(iv) *Fault-plane Falls*— When the upper part of a river is uplifted by earth movement and a fault develops in its course then the river forms a waterfall along the fault scarp because of the sudden change in level. Examples of this type can be seen in the ancient uplifted plateaus. The Hundru Falls on the Subarnarekha river near Ranchi where the main drop is a nearly vertical scarp of granite-gneiss caused probably by Tertiary faulting and the Sivasamudram Falls on the Cauvery river in Karnataka are good examples in the ancient plateau of Peninsular India. Tamsa river (South Tons) descends through several fault plane falls while coming out of the Vindhyan plateau, the most striking of which is the Behar falls which is 370 ft. high and has a width of 600 ft. in the rainy season.

(v) *Knick-point Falls*— When there is rejuvenation in the lower section of rivers either as a result of a fall in sea level or uplift of the land, its ability to cut downwards is increased and there develops a marked break in slope at the rejuvenation head or the knick-point. This is especially so if the rejuvenation head happens to lie in an area of resistant rocks. Under such conditions, the rivers develop waterfalls at the knick-points because of the sudden change in slope. Such falls are found in the hard rock areas of the Appalachian mountains where the knick-points are located on the eastern slopes of the mountains and the rivers descend in falls from the old surface of the Appalachian to the coastal plains.

Examples of falls on the knick lines formed by the scarps in Chota Nagpur plateau are the Hundru falls on the Subarnarekha (75m), the Dasam on the Kanchi (40m), the Sadni on the Sankh (60m) etc. In fact, according to Wadia, the waterfalls in the Peninsular Plateau of India including the Gersoppa on the Sharavati (260 m), the Sivasamudram on the Cauvery (90m), the Gokak falls in Belgaum (54m), the Dhurandhar or Bheraghat falls on the Narmada (9m) etc. are all due to minor disturbances caused by earth movement in late geological age.[1]

(vi) *Hanging Valley Falls*— When there is increase in downcutting in the principal rivers on account of rejuvenation, the principal river is able to cut its valley deeper than the tributary stream. The tributary streams are not able to cut their valley as deep either on account of less volume of water or other reasons. In such a situation, the valley of the tributary stream appears to hang above the valley of the main stream at their junction. Such valleys are called *hanging valleys* and this kind of junction between the principal and the tributary streams is called a *discordant junction*. Under such conditions, the water of the tributary stream falls down in the form of a waterfall before meeting the principal stream. A good example in Chota Nagpur is the Rajrappa fall where the Bhera nadi, a tributary of the Damodar river hangs nearly 15 m above the main stream and granite-gneiss rocks dip upstream in the Bhera nadi at the site of the fall.

This type of waterfall may also be formed as a result of some special kind of differential uplift and/or erosion. The fault of the Yozemite river in California provide an illustration. Prior to the recent mountain building movements, the principal rivers of this region flowed towards the west on a relatively gentle slope. Then the earth movements resulted in the formation of the extensive Sierra

1. Wadia, D.N.- *Geology of India*, Tata Mc Graw Hill, N. Delhi, 1981, p. 25

Nevada mts with slopes towards the west. Suddenly the slope in the bed of the principal rivers became very steep, but the change in slope in the case of the tributary streams was much less pronounced. In course of time, the valleys of the principal streams became very deep because of increased down-cutting and the valleys of the tributary streams remained at a higher level. Later on, as a result of glaciation the difference in level of the valley bottoms of the principal and tributary streams was further accentuated.

In fact, hanging valleys are most commonly found in the glaciated mountains, and the hanging valley falls are, therefore, more characteristic of rivers which occupy glaciated valleys.

Some times, hanging valley falls may be formed along the sea coast under the influence of wave action. Where the sea-waves are quite strong under the influence of powerful winds, they cut the coast into sea-clifls. The cliffs tend to retreat because of the constant attack by the sea-water, but the rate of downcutting in the streams is not so high that they may cut the cliff to a lower and level surface. Consequently, the river valleys hang above the sea-coast and the rivers form waterfalls before joinning the sea. Ordinarily, the height of the fall is more in the case of smaller streams and less in the case of the bigger streams. On the north-east coast of the Hawaii islands, such cliffs have been formed along the volcanic hill of Mauna Kea under the influence of powerful trade winds and several small streams fall into the sea in waterfalls, presenting a most picturesque sight.

It has been stated earlier that the presence of waterfalls is indicative of the youthful stage in the life history of a river. In course of time as the waterfall recedes backwards, its height is reduced and it is destroyed with the establishment of grade in the bed of the river. The more rapid the recession of waterfalls, quicker is the establishment of the grade. The top of the waterfall acts as a local base-level for the river above the fall. Consequently, if the rate of recession of the waterfall is slow on account of the hardness of the rocks, the river channel above the fall may become graded, the valley may become wide and the river may show signs of mature or early old age above the fall, but below the fall the youthful appearance is preserved due to the presence of the gorge and the constricted valley.

Waterfalls are important not only because of their scenic beauty and attractiveness but because they are a store house of waterpower. In many parts of the world hydroelectricity generated from the waterfalls is being used on a large scale for lighting as well as for industry.

Landforms formed in the Middle Course

After leaving its deep, narrow and steeply sloping mountainous course the river enters its middle or plain course. Here the slope is less and the speed of the river is slackened. Because of the reduction in speed it is not able to carry the larger rock fragments, and they start getting deposited. First the larger fragments and as the river advances, the smaller fragments and sand are dropped. Thus where the river emerges from the mountains and starts its middle course, rock fragments, pebbles and sand are deposited in the foot-hill zone. This deposit is triangular in shape (i.e. fanning out downwards) through which the river flows in several braided channels. It thus resembles a fan made of palm leaves, and consequently such semi-circular deposits are called *alluvial fans*. Where the slope is relatively more, and smaller streams emerge from the mountains, the alluvial deposit is less and is known as *alluvial cone*. In the upper section of the alluvial fans coarse boulders and pebbles are deposited, while in its lower part, finer material are deposited. Sometimes the neighbouring alluvial fans may meet and coalesce with each other as they develop, resulting in the development of an extensive alluvial plain in the foot-hill or piedmont zone. This is known as *piedmont alluvial plain or Bajada*. The alluvial fans are built mostly during the rainy season, when the volume of water as well as of the sediments is more. They are more commonly found in areas which experience heavy rainfall but are denuded of a natural vegetation. Bajada has generally abundance of ground water supply and is able to support a dense population. An excellent example is the Los Angeles Basin in the U.S.A. where the ground water has been extensively utilized for the cultivation of fruits, vegetables and other crops and there is a dense population. Cones and fans are found at the junction of all the great mountains and the adjoining plains and may be seen in abundance in the sub-Himalayan belt of 'Bhabar' composed mostly of gravels, shingles and coarse sand.

The river, no doubt, does some erosion (especially lateral erosion) in its middle course, but the principal work of the river in this part of its course consists of transportation and deposition.

In its middle course the river normally follows a meandering

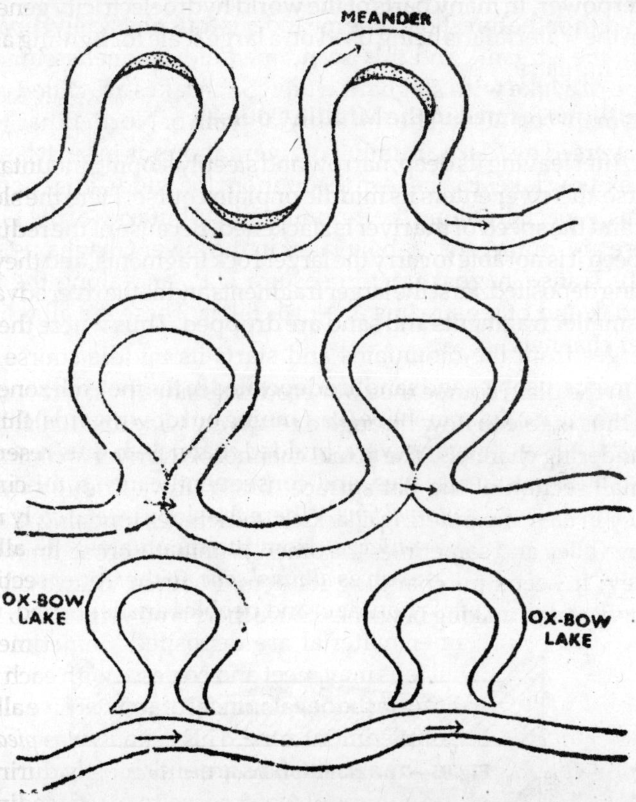

Fig. 84—River Meanders and the formation of Cut-off lakes.

path and not a straight one. Here the land is relatively flat and the speed of the river is less, with the result that wherever it meets with some impediment or obstacle, it turns round the obstacle instead of cutting through it. Such curved channels are called *meanders*. It will be apparent from Fig. 84 that in every meander there is a concave side which is being constantly cut, while sand and silt are being deposited on the opposite side. The side of the meander which is being cut is steep, while the side which receives the deposit is gently sloping. With the passage of time the curvature of every meander goes on increasing and the meander assumes the shape of a horse-shoe. Finally, there comes a time when the two ends of a meander meet, and the water of the river flows through a straight channel instead of curving round the meander. The meander is separated from the main flow and sand and silt are deposited near its mouth or the threshold. The meander is thus physically separated from the stream, and this abandoned old channel assumes the shape of a lake with stagnant water. Such lakes are called *cut-off takes* or *ox-bow lakes*. The flat alluvial plain of North Bihar which is traversed by a large number of rivers abounds in such lakes. In fact ox-bow lakes can be seen throughout the wide Gangetic flood-plains, especially along the innumerable tributaries of the Ganga. There are numerous ox-bow lakes in the lower Ghaghara-Ganga doab, Trans-Ghaghar tract in eastern U.P., and along the Burhi Gandak and other streams in North Bihar. These are all relicts of older channels.

In the plain course the river erodes its sides by lateral erosion and thus is able to flow through a relatively wide and flat valley in meandering channels. The actual channel of the river is confined to a small section of the flat surface but it flows through this flat valley in easy meandering channels sometimes touching one side of the valley and sometimes the other. Within the broad limits of its valley, it keeps on changing its actual course, destroying old meanders and making new ones. Since there is no vertical corrasion,

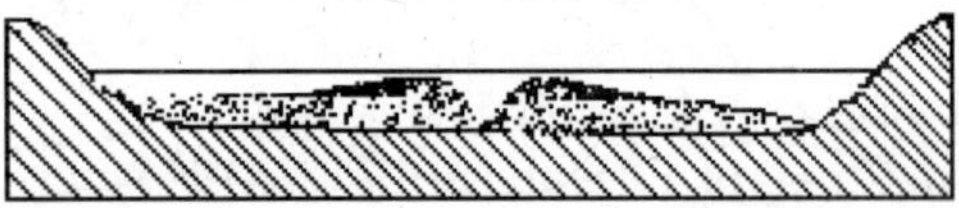

Fig.85—The flood-plain of the river.

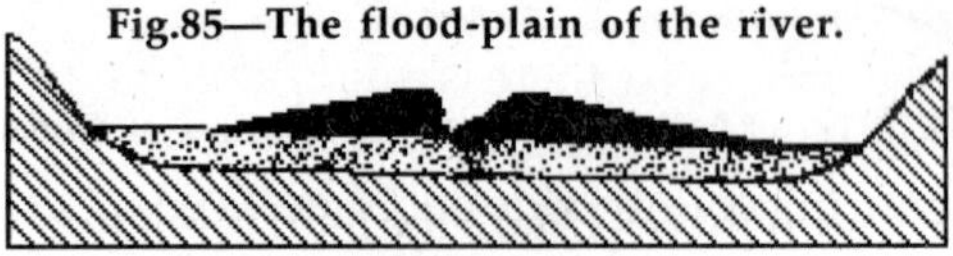

Fig.86—The natural levee of the river.

the river bed is not much below the level of the flat plain and is generally, therefore, shallow.

During the rainy season when the river is in flood, the river water overtops its banks and spreads over the neighbouring relatively flat land. The entire area so affected is called the *flood-plain* of the river (Fig 85). Through this proces the river builds up its alluvial plain. The flow is relatively fast within the confines of its banks, but outside the banks the water is nearly stand-still in the flood-affected areas and consequently the silt brought along with the flood water is deposited along the entire area. In the Ganga plain the inner or the newer flood-plain liable to annual inundation is called *Khadar* or the newer alluvium, while the older or the outer flood-plain subject to inundation only during the highest floods, is called the *Bhangar* or the older alluvium.

The speed of flow in the river is relatively less near the banks. Because of the reduction in the speed of flood water near the banks, sediments tend to be deposited there. When the flood recedes, some of the flood-water drains back to the river and some infiltrates into the ground, but the mud and silt which are deposited near the river banks stay there, and through repetition of this process over the years the river bank is raised above the level of the flood-plain. This river bank deposit is known as *the levee or natural levee* (Fig. 86). If the river contains large quantities of sediments, then portions of the sediments are also depsoited in the bed of the river and in this way the bed of the river also tends to be raised by accretion of sediments. With every new flood fresh silt is deposited along the banks and the height of the levee increases, and concurrently the bed of the river also rises. In course of time the levees become sufficiently high and are noticeably above the general level of the surrounding plain. Excellent exmaples of levee development can be seen along the Ganga and its tributaries in India and along the Mississippi in the U.S.A. There are several rivers in North Bihar which flow above the general level of the plains they have built. In China the Hwangho flows for hundreds of miles above the level of the surrounding plain. In all these areas the natural levees have to be artificially strengthened, otherwise there is a danger of their breaking during the flood. When the levees give way, the river water spreads over the neighbouring countryside and there is considerable loss of life and property. This has happened several times in the Hwangho river of China. That is why the Hwango is called China's sorrow.

Features formed in the Lower Course

In the lower course of the river the slope is even less and the speed of the river is further reduced. Even mud and finer particles start getting depsosited in its bed. The channel of the river is, therefore not deep but shallow and wide. During the rainy season the river water together with the mud and silt easily overflows the banks. As the flood subsides, large quantities of relatively coarse materials are relinquished adjacent to the river channel and the finer materials are deposited further away over the surrounding country. In this fashion a natural levee of silt and sand develops alongside the channel sloping away into bordering clayey deposit, and a flat fertile palin, liable to annual inundation and deposition is built up. With every annual flooding, the natural levee becomes higher and broader and the floor of the river channel is aggraded until it lies at a higher level than the bordering depressions which remian as illdrained, swampy wet lands.

Fluvial deposition aslo takes place at the mouths of rivers, where horizontal acceretion tends to prolong natural levees seaward as sedimentary jetties. They are well developed in areas of limited wave action and current scour, especially in a lagoon environment. Super abundance of fluvial sediment enables protruding jettes and sub-delta lots to develop alongside the extending distributaries on the Ganga and Mississippi deltas where shallow water off-shore reduces the effectiveness of wave and current scour.

Deltas

When the river enters the relatively stationary water of the sea, its speed is very much reduced and ulitmately stops. With the stopping of movement, the entire load of the river is deposited on the sea-bed. In South America it has been observed that the ocean

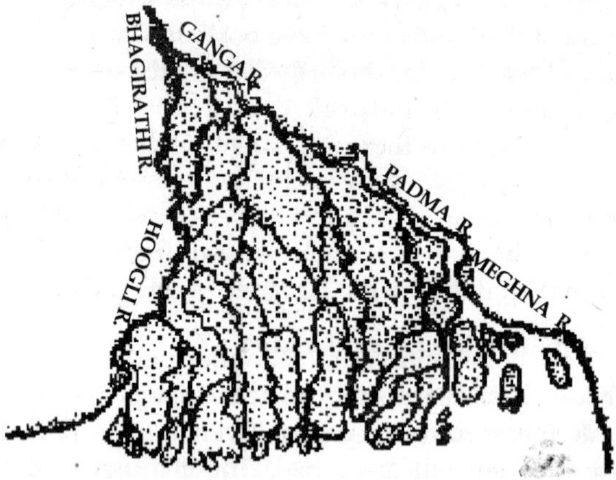

Fig. 87—Delta of the river Ganga

water remains muddy for nearly 200 miles off the mouth of the Amazon river. This distance is quite insignificant considering the vast size of the ocean, and broadly speaking, it would be correct to state that the river deposits nearly all its load near the sea- coast.

As a result of deposition of silt in the river-bed near its mouth, the river easily divides itself into fresh channels. Thus before meeting the sea, the river is divided into several channels separated by its own alluvial deposit. The alluvial deposit over which the several distributaries of the river are spread is traingular in shape with its base towards the mouth (sea-coast). This traingular piece of land resembles the Greek letter delta, and hence its name. Deltas are the products of rapid deposition of stream borne sediments into more or less still standing bodies of relatively shallow water. Their presence indicates the continuing ability of a river to supply and deposit sand, silt and other detrital material more rapidly than they can be removed by currents.

Three things are necessary for the formation of a delta. Firstly, the river should bring adequate quantities of sediments and secondly, the waves and tides should not be so powerful as to erode away all the silt brougt by the river ie the sea should be relatively sheltered at the river's mouth. Thirdly, the sea should be relatively

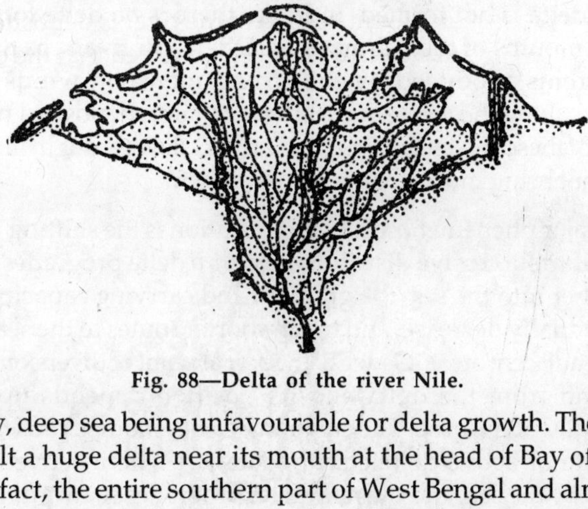

Fig. 88—Delta of the river Nile.

shallow, deep sea being unfavourable for delta growth. The Ganga has built a huge delta near its mouth at the head of Bay of Bengal and, in fact, the entire southern part of West Bengal and almost the whole of Bangladesh consist of the delta of the Ganga (Fig 87). Delta is also found along the mouth of the Mahanadi, the Godavari, the Krishna and the Cauvery on the eastern coast of India, the Indus in Pakistan, the Nile in Egypt and the Mississippi in the U.S.A. The mouths of the Ganga are affected by tides, but the silt brought by the Ganga-Brahamputra river system is so enormous in quantity that only a very small part is washed away by the tidal currents, and the huge lobate shaped delta has been built up. The Indus delta is in a more sheltered sea than the other major deltas of the Indian sub-continent, as it lies beyond the influence of the monsoon winds. The arc of the Ganga delta extends for about 400 kms from the Hooghly to the Meghna. There is a high degree of indentation with north-south marine inlets and distributary channels and marshy lands in between which are covered by sea-water at high tide. The Ganga delta is slowly prograding, especially in the active eastern section which is building at a faster rate and where the different stages of land building into the sea can be clearly seen from the air.[1]

When the river falls into a lake or a sea where the tidal currents are not strong, conditions are favourable for the formation of a delta. The Mediterranean sea is relatively free from tides, and the rivers falling into it make deltas at their mouths. The best example of such a delta here is provided by the delta of the Nile river (Fig.88).

Where the tides are quite strong near the mouths of rivers, nearly all the silt brought by the rivers is carried away by the tidal currents, and the delta is not formed. In India, there is no delta formation near the mouths of the Narmada and the Tapti rivers, as the tides and currents produced by the strong monsoon winds are so powerful along the western coast that no silt is deposited near the mouths of these rivers. Further, these rivers flow through hilly areas and do not bring much sediments with them.

A major phenemenon in delta evolution is the shifting of river courses into successive distributaries. As a delta progrades farther and farther into the sea, the gradient and carrying capacity of the river gradually decrease, and much shorter routes to the sea can be found in adjacent areas. Generlly the actual point of diversion occurs far inland from the delta and its position depends upon the topography of the base. The creation of a new course causes a shift in the site of active deltaic sedimentation. The inactive delta is attacked by the sea, and the new delta progrades seaward rapidly.[2]

Although all rivers do not form deltas, deltas occupy a special place among the landforms created by the depositional action of rivers. In fact, delta-plains have been the birthplace of several ancient

1. Spate, O.H.K. and Learmonth A.T.A.—*India and Pakistan*, Methuen. 1967, p. 589
2. *Encyclopeadia of Geomorphology*, op. cit., p. 255

civilizations. In the deltas formed by large rivers, fine sediments are deposited in enormous quantities. Because of the deposition of sediments, the delta goes on extending towards the sea as well as laterally on two sides of the river. The river gets divided into several distributary channels and follows a braided course. In this way in course of time an extensive low and flat deltaic plain is built up along the sea-coast through which the rivers flow leisurly in several distributary channels and finally meet the sea. Such extensive delta deposits are found along the mouths of the Nile, the Mississippi, the Ganga, the Yangtze, the Rhine and many other rivers.

Though commonly the deltas are triangular in shape and resemble the Greek letter delta, but in reality their shape shows great variations. On the basis of the difference in form, deltas are classificed into four major types:—

1. Arcuate delta
2. Digitate or Bird-foot delta,
3. Estuarine delta,
4. Cuspate or tooth-shaped delta.

Arcuate Delta— This is the most common form of the delta which resembles the Greek letter delta. Its most notable example is the delta of the river Nile whose shape inspired Herodotus in 500 B.C. to use the term delta for the alluvial plain enclosed between the diverging distributaries of the Nile and the sea. Its shape is like the arc of a circle and resembles a bow. Below Cairo which is located at the apex of the delta, the Nile divides into a number of radial channels. As the deposit of sediments is maximum in the neighbourhood of the main channel, the central part tends to project into the sea. The sediments which are carried down to the Mediterranean sea by the principal channels, form curved bars along the coast under the influence of wave-generated littoral currents, behind which there are shallow lagoons (Fig 88).

Such deltas are fan-shaped or arcuate in form with the arcuate base of the triangle lying along the sea-coast. Other examples of delta of this type are the deltas of the Ganga, the Rhine, and the Hwang Ho.

Digitate or Birdfoot Delta— Delta of this type resembles the fingers of a human hand or the foot of birds and hence it is called digitate or bird-foot delta. Its best example is the delta of the Mississippi in the U.S.A. This river flows through areas of limestone and other

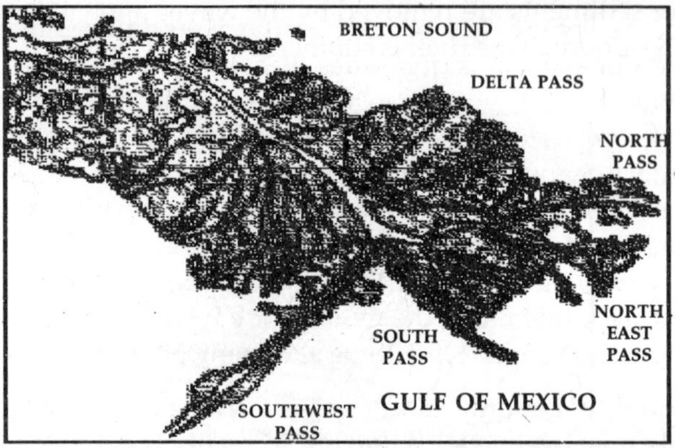

Fig. 89— Bird-foot delta of the river Mississippi.

fine-grained sedimentary rocks. It, therefore, carries considerable quantitities of limstone and other finegrained particles. The fine sediments enter the sea for considerable distance on account of the speed of the river, but on account of the saline water of the sea tend to get deposited on both sides of the stream channel and thus give rise to long, finger-shaped deposits. The Mississippi river flows into the Gulf of Mexico for some distance in a single channel and then divides itself into four distributaries with deposits along the sides of the distributary channels, resulting in the formation of the finger-shaped delta.

Estuarine Delta— Where the river enters the sea through a single mouth or estuary, and the estuary is filled up with sediments, a long and narrow delta is formed. Normally the estuary is open and

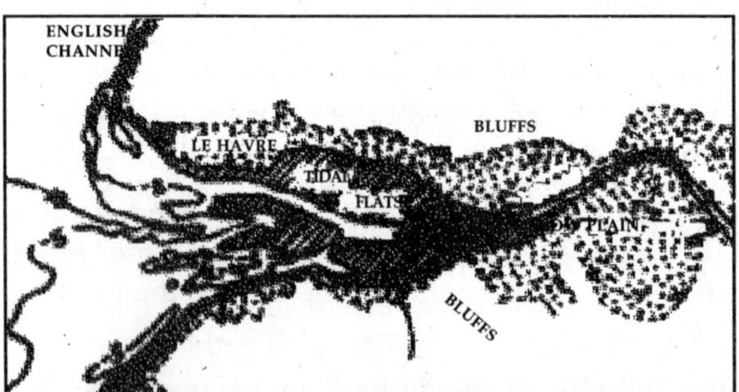

Fig. 90—Estuarine delta of the river Seine

the river sediments are removed by the waves and currents. But when the mouth of the river is submerged below the sea, a narrow

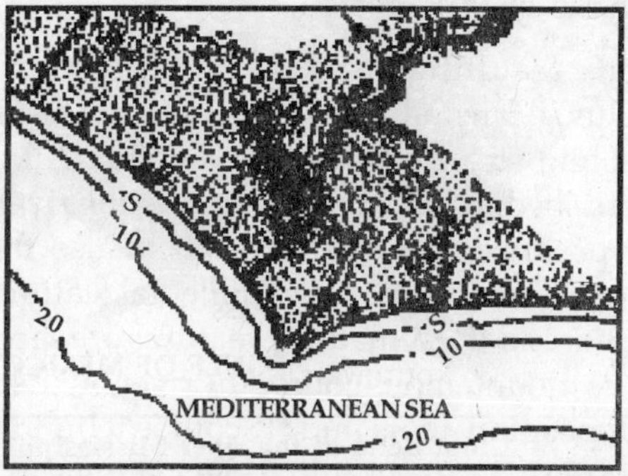

Fig. 91—Cuspate Delta of the river Tiber.

linear delta is formed as a result of the deposition of the sediments. Examples of this type can be seen seen along the mouth of the Seine river in France and the Sasquehana along the eastern coast of the U.S.A.

Cuspate or Tooth-shaped Delta—A good example of this type of delta is the delta of the Tiber river in Italy. In this type the deposit is along the main channel and the delta projects into the sea, but the sedimets brought by the river are deposited and spread on both sides of its mouth by the waves. Thus on both sides of the projected mouth of the river are formed curved deposits which are concave towards the sea.

Landforms due to River Action

We thus find that we may divide the landforms built by river action into two major categories—erosional landforms and depositional landforms. Of the landforms due to erosion the following are noteworthy—gorge and canyon,. V-shaped valley, waterfalls and rapids, pot-holes ,structural benches, river terraces, and peneplain. Of these gorge and canyons, rapids and waterfalls, pot-holes and meanders have been discussed in this chapter and the remaining lanforms wil be discussed at the appropriate place in the three subsequent chapters. Of all the landforms carved by river erosion the most basic and by far the most important is the river valley and the other featrues are secondary. The development of the river valley is, therefore, discussed at some length in the next chapter. The end product of river erosion is the development of the peneplain or a featureless lowlying surface which is analysed in the chapter on the cycle of erosion.

Landforms resulting from river deposition include the following—alluvial fans, alluvial cones, natural levees, floodplains, ox-bow lakes, mid-stream deposits and deltas. All these features have been described in this chapter. Just as the river valley is the basic feature produced by river erosion,so also the flood-plain (including the delta) is the most fundamental feature produced by river deposition, and the remaining features are secondary. In the lower part of its mountainous course the river builds alluvial cones and fans by deposition, in its middle course it builds an extensive flood-plain, while in its lower course depositional action results in the formation of flat deltaic plains. All these depositional features are of great human significance, as they provide fertile land for cultivation and from the earliert times have provided the most attractive sites for human settlements.

Questions

1. Write a concise account of the work which a river does in the different parts of its course.
2. Give an account of *either* the erosional *or* the depositional landforms produced by river action.
3. Explain the formation of the different types of waterfalls, giving examples Why do waterfalls recede?
4. How are deltas formed? Classify deltas according to their shape and give examples.
5. Write short notes on :—Canyons, Alluvial fans, Retreat of waterfalls, Hanging valley falls, Natural levees and Flood plains, Load of a river, Overland flow. River regimes, The transporting power of a river.
6. Discuss the mechanism of fluvial erosion and processes.

Chapter 17

River Valley Development and Channel Morphology

The evolution of the river valley may be studied under two broad heads—(a) the development of the valley of a single river, and (b) the evolution of the drainage system of a region in which several river valleys develop as a unified system. In nature, these processes are co-existant and cannot be separated from each other. But for the sake of convenience we shall isolate the two, and study first the evolution of a single river valley in this chapter and analyse the evolution of the drainage system in the next chapter.

We have already seen the landforms built by the river in its upper, middle and lower courses. We shall now turn our attention to the development of the river valley proper with special reference to the valley form. We shall, therefore, proceed to examine the cross-profile and the longitudinal profile of the river valleys, as it is these two elements which really determine the form of the valley.

The Cross-Profile of River Valleys

By the cross-profile of a river valley is meant the bank to bank profile across the width of a river. It represents the outline of the river bed between its two banks. The crossprofile thus gives us an idea of the width of the valley and its varying depth. Two kinds of processes are involved in the development of the river valley—valley deepening and valley widening. In the initial stage of valley formation the river principally cuts downwards, i.e. the river deepens its bed by vertical corrasion or down-cutting. Where this process is dominant, the valley is deep and narrow like a gorge. But under normal climatic conditions,

Fig. 92—Valley widening results in the removal of a much larger amount of material than valley deepening

after a period of time the rocks on either banks of the river start breaking by weathering, rain-wash, gullying and slipping and then the mouth of the gorge starts opening up and the valley of the river becomes V-shaped. It is clear from Fig. 92 that the river has to cut and remove a much larger amount of material in widening its valley than in deepening it. Therefore in the initial stage the river only deepens its valley by down cutting. At this stage the amount of water and the load of the river are both small and tributary streams are either absent or in their formative stage. Therefore the river is incapable of cutting its banks to any appreciable extent. In other words, lateral corrasion is either absent or nominal and the river only deepens its valley by vertical corrasion. The vertical corrasion is performed mainly with the help of angular fragments of rock in the river which on account of the velocity of river water start revolving in an eddy and act like drillers. This results in the formation of circular depressions in the river bed which are called potholes. Besides pot-hole drilling, hydraulic action and in special circumstances solution of rocks also assist in valley-deepening. On account of the preponderance of vertical corrasion in the early stage young rivers flow through narrow valleys with steep sides which resemble a gorge or a narrow V. The chief agents of valley widening are weathering and rainwash, as under their influence the side rocks disintegrate and fall into the river. In course of time, the volume of water in the river as well as the load of the river also increases, and then the river is in a position of cut its banks. With further passage of time a stage is reached when down-cutting becomes relatively slow, but the processes of lateral corrasion and valley widening continue and become more important. Thus the width of the valley goes on increasing, and in mature stage the river valley becomes open and wide and relatively shallow. We thus find that the cross-profile of the river is indicative of the stage of its development. With age there is a gradual and systematic change in the form of the valley, and in the long run the process of valley

Fig. 93—The process of Valley widening

widening becomes far more important than the process of valley deepening (Fig.94).

We have to, however, keep in mind that even in the initial stage when the river is primarily engaged in deepening its valley by vertical corrasion, lateral corrasion is not totally absent. When the river initially carves its course following the slope of the land and the irregularities on the surface, even at that initial stage the course of the river frequently tends to be winding and is rarely straight. Where the river bends, the curve of the bend tends to increase by lateral erosion, because the currents are faster on the concave side of the valley and as they impinge, there is lateral erosion and cutting of the lower portion of the banks. Thus by the process of differential erosion on two sides of the valley, in course of time a chain of meanders with inter-locking spurs projecting into the valley at alternate intervals, comes into existence.

Fig. 94—Development of the river valley with age.

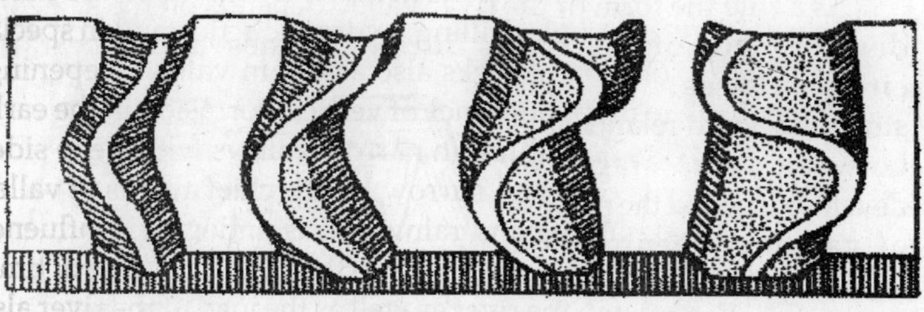

Fig. 95—Valley widening by lateral erosion and down-valley migration of meanders.

Thus *inter-locking or the overlapping spurs* are the projecting portions of rocks on the convex side of the meander which are not cut by the river, as on this side the velocity of the river is less and there is little erosion. But weathering and rainwash continue to be active on the convex side with the result that rock fragments break off from the spur and collect at its bottom. On this bank, therefore, the slope of the spur becomes gentle and is known as *slip-off slope*. On the other hand, on the concave side of the meander erosion is rapid and consequently the slope is steep and *cliff-like*. This is called *river-cliff*. By lateral erosion the concave side is thus pushed further away, and on the convex side as a result of deposition of rock fragments from the spur a relatively flat surface is built up below the spur which is covered by water only when the river is in flood. Simultaneously, as the spur is also being eroded, and ultimately when the spur is destroyed and levelled down, the river flows through a broad and relatively flat valley, with steep cliff on both sides. The flat valley floor is really the flood-plain of the river. When this is first built, it is limited to the meander belt of the river. But with the passage of time, the size of the meander increases and the lateral cliffs also start getting eroded. In this fashion the floodplain widens and ultimately becomes wider than the meander belt. With the removal of the obstruction placed by the spurs, the meanders begin sliding downwards and down-valley migration of meanders begins (Fig. 95). Lateral erosion thus increases and the floor of the river valley becomes in the process wider and flatter. The flood-plain built by this process is thus an erosional feature, though there is also alluvial deposition. This is different from the wider flood plain built by the river in its middle and lower courses which is essentially a depositional flood-plain.

As a rule the form of the river valley depends on the stage of development of the river. But sometimes under certain circumstances, the straightforward relationship between the stage of development and the process of valley deepening and widening is disturbed and does not operate. For instance, in the arid regions the process of valley widening is extremely slow, because in the absence of rainfall lateral cutting by weathering and rainwash is non-existant. Under such conditions, the youthful character of the river valley is maintained intact

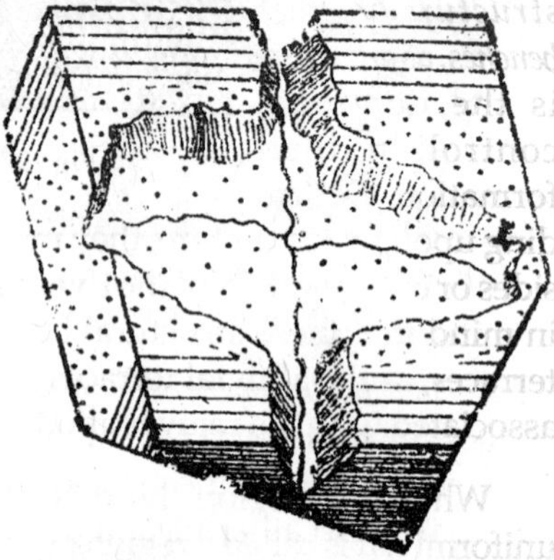

Fig. 96—Influence of rocks on Valley form.

for a long time. A good example of this is provided by the Grand Canyon of the Colorado river in the U.S.A. There is rainfall in the source region of the river, but it passes through the Arizona desert before it meets the sea, and it is during its journey through the desert that this canyon, one mile deep and 300 miles long, is found.

The valley form is also influenced by the nature and structure of rocks. In the areas of soluble limestone rocks, the form of the valley tends to become rounded instead of being V-shaped, as together with corrasion solution also becomes important. Where the river passes through an area of hard and more resistant rocks, the valley is narrow and has steep sides, but in areas of soft rocks, the valley becomes more open and wide. In other words, in the same river valley we come across portions which are open with gently sloping sides or narrow with steeply sloping sides, depending upon the nature of rocks (Fig. 96)

Similarly in the young stage of a river, the cross-profile of a river varies on account of the differential erosion of the hard and the soft rocks. Where the hard and soft rocks lie in alternate horizontal layers in the valley floor, the valley sides become stair-like, as the soft rocks are eroded easily while the hard rocks are eroded less and project as benches. These are called *Structural benches or structural rock benches,* as structure is the dominant control in their formation. Depending upon rock structure, they may be found locally either on both sides or only one side of the river valley (Fig. 97). It should be borne in mind that these structural benches are different from alluvial terraces, as the alluvial terraces are features which are essentially associated with the rejuvenation of rivers.

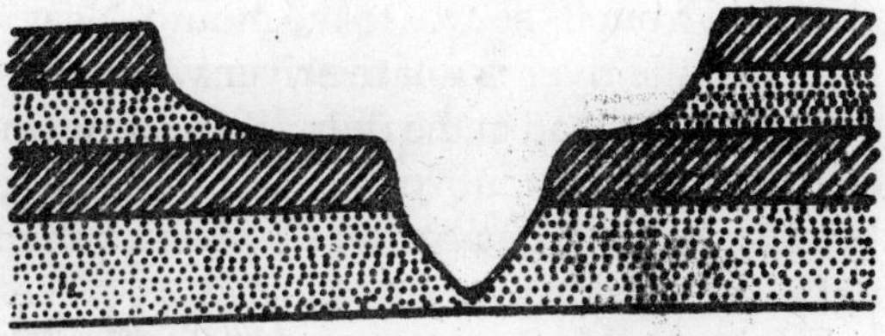

Fig. 97—Structural rock benches.

When in a region the rock strata dip in the same direction uniformly, it is called a region of unilclinal or homoclinal structure. Where the dip of the rocks cuts across the direction of the river i.e. where the river follows the line of the geological strike, the river cuts not only vertically but also laterally on one side and thus slides sideways on account of the dip of the strata. Thus while one side of the valley becomes gently sloping, the other side continues to remain steep-sided, and the cross-profile of the river becomes asymmetrical. This kind of shifting of the river valley in response to the dip of the rocks, is called *uniclinal or homoclinal shifting* (Fig.98).

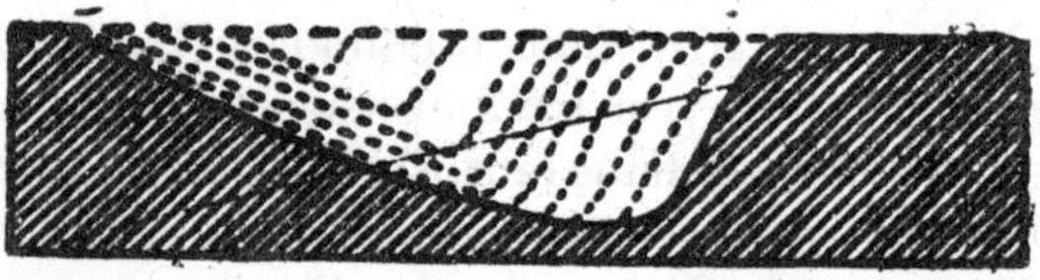

Fig. 98—Uniclinal or Homoclinal Shifting of the river valley.

We have seen earlier that asymmetrical cross-profile of the river valley may also be due to differential erosion of the valley sides where there are meanders. We may thus conclude by saying that for various reasons, departures from the ideal crossprofile are usually found and the cross-profile of a river valley is often found to be asymmetrical.

The Long-Profile of River Valleys

By the long profile of a river valley we mean the profile of the river bed from its source to the mouth. Near its source the erosive power of the river is relatively small because both the volume of water and the load in the river are small. Then in the lower reaches also i.e. near the mouth of the river the erosive power is again low, because the river cannot cut below sea level, and further on account of the very gentle gradient the velocity of the river is low while its load is heavy. Maximum erosion, therefore, takes place in the region mid-way between the source and the mouth. The result is that quite early in the life of a river, its course is marked by a rough concave profile of the river bed.

In the initial stages of down-cutting by the river, irregularities of surface are found in the river bed resulting in the formation of rapids, falls and other related features. These irregularities may be either related to the original irregularities of the surface or they may be caused by differential erosion of rocks of different hardness in the river's bed. But these irregularities in the bed of the river are temporary, and as the river erosion proceeds, the irregularities are destroyed and levelled down. The irregularities which result from differential downward corrasion are relatively more important and persist for a longer duration of time. But even these are destroyed in the long run. Striking examples of features caused by such differential erosion of hard and soft rocks are provided by rapids and water-falls which have been discussed earlier. We have seen how the waterfalls recede backward and there is formation of gorge in front of the fall, a good example of which is provided by the Niagara Falls on the U.S.A. - Canada boundary. The fall is receding at the rate of 1 to 5 ft. per year and the present length of the gorge in its front is nearly 7 miles. If the rate of recession be taken as 5 ft per year, then it must have taken 7,000 years for it to recede seven miles, and very likely the actual time taken has been much more.

Another kind of irregularity found in the course of a river in its young age is the existence of shallow lakes, but slowly their mouths are cut and levelled down and ultimately they disappear.

Graded Curve or Profile of Equilibrium

If there is no interruption of any kind by tectonic movements, then in course of time the river destroys all the irregularities in its bed and develops a smooth concave curve from its mouth to the source which is called *the curve of water-erosion* or the *graded curve of water-erosion* (Fig.99). This stage in the life of a river is reached when the river has degraded its bed to the maximum and no further deepening is possible This is that state of equilibrium when deposition equals corrasion and at every point in its course the strength of the river is just sufficient to enable it to carry its load. In other words, when a balance is reached between the transportation capacity of the river and its load, then the river is said to be *graded or at grade*. Thus when a smooth concave curve develops in the river-bed from the source to the sea, then this is called the graded curve. In reality, grade is a state of balance towards which the river constantly tends and in practice may come very close to this state. The longitudinal or long profile of a river which has reached the graded stage is called the *profile of equilibrium*. This concept owes its origin to the famous American geologist and geographer William Morris Davis (1902) and is considered to be fundamental concept in the study of landform carved by river erosion. According to Davis this stage is indicative of the beginning of the mature stage of a river. This concept has been criticized by several people, the most prominent of these being G.H. Dury (1966), according to whom it does not in any way help in the analysis of landscape, But there are others who have supported his concept, the most notable of these being Mackin (1948). According to Mackin, graded river is one in which in a period of time the slope of the river becomes such that

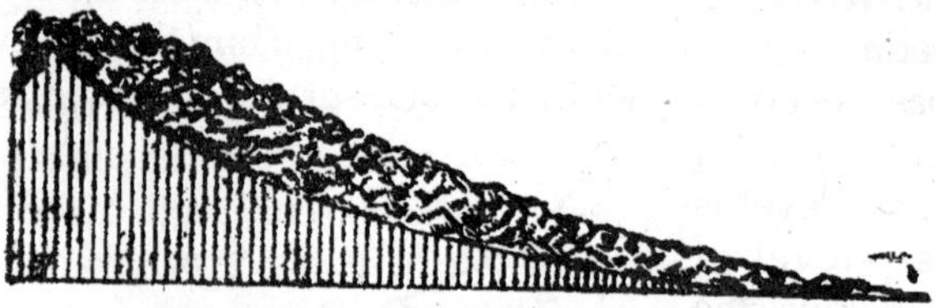

Fig. 99—Curve of water erosion

the velocity of the river is just sufficient for the transportation of its load. The graded river is a system in equilibrium. Any change in the factors controlling it (velocity, load, volume of water, slope) disturbs the state of equilibrium, but soon a process starts that tends to absorb the effect of change and the disturbed equilibrium is re-established.[1]

Three variations or types of profile of equilibrium have been identified. One is the ideal concave profile, typical of relatively small rivers. The second is the concave profile broken by gentle steps characteristic of large rivers with tributaries and flowing over homogenous rocks, such as the Ganga. Below the cofluence of each tributary stream, the discharge and competence of the main river increase resulting in increased erosive capacity which steepens the gradient of the profile of the main river below the confluence and stepped profile results. The third is the Penckian or Treppen type, marked by steep falls at the knick-points making the junction of higher pediment with a lower and younger pediment or occasioned by subsequent uplift and faulting of the old erosion surface. The profile or thalweg of the principal rivers of the Chota Nagpur plateau are of the latter type.

1. "A graded stream is one in which over a period of years, slope is delicately adjusted to provide, with available discharge and with prevailing channel characteristics, just the velocity required for transportation of the load supplied from the drainage basin. The graded stream is a system in equilibrium; its diagnostic characteristic is that any change in any of the controlling factors will cause a displacement of the equilibrium in a direction that will tend to absorb the effect of the change. "Mackin, J.H. Quoted by Thornbury, W.D.-op. cit, p. 107

Base-level of Erosion

There is a limit up to which a river can cut and degrade its channel. This limit is known as the *base-level of erosion. The* base level concept was first proposed and elucidated by Powell in 1875. Since then it has been generally accepted that there is a lower limit of deepening of a river valley. According to him the sea level is the grand base-level below which the land surface cannot be degraded. For rivers which join the sea, sea-level is the base -level of erosion, because the river valley cannot be cut deeper than the sea-level. But for rivers which do not fall into the sea but join other rivers, the level of confluence is the base-level of erosion. Similarly, for rivers which drain into a lake, the level of the lake is the base level of erosion.

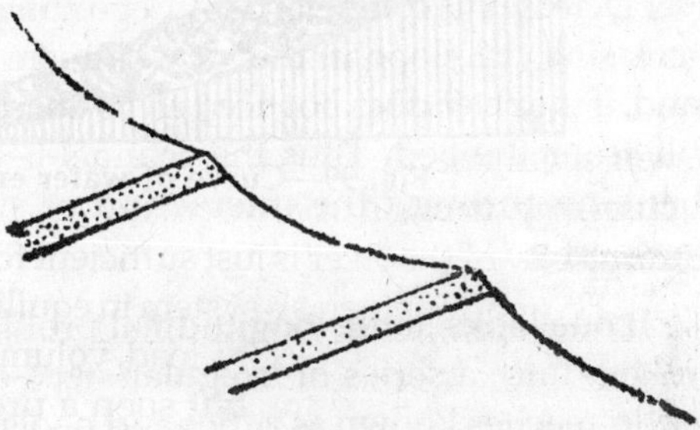

Fig.100 —Effect of local base-levels on the evolution of graded curve.

Besides the sea-level, it is possible for the river to have local or temporary base levels in its course. The grading of the river bed from its source to the mouth takes a long time. But before this, establishment of grade is possible in portions of the river's course under the influence of local base-levels (Fig 100). Where an outcrop of hard strata is found in the river course, every such hard layer acts like a local base-level of erosion. In other words, the slope of the river bed above the rock is determined with reference to that hard layer of rock. In course of time, these local base-levels are destroyed and then a graded concave curve is established in the entire course of the river. In any case, it is important to remember that the development of the graded curve starts from the mouth of the river i.e. from the base-level of the sea, as the vertical down-cutting is the least here, and with the passage of time the curve is extended upwards until the entire curve becomes graded.

Channel-bed Topography

The river tries to maintain a channel morphology which is most efficient for the transportation of the load supplied to it. When there is an imbalance between sediment input and output through a stretch, the channel morphology has to be adjusted to correct the disequilibrium. Stream capture, headward erosion, large scale deforestation of catchment, deltaic growth, meandering etc. may bring about changes in the river profile. However, adjustment is usually made by the river in terms of scour or fill in the channel bed to lower or raise its gradient. Where the load cannot be moved, the river increases its competence by depositing the material and thereby increasing the slope in the downstream direction. On the other hand, if the river does not need all its energy to transport its load, it will scour the bed. Thus the channel-bed topography keeps on changing through the interaction of stream flow and debris transport.

If one looks at the longitudinal profile of a river in detail it is broken into a series of irregular steps of alternating steep and gentle reaches known as riffles and pools respectively. A *riffle* is a topographic high area produced by the accumulation of relatively coarse sediment, and a *pool* is a topographic low, usually characterized by finer material. Variations of width, depth and velocity along a channel occur because of pool/riffle sequence. Riffles are usually wider and shallower than pools. The velocity in pools is less than that in riffle. Pools produce an asymmetrical cross-section while riffles have a more symmetrical cross section. The existence of pools and riffles in both straight and winding channels suggests that such irregularities in the channel-bed are a kind of self-adjusting mechanism to balance energy expenditure along a stream.

Channel Pattern

The channel pattern is the map view of a river. The channel pattern is usually classified as straight, braided and meandering (Leopold and Wolman, 1957). Two other types have also been recognized—sinuous and anastomosing. Sinuous pattern is intermediate between straight and meandering. The term anastomosing has been proposed for braided patterns where the islands dividing the channel are stable (Schumm, 1968). These various patterns have been shown in Fig. 101.

Leopold and Wolman based their classification on two criteria, namely sinousity and channel multiplicity. They defined sinousity index (SI) as Channel (thalweg) length ÷ Valley length for a given reach. Brice (1964) defined SI as Channel length ÷ Length of meander axis belt, and used the sinousity index to separate straight from sinuous and meandering channels. If the sinousity is less than 1.05, the channel is straight, an SI between 1.05 to 1.5 means a sinuous channel and when SI is greater than 1.5, the pattern is meandering. *Meandering* is believed to result from the hydraulics of flow and the requirements of capacity and competence to carry the load supplied from upstream. Pools, riffles and asymmetrical crosssection at bends are characteristic of meandering streams. Meander patterns change and the bends migrate over time.

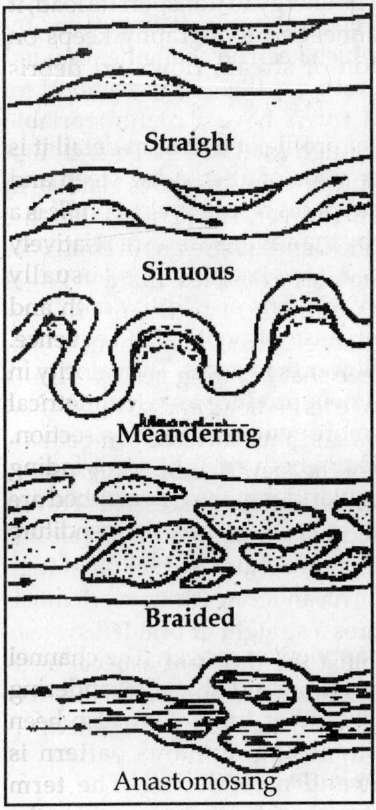

Fig. 101—Types of Channel Patterns.

The cause of meandering is not fully understood. Three types of hypotheses have been proposed:

(1) Meandering results from the hydraulics of flow;

(2) Meandering results from the requirements of capacity and competence to carry the load supplied from upstream,

(3) Meandering represents a least work tendency and is an attempt to balance more energy dissipation by friction at a curve with less dissipation over the pool. It has been suggested that "meanders, pools and riffles are just aspects of the total river morphology that are adjusted in the effort to carry the load under the environmental constraints imposed on the river. Meandering results from over competence or lack of bedload. The river meanders to decrease slope and velocity and increase resistance to flow." (Morisawa, 1985, p. 103).

S.A. Schumm has distinguished between braided and anastomosing patterns on the basis of stability. A *braided pattern* is one where the various divided channelways are always shifting. An *anastomosing pattern* is one where there are many channels but they are stable and retain their identities with changing discharge and time. Both types are characterized by many channelways separated by bars or islands. Braided rivers are most typical of alluvial fans and deltaic regions and outwash. Braiding has been attributed to many causes such as erodible banks, steep river gradient, abundant and coarse bedload, and a lack of river competence. Morphology of braided pattern is distinguished by instability of bars and channels which keep on changing. Braided rivers have a high width/depth ratio i.e. their cross-sections are wide and shallow. The braided rivers have a highly variable discharge with high flood peaks, high total discharge and high monthly variability. The Brahamputra a few km above Gauhati in Assam is a good example of a braided river. The Kosi in Bihar, the Ganga below its confluence with the Gandak, streams draining the slopes of the Siwaliks are other notable examples. Braided rivers carry a heavy bedload. The Brahamputra carries a load of 7 million tons/day during its peak discharge period (Coleman, 1969). Knighton (1972) and others believe that braiding occurs when a river is locally incompetent to move its bedload and so deposition takes place to increase the gradient and thus increase competence. Leopold and Wolman (1957) are of the view that braiding results from lack of competence and is not indicative of excessive bedload since aggradation can take place at constant slope without braiding. This view is in agreement with that of Schumm (1968) that high suspended loads are transported in meandering (sinuous) channels whereas bedload transport requires a straight or braided system. The river continually tries to adjust the different aspects of its morphology to changing environment and hyraulics, and there is a continuum of patterns from straight to meandering to braided.[1]

Effects of Rejuvenation on Valley Form

By rejuvenation we mean that state in which for any reason there is change in slope resulting in an increase in the erosive capacity of the river. This happens primarily either when there is

1. This section is based on Morisawa, M.—*Rivers*, Longman Group, London and New York, 1985, pp. 90-114.

a fall in sea-level or there is an upliftment of the land. Since sea-level is the ultimate base level of erosion by rivers, there is change in the base-level of erosion whenever there is a change in the sea-level either by rise or fall. In such a situation, the river starts the regarding of its course in relation to the new base-level, and starts building a new curve of water erosion in place of the old one. With a fall in the sea-level, the base level is also depressed, and the river by its increased power of erosion, starts building a new curve in conformity with the new base level. As a result of fall in sea-level the slope near the mouth of the river increases and vertical down-cutting begins and is extended upwards to the top, and in this manner the new curve of erosion from the mouth to the source begins to be refashioned, starting from the mouth of the river. At any point of time, where the new curve cuts across the old curve, there is a distinct break in the slope. This point is called *rejuvenation-head* or *knick-point*. The knick-point tends to migrate towards the river source. The migration of the knick-point towards the river source is, however dependent upon the nature of rocks, and this process can be extremely slow in areas of hard rocks. In course of time when the river is able to carve the new curve from its mouth to the source in accordance with the new base-level and balance is achieved between its transporting power and its load, then the river is said to be *regraded* and its curve of erosion is called the *regraded curve or regraded long-profile*.

Knick-points can be identified in the course of a river by the presence of rapids or waterfalls, especially in places where hard rocks are present. A good example is provided by the Fall-line or Fall-zone of North-America. After the uplift of the Atlantic coastal plain, the pre-existing rivers started regrading their courses, and their knick-points are found in the areas of hard rocks of the Appalachian mts., where the rivers descend into waterfalls from the old surface to the new surface of the coastal plain.

Sometimes an examination of the long-profile of a river reveals the presence of several knick-points which shows that rejuvenation has taken place more than once. This kind of longitudinal or long-profile in which there are distinct breaks in slope, is called *composite profile* and this kind of valley form is described as *composite valley form* (Fig. 102). In such valleys, together with the composite long profile, composite cross-profile is also found and besides the sudden break in slope in the long profile, the cross-profiles show distinct valley-side facets such as river terraces, which are separated by breaks of slope.[1]

In such river valleys when we move from the mouth towards the source, we come across a new valley below the knick-point and portions of the old valley above it, and as we move upwards, the slope of the river sides becomes gentler. The form of the upper valley is one of mature or old age, while the form of the lower portion of the valley is youthful. In course of time, as a result of the increase in down-cutting in the bed of the old valley, the new valley is extended upwards, and in this fashion there is the development of *valley in valley or two-story valley or two -cycle valley.* The chief characteristic of the cross-profile of such a river valley is that on both sides of the valley, a terrace-like relatively flat surface is found between the old valley above and the new valley below. This is known as *river terrace.* River terraces are of two kinds—alluvial terraces and bedrock terraces. The difference between them is that while the alluvial terraces are wholly made of alluvium, in the bedrock terraces the rocky surface is covered with a very thin layer of alluvium. With repetition of this process by subsequent rejuventations, several terraces at different levels are built up in the river valley. These river or bed rock terraces are wholly different from the structural benches or terraces which are formed due to the presence of strata of hard rocks. The bed-rock terraces, on the other hand, are remnants of the old base level of erosion and have no relationship with rock structure. An important thing to remember is that the river terraces are formed by erosion and not by deposition.

Fig. 102—Composite Valley form.

The other effect of rejuvenation is to be seen in the formation of *incised or inclosed meanders.* When within a normal meander, a new

1. "The composite long profile of rivers can frequently be correlated with composite tranverse profiles showing a series of distinct valleyside facets, seperated by breaks of slope." Wooldridge, S.W. and Morgan, R.S.., Op., cit., p. 221

narrow and deep meander is formed by vertical down-cutting as a result of rejuvenation of the river, such meanders are called incised or inclosed meanders. Incised meanders are of two types— (i) *Intrenched or entreched meanders* where at the place of the meander the form and slope of both sides of the valley are similar or identical, and (ii) *Ingrown meanders* where the cross - profile of the valley is asymmetrical. In the first type of incised meanders the crossprofile is symmetrical, and the new meander is formed by rapid down-cutting in the old meander. But in the second type i.e. in ingrown meanders, the down-cutting is not so rapid and vigorous and the meander gets enough time for lateral erosion, with the result that on the concave side there is development of *undercut slopes* or vertical bank by lateral cutting and on the convex side there is development of *slip-off* slope or a gently sloping extended slope on account of deposition. To a very large extent it would be true to say that where the rate of uplift is high, there is development of intrenched meanders, and where the uplift is either slow or gradual, ingrown meanders develop. But the form of the meanders is also considerably influenced by the nature of the rocks, and sometimes both the types of meanders may be found in the same valley within a relatively short distance. Where the bed-rocks are hard, the rate of down-cutting is slow and the river gets time for lateral shifting, and conditions are more favourable for the formation of ingrown meanders. On the contrary, where the rocks along the valley sides are hard and consequently lateral cutting is slow, vertical down-cutting is more active and intrenched meanders are likely to be formed[1]. But in reality the problem of the formation of these two types of meanders is extremely complex, and it is not clear why in one area together with down-cutting there is also lateral erosion, while in another area there is only vertical down- cutting.

---

1. Moore (1926) in discussing the inclosed (incised) meanders of the Colorado plateau concluded that whether entrenched or ingrown meanders develop depends largely upon the character of the inclosing rock and the amount of stream load. He found entrenched meanders in hardrock areas and ingrown meanders or open valleys in areas of weak rock. He attached considerable significance to whether the stream carried a large or small load and concluded that heavily loaded streams are more likely to cut laterally as they incise their valleys than those with lesser loads and would thereby develop ingrown meander rather than enrenched. Thornbury, WD.—op. cit., p. 143

Questions

1. How does a river develop its valley and what are the factors influencing this development ?
2. Give a reasoned account of the evolution of *either* the cross-profile or the long-profile of a river valley. Draw diagrams wherever necessary.
3. Discuss the influence of rejuvenation on the form of river valley.
4. Write short notes on the following :-

   Base-level of erosion, Profile of equilibrium or the graded curve of water erosion, Structural rock benches and river terraces, incised meanders, River channel patterns.
5. Trace the evolution of the long profile of river valleys, bringing out clearly what you understand by the base level of erosion and the profile of equilibrium.
6. Explain Channel- bed morphology and channel-pattern.

Chapter **18**

River Systems and Drainage Patterns

We have considered in the previous chapter the changes which a river brings about in the form and size of its vailey. But in a region there are usually more than one river and then each river has a number of tributaries. It is important to study their mutual interrelationship and organization together with the changes they produce in the landscape.

River System

A number of rivers join together to form a river or a drainage system. Genetically it is possible to recognize two distinct types of drainage systems— the *consequent* and the *inconsequent*, depending upon whether the rivers show adjustment to structure or lack of adjustment to structure. In the consequent drainage system we find accordance or conformity between structure and the development of drainage, giving rise to a whole range of drainage patterns which are dependent on the geological structure of the region and the resistance of the rocks to erosion. On the other hand, the inconsequent drainage system is characterized by discordance between structure and drainage in which the rivers are not fully adjusted to the geological structure and topography and show no relationship to the present structure of the rocks over which they flow.

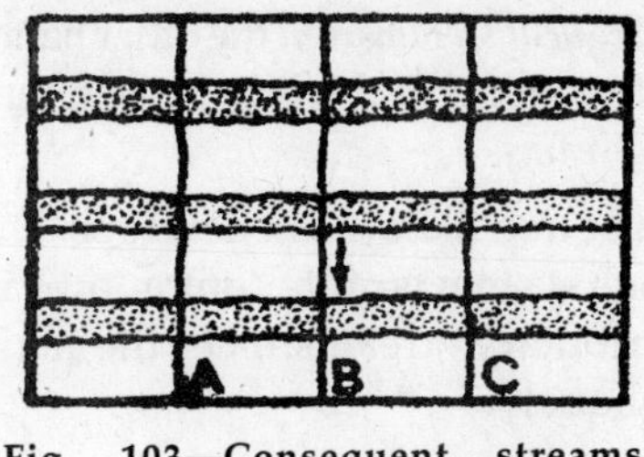

Fig. 103—Consequent streams. A,B,C, are consequent streams which follow the original slope.

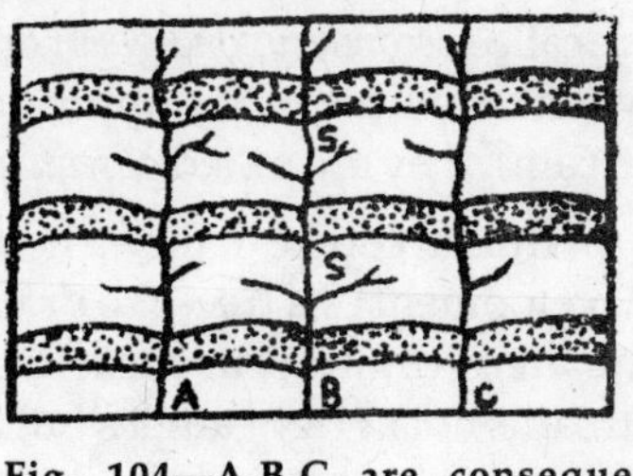

Fig. 104—A,B,C are consequent streams, and S, S are subsequent streams

Consequent Drainage

River systems that develop as a direct result of, and in harmony with the pre-existing land surface are called consequent. On a new land surface the streams start flowing in the direction of the initial slope. If the landmass is upraised in the middle and is sloping on two sides, then on each side a number of parallel streams will start flowing down the slope. The streams which follow the slope are called consequent streams and the system of such streams is collectively described as a consequent drainage system. J.W. Powell (1875) was the first to use the term consequent to denote a river whose course was determined by the original slope of the land surface. Based on this a genetic classification of streams was later on developed by W.M. Davis (1894-95). The higher land separating the rivers on two sides is called the *watershed* or *water-parting* or *divide*. After sometime when the rivers have excavated their valleys, tributary streams following a zone of weakness (such as joint, fault or easily erodible rock layer) develop and join the main river from both sides. These streams are called *subsequent streams* (so named by J.B. Jukes in 1862), as they develop subsequent to the development of the main consequent streams (Fig. 104). As the consequent streams flow in the direction of the dip, they are also known as *dip streams*, while the subsequent streams flow in the direction of the strike and are hence called *strike streams*. Since the softer rocks are more easily eroded than harder rocks, the tributary streams are formed in the areas of softer rocks, while the harder rocks form divides between the subsequent river valleys. After some time even the subsequent streams come to have tributaries of their own which meet the former from both sides. These tributaries follow either the direction of the first consequent streams or flow in just the opposite direction. When their course is in the direction of the first consequent stream they are called *secondary consequent* or *resequent streams*. On the other hand, when they flow in the opposite direction (of the first consequent stream), they are called *obsequent streams*.

Thus in course of time a netted or tree-like system develops. The development of a tree-like drainage system is quite common, as in the areas of uniform rocks the tributary streams meet the main streams not at right angles but obliquely.

Glock (1931)[1] divides drainage network development into two

1. Glock, W.S.—The development of drainage systems: a synoptic view. *Geog. Rev.* 21, 1931, pp.74-83

stages. In stage one there is initiation of channels, followed by headward growth and elaboration by addition of tributaries. After maximum development, there comes stage two when there is integration of the network. Loss of tributaries occurs by abstraction or piracy and tributary loss may also occur as the main river widens its flood-plain and meanders. Another important change in the network with growth is in the angle of junctions between two streams which decreases with time (Schumm, 1956). There is also a down-valley movement of junctions with development.

Migration of Divides and River Capture

In the initial stage of the development of river systems, the divide separating the two river systems is not very clear. But as the river system develops and the rivers cut back towards their source, the divide becomes more clearly marked. This type of erosion towards their source by the rivers is called *headward erosion*. One of the two rivers normally erodes more than the other, with the result that one slope of the divide is cut and eroded more rapidly than the other slope, and the divide tends to migrate towards the weaker stream which is able to erode less. The divide crest line thus recedes or moves further and further away from the more rapidly cutting stream. This process is called the *migration of divides* (Fig. 106).

Migration of the divides between two tributary streams flowing in opposite directions leads ultimately to the capture of the weaker stream by the stronger stream. This phenomenon is called *river capture or river piracy*, and can be easily understood with the help of Figs 105 to 108. One tributary stream erodes headward so rapidly and the divide is pushed back with such speed that it is ultimately destroyed, and the stream joins the tributary stream on the other side of the divide and captures its flow. The tributary stream which contains more water or flows over a steeper slope and along weak rock belts, is usually the stream which captures the other stream by more vigorous headward erosion. As the more rapidly cutting stream has a deeper valley, the entire water of the other stream starts flowing into the former. This is called *river capture or river piracy*. This has been shown in Fig. 107. The river which captures the flow of the other river is called the *pirate river*, and the stream whose flow is captured and diverted is called *the beheaded stream*. In the beheaded stream the volume of water below the point of capture naturally becomes less and it now flows as a much smaller stream and its valley looks much bigger than the amount of water it carries. The shrunken river thus does not quite fit into its valley and is, therefore, called a *misfit* river. Sometimes the volume of water in the misfit river is so much reduced that its source migrates downwards away

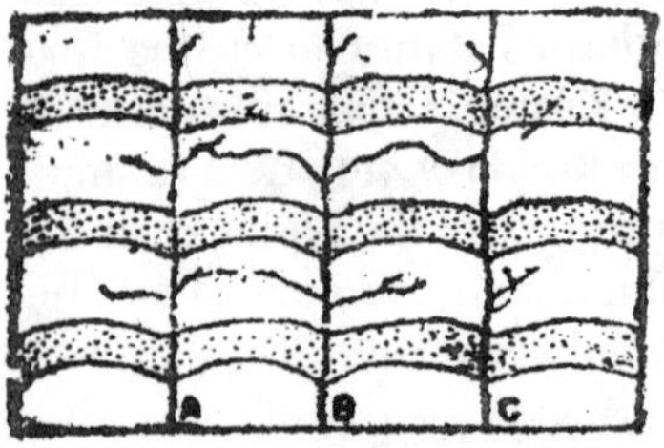

Fig. 105—The situation just before river capture.

Fig. 106—Migration of divide.

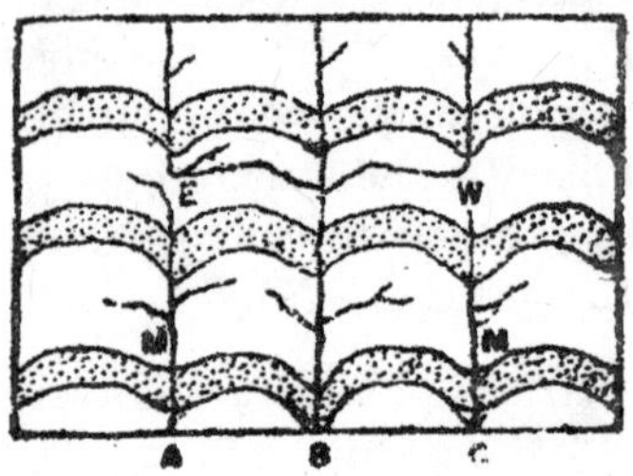

Fig. 107—River capture. E=Elbow of capture, W=Wind gap, M=Misfit

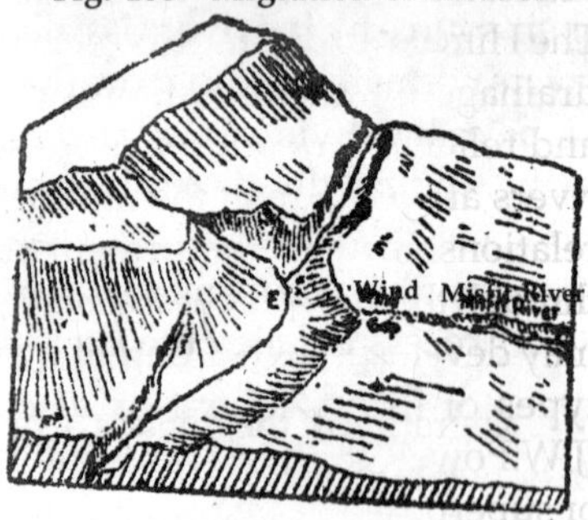

Fig. 108—Elbow of capture, wind gap and misfit river. E=Elbow of Capture.

from the point of capture and the intervening portion of the valley above its new source becomes dry and is called *wind-gap or dry gap*. There is a point at which the course of the beheaded stream suddenly changes on account of the capture, and this sharp bend in the course of the river is called *the elbow of capture* (see Fig. 108) The sudden diversion of a greatly increased volume of water through the capturing system causes rapid entrenchment usually with a waterfall or rapids that migrate rapidly up the captured segment from the point of initial capture.

Many of the Himalayan rivers, in their higher courses exhibit the phenomena of river-capture. This has happened through the rapid headward erosion of their main transverse streams, capturing or beheading successively the secondary laterals belonging to the Tibetan drainage system on the northern slopes of the Himalayas, the rivers being greatly helped in this task by accelerated erosion on account of the successive uplifts. The Kosi river has captured the waters of the Arun river which originates to the north of the Himalayan watershed. Other examples of river-capture are furnished by the Bhagirathi and other tributaries of the Ganga, the Tista river in Sikkim and the Sind river in Kashmir. Examples of river capture are also found in the Son Valley region. A very clear example river capture can be seen where a tributary of the Banas river has captured the head waters of the Kushmahar, a tributary of the Son.

Inconsequent Drainage

As already pointed out, inconsequent drainage is one which is unrelated to the structure of the rocks over which the rivers flow. The Himalayan drainage system is in a large measure inconsequent drainage, i.e. its formation is not consequent upon the structure and relief of the country as we see them. Many of the Himalayan rivers are older than the mountains they traverse. They show no relationship with the present structure and relief and cut right through the line of highest elevations. Lack of adjustment to structure may develop in two different ways, and accordingly two principal types of inconsequent or discordant drainage are recognised (J.W.Powell,1857)—(a) antecedent drainage, and (b) superimposed or superposed drainage.

Antecedent Drainage

If the river is able to maintain its course as a mountain range is elevated across its path, it is called antecedent. Antecendent drainage is thus a form of discordant drainage in which the stream existed there before the mountain began to be uplifted. The rate of uplift is sufficiently slow so that the streams are able to downcut their beds at approximately the same or accelerated rate and more or less keep pace with the uplift and so maintain their course across the tectonic ridges. Such a drainage system where streams have maintained their valleys across tectonic mountains and cut across them in deep gorges, are said to be antecedent, because the river is older than or antecedent to the deformation responsible for the present structure and relief. The antecedent drainage is mostly found in regions of recent mountain movements. Since the rivers cut through the high mountains which rise above the level of their source, it is clear that the uplift in the middle of the course was much greater than at their sources.

The best examples of antecedent drainage are furnished by some of the great Himalayan rivers, more particularly the Brahamputra, the Gandak, the Kosi, the head waters of the Ganga (Bhagirathi and Alaknanda), the Indus and the Sutlej. All these rivers have characteristics which favour their pre-Himalayan origin. They have their sources in Tibet well to the north of the Himalayan mountains and flow for long distances in structural trough parallel to the main range. Then they suddenly change their direction, turn south towards the mountain ranges, cut across the high Himalayan ranges through deep and precipitous gorges and finally enter the northern plains of India. For instance, when the Indus river comes out of Kashmir near the Nanga Parbat, the walls of the gorge are about 5,200 m high above the river bed. It was possible for these rivers to cut across the high Himalayan ranges, only when it is assumed that they are older than the Himalayas. The rate of uplift of the Himalayas was slow. As the Himalayas were uplifted, the rivers continued to cut downwards and degrade their course at a faster rate, and were thus able to maintain their pre-existing channel.

In North America the Columbia river which passes through the Cascade mountains in a deep gorge is also an example of antecedent drainage. The transverse rivers of the Appalachian mountains present yet other example of antecedent drainage. They rise on the eastern slopes, cut across the Appalachians and flow westwards towards the Mississippi river. There is, however, some difference of opinion about their origin, and some scholars believe that they have been formed by river piracy or by superimposition. The antecedent rivers may have their tributaries, but the smaller tributary streams will not be able to downwcut as rapidly as the main stream. As a result they generally get seperated from the principal river. As the process of mountain building continues, they are either destroyed or sometimes may be traced in the form of hanging valleys.

Superimposed or Superposed Drainage

In this type of drainage too there is lack of correlation between the direction of river flow and the geological structure of the region. A superposed or superimposed drainage extends across a geological structure that antedates the river valley and which was buried beneath a cover of sedimentary rock, glacial drift or lave flow. The streams originated on the cover-mass as consequent streams and their courses were determined by the slope of the cover-mass. But in course of time when the upper cover of rocks has been removed by denudation, the river starts flowing on the underlying older rocks and cuts its valley into them. The underlying old rocks may be wholly different in structure, but by this time the organization of the river system is so well established that the rivers continue to flow in their old courses in spite of the change in the nature of rocks and the geological structure. This type of drainage is called *superimposed or superposed or superinduced drainage*, as the drainage pattern, as we see it today, has been laid over i.e. superimposed or superposed on the older rocks as an inheritance from the vanished cover.

An excellent example of superimposed drainage is provided by the Lake district in Britain. The older rocks which are now mostly found on the surface are folded. The rivers radiate in all directions from the centre, and bear no relationship with the disposition of the surface rocks. The overlying cover of relatively newer rocks which has been removed from the greater part of the region, is present only on the margins. At one time these newer rocks covered the entire area and formed a dome. The streams started from the domal top and flowed in all directions. Long period of erosion has removed the younger strata and the dome, but the directions of the rivers had been determined by them and as they have down-cut their valleys, they have continued to flow in their old courses and are incised into the older, more resistant rocks, with strong structural features trending across the river courses like barriers. In India, examples of superimposed drainage are provided by the Chambal and Banas rivers of Rajasthan. The course of the Chambal was determined when the Deccan lava extended from the Malwa region upto Kaimur and now this river cuts across the Vindhyan scarps, and unmindful of structure flows through a 32-km long gorge in the successively rising plateaus of Kaimur, Rewa and Bhander, before reaching the town of Kota.

The Appalachian rivers of U.S.A. are not interpreted as examples of superimposed drainage, where the rivers have cut their valleys into the foundation rocks but their pattern was largely determined by the sedimentary cover of rocks which has vanished except along the coastal plain. The present course of several of these rivers is the result of a combination of superimposition and antecedence, for which the term *anteposition* has been suggested.

Types of Drainage Pattern

Drainage pattern refers to the plan or design which an aggregate of rivers forms in a region. The classification of drainage pattern usually followed is by geometric form of the spatial arrangement of the streams and is non-genetic. But drainage patterns reflect the influence of structural control, inequalities in rock hardness, initial slopes, recent diastrophism and geomorphic history, and consequently are extremely useful in geomorphic analysis.

E.R. Zernitz (1932) summarized seven major types of drainage pattern and their significance—dendritic, trellis, rectangular, radial, annular, parallel and irregular[1]. In 1967 A.D. Howard compiled and classified drainage pattern into categories of basic pattern, modification to the basic pattern and varieties of the modified patterns. The basic drainage patterns follow regional structural trends and cover a wide range of scales. They are subjectively defined and are transitional from one to another[2]. His basic patterns include all the types of Zernitz except the last (i.e.irregular) plus two more, namely, multi-basinal (which is very similar to irregular) and contorted. The principal types of drainage pattern which are commonly recognized are described below:

Dendritic Drainage Pattern: This pattern is named after the Greek word 'Dendron' meaning tree, as it has a tree-like appearance. In areas of uniform rocks, the tributary streams (called insequent by Davis, as they follow primarily the regional slope) meet the principal ones not at right angles but obliquely so that the principal consequent stream resembles a big tree and the tributary streams look like its branches. As each insequent stream develops its own valley it receives in turn a second generation of tributaries. This branching tree-like drainage pattern has been called dendritic, the term being first used by I.C. Russell in 1898. This is the most common type of drainage pattern and is especially characteristic of regions having

---

1. Zernitz, E.R. (1932) : "Drainage Patterns and their significance", *Jour. Geo.* 40, 498-521.
2. Howard, A.D. (1967) : "Drainage Analysis in Geologic interpretation : A Summation". *Am. Assoc, Petrol, Geologists Bull.*, V. 51, pp. 2246-59

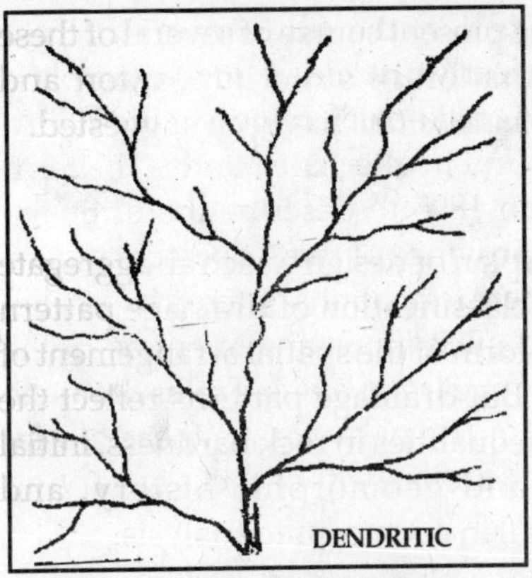

Fig. 109—Dendritic drainage pattern.

horizontally bedded sedimentary rocks as well as massive igneous rocks. In the Chota Nagpur plateau where the prevailing igneous and metamorphic rocks are almost equally resistant, we find the development of this type of drainage pattern. Modified dendritic drainage has been described as pinnate, sub-parallel and anastomatic. The pinnate pattern is feather-like, with tributaries almost parallel to and joining the principal stream at acute angles. In the sub-parallel dendritic type, the acute angle formed at the junction of the principal stream and tributaries is so small that the major and secondary streams look nearly parallel. The anastomatic pattern is characteristic of flood plains and deltas and consists of braided streams separated by channel bars and ox-bow lakes.

Trellised Drainage Pattern : In the trellised drainage pattern the rivers form a net-like system and the tributary streams of each class flow roughly parallel to each other. This pattern of drainage develops in areas where the hard and soft rocks occur in roughly parallel bands. The principal rivers follow the slope and flow in the direction of the dip, and the tributary streams flow in the direction of the strike and meet the principal streams at right angles. The secondary tributaries are commonly elongated, parallel to the master stream and join the primary

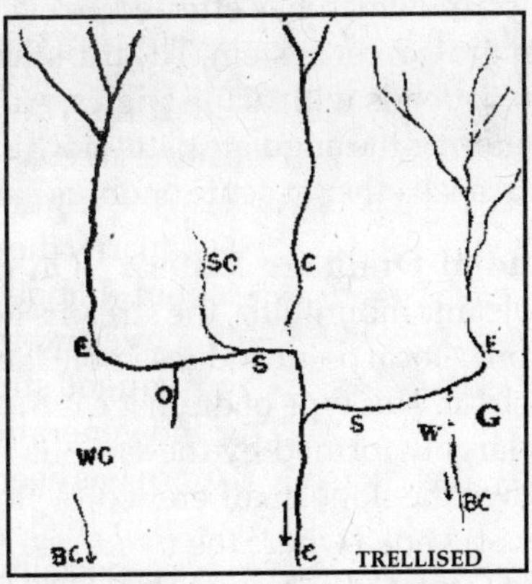

Fig. 110—Trellised drainage pattern. C—Consequent, S - Subsequent, SC - Secondary consequent, O - Obsequent, BC - Beheaded consequent, E - Elbow of Capture, WG -Wind gap.

tributaries at right angles. In this pattern the influence of structure on the course of the rivers is very clearly marked. A good example is provided by the drainage of the Appala-chian region in the U.S.A. where hard and soft rocks are found in alternate bands. The term was first used by Bailey Willis in 1895 to describe the drainage pattern of the Ridge and Valley province of the Appalachians, an area of folded and tilted strata, where formations outcrop in parallel order of drainage. Willis named such an arrangement as trellis pattern because of its resemblance to garden trellis. Trellis drainage may develop in glacial regions, in the presence of linear glacial features as well as regions where parallel belts of rocks of varying resistance outcrop.

Rectangular Drainage Pattern : This pattern is very similar in form to the trellised pattern because the tributary streams meet the principal ones at right angles. Their rectangular form, however, is not related to the surface slope but is determined by the rectangular nature of rock joints and the presence of right angle faults. This pattern develops in regions where the rocks have rectangular joints and faults. Examples of this pattern can be seen along the Norway coast and in the Adirondack Mountains of the U.S.A.

A modification of this is *barbed drainage pattern* characteristic of headwater portions of drainage systems and is the result of river capture which has effected a reversal of the drainage of part of a separate river system. The tributaries join the main stream in boat-hook bends with acute angles. Another variant of the rectangular pattern is the angulate pattern which develops where faults or joints join each other at acute or obtuse angles rather than at right angles.

Radial Drainage Pattern : On domeshaped mountains or on volcanic mountains, the streams generally radiate in all directions from a focal point and resemble the spokes of a wheel or the radii of a circle. This type of drainage is therefore called radial. This pattern is largely formed by the various consequent streams which flow down the slope in different directions. A good example is provided by Sri Lanka where the rivers rise and radiate in all directions from the Central Highlands (Fig-111). Other examples are met with in volcanic mountainous regions, Henry Mts. of Utah and the Central Plateau of France.

This is really the radial outward pattern where the drainage

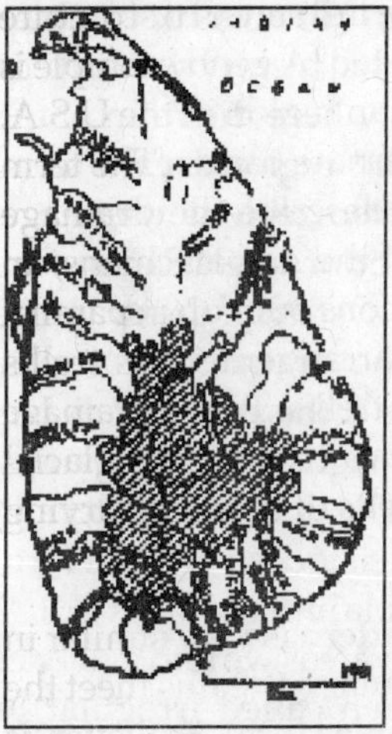

Fig. 111—Radial Drainage pattern of Sri Lanka.

lines radiate outwards from a common centre. The converse of this is the centripetal or radial inward type of drainage where the streams flow into a common centre from circular basin walls such as, in structural basins, calderas etc. This commonly occurs in areas of inland drainage where the rivers instead of joining the sea fall into some inland depression or lake. Apart from basins of inland drainage, a beautiful example is provided by the Kathmandu Valley in Nepal. The valley consists of deltaic low deposits and forms a sort of amphitheatre surrounded by spurs of the Middle Himalayas and on the south by the Mahabharata range. The rivers rising in the surrounding mountains flow towards Kathmandu in the valley centre and meet the Baghmati river there, after which the Baghgmati flows southwards towards the Ganga, cutting a deep gorge through the Mahabharat range.

Annular Drainage Pattern : On maturely dissected domal mountains where hard and soft rocks are found in alternate bands on all sides, a ring-like or annular pattern develops as a result of the capture of the radial streams. As breaching proceeds, the initial radial streams disappear and concentric arrangement of streams on the least resistant formation develops. On these domes the tributaries of the principal consequent flow in a circular fashion round the dome following the areas of softer rocks. Examples of this pattern can be seen in the Henry Mountains and the Black Hills of the U.S.A. and in the Wealds of England. (Fig 111 A)

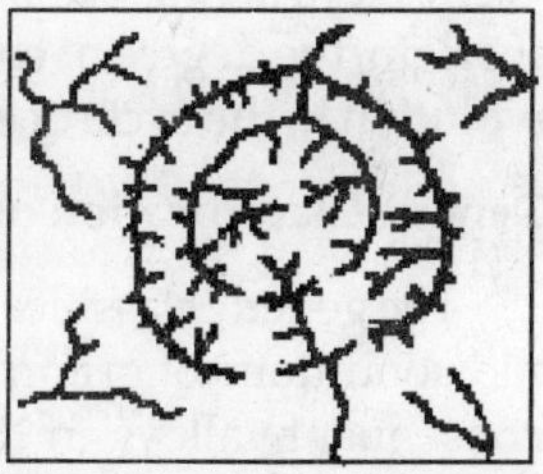

Fig. 111 A — Annular drainage pattern.

Parallel Drainage Pattern : This pattern consists of parallel master and tributary streams. Pronounced slope, parallel faults and parallel topographic features such as drumlin and lateral moraines are favourable to the development of this drainage type.

Complex or Irregular Drainage Pattern : In areas of complex geological structure and complex geomorphic history we find the development of complex drainage pattern where the form of the drainage is not uniform but shows marked regional variations. Examples of this can be seen in the glaciated regions where the drainage system developed in the period prior to glaciation has been partly or wholly destroyed and a new system of drainage has not yet fully developed. In such areas the drainage system is disorganized; abandoned valleys may form if the stream flow has been blocked by a moraine, the rivers flow through lakes and swamps, and tributary streams are small and few. The drainage pattern of Finland is a good example. In fact, in the Pleistocene period North and Central Europe and large parts of Canada and northern U.S.A. were covered by glaciers and in all the above areas we find evidence of disorganization of drainage. It has been called irregular or deranged drainage, as it is largely produced by blocking or displacement of an earlier river system. Mature or rejuvenated shields may also develop an irregular pattern which is termed complex drainage pattern.

Classification of River Valleys

We may now briefly consider the classification of river valleys, as this is closely related to the classification of drainage system. The river valleys may be classified on the basis of their mode and condition of origin, the degree of control exercised by structure or the stage of evolution in the cycle of erosion.

Genetic Classification of River Valleys

The genetic classification of river valleys is based on their mode and conditions of origin. J.W. Powell was the first to use the term 'consequent valleys' in 1875 for valleys carved by streams flowing along the original slope. Later on Davis presented a more detailed classification based on the sequence of their development and divided river valleys into consequent, subsequent, obsequent, resequent and so on. In an area of consequent drainage system, after the formation of the principal river valley, there is a gradual development of smaller tributary valleys with the rise and evolution of the tributary streams of different order. The valleys have been named after the type of the related streams and their sequence of origin. The consequent valley has also been called dip valley and the subsequent valley has been referred to as strike valley.

Similarly the valley formed by the two types of inconsequent drainage have been called antecedent valleys and superimposed valleys according as they are formed by either antecedent or superimposed rivers. While an antecedent valley antedates the structure across which it is cut, the superimposed valley extends along a structure which antedates the valley but was not exposed at the time of the initiation of the valleys. Both these types of valleys are discordant to and have developed transverse to the prevailing structure.

Classification of Valleys based on Controlling Structure

Sometimes the river valleys are classified on the basis of the type of structures which have determined their direction and disposition. In areas of homoclinal structure, the layers of hard and soft rocks are inclined at the same angle and in the same direction. Here the subsequent streams flow along the strike through areas of soft rocks. Their valley are called homoclinal valleys. Such valleys are found along the margins/sides of the folded Appalachians, the Jura mountains and the Rockies. The valleys developed along the axes of breached anticlines have been called anticlinal valleys. The valleys formed by rivers occupying down faulted basins are referred to as fault valleys, while those carved along fault lines have been called fault-line valleys. Where the direction of the valleys has been determined by rock joint systems, they are called joint valleys.

Classification based on Stage in the Geomorphic Cycle

This classification is based on the gradual evolution of river valley in the time sequence. The valley forms differ in the different stages of the erosion cycle and also where more than one cycle has been in operation. Accordingly, it is possible to identify four different types of valleys, based essentially on their age and stage of development.

(a) *Young Valleys*—Such valleys are characteristic of the first stage of the erosion cycle. There is predominance of down-cutting, and in the early stage the valleys are deep and narrow. Gradually, however, as result of rain-wash, weathering and lateral erosion the valley tends to be widened and approaches the V form.

(b) *Mature Valleys*— Mature valleys are formed in the second or mature stage of the erosion cycle. Here, on account of the increasing importance of lateral erosion the mouth of the V becomes wider.

Simultaneously, the long profile of the river valley approaches a graded concave curve with the removal of irregularities in its course.

(c) *Old Valleys*— Old valleys are characteristics of the last stage of the erosion cycle. Such valleys are characterized by increasing lateral erosion, increasing deposition and meandering. The valleys are very wide, shallow and extensive and the rivers form flood-plains and deltas.

(d) *Rejuvenated or Polycyclic Valleys*—We have seen that rivers may undergo rejuvenation as a result of change in their base-level of erosion. This is made possible either by uplift of the landmass or by a fall in the sea-level. As a result of rejuvenation, the valleys, whether in mature or old stage, develop signs of youth. Rate of vertical down-cutting increases, and the rivers start cutting new and narrower valleys within their old valleys. Thus valleys within valleys are formed. Further new irregularities develop along the long profile of the valley and knick-points are found. The streams cut down and back from the base-level into the previous flood-plain and carve out a series of paired terraces along the valley sides, while a new flood-plain is gradually formed adjusted to the new base-level. Such valleys are referred to as polycyclic or multi-cyclic valleys and are quite common due to changes in sea level and diastrophic movements.

Questions

1. Explain the mode of development of river systems.
2. Describe with examples the evolution of the different types of drainage patterns.
3. Explain with examples, what you understand by the following :

   Consequent drainage system, Antecedent drainage, Superimposed drainage, Dendritic Drainage pattern.
4. Describe the different stages in the development of a dendritic drainage system and discuss with examples the conditions under which superimposed and antecedent drainage may develop.
5. Write short notes on:

   Subsequent streams, Migration of Divides, River Piracy, Radial Drainage Pattern.
6. Classify river valleys on the basis of
   (a) Stages in the erosion cycle,
   (b) Conditions and order of genesis, and
   (c) Controlling structures.

Chapter 19

The Cycle of Erosion

The Concept of Davis

The originator of the concept of the cycle of erosion is the famous American geomorphologist William Morris Davis (1850-1934).He was Professor of Physical Geography at Harvard University and founded the Association of American Geographers in 1904. He published this concept in 1899 in a well-developed form and called it the geographical cycle, though he had put forward his ideas, earlier at the mectings of the American Association for the Advancement of Science in 1884. According to this concept a landscape has a definite life history, and as the processes of land sculpture operate on it the surface features are marked by several changes in its life-time. The changes in the surface features follow a definite sequence where the initial forms pass through sequential forms and finally reach the ultimate form. Thus the evolution of the landscape passes through a cycle, and the cycle follows a definite sequence of development. The gist of the cyclic concept is that commencing with an initial landscape, generally assumed to be one of low relief which is uplifted rapidly, the land surface passes through an orderly sequence of forms eventually to return to a surface of low relief, to an assemblage of forms which,to all intents and purposes is identical to the original.

The concept of the cycle of erosion is an important landmark in the history of geomorphology. Though it is no longer accepted wholly in its original form it was so widely accepted in the English speaking world that its validity could hardly be questioned for more than half a century. Developed to cope with the sequences found in humid temperate regions, the cycle should be divided into three components, a cycle of river development, a cycle of slope development and a cycle of landscape development. The successive stages of development sequence can be divided into three parts and may be identified as youth, maturity and old age. Davis's presentation of the scheme was both vigorous and vivid and his colourful analogy of the human life and landscapes both passing through the stages of youth, maturity and old age, caught the imagination of the scientific world. In a very real sense the cyclic concept was a product of its times, for it is evolutionary in character and was conceived in an age when the scientific climate was dominated by the Darwinian theory. In order to understand the evolution of a particular landscape it is extremely important to know the stage of development. But the process of sculpture , geological structure and the nature of rocks also exert an important influence on the fashioning of the landscape. Keeping the above facts in mind Davis states that landscape is a function of structure, process and stage. These are what Davis called 'the trio of geographic controls'- the three principal controlling factors in the evolution of any landscape, as it passes from an initial state via sequential forms to an ultimate state. In the absence of complicating factors, such as tectonic activity, Davis believed that humid temperate landscapes evolve through a fixed sequence of forms that reflect the interaction of the above three factors.

Structure

The word structure has been used here in the larger sense and includes not only regional geological structure, folds, faults and such other broad features but also the nature of rocks and their characteristics, their physical and chemical properties, their permeability and solubility, the nature of joints etc. Thus on a very permeable rock, there will be minimum run-off and so erosion by running water will be slower than on an adjacent impermeable rock. Soft shales erode more readily than massive quartzite, while a bed of limestone or gypsum will undergo solution by underground water in addition to being modified by surface stream. As a rule, the influence of rocks on landscape is quite marked, and it would be wrong to conclude that where this is not discernible, rock structure has had no influence on topography. Sometime the influence of structure is so pervasive and uniform over a relatively large area that this is not apparent from the landscape.

"In general, structural features of rocks are much older than the geomorphic forms developed upon them,."[1] The structure, as a rule,

1. Thornbury, W.D.—op.cit., p. 18,

is older than the landscape and provides the base for the operation of the various gradational processes which give rise to new sequential landforms.

Process

The word process includes the action of all the forces which bring about changes in the configuration of the earth's surface. As we have seen, some of these forces are endogenic or internal and express themselves through volcanicity and diastrophism and produce irregularities on the earth's surface by building and uplifting mountains, plateaus and hills. On the other hand, there are exogenic forces or external processes which tend to level down the earth's surface. The agents of these external processes of denudation are running water, groundwater, glaciers, wind and sea-waves. These tend to bring down by subaerial erosion the land uplifted by the internal forces, and simultaneously they also transport the eroded materials and deposit them elsewhere, the total effect being a general lowering and levelling down of the earth's surface. As a result of the operation of these agencies over a long period of time, the uplifted parts are gradually reduced in height and in the end are changed into a featureless and level surface. Process is significant in the evolution of landscape, because the landscape changes with a change in the process. In other words, a particular process gives rise to distinctive landforms. For instance, alluvial fans, flood-plain and deltas are characteristic of regions subject to erosion by river, underground streams and caverns are formed by the action of groundwater, and moraines, drumlins and such other features are produced only by glacial action. Every landscape bears the unmistakable impress of the process which has shaped it, and it is possible to classify landscapes on the basis of the dominant process which has sculptured and fashioned them. Davis considered running water to be most universal process and initially based his cycle concept on fluvial erosion, but subsequently extended it to include other processes.

Stage

When the different agents of erosion work on the earth's surface, there is a gradual and systematic evolution of the landscape. Broadlly, three stages of this evolution may be identified. These three stages of the cycle of erosion, according to Davis, are youth, maturity and old age. Just as each process gives rise to a distinctive landscape, so also each stage of the erosion cycle is characterized by distinctive landforms with the help of which it is possible to identify the stage of evolution of the landscape. There is a direct relationship between the stage of development and the character of landforms and Davis lays great emphasis on the stage of development in his concept of the cycle of erosion. Stage is the most important determinant of landscape form (as marked by passage of time) and the age of the landscape can supposedly be read from its form. Landforms are thus 'time -dependent.'

Although these stages are related to the sequence of time and time broadly defines the stage reached in the cycle, it would be wrong to conclude that the three stages have equal life-span or that

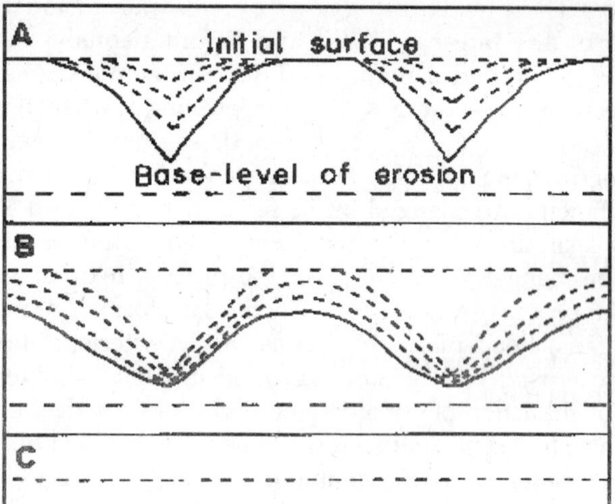

Fig. 112—Davisian cycle of erosion. A-Youth, B-Maturity, C-Old Stage.

where the landscape in two separate regions is said to be in the same stage of development, it has taken equal time for the two regions to reach that stage. Further, two regions having reached the same stage of development may still show differences in the details of their landscape. The important thing to remember is that the division of the erosion cycle into stages is intended to emphasize the orderly evolution of landforms. For example, if we take a landmass which has just been uplifted above the sea-level, it is possible to ouline the sequence of development of the landforms on this surface by the processes of erosion (Fig. 113). In the initial stage the uplifted landmass will not have diversity of relief, but under humid temperate conditions down cutting of the valley by rivers will begin vigorously which would be deep and narrow. The inter-stream areas or the inter-fluves will be relatively wide and flat and upraised and will consist of the original land surface. As the rivers incised their beds, the amplitude of relief would increase. This stage of deep narrow valleys with remnants of the initial surface preserved high in the inter-fluves and increasing relief amplitude, was termed 'youth' by Davis. With the passage of time the valley will widen as lateral erosion proceeds and with the widening of the valleys, the interfluves will become narrower. Even after vertical down-cutting by the river has stopped i.e. even after grading, the process of valley widening continues, and in this manner the inter-fluvial original surface is narrowed to knife-edge crests and ultimately is completely destroyed. This is the end of youth and the beginning of maturity. Thus relief goes on increasing in youth until maximum relief is reached towards the end of youth or the beginning of maturity. After this, on acount of the further erosion of the interfluves and deposition in the valleys, amplitude of relief would start decreasing. In fact the entire stage of maturity is a period of decreasing relief.[1]

How soon youth will turn into maturity depends to a large extent on the texture of the drainage. The number of streams and their relative distance will determine the speed with which the inter-stream areas are being denuded. During this period there will also be an extension and evolution of the drainage system with the development of tributary streams, river piracy, migration of divides and other related phenomena. It is, however, important to note that the end of youth and the first grading of the drainage system are not coincident, as the grade is not established in the entire river system at the same time, and more especially the grading of the head streams

1. While youth is a period of increasing relief, maturity, at least in its later stages, is a period of decreasing relief. Sparks, B.W.,—*Geomorphology*, Longman, London, 1967, p. 16-17

takes a much longer time. Thus the work of grading of the streams continues even during the period of maturity, and at the same time the rivers build flood-plains and their courses are meandering.

In youth the slope of the valley is steep and the rockwaste falls easily down the slope. The valleyside rocks are therefore bare and free of debris. Later on, when the slope becomes gentle, the debris start collecting and move down only slowly. There is thus grading of the valley-side slopes. This is the reason why whereas the water-divides are steepsided in youth, they become rounded or covex in maturity, when the curve of water-erosion or the graded long profile of the river becomes concave. In this way in the middle and late stage of maturity, the processes of both lateral and vertical erosion continue. Simultaneously the rounded water-divides are also lowered. Maturity imperceptibly changes into old age, and the down-wasting of the divides would continue until the entire area is reduced into a low, nearly level and featureless plain with only a few broad convex swells, called the *peneplain*. Thus slope angles decline through the cycle and ultimately a peneplain, a low relief plain close to river base level is developed.

The peneplain is not really a plain with a level surface, but is a rolling lowland where the different rivers are seperated from one another by slightly higher ground and flow slowly down very gentle slopes. In this stage the influence of rock differences on the landscape (topography) is almost entirely lost, but here and there some resistant rocks may remain as high ground or form residual hills called *monadonocks*. Davis called such residual hills by this name after Mt. Monadonock in New Hampshire.

Thus an uplifted landmass after passing through the different stages of erosion eventually changes into a featureless nearly level plain. This is the basic principle of the concept of cycle of erosion. The cycle of erosion has been called the geographical cycle by Davis, but it is better known now as the *geomorphic cycle*. According to some scholars there is a difference between cycle of erosion and geomorphic cycle. The cycle of erosion is that duration of time in which an uplifted landmass changes into a featureless and nearly level plain through the processes of erosion, whereas the gradual evolution of landforms with the progress of the erosion cycle is included under the term 'geomorphic cycle'. According to Worcester the geomorphic cycle is the landscape which has evolved in the different stages of the erosion cycle. But in acutal use no such distinction is made between these two terms and both are used as if they are synonymous.

Since the river or running water is the most powerful and pervasive agent of denudation the cycle of erosion as evolved under fluvial erosion has been called the normal cycle of erosion by Davis. Some other scholars do not quite agree with this view and consider erosion performed by other agents like galciers or wind to be equally normal. Later on Davis did develop further cycles to deal with landform development under climatic conditions other than humid, but he considered them to be landforms produced by climatic accidents.

The principal characteristic of the old age of the cycle of erosion or the geographical cycle, according to Davis, is the existence of the peneplain with stray monadonocks. Though there is considerable difference of opinion about the concept of peneplain among scholars, it can be stated that when the uplifted landmass is so much eroded that it reaches the base level of erosion and changes into a low, nearly flat surface with a few residual hills and hillocks then that lowlying nearly level surface is called a peneplain.

Graphical Representation of the Cycle of Erosion

Davis's concept of the cycle of erosion can be easily understood with the help of the graphical representation given below. The uplift of the landmass has been shown by broken lines. A represents the average initial height of the uplifted surface and B represents the

Fig. 113—Graphical representation of Davis's Concept of the cycle of erosion

average height of its lower parts (Fig. 114). Thus the distance AB represents the average initial relief. In the first stage of erosion the floor of the valley is rapidly degraded as a result of vigorous down-cutting, but there is very little erosion of the ridge tops or the divide tops. Consequently the relief tends the increase which is shown by the divergence between the two curves in the diagram. By the time the stage of maturity begins, the relief has become maximum which is shown by the straight line CD. After that, vertical corrasion starts decreasing, and in place of the valley floor the higher parts start getting denuded with greater rapidity. Consequently there follows a period of decreasing relief, and the two curves tend to come closer together. In the later stages of erosion this tendency further increases, resulting in increasing convergence of the two curves. This process continues for a very long time, and ultimately in the old age the landmass is reduced to a low featureless surface.

Criticisms of the Erosion Cycle Concept

For the sake of simplicity, Davis has assumed in his theory that erosion begins only after the uplift has taken place and that uplift occurs very rapidly and then ceases for the rest of the period of time in which a cycle would take place. His critics have accused him of unnecessarily simplifying the concept of the cycle of erosion. In fact, uplift and erosion go on together and the reality is that uplift is a long and slow process and it is only rarely that the uplift is so rapid that erosion starts only after the completion of the uplift[1]. As soon as a landmass begins to be formed by uplift erosion immediately starts working on it. But it would be unfair to Davis to say that he has assumed rapid uplift followed by long stillstand to be the normal situation and the normal process. As stated earlier, he has made this assumption only for the sake of simplification of presentation of his ideas, and has not completely overlooked the possibility of variations in the kind of uplift resulting from earth movements.[2] Further, broadly speaking it would be true to say that uplift followed by stillstand and erosion is the normal sequence and the other possible variations of the sequence are rather exceptional. Though the cyclic concept offers a guide to what

1. " In general, the cycle should not be regarded as beginning with the erosion of an already uplifted mass. Erosion and uplift inevitably overlap to some extent, and the cycle really begins with the first emergence of a sea floor, or the first movement of deformation of a landmass". Wooldridge, S.W. and Morgan, R.S.—op. cit. p. 178.
2. "Although Davis may have overworked the idea of rapid uplift followed by stillstand to permit the geomorphic cycle to run its course, he did at least recognize that there are many possible variations from this simplified case". Thornbury, W.D.—op.cit., p. 204,

happened and in what sequence, the assumption that the initial uplift of the land is rapid is not considered to be justified and the effects of erosion during uplift also remain uncertain.

The other criticism advanced against this concept is that the cycle of erosion can never run its full course and that the "peneplain" is an unrealized and unrealizable abstraction.[1] Such a long duration of time is needed for the completion of the cycle and for the formation of the peneplain that it is impossible for a landmass to remain stable and absolutely unaffected by earth movements for such a long time. Some critics have, therefore, argued that the peneplain is essentially an imaginary surface, as no peneplain has ever been formed in the past, nor there is a possibility of its being formed in the future. They substantiate their argument by pointing out that we cannot find a single example of an extensive peneplain in any continent today with reference to the existing sea-level. But we have to remind ourselves that the Pliocene- Pleistocene diastrophism and uplift are relatively recent events in the perspective of geological time and the landmasses of the globe are at present passing through a period of active degradation. If we look back to more ancient geological times, we shall find that we do come across surfaces of low relief which were probably peneplains but have been raised to form plateaus, or they have been covered and masked by later sediments and can be identified only as planes of unconfirmity of the geologists, and in some places where the upper cover of rocks has been denuded, the former have been exposed on the surface.

Some of the other important criticisms of the Davisian cycle concept have already been referred to earlier and may be seen on pages 18 to 19.

General Systems theory and the Cycle concept

R.J. Chorley (1962) has made it clear that the erosion cycle obeys the laws governing *general sytsems theory*. According to the systems theory there are two main kinds of models which we may develop. The first is the *closed system* in which there are clearly defined boundaries across which no import or export of materials or energy may occur. Another characteristic is that the system moves in the direction of maximum entropy i.e. there is minimum of free energy left in the system. Davis's concept is now regarded as an example of

1. Wooldridge,. S.W. and Morgan, R.S.—op. cit., p. 184.

a closed system. Uplift starts erosion and the erosion continues until the land is reduced to a peneplain. When this stage is reached the system will have minimum free energy. The entire sequence of events from uplift to the production of peneplain constitutes one cycle and it is assumed that this cycle would go on its way relentlessly until the changes were completed. Davis later allowed for modifications of the cycle by other factors, but he did not change his approach to embrace the open system approach. Chorley (1965) has pointed out that the basic fallacy was the supposition that from the present landform one can deduce its history. The Davisian ideal cycle can scarcely ever be achieved because of climatic and eustatic changes as well as isostasy and other tectonic factors.[1]

In the case of the *open system*, an energy supply is essential to its maintenance and preservation. It is maintained by a constant supply and removal of material and energy. An advantage of this concept is that the system may achieve a steady state so that any necessary adjustments of energy within the system may be effected by an adjustment of the form of geometry of the system. Considered from this point of view, most of the cycles considered in geomorphology are better suited for treatment using the open system analogy than the closed system models.

Another characteristic of the open system is that it behaves equifinally i.e. different initial conditions lead to the same end result. This was not anticipated by Davis. For example, he interpreted all knick-points as being produced by a change in the base level in the middle of the cycle, and ignored the production of knick-points by climatic or vegetational changes causing an increase in run-off. Such changes can be explained only by using the open system approach.

The Davisian concept using the closed system models has also been applied to a cycle of shore line development, especially by Johnson and Cotton. However, recent studies indicate that shore lines may be understood better by considering them as open system since climate, wave action, ocean currents, sea-level and the nature of the material being eroded and providing beach sediments vary so much with time. The cyclic concept has also been applied to

1. Chorley, R.J. (1965): "A Re-evaluation of the Geomorphic system of W.M. Davis "in *Frontiers in Geographical Teaching* edited by Chorley R.J. and Haggett, P., Methuen, London, pp 31-38.

karst regions, but the diversity of theories to explain the development of karst topography suggests that the use of an open system would be more fruitful.

Open system models have been developed in a few cases. One is the hydrologic cycle. It is likely that more open system cycles will be postulated in the coming years and will gradually replace the closed system cycles. Davis's concept of a closed system may not work in detail but his concept of structure, process and stage is of fundamental importance in the more practical open system models as well which are taking its place.

Penck's View of the Erosion Cycle

German geologist Walther Penck has been the most prominent critic of Davis. Penck has criticized Davis mainly for the latter's assumption that erosion starts only after complete uplift and that uplift is short-lived while the landmass stands still during the long period of erosion. Penck has also expressed disagreement with the stage concept of Davis, and in his view the cyclic evolution of landscape on a still-stand landmass for such a long duration of time can be considered only exceptional.[1] According to Penck the landscape is the result of the relative intensity of the degradational processes and the phase and rates of uplift. In other words, the landforms were determined by the rate of uplift and the rate of erosion. Penck talks of three phases in place of the three stages of Davis. These three phases are related to the rate of uplift. As the rates of uplift and degradation are different in the three phases, different kinds of slopes developed in them resulting in the formation of different landscapes.[2] The uplift was rapid in the beginning, moderate in the middle, and decreased in intensity towards the end. After sometime the uplift stops altogether and from that point onwards there is continuous degradation of the landscape.

---

1. Penck's book in German was published in 1924 but its English translation under the title of "Morphological Analysis of Land forms" came out only in 1953.
2. "The earth's surface was regarded as the 'Zone of Conflict' between external and internal forces and he believed that the land forms reflected the relative intensity of the endogenetic and exogenetic processes. Thus he believed that slope gradients were largely determined by the intensity of erosion as influenced largely by crustal movements". Thornbury, W.D.—op.cit,1969,p.201

Penck has assumed a low, featureless plain called *primarrumpf* by him for the initiation of the erosion cycle. The primarrumpf closely resembles a peneplain in which there is neither appreciable height nor appreciable relief. Initially when the land rises above the sea-level the uplift is so slow that uplift and degradation are almost equal, and irrespective of the geological structure, a low surface without relief is formed which is called the *primarrumpf*. According to Penck, this is the universal initial geomorphic unit on which the landforms develop as a result of the differential rates of uplift and degradation. Davis too has conceived of a similar initial surface in areas of extremely slow uplift and has called such a surface "old from birth peneplain". Penck calls the terminal plain which is formed at the end of degradation by the name of *endrumpf*. Endrumpf is the last stage or phase of degradation and the ultimate landform of the cycle, which closely resembles the peneplain of Davis.

Graphical Representation of Penck's Concept of Erosion Cycle

In the diagram given below (Fig.115) the upper curve represents the average maximum height, and the lower curve represents the average height of the valley floor. The entire cycle has been divided into five stages, but these relate to the changing conditions of the evolution of relief which are related to the rate of uplift and the rate of degradation. In the initial stage when erosion begins with the uplift of the primarrumpf, erosion is less than uplift. Consequently, there is an increase both in altitude and relief. In the second stage with the widening of the river valleys, the inter-stream areas

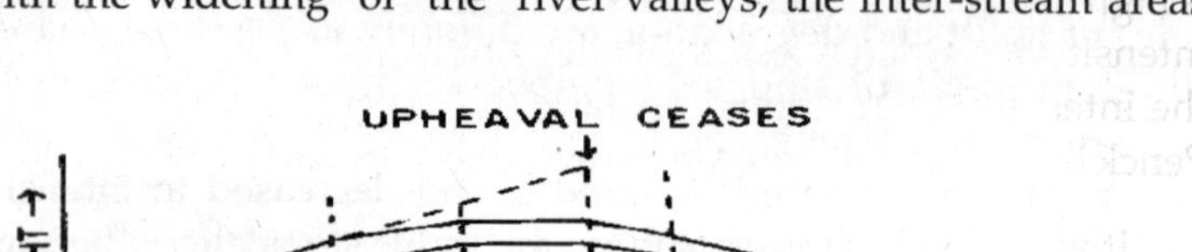

Fig. 114—Graphical representation of Penck's concept of the cycle of erosion.

become sharp-edged but there is no marked difference in relief as the rate of erosion in the valleys as well as in the upper peaks is more or less the same. But since the rate of uplift exceeds the rate of erosion, the relative height goes on increasing. In the third stage, the uplift of the landmass still continues, and under the stimulus of uplift the rivers continue their downward cutting. The rate of erosion, however, more or less balances the rate of uplift. Consequently both the relative height and relief are stationary. Early in the fourth stage the uplift ends, but erosion continues to be active along both the curves. Due to the increase in erosion, there is reduction in height but relief is stationary, i.e. the two curves run parallel to each other. In the fifth and the last stage vertical downcutting of valleys slackens, and lateral erosion of the valleysides is more active. The peaks and ridges of the inter-fluvial areas become rounded by erosion, and their height also suffers reduction. Consequently, there is a continual decrease both in absolute height as well as relief.

The conditions outlined above are possible, though not essential and absolutely natural and are basically dependent upon the mutual relationship between the rate of uplift and the rate of degradation. Relief increases only in the first stage, remains stable in the second, third and fourth stages, and starts decreasing in the last stage. In the first stage the rate of uplift exceeds the rate of erosion, and in the second and third stages both are almost equal. The uplift stops by the beginning of the fourth stage and thereafter only erosional or degradational activity proceeds.

According to Penck, landforms are built by slopes and the form of slopes depends on the intensity of erosion, which in turn is mainly determined by the nature of uplift i.e. by the nature of earth movements. When under conditions of rapid uplift the erosion rate is high, the valley slopes become convex. On the contrary when the intensity of erosion is less, the valley slopes are concave, and when the intensity of erosion is constant, the slope is uniform. In fact, Penck has laid great stress on the evolution of slopes in his concept.

It will not perhaps be true to say that the concept of Penck is more practical or effective than that of Davis. Penck has undoubtedly drawn our attention to some of the weaknesses of Davis's theory, but there are practical difficuties in accepting Penck's concept as an alternative to the concept of Davis. This much, however, must be said to the credit of Penck that he has put forward a new viewpoint and has indicated a new possible direction of investigation. But the concept of Davis with all its faults and limitations is more significant and important and has acquired a special position in the study of geomorphology. The entire geomorphic thought has been influenced by the concept of Davis. The essential significance of his concept lies in the fact that it lays stress on the gradual evolution of landforms and provides a basis for their genetic classification.

King's Cycle of Pediplanation

In 1948 Lester King proposed a new cycle of landscape evolution called the cycle of pediplanation based on his study of the planation surfaces dotted with isolated hills (inselbergs) found in the arid and semi- arid parts of South Africa. His ideas may be considered to be a blending of the ideas of Davis and Penck. Like Davis he thought that an initial uplift starts a new cycle of landscape evolution in which the landscape passes through the sequence of youthful, mature and old age forms. But unlike Davis, he argued that the landscape evolves by parallel retreat of slopes (and not by slope decline) and culminates in a pediplain consisting of low-angled pediments (multi-concave surface) with residual hills called inselbergs representing small vestiges of the initial surface. King claimed that pediplanation is the basic process of landform evolution and is true under all climatic conditions, but it has been found that his theory is not universal in application and is essentially applicable in arid and semi-arid regions. This theory is discussed in greater detail in Chapter 20 on slopes.

Interruptions in the Erosion Cycle and Multi-cyclic Landscape

Usually the cycle of erosion is not able to run its full course and remains incomplete on account of the various kinds of interruptions that it may have to face in the course of its progress. For various reasons, interruptions may be caused at any stage in the cycle and the landscape may revert back to the first or intital stage before the completion of the cycle. This explains why examples of the completed cycle are so rare and partial cycles or interrupted cycles are more commonly met with. Such interruptions cause several incomplete/ complete cycles to run one after the other in the same region. Such cycles are called poly-cycle or multi-cycle and landscapes in which there are preserved evidences of more than one geomorphic cycle are called multi-cyclic or polycyclic landscape. "The multi-cyclic evolution of landscapes is more common than the monocyclic development"[1]. Monocyclic landscape is one on which we find evidence for the completion of only one cycle of erosion, while multi-cyclic landscape shows evidence of more than one erosion cycle.

1. Thornbury, W.D.:Ibid., p. 139.

The main cause of interrruption in the orderly progress of the erosion cycle is rejuvenation. As a result of uplift and rejuvenation, the rivers which have reached the base level of erosion, again become capable of deepening their valleys. According to Davis, any change in sea-level, which is the base level of erosion, will give rise to a new cycle, irrespective of whether the first cycle has been completed or not. Since the permanent base-level of most rivers is the sea level, any change in the sea-level influences the work of rivers. In ice ages, a large proportion of the ocean water is deposited on the continents in the form of ice-sheets and the sea-level falls. The fall in sea-level leads to the rejuvenation of the rivers and their ability for down-cutting increases. On the contrary, in the warmer periods the ice melts and the sea-level rises. The rise in sea level reduces the down-cutting by rivers and flood-plains start building up. Similarly, the uplift of a land-mass by earth movements will cause interruption in the erosion cycle, the rivers are rejuvenated and the cycle again reverts to the youthful stage. In the bed of the old river valleys, new valleys are formed as a result of increase in vertical downcutting. Such features are called *valley in valley* or *two-cycle valley*. Rejuvenation brings about significant changes in the long and cross profiles of river valleys. There is formation of knick-points in the long profile of the river valley and of river terraces along the valley sides (see pages 289/92).

Sometime a catastrophic event may lead to the sudden termination of the cycle. Davis has used the term 'accident' for such interruptions. He has pointed out that such sudden termination of the cylce even before it is completed, may be brought about by volcanic and seismic activity or by climatic changes. The cycle may also end abruptly by complex diastrophism or by marine transgression of the landmass.

Dissatisfaction with the Cyclical Theories

In the 1950s and 1960s dissatisfaction with the cyclical theories led to a shift from the denudation chronology approach to landscape evolution approach based on detailed quantitative investigation of form and process in small drainage basins. Dissatisfaction arose mainy for two reasons. Firstly, it was found almost impossible to verify in the field the denudation chronologies dervied from the cyclical theories. Secondly, it was felt that Davis's and King's assumption of initial rapid uplift followed by a long period of standstill and tectonic quiet was unrealistic. Thirdly, even under conditions of tectonic stability denudational unloading would result in isostatic uplift as a nessesary consequence. Coupled with the above reasons was the emergence of the quantitative approach which hastened the process of shift from the cyclical approach to detailed quantitative investiagtion of form and process in landscape evolution of small areas. One such non - cyclic theory that deserves mention is that of John Hack which draws heavily on the ideas of Davis. Contemporary G.K. Gilbert had emphasized the idea of dynamic equilibrium form that balances erosional energy and lithological resistance but his ideas were overshadowed by the forceful writings of W.M. Davis.

Dynamic Equilibrium Concept

The latest challenge to the concept of erosion cycle of Davis has come from. J.T. Hack, supported by Strahter, Chorley and others. By his study of the Appalachian and the neighbouring highlands in the 1960s, he has tried to prove that there is no evidence of cyclical erosion surfaces in this region, and it is possible to provide a satisfactory explanation of the present landscape with reference to the existing processes acting on the different kinds of rocks. He is not prepared to accept that there is a cyclic and orderly development of the landscape or that stage and time are important in the evolution of landforms. While rejecting the concept of the erosion cycle, he has put forward a new concept called the Dynamic Equilibrium Concept. "As applied to landscape, the principle of dynamic equilibrium states that when in equilibrium a landscape can be considered a part of an open system in a steady state of balance in which every slope and every form is adjusted to every other. Changes in topographic form take place as equilibrium conditions change, but no particular cycle or succession of changes occurs through which the forms inevitably evolve, as was assumed by Davis and most later workers in geomorphology. Differences in form from place to place are explained by differences in the bedrock or in processes acting upon the bedrock. Changes which take place through time are a consequence of climatic or diastrophic changes in the environment or of changes in the pattern and structure of the bedrock exposed as the erosion surface is lowered."[1]

---

1. Hack, J.T. (1965) : Quoted in Thornbury, W.D.—Ibid., p.198.

The dynamic equilibrium theory can be understood by reference to Fig. 116. It shows an area with three horizontal layers of rocks. The

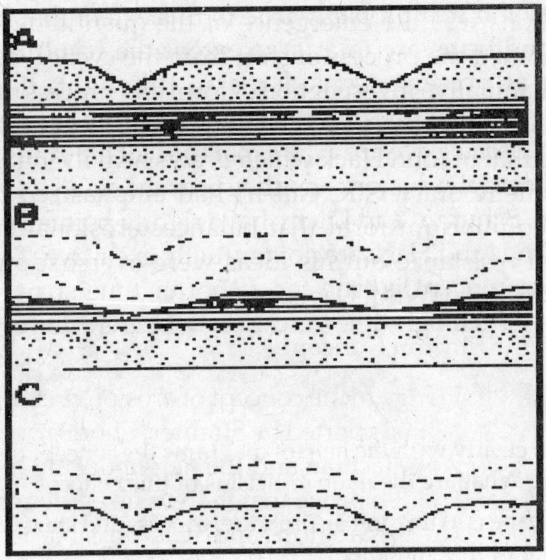

Fig. 115—Landform development on horizontal rocks according to dynamic equilibrium theory.

middle layer is weak clay while the upper and bottom layers comprise resistant sandstone. It is assumed that all controlling factors other than rock type remain constant. Landform will first develop on the upper sandstone (A). Because of resistant rocks, valleys will be narrow, valleysides steep and relief considerable. When the sandstone is removed by erosion and slope retreat and the rivers start cutting into weak clay, the valleys will become broader, valley-side slopes will become gentle and relief will be reduced (B) When the claey layer is removed and the lower sandstone is exposed, the landform will again revert to narrow and steep sided valleys and considerable relief (C). This sequence of landform changes is non - evolutionary. The landscape merely adjusts its form to maintain a state of equilibrium in relation to the controlling factors, in this particular case being the rock-type. Similar effects might be produced in an area of homogeneous rock by change in climate or rate of river erosion. In reality several factors may operate concurrently which may counter - balance each other, and the form of land may remain the same.

It appears that the concept of Hack may probably be accepted as a complementary and not as an alternative concept to that of Davis. The concept of Hack lays stress on the present processes and conditions, and it is probably true to this extent that most of the existing landforms are to a large extent the result of processes operating at present. Considered in this light, the dynamic equilibrium concept does not consider the past and is incomplete to that extent, because we are aware that most of the landscapes have a long and complex history and they bear the stamp of this history upon them. Schumm and Litchy have rightly pointed out that the views of Davis and Hack are not mutually exclusive. The dynamic equilibrium concept is true for a shorter time span, but in the perspective of geologic time the concept of the erosion cycle is more relevant.[1]

Questions

1. Explain clearly with the help of diagrams the concept of the cycle of erosion. What are the main criticisms of this concept ?
2. "Landscape is a function of structure, process and stage". Explain the meaning of this statement.
3. Discuss critically the Davisian concept of the Cycle of Erosion.
4. Explain with the help of a diagram Penck's conception of the erosion cycle.
5. Compare the views of Davis and Penck about the cycle of erosion.
6. Evaluate the significance of structure, process and stage in the evolution of a landscape. Give suitable examples.
7. Analyse the causes and consequences of interruption in a normal cycle of erosion.
8. Write short notes on :—

   Peneplain, Polycyclic landscape, Normal cycle of erosion, Primarrumpf and Endrumpf, Dynamic equilibrium concept.

---

1. Landscapes can be considered......either as a result of past events or a result of modern erosional agents. Depending upon one's view point the landform is in one stage in a cycle of erosion or a feature of dynamic equilibrium with the forces operative. These views are not mutually exclusive. It is just the more specific we become the shorter is the time span with which we deal and smaller is the space we consider. Conversely when dealing with geological time we generalize. The steady state concept can fit into the cycle of erosion when it is realized that steady state can be maintained only for a fraction of the total time involved. Schumm, S.A. and Litchy, R.W. (1965). quoted in Thornbury. W.D., Ibid. p. 200.

Chapter 20

The Development of Slopes

Importance of Slopes

A major proportion of the earth's land surface consists of slopes. Barring the level plains, slopes are present almost every where and their maximum development is found in the hilly areas. The physical landscape consists essentially of an assemblage of different kinds of slopes, and the morphological characteristics of a region can be analysed only on the basis of slope forms. Slopes, in fact, provide the key to the study and analysis of landforms. Slopes also exercise an exceedingly important influence on soils and natural vegetation. From the point of view of human activities too, the influence of the physical landscape is largely exercised through slopes, and land utilization, agriculture, settlements, transport and urban activities are all not only visibly influenced by slope but to a very great extent are controlled by it. Ordinarily only land having a slope of less than 10^0 is cultivated, and it is only in areas of very dense population (such as Java, China or Nepal) that lands with a slope of 25^0 or more have been cut into level terraces and brought under cultivation. The increase in slope results in an increase in the danger of soil erosion. Generally there is an inverse correlation between the angle of slope and the depth of the soil. Consequently in land classification schemes, together with soil characteristics the slope angle has to be given due consideration. For all these reasons, scientific analysis of slopes is being given special emphasis in geomorphological studies in recent years.

Genetic Classification of Slopes

Slopes can be produced either by endogenetic or by exogenetic processes, and accordingly two broad types of slopes are recognized—endogenetic slopes and exogenetic slopes.

(a) *Endogenetic (or hypogenic)* slopes are those which owe their origin to processes which originate within the earth such as, earth movements resulting in uplift, folding, faulting and tilting, or volcanism which may produce entirely new constructional slopes by accumulation of lava and pyroclastic materials. The slopes produced by earth movement are referred to as *tectonic slopes* and are directly related to geological structure. Fault scarps which are found as steep vertical slopes on either side of a rift valley are a good example of this type. The *volcanic* cones, domes and plateaus are *slopes* which are formed essentially by the deposition of materials ejected from volcanic vents and fissures and constitute yet another sub-type of endogenetic slopes. It is important, however, to remember that all tectonic and volcanic slopes are in course of time modified by the agents of sub-aerial denudation.

(b) *Exogenetic or epigene* slopes result from the action of external atmospheric processes which originate at or near the earth's surface such as, weathering, mass-wasting, erosion, transportation and deposition by the denuding agencies. Exogenetic slopes may be produced either by degradation or by aggradation and accordingly there are two sub-divisions of exogenetic slopes—*degradational or erosional slopes and aggradational or depositional slopes*. (i) Erosional slopes include slopes created by erosion by running and ground water, glaciers, wind and waves. In the course of erosion rivers and glaciers impart different forms to their valleys, and even in arid climate different kinds of slope are formed by fluvial erosion. These are essentially valley slopes. Degradation slopes thus include escarpments, watersheds of streams, river channels and river-cut terraces as also the various forms carved by glacial erosion. In the coastal areas cliffs are formed by marine erosion (ii) *Depositional or aggradational slopes* include alluvial, colluvial and talus slopes as well as depositional slopes built by glaciers, winds and waves. Rivers build alluvial cones and fans and flood-plains by aggradation. Glaciers build different glacial depositional slopes and in the desert depositional slopes are built on a large scale by wind, the most prominent of these being the various kinds of sand dunes. Similarly, along the sea-coast beaches, various kinds of spits and bars give rise to a variety of aggradational slopes through marine deposition.

Elements of Slope

Slopes have a variety of forms. A commonly observed slope profile consists of a convex 'crest' or 'nose', a rectilinear 'side slope' or 'mid-slope' and a concave foot-slope. In the convex slope angle constantly increases downslope, in the concave slope angle constantly decreases downslope, while in the rectilinear or straight slope the angle remains unchanged. A slope, of course, may be complex with great variation of slope angle in a short distance. According to L.C. King (1957-1962) and A. Wood (1942), a composite slope profile consists of four distinct parts or elements, and all the four elements are present in a standard hill-slope (Fig. 117). In the uppermost part or crest the slope is convex. Below this is the scarp

Fig. 116—Four slope elements of composite slope profile according to Wood and King.

or free face slope. Further down is rectilinear slope and at the bottom is concave slope. It is obvious that this division is based essentially on the most widespread composite slope profile form but in reality several combinations of these slope forms or elements are possible depending upon structure, the nature of the ground surface, regolith and the processes involved.

The four main types of *simple slope forms or elements* are shown in Fig. 118 and are described below. All the four elements may not be always found on the same slope and one or two elements may be missing. The most commonly found elements are convexity at the top and concavity at the base of hill slopes, occasionally with a rectilinear section in between. In arid areas of hard crystalline rocks, the summital convexity may be altogether absent and a composite slope form may consist of an upper steep free face (with 40° or more angle), a central boulder controlled rectilinear section and a basal concavity or pediment. In areas of alternating resistant and weak rocks where valleys are deep and weathering quite active, the composite profile may comprise numerous free faces associated with the massive and resistant strata separated by rectilinear sections of the easily weathered weak rocks, with only small stretches of summital convexity and basal concavity which in some cases may be altogether absent.

(i) *Convex Slope*—Sometimes the whole of the slope may assume a convex form, but it is more usual for the convexity to be developed only on the upper part of the slope. In the uppermost portion of hillslope called the crest or summit the slope is commonly convex to the sky i.e. the slope angle constantly increases downslope from the summit. This is commonly referred to as 'summital convexity'. Two processes are particularly responsible for the formation of the convex slope—weathering and soil creep. It is, therefore, also referred to as upper wash slope (Fig. 118 D).

(ii) *Free face slope*— Just below the summital convexity is found a steep wall-like slope called the scarp or free face slope. It consists of bare rock and on account of the steepness of the slope no rockdebris or regolith is found on it. Along this slope the weathered material moves down with water or rock fragments fall down due to land-slides but the slope itself remains free from all kinds of debris or regolith. The simplest example of free face slope is the cliff, developed along the coast by wave under-cutting, or the scarp in deep river gorges or in faulted structures. Here the cliff is so steep (40^0 or more) that the products of weathering fall immediately to the base and the weathered material accumulates at the foot of the cliff in the form of scree slope (Fig. 118A).

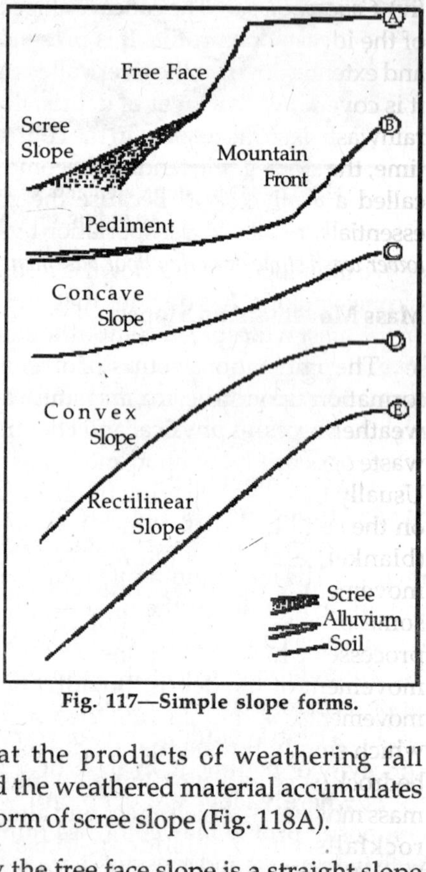

Fig. 117—Simple slope forms.

(iii) *Rectilinear Slope*—Below the free face slope is a straight slope where the slope angle remains constant. This is called rectilinear slope as it is straight in profile. Here the debris which fall down the free face slope are deposited. The angle of this slope is determined by the angle at which the coarse rock debris can come to rest. These debris become finer with weathering and the finer debris are carried downwards by the rain water. The slope angle of this section, therefore, remains constant. It is therefore sometimes referred to as *constant slope.* Since the slope angle is determined by the balance between the debris supply from above and the rate of its removal downhill, it has also been called *debris controlled slope.* This straight segment is usually a talus where the bedrock surface is veneered with a layer of sliderock that is creeping (Fig. 118E).

(iv) *Concave Slope*—The concave slope is found in the lowest portion of the ideal slope profile. It is present at the bottom of a hill slope and extends down to the river valley or the alluvial plain. Although it is covered with a layer of debris, its form is really controlled by rainwash or surface wash. In the course of its development through time, the slope angle tends to become less and less. It is, therefore, called a *waning slope.* Because the concavity of the footslope is essentially related to transportation by flowing water, it is also called *lower wash slope or valley floor basement slope* (Fig. 118 B and C).

Mass Movement on Slopes

The two main groups of processes responsible for slope formation are weathering and surface transport. The prime role of weathering, both physical and chemical, is the production of rock waste or debris lying upon and being transported down the slopes. Usually there is a covering of unconsolidated materials or debris on the solid bed-rock, which is called *regolith.* (from Greek *regos* (blanket) and *lithos* (stone). Normally there is a donwslope movement of the rock debris and soil through various processes, some of which are slow but continuous while there are other processes which are responsible for larger and intermittent mass movement of the debris. Broadly, however, two classes of mass movements may be said to affect slopes—(i) surface mass movements which simply transfer debris downslope in a layer overlying the bedrock (such as creep, slump, solifuction etc), and (ii) deep-seated mass movement which affect and alter the bedrock surface (such as rockfalls and landslides). Mass movement is often used interchangeably with mass wasting, though there are some who would like the term mass movement to be reserved for movement of large masses as a unit. It is important to remember, however, that the present processes which may have been operating on a slope slowly for a fairly long time, may not be very significant as it is possible that the form of the slope has been primarily determined by large and intermittent movements. Further, it is also possible that the form of the slope has been primarily determined by a process which was active in the recent past when the climate was different (e.g. Pleistocene glaciation) and the present processes have not had time to exercise any significant influence on the slope form.

Classification of Mass Movements

Sharp (C.F.S. Sharpe, 1938) has made a detailed study of the different forms of mass movements. His classification is essentially based on the amount of water present in the debris, because water reduces the cohesive strength of fine-grained materials and acts as a lubricant in the downslope movement of debris. As the proportion of water present in the debris rises, the slope angle required for the transportation of materials becomes less and less. In other words, as the amount of debris in proportion to the water present increases, steeper and steeper slopes are needed for the transport of the materials. Mass movements can be divided into two broad categories on the basis of the proportion of water present in the debris—(i) Flowing, and (ii) Sliding.

Flow Movements

When water is present in the debris then the debris tends to flow down with the water. When the amount of water present is relatively small, the movement of debris is slow, and when the amount of water is more, the movement of debris is faster. Consequently, it is possible to identify two-sub-divisions of flow movements, namely, (a) Slow flow movements, and (b) Rapid flow movements.

(a) **Slow Flow Movements**—These include soil creep, talus creep, rock-glacier creep and solifluxion. Of these *soil creep* is the most important. This is the slow movement of soil materials down the slope under the influence of gravity. While the other three kinds of creep (namely, talus, rock and rock-glacier creep) involve rock debris, soil creep concerns only fine-grained particles. Soil creep is observed both in tropical and temperate climates. It is on account of soil creep that telegraph poles and trees are found bent downwards on slopes, and just below the upper layer of the soil we find lines of rock fragments which are related to the types of rocks found in the upper parts of the slope. Soil creep occurs as a result of the combined influence of several processes, the transporting capacity of each of which may be very small. Fine particles are transported down slopes with rainwash and the impact of raindrops, and where the fine particles from underneath the larger rock fragments have been removed, the downward movement of the larger fragments also becomes easier. Soil creep is also aided by frost heaving. In the process of frost heaving rock fragments of debris rise at right angles to the slope, but they fall perpendicularly downwards under the influence of gravity, and in this way they slide down along the slope. Further when a rock fragment expands by heating on a slope, its downslope side expands more than its upslope side, because on its downslope side the force of gravity also works along with expansion. On the contrary, when there is contraction by cooling, the contraction is more on the upslope side and less on the downslope side on account of gravitation. The net result of the alternate heating and cooling of rocks is that the rock fragments slowly move downwards. Soil falls from above into the cracks on the slope and aids in soil creep. Soil creep is also facilitated by ploughing and the treading of animals on slope.

Soil creep in tundra and colder regions and at high altitudes grades imperceptibly into *solifluxion*. The term is used to describe naturally occurring downslope movement of soil materials resulting from frost action. Solifluxion is particularly important in peri-glacial regions of the treeless tundra where on account of annual freezing and thawing, there is mass movement of the upper layer of over-saturated and disturbed debris in summer over the frozen permafrost even on relatively gentle slopes under the influence of gravity. Movement may be confined from a few cm depth to about 75 cm depth, depending upon the depth reached by annual freezing and thawing. Solifluction is largely absent from cold, frosted regions where the presence of tree roots stabilizes the soil.

In talus creep, rock creep and rock glacier creep there is mass movement of relatively coarse and dry debris under different conditions. According to Sharpe, *talus creep* is the downhill movement of material in screes; *rock creep* is the movement of jointed blocks, partly as a result of soil creep and partly as a result of sliding; while *rock glacier creep* consists of streams of boulders with interstitial ice and little soil.

(b) **Rapid Flow Movements**—In rapid flow movements the amount of water present in the debris is more, so that there is thorough lubrication of the debris and the mass movement takes place principally or wholly in the form of flow. Sharp (1938) mentions two principal examples of rapid flow movements—*earth flow* and *mud flow*. A *mud flow* is a flowing mass of predominantly fine-grained earth material that possesses a high degree of fluidity during movement. In mud flow the transported material is partially in a solid stage. The velocity of mud flows is at least 10 metres per hour, and in earth flow it is usually less. In both there is movement of wet debris but in mud flow the amount of water is more and the debris possesses a high degree of fluidity and lubrication and the slope angle is also more. Consequently the velocity of mud flow is more than earth flow. Mud flows occur most frequently in areas of steep slopes and sparse vegetation especially in environments that provide abundant incoherent fine-grained debris and intermittent heavy rainfall. The most favourable conditions for mud flow is found on mountains in arid and semi-arid regions where the natural vegetation is absent, there are periods of intense rainfall or local cloud-bursts after long intervals and there is accumulation of loose debris on steep slopes. In humid alpine environments mud flows usually result from the saturation of glacial debris by water from melting snow, and enormous mud flows have been observed in the Himalayas above the timber-line.

In *earth flow* the movement is so slow that usually it is imperceptible and the turf cover remains intact. Near the place of origin the surface sinks as a result of slumping, and towards the bottom of the slope where the earth flow stops or slackens, the ground swells at intervals. Clay soil is the one most likely to be affected by earth flow.

There is a third kind of rapid flow movement called *debris avalanches* in which the movement is so rapid that in the process of movement the mass of debris breaks. Like mud flows, debris avalanches also move in long and narrow paths and usually follow the gullies present along the hill slopes. There are some who believe that valleys may be carved on hill slopes by this process, but mud flows in any case appear to have limited erosional ability. Debris avalanches are also found on slopes on which there may be no gullies and under such conditions the entire cover of regolith may be removed from the slope. It is not necessary to have a thick cover of regolith for the formation of debris avalanche and in most cases this has been found to be only 20 to 100 cm deep. Examples of debris avalanche are met with in areas of soft sandstone mixed with clay as well as in areas where boulder clay and peri-glacial deposits are found on steep slopes.

In earth flow, mud flow as well as debris avalanche the regolith is completely wet, and they generally occur after an unprecedented heavy rainfall. But the onset of movement is due to lubrication, since only a thin film of water, such as is nearly always present, is required to produce the maximum decrease in cohesion between particles The cause is a rise in the water table to a position above the material affected which produces an increase in pore water pressure, and a consequent decrease in shearing strength.[1]

Sliding Movements or Landslides

In sliding there is a rapid downslope movement of principally dry materials. There are different kinds of sliding movements such as, rock slide, debris slide and landslide. Of thesee *landslide* is by far the most important. In landslide a mass of rock and debris separates from the underlying rock along a shear plane and moves rapidly downslope under the influence of gravity as a result of failure along the shear plane. The movement is often lubricated by water and tends to occur suddenly. In the beginning the movement is through slip and there is no deformation of the rocks above the failure surface. But after sometime the mass of rocks starts at least partially breaking and together with slip there may be also flow. In *rock slide* and *debris slide* the failure surface is shallow and frequently parallel to the surface, and when there is mass movement of consolidated rocks it is called rock slide, and when there is mass movement of unconsolidated debris or regolith, it is called debris slide. On rectilinear slopes with homogenous material the failure surface is concave to the slope. This is the ideal condition of slump or rotational slip. But where the rocks are not homogenous, the failure surface may take various forms.

1. Young, A—*Slopes*, Longman, New York, 1972, p. 76.

On the basis of Depth/Length or D/L ratio i.e. the ratio of the maximum thickness of the moving mass to its length up the slope, expressed as percentage, A.W. Skempton (1953) has divided landslides into two types—*surface landslides* and *deep landslides*. In surface landslides the D/L ratio is 2 to 5% while in deep landslides this ratio commonly ranges from 10 to 30%.

Causes of Landslides

Landslides result from various causes under different conditions. Broadly speaking, we may divide these causes into two major sub-divisions—passive causes which present favourable conditions for landslides, and active causes which are really responsible for the movement. Among the passive causes the following are noteworthy:-

(a) *Structure and Lithology of Rocks*—Certain lithologies and disposition of strata which make for low shearing strength, are usually unstable and the possibility of landslides in them is greater. Sedimentaries composed substantially of layered clays and shales, porous volcanics such as tuff and breccia, unconsolidated, weak or loosely coherent rocks, foliated rocks and permeable rocks underlain by impermeable strata dipping downslope are more prone to landslides.

(b) *Steep Slope*—It is necessary to have cliff or steep slope for the landslide to occur. Once the lansdlide starts it continues till the slope becomes so negligible that the mass of rock and debris comes to rest.

(c) *Hydrological Conditions*—On steep slope even a small amount of water is sufficient to lubricate the debris and initiate landslides. Thus conditions which favour the accumulation of water either on the surface or just below the slope surface are helpful in the causation of landslides. Apart from the water present in the debris, water enters the cracks and joints of rocks and freezes in winter in the cold regions. When the frozen water melts inside it helps landslide.

(d) *Heavy Rainfall*— Chances of a landslides are more in areas of heavy rainfall, especially in places where the forests have been cut down. Climates which have high intensity rainstorms present especially favourable conditions for landslides.

Of the *active causes,* three are particularly important—steepening of slopes by erosion, earthquakes, and unprecedented heavy rainfall especially when it is preceded by a fairly long wet period as a result of which there has been a marked rise in the water-table. The heavy rainfall saturates the slope and makes it unstable and most of the landslides are caused by prolonged and heavy downpour of rain. In the Himalayas landslides are common in the rainy season on steep slopes denuded of forests. Landslides are especially important in tectonically active zones subject to earthquakes in mountains in humid lands, cliffed coastal belts where marine erosion is active and in areas where there is undercutting at the toe of a slope by river erosion.

Geomorphological Effect of Landslides

A landslide leaves a basin-shaped scar on the slope from where the material has slid downwards and a mass of debris at the foot, often projecting forward as a tongue. As a rule landslide leads to a reduction of slope angle by transfer of rock and debris from upslope to downslope. Every subsequent landslide results in the formation of a gentler slope and even where steeper slopes are formed, they are of lesser height than the pre-existing one. The net geomorphological effect of landslides is, therefore, reduction in the steepness of the slope and to bring it to an angle at which it may attain long term stability. When a slope attains the stage of long term stability, landslides cease to occur on such a slope.

Landslide or any other rapid mass movement causes essentially catastrophic denudation which is different from the normal continuous denudation. Catastrophic denudation generally follows a heavy rainfall and is accomplished through rapid mass movement. At one time catastrophic denudation affects a limited area and a relatively large area is affected only in the long run. On the contrary, continuous denudation influences the entire slope or a major part of it simultaneously and continuously, though with varying intensity.[1]

The amount of denudation by landslides in a short time equals denudation done by the continuous process over thousands of years.

1. Catastrophic denudation is spatially localized in the short term and highly irregular in time, while continuous denudation acts more uniformaly in space and is less variable in time. Ibid, p. 85.

Some people are, therefore, of the view that on slopes above the angle of long-term stability as well as in other areas where there are clear evidences of mass movement, rapid mass movement is the principal mechanism of denudation. But this view cannot be advanced with any firmness, because rapid mass movement gives rise to dissected and irregular topography while in fact most of the slopes are not dissected and irregular.

Different Approaches in the Study of Slopes

The study of slopes involves primarily a consideration of the following matters— changes in the form and gradient of slopes with time, the relationship between slopes and the different processes of denudation and mass wasting, the influence of rocks and structure on slopes, the relationship between the slope-form and climate etc. The attempts that have been made to study the above problems can be broadly grouped under two main approaches- (i) the slope-evolution approach, and (ii) the process- form approach. The first approach lays emphasis on the evolution of slopes with time and the different stages in the evolution of slope from its initial form to its present form are analysed, while in the second approach the dynamic role of the processes is emphasized and the varied slopes produced by the processes of erosion in the different climates and over different rocks are studied.

(i) **Slope-evolution approach**: In this approach the evolution of slopes is studied from the historical point of view. The theories of slope evolution put forward by Davis and Penck belong to this class and are discussed at a later stage in this chapter. But there are several problems in this approach which cannot be easily solved. (a) Firstly, for the study of an existing slope it is necessary to have an idea of its initial form which is extremely difficult to find. Most of the scholars have assumed the initial form of the present slope to be a vertical cliff. (b) The other problem relates to the dating of the slopes. The slopes are kept in proper time-sequence on the basis of the study of slope profiles and their relative dates. But this is possible only in those places where there has been no interruption of any kind in slope decline. Further, it is extremely difficult to date the slope correctly where they have a cover of new deposits.

(ii) **Process-form approach** : In studies of this kind a close relationship between process and slope form has been indicated.

Process is the most potent factor in the determination of slope form, and in a particular area which denudation process will be most effective depends upon climate, rock type, structure and relief and the vegetation cover. Several problems have to be faced in this approach as well. (a) The processes determining slope form operate extremely slowly. Consequently their influence can be traced only by highly sophisticated methods. (b) In reality the present form of most slopes is not the result of present processes but is the relict of old slopes. (c) Which process influences slope form under what conditions has remained a controversial subject. (d) If it is correct that slope form is determined by process, then it is clear that different climates will have different slope forms, because the processes and their relative effectiveness change with climate. But there are some who are of the view that it is not always possible for the slope form to be of a distinct type in a particular climate. In fact, with the exception of glacial conditions the four principal elements of slope, namely, convex, free-face, rectilinear and concave, may be found in different proportions in all climatic regions.

Process - Form Relationships

In the earlier studies of process-form relationships only one particular process has been considered to be especially active in the determination of slope form, but more recent studies emphasize the fact that in the evolution of slopes not one but several processes act simultaneously. It is thus possible to divide the theories regarding process-form relationship into two broad classes—(a) Monoprocess or univariate approach and (b) Poly-process or multivariate approach.

(a) **Mono-process or Univariate approach :** In the earlier studies the evolution of normal slope i.e. the evolution of convexity and concavity has been ascribed primarily to soil creep and rainwash respectively. In 1908 Fenneman tried to explain slope form only with reference to running water. According to him, on account of the restricted area of the slope summit the amount of rainwater received is small and there is, therefore, limited erosion, but as we move down from the summit, both the surface flow and the debris increase and there is more erosion resulting in concave slope. Convex slope at the summit was thus a result of the action of rillwash which concentrated down-slope to produce gullying and concave slope. In 1945 Rebort E. Horton put forward similar views in his important quantitative and experimental study of the nature of overland flow and its influence upon slope form. He emphasized the erosional capability of running water (surface run-off) and concluded that the form of the concavity represents the minimum slope across which material brought in from upslope can be transported. Eroding force generally increases downslope from the divide due to increased run-off and the changing pattern of flow. In the upper convex part of slope there is little erosion because the eroding force of the rainbeat is less than the resistance of the slope material to erosion, after a certain critical length eroding force becomes equal to resisting force of slope material and below this erosion by running water begins. The middle concavity is therefore due to erosion by sheet flow, and there is lower concavity where the slope debris is accumulating. Though Horton's theory remains an authoritative explanation for many slope phenomena, the principal weakness of his theory as well as that of Fenneman is the lack of attention to the role of soil creep in slope development which is an active agent in humid regions.

G.K. Gilbert (1909) in his paper on 'The Convexity of hill tops' considered convexity to be principally the result of soil creep. Soil creep is minimum at the slope summit and increases downslope. As the amount of debris to be transported increases downslope, increasing slope is needed for the downward movement of the increased debris, and the slope becomes convex. Gilbert has not thrown light on how this becomes possible, because increased soil creep will ultimately lead to accumulation of debris in the lower part of the slope and then the form of the slope will become concave there.

A.C. Lawson (1932) put forward the view that the change from convexity to concavity represented a change in the nature of flow of water from eroding flow to aggrading flow. Rainwash is most effective in the upper part of the slope and less effective in the lower part. The amount of debris goes on increasing in the surface run-off as it moves downslope and erosion is less and finally in the lowest part there is deposition in place of erosion. Thus the longitudinal profile of a slope has two sections — an upper section in which there is erosion and a lower section in which instead of erosion there is deposition of debris. In this way convex slope develops in the upper part and concave slope in the lower parts. In other words, convexity is the result of erosion and concavity is the result of deposition. The intervening straight profile found on many slopes represents either (a) an equilibrium between erosion and deposition, or (b) a zone along which occurs seasonal oscillation of the point of change from the one type of flow to the other. With rain-wash the debris is being transported down the slope from the summit and the radius of the curve of the longitudinal profile goes on increasing, and consequently relief goes on decreasing. The main criticism of Lawson's theory is that there cannot be much erosion at the slope summit because here the surface run-off is small and there is paucity of debris which is the principal tool of erosion. But some distance below the summit the erosive power of water increases as a result of increase both in the amount of water and debris. Further, it cannot be said that the concavity of the lower part of the slope is invariably related to deposition.

(b) **Poly-process or Multivariate approach** : The weaknesses of the mono-processes approach are obvious, because various and complex processes are at work in the evolution of slopes. H. Baulig (1940) has criticized the views of Fenneman and Gilbert and has called the concept of one process one slope to be misleading and wrong. Baulig regarded regolith as the fundamental factor controlling slope form and development and recognized that both rainwash and soil-creep were important slope forming processes. In the upper portion of the convexo-concave slope the erosional effect of rainwash is very limited, as the amount of water is small because of the small catchment area, and most of the flow is in the form of sheet wash. Convex slopes occur at the summit because the debris at the summit and ridge tops is thin, coarse and permeable, run-off is limited and soil creep dominates. The convex form of the summit is therefore mainly determined by soil creep. Lower down the debris become finer, thicker and impermeable, the amount of slope wash increases and becomes concentrated into rills and flows rapidly down in several channels. Erosion, therefore, increases and soil creep does not remain so effective. In this way concavity develops in the lower part of the slope because of the change in the nature of flow. Rainwash and soil creep both are agents of downslope movement of debris and work together. Although in the lower part erosion by running water (rills) is important but as the debris becomes finer and wetter, erosion by soil-creep also increases. It has been found by actual field studies and measurements that in the upper part of the slope the debris received by erosion is not transported down. Consequently the steepness of the slope increases and convex slope develops. Lower down on account of increase in surface flow by rainwash and increase in the mobility of soil creep, there is increase in the mass movement of the debris and this increase is much greater than the increase in the amount of debris. Because of the greater effectiveness of mass movement there is the development of concave slope. Schumm (1956) confirmed that both soil creep and overland flow (rainwash) are important in determining hill slope form.

For temperate humid climate the theory of Baulig is true to a large extent and as a rule convexity is found in the upper part of the slope and is the result of concentrated rill action. Sometimes in between these two is found a rectilinear or straight section in the slope profile where it would seem that the downslope increase in load balances the downslope increase in transporting power.

It is now widely accepted that convexity is mainly the result of soil creep, while concavity is essentially produced by the action of rain-wash. But it is important to bear in mind that the effectiveness of these processes and their influence on slope form vary with climate, lithology and structure, nature of the regolith and age. What proportion of the slope will be affected by soil creep and what proportion by rain-wash will vary with climate. In humid climates the areas affected by rainwash is likely to be bigger, and concave form will predominate. On the other hand, in arid climates convex slopes are likely to develop. The influence of impermeable rocks is similar to the influence of heavy rainfall, and concave slope mostly develops in areas of impermeable rocks. On the contrary, slopes developed upon chalk are typically convex due to its high permeability. Rock types also influence slopes by determining regolith characteristics, particularly its texture and infiltration capacity. Granite is not permeable rock but the coarse debris produced by it is permeable and is conducive to the production of convexity. For various reasons, the concave portion of the slope becomes increasingly important with age and the convex portion tends to decrease. Further, youthful slopes may be irregular and steep, but with age the slopes tend to become smooth and gentle.

Theories of Slope Evolution

In slope evolution we study the change of slope form with time.

A basic question is, under certain given conditions and after the

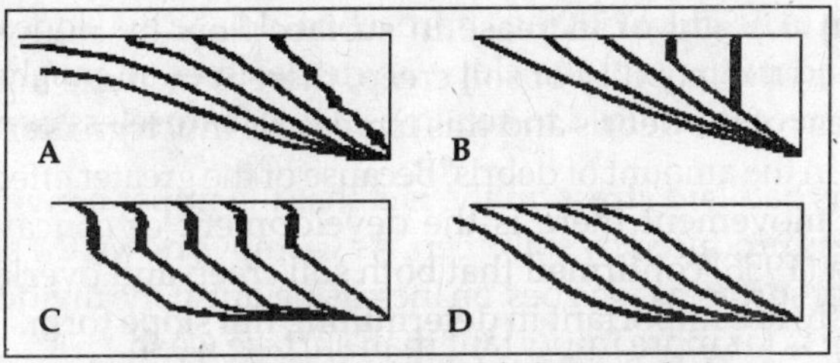

Fig. 118—Three concepts of slope evolution.
A-Slope decline, B-Slope replacement,
C-Parallel retreat with free face, D-Parallel retreat without free

processes have worked for a period of time, whether the slope angle increases or decreases or remains constant. In this connection three important theories have been put forward:

(i) Slope decline theory of W.M. Davis,
(ii) Slope replacement theory of W. Penck, and
(iii) Parallel retreat theory of L.C. King.

In *slope decline* the angle of the steepest part of the slope goes on progressively decreasing and there is the development of convexity and concavity. In *slope replacement* the decline of the maximum slope is achieved through the establishment of gentle slope from below until the entire slope becomes concave or is divided into concave sections. In *parallel retreat* the maximum angle remains constant, the length of the concavity goes on increasing but the slope angle of the concave part goes on decreasing. Parallel retreat can take place with or without free face (Fig. 119).

Slope decline theory of W.M. Davis

Davis has put forward his view regarding slope evolution in the context of his geomorphic cycle. According to him as the cycle of erosion advances the slope form changes with time and ultimately the slope declines. The slope form is dependent upon time or the stage of the erosion cycle. In the youthful stage of the cycle there is the development of steep convex slope on account of vertical down-cutting by rivers. In the mature stage, lateral erosion becomes more important and the summit of the divides is also subjected to erosion. There is thus a progressive decline in the slope angle as a result of the wasting divides, and the form of the divides tends to become rounded or convex and simultaneously, the rounded divides suffer reduction in height. With the approach of old age the slopes become flatter, and in the end the slope angle declines so much that the entire surface is reduced to an almost level featureless peneplain.

Davis had laid stress on the fact that the upper convex part of the slope is produced by soil creep. As we move down the slope, the amount of surface wash goes on increasing but at the divide summit soil creep is far more important than surface wash.

Davis has also put forward the concept of graded slope or graded valley sides and graded waste sheets. In his view the river as well as the mobile debris sheet represents basically a moving mixture of rock debris and water in different proportions. The establishment of the grade starts from the lowest part of the slope and extends from below upwards. From any point on the graded slope the debris is being transported down into the river. In fact he equates a graded slope with a graded waste sheet. A graded slope is one having a continuous soil cover, while a graded waste sheet is one on which there is equality between the supply and removal of debris.[1] To use his own words, 'just as graded rivers slowly degrade their courses after the period of maximum load is past, so graded waste sheets adopt gentler and gentler slopes.... When the graded slopes are first developed they are steep and the waste that covers them is coarse and of moderate thickness.. In a more advanced stage of the cycle the graded slopes are moderate, and the waste that covers them is of fine texture and greater depth than before.' (1899) 'The retreat of a valley side is usually accompanied by a decrease in the steepness of its slope as well as by the development of a convex profile at its top and a concave profile at its base'. (1932).

---

1. " In the youthful stage of the cycle the rocky cliff and ledges that often surmount graded slopes are not yet graded if waste is removed from them faster than it is supplied by local weathering and by creeping from still higher slopes, and hence the cliffs and ledges are left almost bare....As late maturity passes into old age, even the ledges on ridge-crests and spur-fronts disappear, all being concealed in a universal sheet of slowly creeping waste. From any point on such a surface a graded slope leads the waste down to the streams. At any point the agencies of removal are just able to cope with the waste that is there weathered plus that which comes from further uphill". Davis-*Geographical Journal*, 1899. Quoted by Young, A.-op.cit., p.26

Fig. 119 A — Slope decline theory of Davis, B-Parallel retreat theory of King, with free face and without free face.

After this Davis comes to his theory of slope decline. As the cycle of erosion progresses and relief gradually decreases, slope profiles change from steep and straight during early maturity to convex-concave by late maturity. As the cycle advances, both the convex and concave elements become longer and the radius of curvature goes on increasing. Slopes decrease in declivity as relief is reduced and there is a reclining slope retreat and a decline in both height and inclination. His theory is empirical and based upon direct observation in nature and topographic maps and pure deduction. He expressed this theory in a composite profile drawing lacking units of measure and without a time-scale (Fig. 120 A), but the decline of slopes with time was made clear enough. According to Davis the principal reason for the decline of slopes is that debris is being transported from the upper convex part of the slope to the lower part. As the cover of debris increases in the lower section, the rocks in the lower part are protected from erosion. In his opinion, slope decline prevails in the humid regions and even in the arid regions there is decline of slopes with time, though in his later studies (1930) he recognized that parallel retreat occurs in arid climates. The principal merit of Davis's theory is that in the context of geomorphic cycle it emphasizes the fact that slope form is time dependent

Slope Replacement theory of W. Penck

The views of Penck are completely at variance with those of Davis. He did not agree with the Davisian assumption that during an erosion cycle uplift is complete before erosion which in turn is largely determined by the nature of the uplift. When uplift is rapid

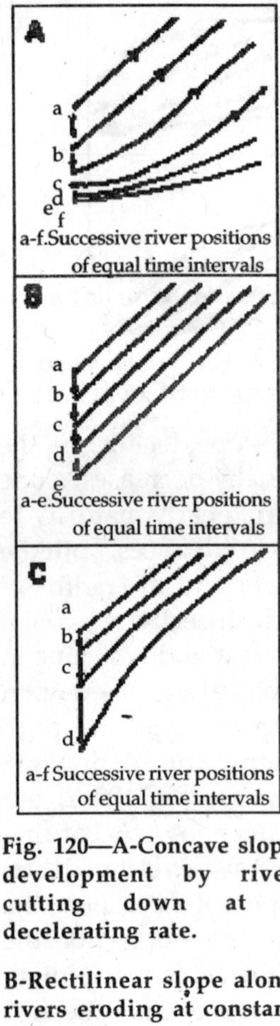

Fig. 120—A-Concave slope development by river cutting down at a decelerating rate.

B-Rectilinear slope along rivers eroding at constant rate.

C-Convex slope along rivers cutting at accelerating rate.

down cutting by streams would be more pronounced and the valley slopes would be convex. In a state of constant uplift i.e. where rate of uplift is equalled by erosion the streams would cut down only as fast as the land rose, and the slopes will be straight. In a state of decreasing uplift or stability, the rate of erosion will be less and concave slopes will develop. Penck made further deductions which have not been generally accepted but the theoretical development outlined above has met with a fair degree of acceptance. Undercuting by streams does commonly steepen adjacent valley side slopes, resulting in local convexities. Straight slopes are associated with high relief which may result locally from the rate of uplift being equal to rate of erosion.

Penck's main thesis was that slopes don't necessarily change form in an evolutionary manner with the passage of time as assumed by Davis but that slope forms (and the way they are altered) are determined by one overriding control—the rate of downward erosion of the river flowing at their bases. According to Penck (A) concave profile develops above rivers which are cutting down at a decelerating rate, (B) rectilinear slopes develop above rivers thatare eroding at a constant rate, and (C) convex slopes are formed above rivers cutting down at an accelerating rate (See Fig. 121). In his view, the precise rate of river erosion was of the utmost importance. A river cutting down at a rapid constant rate will give rise to a steeper rectilinear slope than a river cutting down at a slow constant rate.

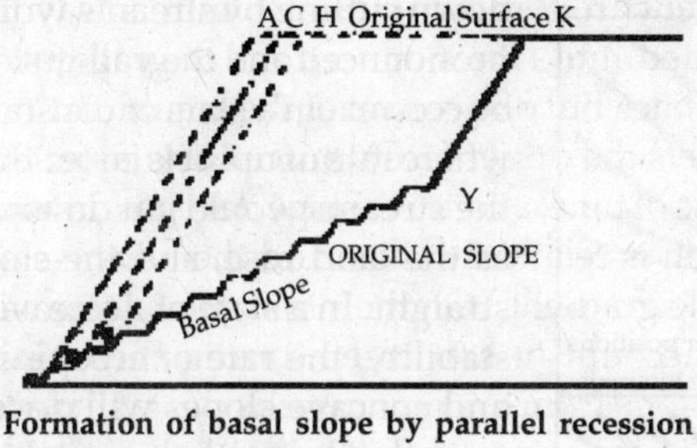

Fig. 121 — Formation of basal slope by parallel recession according to Penck.

As an illustration of the methods of analysis of Penck, the evolution of the concave slope is discussed below (Fig. 123). For the explanation of his theory, Penck assumes a straight rock slope on which the rate of weathering is the same throughout. The weathered rock fragments fall downslope under the influence of gravitation and collect near the base of the slope where the stream is flowing.

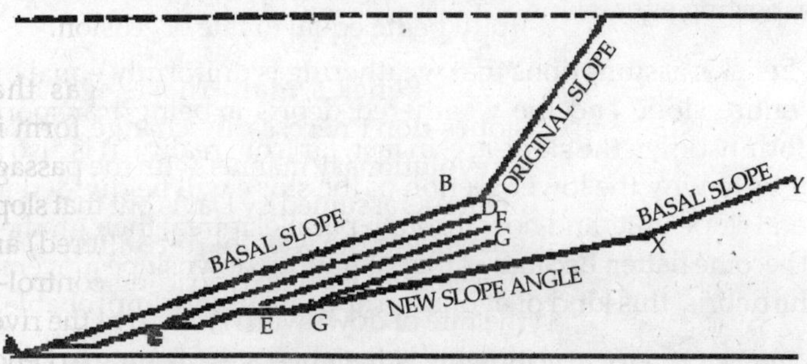

Fig. 122 —With slope development, the original steep slope changes into a basal slope of lesser gradient.

In this way after sometime the slope profile moves back from AB to CD. As a result of the continuation of this process, the slope profile moves back further from C D to EF, but the lowest part of the slope remains stationary, because here the slope is insufficient for the rock fragments to move further down. In course of time, the original slope moves back to X Y through this process, and the lower section of the slope assumes a terrace-like form. On this basal slope there is weathering, but as the gradient is small the debris is not transported till it has weathered into fine particles. When this debris is transported down it stops near the base in the absence of slope, and in the lower section of the new basal slope, facets of gentler gradient are formed (Fig. 123). In this way the new basal slope is divided into facets or sections of gradient. These facets slowly move back but on account of the accumulation of debris on the lower facets the upper facets move back faster than the lower ones. In course of time, a concave slope develops in which the upper steeper section is relatively short and the lower section is longer and has gentle gradient.[1] On the slope summit convex slope may develop on account of soil creep, but in the opinion of Penck flattening of slope from top downwards can take place only up to a limited extent, and basically flattening of slopes takes place from below upwards.

According to Penck the form and angle of slopes are determined mainly by the rate of erosion by the river which itself is controlled by the rate of uplift. Penck does not attach much importance to slope processes, amount and calibre of the debris or volume of surface water. In his view the slope form is not dependent upon the processes operating upon it and the processes act merely as transporting agents.

Penck's assumptions that weathering is uniformly equal over the entire slope and the weathered debris is being transported uniformly down the slope are, in fact, far from reality. It is also not clear as to why the lower section of the slope will be divided into several slope units and how through parallel retreat their gradients will become flatter. Because of the continuous downslope movement of the debris, this kind of evolution appears highly improbable.

Penck's theory was initially called the theory of parallel retreat. In the opinion of some people this is incorrect, as according to Penck parallel retreat takes place only on the initial rectilinear slope. At a fairly early stage in its evolution, this changes into a concave slope, and after that the evolution is through slope replacement from below upwards. The theory is therefore more appropriately called the theory of slope replacement.[1]

Parallel retreat theory of L.C King

King's theory of slope evolution is a part of his more

---

1. A steep rock face left to itself, moves back upslope, maintaining its original gradient, and a basal slope of lesser gradient develops at its expense. Penck, W.—*Morphological Analysis of Landforms*, Engligh translation by H. Czech and K.G. Boswell, Macmillan, London, 1953, p.135

comprehensive theory of landscape evolution. His concept of pediplanation cycle is based on the parallel retreat of slopes. While Davis's slope decline produces peneplains, King's parallel retreat gives rise to pediments and pediplain. In the initial stage of pediplanation steep slopes develop as a result of vertical downcutting by streams. When the angle of these slopes stabilizes, the slopes tend to move away from the river by parallel retreat, and the pediment between the river and the slope is thus extended. In the course of parallel retreat the gradient of the slope remains constant. In course of time the slopes start disappearing as a result of the extension of pediments of neighbouring streams, and at the end of the cycle a multi-concave landscape is formed. But it is likely that at the slope summits convex slopes may develop on a limited scale due to weathering and creep.

King accepts the hillslope elements of Wood and states that these elements are part and parcel of slope evolution. Full development is dependent upon local conditions in which the hard bed rock and adequate relief are especially important. If one of the two conditions is not present then both scarp as well as debris slope will be absent and only a convex-concave slope will develop. But in the opinion of King, the existence of scarp in a slope is a normal feature, and in the evolution of slope it is the scarp element which wears back parallel to itself and controls the evolution of the slope as a whole. As a result of the retreat of the upper element, the pediment extends in length "Since the concave curve which the pediment acquires early in its growth would, if extended upslope, progressively eliminate the upper elements, it must be that the pediment is regraded; that is, as it extends in length its angle is slightly but continuously reduced."[2] King contended that slopes develop similarly virtually in all climates, and his theory was true for all climates except for the very cold and the very arid climates.

---

1. In Penck's system parallel retreat only takes place on an initial rectilinear slope; such a slope is transformed into a concave form relatively early in its evolutionary history, and development then proceeds by successive replacement from below. If this replacement is not in discrete stages but continuous then no part of the concave slope retreats parallel to itself. This system of slope evolution is therefore more appropriately referred to as the hypothesis of slope replacement. Young, A.—op. cit., p. 34.
2. Young, A.—op. cit., p. 36

Davis (1930) admitted that humid and arid slopes might sometimes be similar in form but he did not accept that they were similar in origin. Despite the insistence of King, most geomorphologists seem to believe that slope forms and processes do vary with climate. However, there are many other factors which control slopes. Further, there are some slopes whose origin is related to past climatic conditions.

One of the principal weaknesses of King's theory is that it is not based on field observation and nowhere do we find a measurement of either form or process. But even more important than this is the criticism that King has not given adequate evidence in support of his views, and has in fact sometimes ignored adverse evidences, and has employed repeated assertion in place of evidence as a means of establishing his hypothesis.[1] There is lack of objectivity in his writings. But the value of his work lies in the fact he has put forward some valuable hypotheses for testing and analysis, though he himself has failed to prove them.

The Concept of Wood

In order to put forward his hypothesis, A. Wood (1942) starts with a debris-free cliff slope. He calls it free face. The retreat of the cliff starts on account of weathering and the debris received from the weathering of the free face starts collecting near the base of the slope in the form of talus or scree (Fig. 124 B) Gradually the talus is extended upwards (Fig. 124 C) and the debris is deposited at a fixed angle in the lower section of the free face slope. This has been called constant slope. In the early stages the talus grows from below

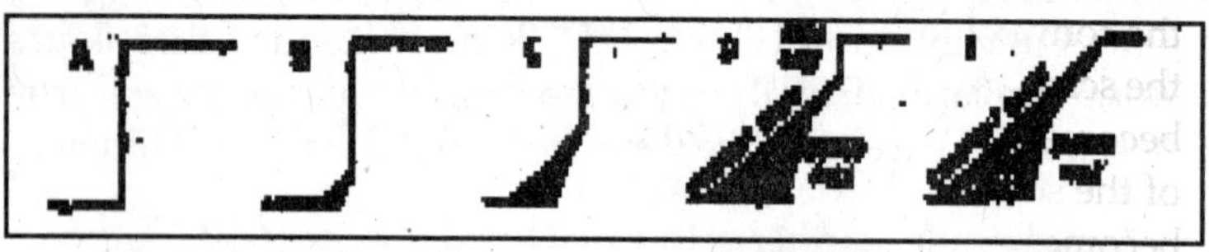

Fig. 123—Development of constant slope according to Wood.

upwards at a rapid rate because it receives the debris from the weathering of the entire free-face. But with the passage of time as the lower section is covered more and more with talus, the uncovered upper section of the free face is proportionately reduced and the

1. Ibid, p. 37

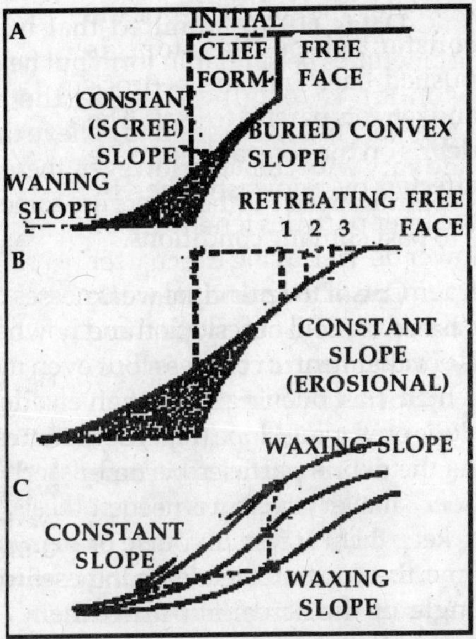

Fig. 124 — Development of slopes according to Wood.

amount of debris received from above consequently decreases. Thus the rate of vertical growth of the talus is greatly retarded as the area of the free-face is reduced. As each higher portion of the free-face is exposed to longer periods of weathering than the portion immediately below it and suffers erosion corresponding to those periods, there is the development of convex rock slope below the cover of debris. Through this process ultimately the entire slope is covered by talus and forms a constant slope and udnerneath the debris is the convex buried face as shown in the diagram (Fig. 125).

This is an ideal situation and Wood admits that for various reasons there may be variations in this. The first important fact to bear in mind is that the volume of scree is more than the volume of the weathered rock, because there is interstitial space between the rock fragments. The vertical growth of the talus is therefore normally much faster than assumed in Wood's model and the steepness of the convex buried face is also more marked. If the finer materials in the scree are transported down by rain-wash or the debris particles become smaller by chemical weathering, then the upward growth of the scree is slowed down and the slope of the buried face also becomes gentle. Further, if the weathered rock fragments are coarse, the talus growth will be more rapid, and if the nature of rocks is such that weathering produces finer fragments the growth of the scree will be slow. In the first case i.e. where the debris is coarse the slope of buried face will be steeper than in the second case where the debris produced by the parent rock are fine.[1]

1. Sparks, B.W. : op. cit., 1972, p. 68

After the formation of the constant slope, the slope is pushed back into the bedrock and retreats parallel to itself. The debris on the constant slope are affected by rainwash whereby the finer particles tend to move lower down leaving the coarser fragments above, and near the base of the constant slope there is development of concave slope, which has been called wash slope or waning slope (Fig. 125). As the debris particles become finer, smaller gradient is needed to keep them at rest. In course of time, the slope, still retaining the angle of the scree, is pushed back into the solid rock behind the scree and continues to retreat parallel to itself.[2] In this way the convex rock slope underneath the debris is transformed into a concave slope.

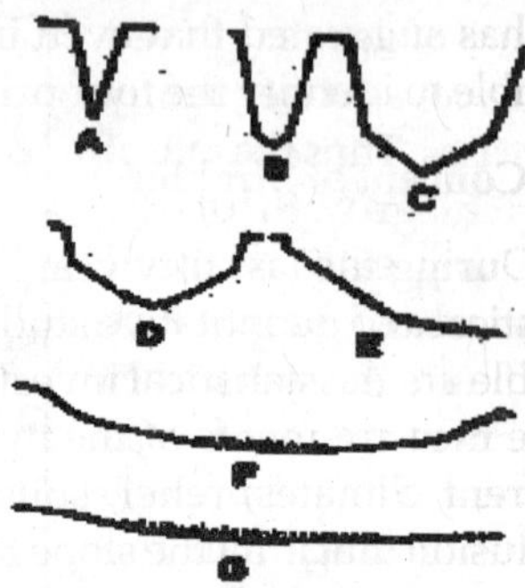

Fig. 125 — Slope cycle according to Wood
A-Free face only, B and C-Constant slope forms; D and E-Waning slope develops, F-Waxing slope with alluvial filing; G-Slopes flatten to approach peneplain

On the cliff summit, under the influence of weathering the form of the initial angular intersection becomes rounded as a result of which there develops a summital convexity. This is known as *waxing slope*. Thus along the logitudinal profile of the entire slope we find three major elements-summital convexity, basal concavity and middle rectilinearity. In the course of evolution, both the concave and convex sections tend to be lengthened and the length of rectilinear section is reduced. Finally, the rectilinear section disappears altogether resulting in the development of a convexo-concave slope which tends to flatten with time. Wood has emphasized the fact that all slopes do not develop in a uniform manner but their evolution varies with structure, climate, tectonic activity and other factors. Under specific conditions, one or more of the factors may be absent, and in the course of evolution the form and character of the four slope elements keep on changing (Fig. 126).

The principal contribution of Wood lies in the fact that in his hypothesis he has determined and defined the elements of slope,

1. Ibid., p.37

and has suggested that even in apparently dissimilar slopes it is possible to identify the four principal elements.

The Concept of Strahler

During the last fifty years attempts have been made to make statistical and quantitative studies of slopes. Among these the most notable are the statistical investigations of A.N. Strahler (1950). He made measurements of the maximum slope angles in regions of different climates, relief, soils and vegetation and came to the conclusion that if all the slope forming processes and the age of the slope are similar, the average maximum slope angle is the same and the variation in the maximum angle of the different slopes is negligible. Strahler has explained that the development of the same gradient in all the slopes is due to the fact that they may be able to transport their debris easily downslope. Such slopes are in a state of equilibrium and may be called *euqilibrium slopes*. But the equilibrium slopes are governed by the various slope forming processes. Change in the nature of the processes and their activities will result in readjustments of the equilibrium. According to Strahler, in all areas the maximum angle of retreating slopes is similar, even though they are of different age. On slopes having similar maximum angles, parallel retreat is more important than slope decline.

Strahler has also shown on the basis of his field investigations that maximum slope angle and channel gradient are inter-related. This confirms the fact that channel slope is adjusted in accordance with the debris obtained from the valleyside slopes. In other words, the entire system is in equilibrium. Where the valleyside slope is steep, the channel slope is also steep so that it may transport the maximum amount of debris. On the other hand, less debris is available from gentle valleyside slope and consequently the channel slope is also gentle.[1] There are however, some exceptions to this general rule. For instance, even where the valleyside slope has similar gradient, the river receives less debris if the slope has a cover of vegetation than in the case of a naked slope, and consequently the channel gradient is also less.

Strahler has also shown that where the stream is close to the

1. As the gradients of streams decrease towards their mouths, so do the side slopes flatten correspondingly, a state of affairs visualised in Davis's cycle of erosion. Ibid., p.90.

foot of the slope, the slope is steep. In such a situation the stream transports rapidly the debris obtained from the valleyside slope. But where the stream is away from the foot of the slope, the slope is gentle and the river is not able to transport all the debris. There is accumulation of debris on the slope which is thus protected from erosion.

The Concept of Savigear

R.A. Savigear (1952) studied the development of profiles of slopes located at the head of Carmarthen Bay in south Wales and came to the conclusion that both parallel retreat and slope decline are important in the evolution of slopes, and sometimes both the processes may be operating simultaneously in a particular region. In the eastern part of his study area he found slopes which have a basal free face, middle rectilinear slope and limited convex slope at the summit. In the western part of the study area, on the other hand, there is extensive convexity at the summit, limited rectilinear slope in the middle and concave slope at the bottom. The western part has receded long ago from the sea and the eastern part has receded from the coast recently. Consequently the slope becomes flatter as we move from east to west. Not only are the slopes steeper in the eastern part but are predominantly convex whereas in the western part the slope is mostly concave. From this it may be concluded that where the debris are tranported rapidly by the seawaves, the sea-cliff is vertical but after recession of the sea the gradient of the slope starts declining i.e. the slope becomes more gentle. In other words, there is a close correlation between slope angle and the active agent at the foot of the slope. This active agent can be waves on the sea-coast or a stream on the land. Thus it is necessary to have an active agent at the foot of the slope for parallel retreat. In the absence of an active agent the debris begin to be deposited which provided protection to the lower section of the slope, while retreat continues in the upper section as a result of continued weathering. In such a situation there is decline in the slope angle and the slope becomes flatter. Savigear has also suggested that the summital convexity develops at quite a late stage. His analysis appears very close to reality

Conclusion

It is clear from the above discussion that the slope-forming processes are extremely complex and inter-related, and our knowledge about them is incomplete and rather general. Even so it is possible to imagine a general cycle of slope development. In the initial stages of development there is increase in slope gradient with time. In the middle stages there is parallel retreat under the influence of active streams and constant basal erosion, and in the later stages there is decline in steepness as the active eroding agent gets separated or becomes far removed from the foot of the slope.

Questions

1. What are the elements of a slope and how do they develop ? What is the importance of slopes in geomorphology?
2. Classify mass movement on slopes and discuss their significance.
3. Discuss three important theories regarding slope evolution.
4. Explain what you understand by slope decline, slope replacement and parallel retreat and show the significance of these processes in slope evolution.
5. Discuss critically the views of Davis, Penck and King regarding slope evolution.
6. Explain the view of Wood and Savigear regarding the evolution of slopes.
7. Analyse critically the different approaches to the study of slopes.
8. Write short notes on :—
   (a) Landslides—their causes and geomorphic effect.
   (b) Slow flow mass movements.
   (c) Gilbert's hypothesis of upper convexity.

Chapter 21

Planation or Erosion Surfaces

G.K Gilbert (1877) defined planation as the 'process of carrying away the rock so as to produce an even surface, and at the same time covering it with an alluvial deposit'. The term planation surface is employed to describe the geographically plain surface or surface of faint relief which is the end-product of all processes of planation by erosion, and which might have been produced either by a completed or incomplete cycle of erosion. It is synonymous with the term 'erosion surface' frequently used by the British geographers. Each one of the processes of erosion mentioned earlier is, theoretically speaking, capable of producing a planation surface. A planation surface may be formed by sub-aerial erosion, marine erosion, pediplanation and possibly other processes and the difference in process may result in some difference in its form. Whether the planation surface has been completely formed or not, depends on the duration for which the process has been active. In the recent geological times the crust has been unstable and consequently planation surfaces are found at a height in the present day landscape. Where the warping in the earth's crust is small or limited in area, it is possible to identify the traces or remnants of the old erosion surfaces. But where the deformation is more marked, it becomes extremely difficult and sometimes impossible to identify the old erosion surfaces and to establish any correlation between them.

The surviving fragments of the old planation surfaces at different altitudes provides clues of their approximate age and the number of times the denuding agencies have tried to base level the surface. They help in understanding the historical evolution of the landscape and its present form. A study of the altitudinal levels and origin of the planation surfaces gives an insight into the old processes and base levels of erosion. The gentle warping of the planation surfaces furnishes facts and clues regarding recent tectonic movements. It must however, be admitted that the correct identification of planation surfaces and their extent is an extremely difficult task.

There are various types of planation surfaces. Planation surface is an omnibus term and includes peneplain, panplain, pediplain, plain of marine origin, as well as glacial and periglacial erosion surfaces. In addition, it also includes exhumed surfaces i.e. an ancient surface which was buried under a cover of sediments and has been subsequently exhumed or exposed on the surface as a result of the erosion of the overlying sediments. It is, of course, not necessary for the exhumed surface to be always a former erosion surface.

Peneplains

Peneplain is an almost featureless plain which characterizes the final stage of a fluvial cycle as conceived by W.M. Davis (1889, 1895). The concept of peneplain has already been discussed at some length in Chapter 18. Davis was not the originator of the idea of peneplain, he only gave it a name and a clear definition. The concept itself came from the pioneer work of Powell and Dutton. Powell (1875) introduced the concept of the base-level of erosion below which dry lands cannot be eroded. Dutton gave greater precision to the idea and wrote that if time be long enough, every region, in its turn, will approach nearer and nearer the base levels of erosion and at last sensibly reach it (1882). This goal is the peneplain state of Davis.

Peneplain is not a plain of flat surface but is a lowland of faint relief where the rivers flow on an extremely gentle slope through wide valleys and the divides separate the river valleys in broad gentle swells. On such a plain the slope is minimal and just sufficient for the transportation of the debris. The slope is determined by the principal rivers and these rivers have the minimum necessary slope for debris transport. In the smaller streams the slope is more than in the principal rivers. It is not easy to define the form of the peneplain. It undoubtedly represents the penultimate stages of the erosion cycle. But it cannot be said that this is the final form (state), because even this continues to undergo slow degradation. Neither it is a completely flat plain nor are the divide summits of the same height, as the streams continue to have some slope. Consequently the height of the peneplain increases inwards away from the coast.

Further, the process of peneplanation starts from near the coast as it is here that the river first attains its base level of erosion. Gradually the peneplain is extended up inland away from the coast where the surface is already in the late stages of maturity, although there is no clear-cut dividing line between the two.

Examples of peneplain are extremely rare, especially those which can be related to the present base level. Some of the critics, therefore go to the extent of saying that the peneplain does not in fact exist and is a mere abstraction. Davis has, of course, pointed out that for the formation of a peneplain a long geological period, such as the Cretaceous or the Tertiary is needed. It seems improbable for a land surface to remain standstill for such a long time so that an extensive peneplain could develop on it. The present sea-level is only about 5,000 years old and the time available is too short for peneplanation to occur. There are, however, extensive areas of weak rocks such as the Eocene clays of the London Basin and the Cretaceous clays of the Weald of south-eastern England which have been reduced to the peneplain form during the last one million years. Davis described peneplains in two radically different situations. In the first the peneplain is buried beneath later deposits. In the second, the peneplain has been uplifted or even warped and then incised by rivers so that it appears in the landscape as an upland plain or plateau. There are numerous examples of such plateaus.

Peneplain may be recognised on the basis of the following characteristics—(a) form of the peneplain, (b) nature of deposits on the peneplain, (c) the nature of drainage on the peneplain, and (d) accordance of summit levels of the surviving fragments of the peneplain.

(a) **Form of the Peneplain** — The principal character of the form of the peneplain is the predominance of lowlying surface and faint relief. This, however, is not of much help in identification, as the peneplain is rarely, if ever found on the original level of its formation. More commonly they have been uplifted and sometimes warped and are found on the summit of mountains and plateaus where they have been also dissected. It becomes impossible to determine whether the surface of low relief is a fragment of the old peneplain or whether it is the product of subsequent sub-aerial erosion. They, however, truncate the geological strata which underlie them and may occur on soft and resistant rocks alike, showing little sympathy with structure.

(b) **Nature of Deposits** —The nature of the deposits on the peneplain helps to some extent in its identification. As a rule there should be a deep cover of weathered debris and soil. The clay with flints on the Chalk plateau of southern England is such a peneplain deposit and comprises the insoluble residue of clay and flints from the chalk and the once overlying Eocene strata. The pediplains developed in tropical climates have also a deep cover of debris. In pediplains this is formed by the process of laterization, but in peneplain the deposit is alluvial. Even so on account of the uplift and dissection, it is but rarely that the nature of the debris provides a direct clue as to the existence and origin of erosion surfaces. Normally the debris cover is so light that it is not possible to come to any definite conclusion with its help.

(c) **The Drainage Pattern** — It is an important characteristic of the peneplain that the drainage network is well adjusted to the geological structure of the underlying rocks. This kind of adjustment is not found in plain formed by marine erosion. A high degree of adaptation of drainage to structure is a diagnostic characteristic of a peneplain and helps us in identifying a peneplain.[1]

(d) **Accordance of Summit Levels** — Most peneplains are represented today only by their surviving fragments, and these fragments reveal their ancient unity in a general accordance of their summit levels. Where over an extensive area well adapted drainage pattern is found together with accordance in summit levels, it is highly probable that the surface represents a remnant of a penenplain. Accordance of summit levels and accordance of inter-stream levels are useful indicators of an uplifted peneplain. But accordance of summits may also arise from other causes. In the Alps mountains accordance of summit level is found in extensive areas, but it would be incorrect to conclude from this that the peneplains were uplifted to that height. This may arise from severe frost shattering above the treeline which will tend to reduce an initially irregular series of summits to a rough accordance of level.

1. An uplifted and dissected surface with a drainage pattern adapted to the details of the structure is more likely to be a peneplain than any other form of erosion surface, except perhaps a panplain. Sparks, B.W.—Ibid, p. 461

In areas of homogenous rocks where the valleys are almost equidistant and have similar depth and slope angles there is possibility of accordance of summit level.

In several areas of the world a staircase pattern of erosion surfaces is found. The planation surfaces are located at different heights and there is a step-like slope in between two levels. In such cases it cannot be said with any definiteness that these surfaces are fragments of different incomplete peneplains.

While it is true that there is no example of a peneplain where base level may be related to the existing sea level, it is also true that examples of partial or fragmented peneplains do exist in several parts of the world. For example, the area lying to the south and east of the Urals and southern Finland are said to be ancient peneplain surfaces. Examples of ancient uplifted peneplains are to be found in the Appalachians (the Shooley peneplain), Quebec plateau of Canada, central highlands of France, south-eastern England and the Chota Nagpur Plateau (Ranchi and Hazaribagh plateaus) in India.

Panplains

C.H. Crickmay (1933) has proposed an alternative concept of the normal cycle of erosion which in his opinion is nearer to reality than Davis's concept. According to him, lateral corrasion assumes a dominant role in the late stages of the erosion cycle. As a result of planation by lateral corrasion of the inter-stream areas the rivers form broad flood-plains. Ultimately the growing flood-plains become confluent, giving a surface much flatter than that of a peneplain. This plain formed by coalescence or joining together of flood-plains he calls a panplain. These flood-plains are thinly veneered with sediments and the sides of the flood-plains are relatively steep. In the Davisian concept peneplain is primarily produced by the down wasting of the inter-stream areas. But according to Crickmay lateral erosion continues to be important even in the late stage of the cycle, and till the end of the cycle the edges of the flood-plains remain steep-sided while the flood-plains continue to grow. The panplain is much flatter than the peneplain and the residual monadonocks are more steep-sided. Further while in the peneplain there is predominance of weathered debris, in the panplain sedimentary deposits of river alluvium are found.

According to Crickmay the homologues of panplain in a desert are "the rock floor of the desert, known as pediments and panfans". King (1947) thought "the essential condition of panplain is wide areas of river-cut plain", and he wrote of "Savanna with wide panplains." Wooldrige (in Taylor, 1951, *Geography in the Twentieth Century*, p. 170) defined panplanation as "the integrated product of long-continued lateral corrasion by rivers".

There are certain difficulties in accepting Crickmay's hypotheses. His panplain is formed essentially by lateral corrasion and its occurrence and extent depend on the interpretation of the term flood-plain. It does not seem possible that all the land will be reduced to a flat surface purely by lateral corrasion towards the end of the cycle. Further, the flood-plain occupies a relatively small part of the plain-like surface developed in the last stages of the cycle. It does not, therefore, seem possible that the extensive flat surface of the panplain will be formed purely by the growth and joining together of the flood-plains. The concept of the panplain cannot be accepted as an alternative to that of peneplain. At best it can be accepted as a complementary concept applicable to limited areas. Neither Crickmay nor later writers have given any specific example of a panplain.

Plains of Marine Denudation

An account of the plains of marine denudation has been given in the chapter on marine erosion. There are some people who are of the view that long continued marine erosion can result in the formation of extensive planation or erosion surfaces. But for this to happen it is necessary that the sea-level should remain constant for a very long time which appears to be highly improbable. It requires a much longer period than subaerial erosion for the land to be degraded and converted into a low plain by the process of marine erosion, as marine erosion is limited to a narrow part of the coast. Plain of marine denudation may therefore be found in narrow belts along the coasts of continents but not in extensive areas inside the continental masses. In comparison to peneplain, the surface of marine plains is more level and its gradient increases towards the land. On its landward margin the boundary of the marine plain should be sharp and clearly defined on account of the existence of the old cliff. The marine plains are also likely to contain marine deposits, marine fossils and rounded pebbles etc. But where these plains are very old and have been uplifted, the evidence of marine deposition may be destroyed by erosion and weathering.

The relationship between structure and drainage aids in the identification of this type of erosion surface, as plains of marine denudation are likely to have a superimposed pattern of drainage. On marine erosion surfaces with a cover of deposits, the drainage is controlled initially by the slope of the deposits and later is superimposed on the underlying rocks. When the marine erosion surface is deposit-free then the river courses may be determined either by the original beach gradient or by the surface relief. In either case there will be development of superimposed drainage which under normal circumstances will be present for a long time.

Pediplains

In 1948 L.C. King proposed a new cycle of erosion called the cycle of pediplanation to explain the planation surfaces, surmounted by isolated hills and piles of rocky boulders, that characterize much of the arid, semi-arid and savanna landscape of Africa. The end-product of this cycle is pediplain.

Pediplain in an extensive erosion surface of low relief with steep - sided residual hills that has been formed by the coalescence

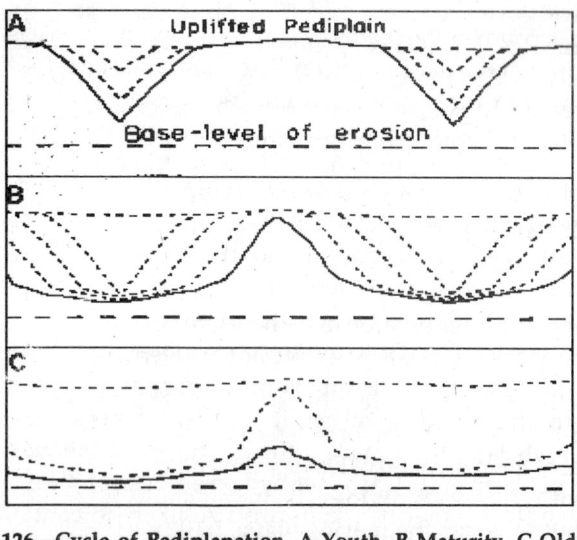

Fig. 126—Cycle of Pediplanation. A-Youth, B-Maturity, C-Old age.

of several pediments. The process by which a pediplain is formed is called pediplanation. Pediplanation is the process of landscape evolution by scarp retreat and pedimentation, and the pediplanation cycle tends to reduce uplifted land ultimately to a landscape of low relief with a concave profile during which the hill slopes retreat more or less parallel to themselves, leaving at reduced level ever broadening pediments extending from the foot of each hill slope to a nearby river bed.

The cycle of pediplanation is illustrated in Fig. 127. It is initiated by the uplift of an old peneplain rather than complex structural movements. (A) The existing streams cut rapidly downwards towards the new base-level. In areas of granite or gneissic rocks such as in Africa, the pattern of valley formed may be roughly rectangular. With time as downcutting becomes less active, small pediments start appearing in the valleys and become more extended as interfluves and uplands are consumed by scarp retreat. The remnants of the original pediplain will, however, exist on all summits, and by late youth many interfluves will have been converted to inselbergs (rounded domes).

(B) The stage of maturity witnesses a progressive reduction in the number of inselbergs as these are weathered into koppies and finally destroyed. The widening pediments of adjacent valleys begin to coalesce. But even during the late maturity the summits of the few remaining inselbergs rising above the surrounding low land frequently preserve vestiges of the former pediplain. Relief which had increased during early youth by valley incision either decreases or remains constant during the mature stage.

(C) In the old age of the cycle when most of the steep-sided residuals have been eliminated under scarp retreat the landscape consists far and wide of low-angled concave pediments and the ultimate form is a multi-concave surface called pediplain. As the relief becomes low, waste tends to collect and scarp retreat fails and a measure of convexity is introduced into the landforms (See Chapter 22).

Whether pediplains are formed by lateral planation or by scarp recession, it is clear that the form of the pediplain is not very different from that of the peneplain of Davis which develops mainly through the down wearing of divides. Pediplains also resemble marine plains, but whereas the marine plains have a seaward slope, the pediplain slopes away from the residual hills. In semi-arid regions or in regions which had semi-arid climates in the past, extensive erosion surfaces are found which are said to be pediplains. It is generally agreed that pediments are typical of arid and semi-arid lands, but L.C. King (1950) considers that pediplains (and not peneplains) are the end-product of the normal geomorphic cycle and all plains moulded by running water are of this type. He has tried to use the concept of pediplain in a comprehensive manner, and many of the erosion surfaces in Africa, Asia, North America, Europe, South America and Australia which were considered to be peneplains are, in his opinion, ancient pediplains. The concept of pediplain is by itself a useful concept and especially applicable to semiaried lands but not many geomorphologists are inclined to agree with its universal application as proposed by King.

Glacial and Periglacial Erosion Surfaces

Several attempts have been made to establish and explain the evolution of the geomorphic cycle under glacial and periglacial conditions, but it is doubtful if under these conditions a sufficiently long enough time is available for the development of extensive erosion surfaces. The Pleistocene Ice Age lasted for about two million years but the climate was not cold during this entire period. If we deduct from this the warm interglacial periods then no more than half a million years are left when glacial and periglacial conditions could be said to exist. Howsoever powerful the agents of erosion, the above span of time is considered to be inadequate for the formation of a peneplain or panplain. It is however, possible that in some limited areas mountains may be reduced to an irregular surface of roughly equal height by recession of cirques. Even more probable than this is that the glacier may bring about changes and modifications in an already existing erosion surface, like the one we find in the Laurentian Shield of Canada.

Peltier (1950) has tried to put forward the concept of a periglacial geomorphic cycle in which an extensive erosion surface of gentle gradient may be built by periglacial processes. This view has been confirmed by the later studies of Demek (1969). The end-product of the periglacial cycle of erosion is said to be altiplanation or cryoplanation surface. While overall reduction in relief is possible by periglacial processes, no example of a planation surface of this kind is available in the world.

Exhumed Surfaces

Exhumed literally means that which is dug up again, after burial. The term is commonly applied to ancient geomorphic features that were buried by such materials as glacial drift, alluvial deposits or trangressive marine sediments and later uncovered by erosion of the covermass.[1] The term resurrected landforms is also used for such features. Exhumed surfaces are thus only a special kind of resurrected landforms or landscape, which were first eroded, then buried and still later exposed by erosion of the overlying deposits. We have so long been considering erosion surfaces which were never coverd with thick deposit of sediments, but there are erosion surfaces which are found buried under hundreds or thousands of metres of later deposits. Under such conditions they are referred to as planes of unconfirmity in the geological parlance. The unconformities are buried surfaces, and where the overlying younger rocks have been eroded away, the exposed unconformity beecomes an exhumed erosion surface. All planes of unconfirmity, however, do not come under the definition of planation or erosion surfaces, as some of the planes of unconfirmity may consist of highly

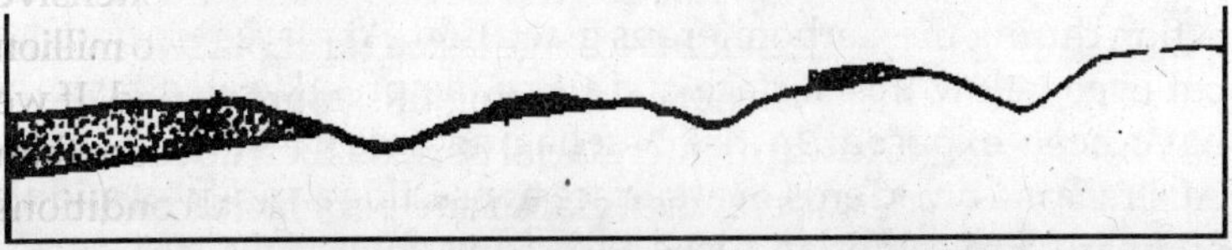

Fig. 127 — Exhumed Erosion Surface.

irregular surface. It is only when a covered planation surface is uncovered by erosion of the cover-mass that we call it an exhumed erosion surface. Exhumed surfaces are more likely to be found where the covered planation surface consists of resistant rock while the overlying layers are relatively soft, such as granite or metamorphic rocks overlain by Tertiary deposits. It is but rarely that the whole of the underlain planation surface is completely uncovered, and most

1. To be considered an exhumed landform, the feature must have been a part of a former landscape that was produced through the operation of normal gradational processes and then later was buried and still alter uncoverd by erosion of the covermass. Thornbury, W.D.—op. cit., 1969, f. 257. Exhumed Plains : This is a plain which has been formed, buried under younger sediment and then exposed by removal of the younger sediment. Ollier, C.B.—*Tectonics and landforms*, Longman, 1981, p.149

of the exhumed surfaces are, in fact, partial, separated by the residual fragments of the overlying deposits. (Fig. 128)

A good example of partial exhumed erosion surface is the Sherman erosion surface in the Larami mts, south-east of the Wyoming Basin in the U.S.A. Some people have called it a peneplain, while others have called it a pediplain. The pliocene sediments which cover extensive areas in the Great Plains further east, covered the entire area of the erosion surface as well. At one time the Canadian Shield and Feno -Scandia were considered to be exhumed Pre-cambrian peneplains. Their landscape has been considerably modified by repeated glaciation, even so Ambrose (1964) is of the view that their landscape is essentially Pre-cambrian, and more especially there has been no significant change in its drainage pattern since the Pre-cambrian times.

Examples of exhumed surfaces are also found in South Africa the most well-known being the Pre-Karoo surface. This surface has been exposed in the Middle Veld and Transvaal as a result of the wearing away of the overlying Karoo rocks. This surface is principally Carboniferous in age and was subjected to extensive action during the Carboniferous glaciation. At places erosion has cut even below this surface and there the pre-glaciated features have been exposed. In the Nuellagine area of north-western Australia is a broad erosional surface transecting a variety of middle and early Pre-cambrian rocks and rising above this surface is a steep escarpment formed by horizontal late Pre-cambrian sediments. Secondary modifications of ancient exhumed surfaces over the central Massif of France by marine fluvial terraces or partial peneplains have been worked out by Baulig (1928) and critically discussed by Sparks (1960).

It is important to remember that most of the exhumed erosion surfaces or exhumed relief features are composed of resistant bedrock which have not been affected by erosion in the process of being uncovered.

Multiple Erosion or Planation Surfaces

In almost every continent we find areas where the landscape consists of planation surfaces situated at different levels. These, in fact, are examples of successive cycles of erosion where there has been uplift after the completion of one cycle and the uplift has been

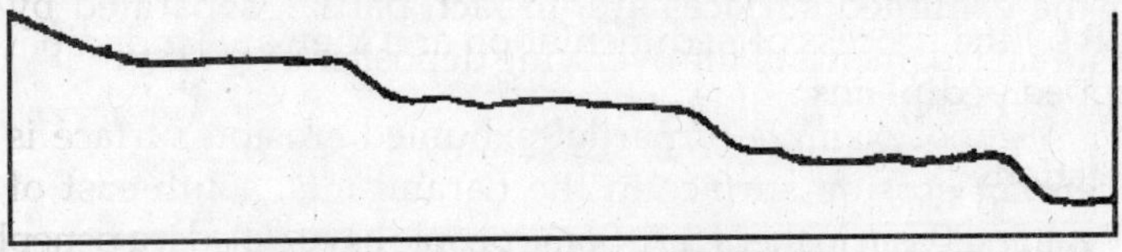

Fig. 128—Multiple Erosion Surface.

followed by another cycle of erosion. Consequently peneplains or planation surfaces built by the successive cycles of erosion are found today at different heights. Such p anation surfaces are together called multiple planation surfaces. The Chota Nagpur Plateau furnishes a good example. The western section of the plateau known as the 'Pat' is higher than the main Ranchi plateau (altitude 2000 ft.) by about a thousand feet and the upper Hazaribagh plateau on the northern side of the Damodar river valley has the same height as the main Ranchi plateau. Nearly a thousand feet below the upper Hazaribagh plateau lies the lower Hazaribagh plateau. Thus in the Chota Nagpur Plateau we find several uplifted flat surfaces at different levels dotted by residual hills. The different flat planation surfaces are separated from each other by steep scarps known as *ghats* which contain some beautiful waterfalls such as the Hundroo and the Hirani. These erosion surfaces were uplifted in the early and late Tertiary times. Similar examples of multiple planation surface are found in the granite areas near Madras, and Budel, (1965) has described two planation surfaces—one 200 to 500 metres high consisting of the Tamilnadu plain near the coast and the other 750 to 900 metres high called the Bangalore plain belonging to Miocene period and containing 1,500 metre high inselbergs. The two surfaces are separated by a scarp slope about 100 metres high.

In the Appalachian highlands of U.S.A. there are evidences of three planation surfaces at three different levels, beginning with the Schooley peneplain at the top followed by the Harrisburg peneplain and further down by the Samerville peneplain.

Davis has tried to explain the multiple planation surfaces with reference to his concept of the erosion cycle and has suggested that the peneplains have been formed in the long period of tectonic quiet which has been followed by a period of rapid uplift. In other words, they are the products of successive cycles of erosion which were interspersed by relatively short periods of rapid uplift. Penck, however, considers such stepped landscapes to have been formed by periodic scarp retreat. King is of the view that they have been built by the process of pedimentation and they are landscapes of stepped pediplains.

Questions

1. What are erosion surfaces ? How do they help in interpreting the landscape ? How can they be identified ?

2. Discuss the origin and characteristics of the different types of Planation surfaces.

3. Explain clearly the distinction between Peneplains, Panplains and Pediplains and give examples.

4. What do you understand by (a) Exhumed surfaces, and (b) Multiple planation surfaces ? Explain with the help of examples.

Chapter **22**

Work of Underground Water and Karst Topography

Water present in the pores and cracks of rocks below the earth's surface is called underground water. The underground water derives its importance from the fact that wells get their water supply from this source and well-water is widely used for drinking purposes. Sometimes the underground water comes out on the surface in the form of springs. Water from springs has been an important source of drinking water from times immemorial and villages have grown up near springs. The underground water is also important from the point of view of landscape evolution, especially in areas of limestone rocks.

Sources of underground water

The underground water may be derived from four principal sources:-

1. **Meteoric water**— The most important source of underground water is water derived from rain and snowfall. This is called meteoric water. When the rain falls or snow and ice melt into water part of it evaporates into the atmosphere, part flows on the surface as run-off, and the balance sinks into the ground through the soil and sub-soil and collects in the rocks below.

2. **Connate water**—When sedimentary rocks are formed, sometimes some water is retained in their pores. This kind of water derived from sedimentary rocks is called connate water, as this water has been present in the rocks from the beginning since their formation.

3. **Magmatic or Juvenile water**—Some minerals also contain water. When rocks melt and the molten magma rises to or towards the surface, the water (in the form of steam) gets separated from the magma and enters the pores and cracks of the adjacent rocks. This

water is called magmatic or juvenile water.

4. **Oceanic water**—Some underground water may be also derived from the percolation of oceanic water underground, especially in the coastal areas.

Of the above sources of underground water, by far the most important is meteoric water, and the other three sources are by comparison relatively unimportant.

Permeability of Rocks

What amount of the water derived from precipitation will percolate underground depends upon several factors. Where the slope is quite steep, almost the entire water flows down the surface as run-off and little is able to find its way to the underground. Then in the hot desert, whatever little rain falls either sinks quickly into the sand or where the surface is rocky and level gets easily lost by evaporation. But more important than climate or slope is the permeability of rocks which primarily determines the percentage of rainwater which will sink under ground. Rocks can be broadly divided into permeable rocks which permit water to penetrate into them, and the impermeable rocks which do not allow water to penetrate. The permeability of rocks is dependent upon two factors: (a) their porosity i.e. their porous and coarse crystalline structure, such as we find in sand, sandstone, chalk and gravel, and (b) the presence of joints and cracks through which water may penetrate, such as we come across in limestone, quartzite and sometimes also in granite. There are no joints or cracks in shale, slate and gabbro and these rocks are impermeable. But there are some rocks which are porous and yet are impermeable. Clay is an example of such a rock. The clay particles are extremely fine. Consequently the pore spaces between the particles are so small that when the rock gets wet, the pore spaces get closed by surface tension and water is not able to penetrate them. Clay therefore does not allow water to enter into it and water collects above a clayey surface. Most argillaceous rocks, though porous, are essentially impervious.

Where there is a layer of permeable rocks underlain by a layer of impermeable rocks, water collects in the permeable layer. The layer of rocks which can hold and permit the passage of water is called an aquifer.

Depth of Underground water

Both the amount and depth of underground water vary from place to place. In areas where shale and some igneous rocks are predominant, no underground water is available up to depths of several thousand feet. On the other hand, in areas of permeable rocks underground water may be available in adequate quantities up to depths of one mile or more. In fact, there is a downward limit below which underground water cannot penetrate. With increasing depth, the pressure of the overlying rocks on the rocks below goes on increasing and finally it becomes so great that all the pores and cracks get sealed. The depth at which this is likely to happen is assumed to be 6 miles below the surface where the pressure of the overlying rocks is 50 lakh Ibs per sq. ft. This pressure is sufficient to close all the pores and cracks in the underlying rocks, and this is therefore, the depth which is considered to be the downward limit of underground water. In reality, it is only rarely that water reaches downwards to such great depths, and most of the underground water is possibly found within a depth of only a few thousand feet.

The Water-Table or the Saturation Level.

Just below the surface, generally speaking, all permeable rocks are wet or saturated. The upper limit of this saturated layer is called the saturation level or the water-table. It is that level below which the pores, joints and cracks of rocks are saturated with water. The water in the wells rises only up to this level. In fact, as we go underground below the surface, we come across the following zones (Fig. 130) :—

The Zone of Non-Saturation—This is the zone just below the surface through which water percolates below to the underlying

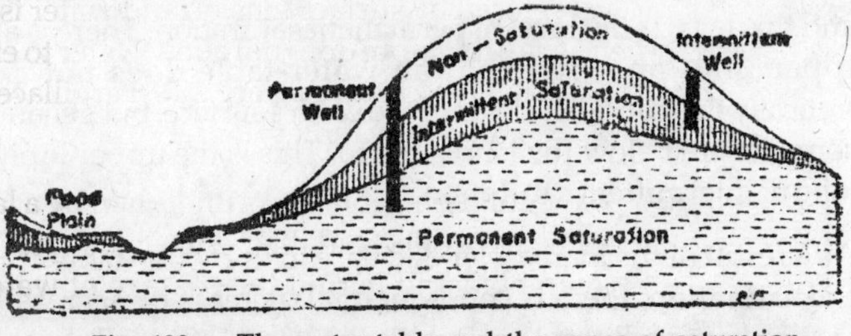

Fig. 129 — The water-table and the zones of saturation.

rocks, but which itself never gets fully saturated. Exceptions are, of course, areas where lakes or swamps may be found on the surface, In this zone of non-saturation, some water is retained by the soils which is used for plant growth.

The Zone of Intermittent Saturation— Just below the zone of non-saturation, is a zone which gets saturated in the rainy season but in which no water is available during the dry season. In other words, the upper limit of this zone is determined by the highest level of saturation attained during the rainy season, while its lower limit is fixed by the minimum/lowest level of saturation during the dry season.

The Zone of Permanent Saturation—Below the zone of intermittent saturation, is the zone of permanent saturation. Its upper limit is formed by the level up to which the underground water rises even in the dry season. This zone extends below upto depths to which underground water can penetrate and cannot go deeper because of the presence of an underlying layer of impervious rocks. This depth is not uniform but varies, but for most areas depths of 2,000 to 3,000 ft. may be taken as the lower limit of the zone of permanent saturation. Wherever the zone of permanent saturation extends upwards right up to the surface, we may find swamps, and sometimes lakes or rivers, though there is no direct relation between surface streams or lakes and the water-table.

We have just seen that the upper limit of saturation is called the water-table or the level of saturation. Normally, the water-table follows the surface relief. It rises below the hills and is depressed in the valleys.

The height of the water-table is not fixed but varies with seasons Its height increases during the rainy season and decreases during the dry season. Normally there is a limit below which it does not fall, and this level is the level of permanent saturation. There is also an upper limit above which the water-table does not rise. Theoretically this can be the level of the earth's surface, but generally it is some where below the land surface. This is the upper limit of the zone of intermittent saturation.

As a result of the increase or decrease in the water-table or the level of saturation, there is consequential increase or decrease in water supply in wells and springs. If the depth of a well does not reach the permanent level of saturation but reaches only the zone of intermittent saturation, then the well will dry up during the dry months. In order to have water throughout the year from a well it is necessary to dig it down to the permanent level of saturation.

Springs

Where water oozes and flows out naturally from the ground we have springs. The water may ooze out rapidly or slowly. Springs are found either at the point where the water-table cuts the ground slope or a little below it. The occurrence of springs is dependent

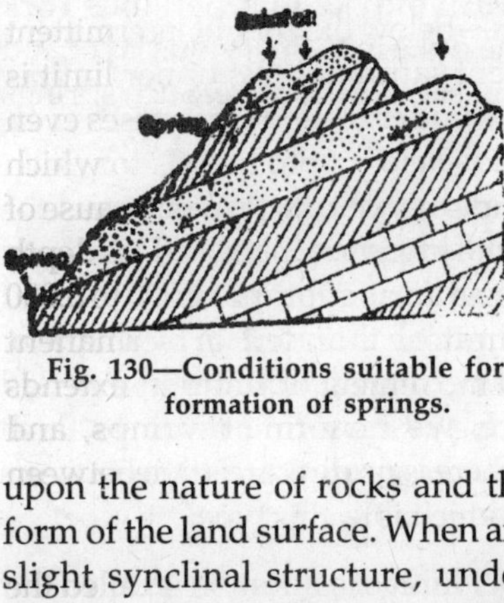

Fig. 130—Conditions suitable for formation of springs.

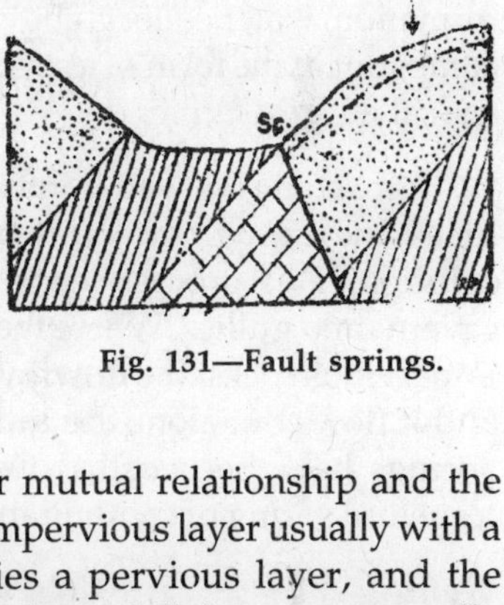

Fig. 131—Fault springs.

upon the nature of rocks and their mutual relationship and the form of the land surface. When an impervious layer usually with a slight synclinal structure, underlies a pervious layer, and the junction of the two layers is exposed along a hill slope, we generally come across a spring. Normally, springs are more commonly found

Fig. 132—Dyke spring

Fig. 133—Vauclusian spring.

in the hilly areas. In Fig 131 a hill consists of an underlying layer of impervious shale overlain by layer of pervious sandstone. When the rain falls, it enters the pervious sandstone but is not able to enter the shale layer. Water thus collects in the sandstone, and oozes out along the hill slope in the form of spring at the junction of the pervious and the impervious layers. If the level of saturation falls down so low in the dry season that there is no adequate water to feed the spring, the spring will dry up during these months. Such springs are called intermittent springs.

Some of the other geological conditions under which springs may originate have been shown in Figs. 132 to 134. In Fig. 132, pervious sandstone has been placed in front of impervious shale as a result of faulting resulting in the formation of a spring along the fault plane. Such springs are called *fault springs*. In Fig. 133, as a result of the presence of an intrusive dyke in a pervious rock formation, water collects above the dyke in the pervious layer and comes out in the form of a spring just above the dyke. Such springs are called *dyke springs*.

Limestone is a well-jointed and soluble rock. Water enters the limestone through the joints and the joints are increased in size by solution. This process leads to the formation of underground caverns and gullies. Where the limestone is underlain by a layer of impervious rocks, the downward movement of water is impeded and it flows out along the hill slope in the form of springs. Such springs have been called *vauclusian springs*, after the famous Vaucluse spring of south France which is of this type.

Wells

We have seen that there is a direct relationship between the water table or level of saturation and the digging of wells. Wells have year-round supply of water only if they have been dug down to the level of permanent saturation. Even in such wells the level of water does not remain constant throughout the year. It rises with the rise in water-table during the rainy months and falls with the fall in the water-table during the dry months. If the wells have been dug down only to the zone of intermittent saturation, they contain water only during the rainy season and dry up in the dry season. Excessive outflow of water through tube-wells may also result in lowering of the water-table.

Artesian Wells

There are some wells in which the water automatically flows to the surface. Such wells are called artesian wells. Long time back a well of this type was first dug in the Artois province of France, and such wells have, therefore, been since known as artesian wells. Geological conditions under which such wells are likely to occur have been depicted in Fig. 135. There is a layer of pervious rock between two impervious layers. All the three beds have been so folded that they form a basin or syncline. The top of the pervious layer reaches the surface and allows the rainwater to enter it. Due to the existence of an impervious layer below it, the water collects in

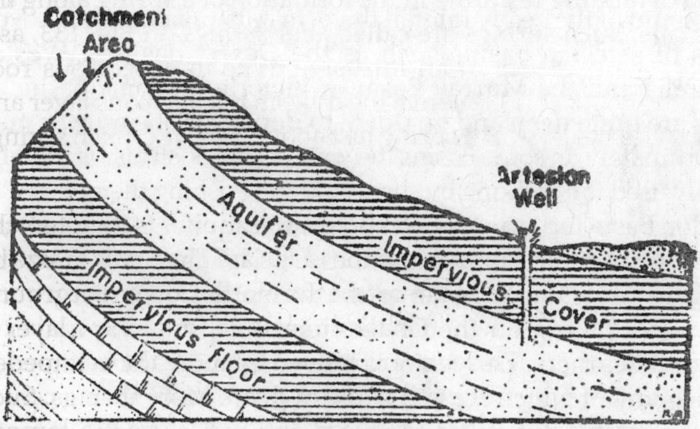

Fig. 134—Geological conditions suitable for artesian wells.

the pervious layer which gets saturated right to the top. In other words, the level of saturation is quite high in this bed. Under such conditions, if a well is dug somewhere in the middle of the basin down to the pervious or acquifer layer, the water in such a well will tend to rise automatically to the surface, as the level of saturation is higher than the surface level of the well. As water always seeks its own level, in such a well the water will automatically overflow to the surface because the level of saturation in the pervious layer is higher than the top of the well. The necessary conditions for artesian wells therefore are — i) an inclined or synclinal water bearing strata or aquifer with impervious rocks on top as well as bottom, ii) Exposure of the aquifer layer at a sufficient height at the rims so that water in this layer has adequate hydraulic head to enable it to rise to the surface where wells are sunk, iii) Sufficient rainfall over the catchment area of the aquifer top for adequate supply of water, iv) Absence of outlet for escape of water except through the wells.

Artesian wells are found in several arid and semi-arid areas of the world especially where a basin-like structure lies alongside a hill where rain water percolates down and collects in adequate quantities in the pervious layer. Examples of this are available from Sahara, Libya and Saudi Arabia, where a number of oases are founded on artesian wells. The most extensive artesian basin is found in Queensland, New South Wales and adjoining South Australia. Here in an area of over 5 lakh sq. miles, aquifer sandstone is found below the impervious surface layer of clay. This sandstone is exposed on the surface in the Eastern Highlands and gets saturated on account of the heavy rainfall there. In Australia there are several areas of artesian basins such as the Desert Basin in Western Australia and the Murray Basin in Vicforia. In some basins the wells are quite deep and go down to depths of one mile or more. Unfortunately, in some basins the water is brackish and is therefore mainly used for drinking by sheep and for washing their fleece. The London Basin too is an artesian basin. The aquifer is the Chalk and the enclosing impervious formations are clay. The Chalk is exposed along the Chilterns in the north and the North Downs in the south. Rainwater falling on the Chalk rims sinks into the basin and accumulates there. This water is tapped by more than a thousand wells many of which are about 200 m deep. More than a century ago the chalk aquifer was saturated and water used to gush out above the surface where wells were sunk. But so much water has been drawn from the chalk reservoir in more recent years to meet the ever increasing needs of the large and growing population that the water taken out far exceeds replenishment by rainfall on the chalk rims and the water level in the chalk has fallen by as much as 120 m in places, and water has to be now raised to the surface by pumps.

The Work of Underground Water

The role of underground water in the formation and evolution of landforms is rather limited but at the same time it is exceedingly important in the areas of limestone rocks where it gives rise to distinctive land forms. Like running water, underground water erodes, transports as well as deposits, but in reality the work of erosion is by far the most significant. Erosion includes four different activities: corrasion, attrition, solution and hydraulic action. While corrasion, attrition and hydraulic action are important in the case of running water, solution is by far the most important process in the case of underground water. Where the underground water passes through well-joined rocks, it separates the loose fragments by solution and carries them with itself. But ordinarily the movement of underground water is so slow that it is not able to perform any erosion merely by hydraulic action.

Solution—Pure water is not a good solvent, but it becomes an active solvent when it comes in contact with carbon dioxide gas. Since the underground water is largely derived from rain, it receives carbon dioxide in adequate quantities from the atmosphere, so that it becomes an active solvent. When the rainwater percolates into the rocks, it dissolves some of the rock particles by solution. Of the soluble rocks, the most notable are limestone, dolomite and chalk. Solution activity by underground water is, therefore, most evident and significant in those areas where these rocks are found.

The principal element of limestone and chalk is calcium carbonate which is easily soluble in rainwater which always contains atmospheric carbon dioxide. Therefore, when percolating water passes through limestone and chalk, some portions of the rock get dissolved by solution and trasported along with the water. In this way small holes are formed on the upper surface. Water naturally moves down through these holes, cracks and joints. The cracks and joints slowly become bigger as a result of solution and thus provide even easier passage for the downward movement of water. After some time the water collects in the passages and starts flowing in the form of an underground river. In course of time solution by underground water is able to produce distinctive landforms both on the surface and in the underground in limestone areas. Thus erosion by underground water is performed mostly through solution.

Transportational Work

The transportational work of underground water is not very important. The underground water moves very slowly. Thus when the rocks are eroded even the small rocks particles start settling down. Only the finest and the lightest particles are transported. Some small rock fragments also roll down with the water, but the speed and extent of such movement are insignificant.

We may, however, mention a special kind of transportation

done by underground water. When surface rocks are loosened as a result of disintegration and decomposition by weathering, underground water may act as a lubricant for the mass movement of the rockwaste down a hillslope. This kind of movement generated by underground water is called mass translocation of rock waste, and may cause land slides along the slopes of mountains.

Depositional Work

As a result of solution minerals mix with the underground water. When the underground water becomes thick with such debris, then the excess materials start getting deposited. As the movement of underground water is slow, the bigger fragments start settling down as soon as they are broken. The finer materials in the solution also start settling down after sometime. This happens particularly when the speed of movement suddenly declines due to some obstruction or when increase in temperature results in partial evaporation of water or decrease in temperature reduces the soluble capacity of the water, or for some reason there is decrease in the amount of carbon dioxide available. Deposition by underground water is done mostly in the underground caverns, resulting in the formation of such features as stalactites, stalagmites and columns and pillars.

Karst Topography

In areas of limestone and dolomite rocks the solution action of underground water gives rise to highly distinctive landforms both on and below the surface which are collectively known as karst landforms. Karst is an Yugoslavian word. On the Adriatic coast in Yugoslavia and the adjoining areas in Italy there is a limestone plateau known as Karst which provides the best example of landforms produced by the solution action of underground water. This region is 500 kms long from N.W. to S.E. along the coast and is about 80 kms wide. The plateau is about 2,500 metres high above sea level. Here the surface drainage is almost completely absent, and the surface is a highly irregular and rocky waste, riddled with innumerable holes, pits and ravines. Below the surface is found an extensive labyrinth of underground rivers and caverns in which several depositional forms present a weird and extremely fascinating picture. This region contains the most developed form of distinctive limestone karst topography which has been studied in great detail by scientific workers. Consequently the terminology relating to karst topography that has evolved is mostly Yugoslavian in origin, although words similar in meaning are available in other languages.

Distribution of Karst Regions

Karst topography has developed in those areas where limestone is present as surface and sub-surface rock. In addition, topography very similar to karst is found in areas containing dolomite, dolomitic limestone and chalk. Besides the Karst region of Yugoslavia, this type of topography has developed principally in the following areas—Causses region of southern France, Andalusian region of Spain, Yucatan and Tabasco in Mexico, Jamaica, northern Pureto Rico, western Cuba, central New Guinea, south-west Celebes, southern China, Vietnam and Malaya, New South Wales and West Australia in Australia and the undernoted areas in the U.S.A.—Salem Springfield region in Missouri, middle Florida, Indiana-Kentucky region, the Great Valley region of Pennsylvania, Maryland, Virginia and Tennessee. Of the above areas the most prominent are the Indiana-Kentucky region in U.S.A. and the limestone mountains of northern and southern Alps in Europe. There are other areas where development of Karst topography has taken place on a lesser scale, such as the chalk areas of England and France, Tanzania in Africa, Jabalpur and the adjacent areas in Madhya Pradesh., Rohtas plateau in Bihar, Kumaun Himalayas near Almora, as well as parts of Jammu and Kashmir and Meghalaya etc. where limestone is found.

Favourable Conditions for the Development of Karst

The following conditions are necessary for the development of a karst landscape:—

1. First of all the rock on and near the surface should be soluble in the ordinary rain or soil water. The most suitable rock for the purpose is limestone. Dolomite is also suitable to a great extent but is less soluble than limestone. Chalk is also soluble, but it does not possess some other attributes which are essential for the full development of this type of topography.

2. The soluble rock should be dense, thinly bedded and well jointed and fractured to allow the easy passage of water. It is not

sufficient for the rock to be porous. In addition to porosity, it is essential for the rock to have a large number of joints and bedding planes. If the rock is porous and permeable throughout, the rainwater soaks into the rock as a whole and is not canalized through specific channels. Chalk is a highly permeable rock and is almost pure limestone but it has no bedding planes and joints. It has capacity to absorb and retain rain water with the result that solution is retarded. This explains why karst topography has not fully developed in the chalk regions of England and France. Underground drainage and karst landforms also develop on soluble rocks like dolomite, gypsum and salt, but these rocks are not widely distributed. Hence karst topography is normally associated with an area of pure, highly jointed and dense limestone.

3. The most favourable condition for the development of karst topography is to be found in areas where deep valleys are found close to a highland and the highland has an extensive basement of limestone. In such a situation, the rain water percolates down the upland into the limestone rock below. Solution is most active in limestone when it acts like a conduit for percolating water. It is extremely important for the water to flow, as it is only moving water (and not stationary water) which can perform solution. The height of the area should be hundreds of feet above sea level to allow free circulation of underground water.

4. The fourth requisite is moderate to heavy rainfall for the supply of underground water. All the karst areas are situated in regions of adequate rainfall. In arid and semi-arid regions karst landscape is generally absent. Wherever examples of karst topography are available today in arid or semi-arid regions, they are merely relict features and were formed when these regions were moister and had a heavier rainfall.

Characteristic Landforms of Karst Regions

In limestone regions various distinctive landforms develop mainly by solution which are not found in other areas. Some of the more important of these landforms are described below. It must however, be remembered that it is not necessary for all these landforms to be present in every region of karst topography (see Fig. 138).

Terra Rosa— When rainwater dissolves part of surface rock and enters the sub-surface, particles of red clay soil are deposited on the surface as well as in the opened joints. This is called terra rosa, and very much resembles lateritic soil. It may not be present on steep slopes but can be seen in areas which are either flat or have gentle slope. Sometimes it may be several feet thick and entirely cover the rocky surface.

Lapies— Where the limestone surface is not covered with terra rosa and the surface is undulating, solution action of water tends to enlarge the rock joints downwards. This results in the formation of a very rough topography where the surface is cut into narrow and deep clefts between thin and needle-like pinnacles. The sharp pin-like vertical pinnacles are roughly parallel to one another, and between them are found narrow and deep trenches which are known respectively as clints and grykes in England. For this kind of serrated landscape, the French word lapies is used at the international level. In Germany this kind of landform is called *karren* (Fig 136). On account of the existence of lapies, the surface becomes so rough and intractable that it is impossible to walk on such a surface. Lapies, however, provides an excellent illustration of how the solubility of rocks, their permeability, joints, bedding planes and other physical and chemical attributes produce differences in the rate and localization of solution. Some scholars hold the view that lapies are not formed on rocks with horizontal stratification and it is essential to have some kind of sloping surface for their formation. Examples of lapies abound in the Dalmatian karst region.

Fig. 135—Lapies or Karren.

Sinkholes and Associated Features—When rainwater percolates into limestone, it enters the rock joints and removes the soluble elements by solution. The joints increase in size by solution and provide even easier passage for the water. Thus prolonged solution results in the formation of holes along the joints which are known as *sinkholes*. Topographically, a sinkhole is a depression, most commonly ranging in depth from 10 to 30ft. and in areas from a few sq. yards to several acres. These small holes are found in large numbers in the karst regions where sinkhole terrain is quite common.

They are usually funnel- shaped with rounded and open tops, though many variations of form are found. Similar dry holes are met with in the chalk areas of Britain which are cylindrical and are called *swallow holes* as these surface openings almost completely swallow the surface drainage during the rains.

Depending on their mode of formation, sinkholes are divided into two classes—one which have been formed entirely by solution and the other which have been formed by slumping. In fact, this distinction is not sharp and clear-cut because even where the sinkholes have been primarily formed by slumping of the ground, there is usually the formation of sub-surface hole by solution into which the upper cover of rocks falls down by breaking or slumping, Sinkholes formed in this fashion are called *collapse sinks*, and their walls are usually steep and rocky. Funnel-shaped sinkholes formed by solution tend to be enlarged downwards because of the repeated action of solution. In the Karst region of Yugoslavia these enlarged funnel-shaped sinkholes are called *dolines.* Here they are usually 6 to 75ft. deep and have a diameter of 30 to 400 ft. Very often the bottom of the doline is closed as a result of the fall of debris from above so that water only percolates slowly instead of flowing down swifty. As a result water may accumulate in the doline and form small lakes called *karst lakes or sinkhole ponds* depending upon their size. These lakes are temporary, and when the debris blocking the passageway are removed, the water of the lake sinks below and the lakes dry up. In this way innumerable small seasonal lakes come into existence during the rainy season when all the rainwater is not able to flow down the dolines and collects in them. Vertical or inclined shafts leading from the sink-holes or dolines to the underground caves are known as *ponor* (Fig. 137).

Fig. 136—Funnel-shaped dolines and vertical or stepped ponors leading to the underground cave.

Dolines or sinkholes that develop slowly downwards by solution without physical disturbance of the rock, as well as collapse sinks resulting from collapse of the roof cover into an underground solution void, may be considered the most primary and fundamental feature of a karst terrain. The water which enters the dolines, goes underground by percolation and down-flow, and eventually takes the form of an underground stream. Where the land separating two adjacent dolines collapses, a portion of the underground stream may become unroofed and the stream can be seen from the surface flowing out of a cavern on one side, across as open space into another cavern on the other side. Such an opening through which the course of the underground stream is visible, is called a *karst window*. The opening may be very small or quite large. Some karst windows may become so enlarged that the stream flows as a surface stream for some distance across alluviated floors. In a region where there are many dolines or sinkholes, they frequently expand in diameter and coalesce and form *compound sinkholes* or compound swallow holes. They usually consist of a large solutional depression, with the collapsed roof and walls of several dolines merged into them. These large depressions or compound sink-holes are also called *uvalas*. An uvala may have a diameter of up to one km. There is, however, some confusion in the use of this term, and sometimes the term uvala has also been used in the sense of a large depression resulting from roof collapse over underground water courses i.e. in the sense of large karst window, but the former is probably the more correct and accepted usage. The development of karst windows and uvalas represents a significant stage in the evolution of karst landscape.

Depressions even larger in size than uvala which owe their origin to solution of down-faulted or down-folded limestone blocks have been called *polje*. It resembles an uvala in appearance, but differs in extent and origin. The area of a polje may cover several sq. kms. The largest polje in Yugoslavia in the western Balkans is the Livno polje which is 70 kms long and 5 to 11 kms wide. A typical polje is an elongated basin with a flat floor and steep sides. It is believed that it has been formed by the solution of down-faulted or down-folded limestone blocks and is thus a structure-controlled solutional form. In Yugoslavia poljes are given extensively to maize cultivation on account of the flat surface and easy availability of water.

Other Karst Features— In limestone regions containing horizontal or gently dipping strata solutionby rain water results in the development of a plain riddled with thousands of sinkholes through which this water sinks and is conveyed through underground routes. These sinkholes are so effective in collecting the surface run-off that hardly any surface stream exists. and almost the entire drainage is underground. It is only the larger streams which rise in the highlands beyond the plain and have cut deep valleys, that are able to flow through the plain. The smaller streams lose their water in one or more sinkholes and are thus called *sinking creeks* and the point at which water sinks below is called its *sink*. In some streams the water disappears in the alluvial materials on the riverbed, and no distinct point of disappearance or sink may be noticeable. The sinking creeks may flow as underground streams for some distance before again appearing as surface streams. The Lost River in southern Indiana, U.S.A. flows underground for 13 kms before it again starts flowing on the surface.

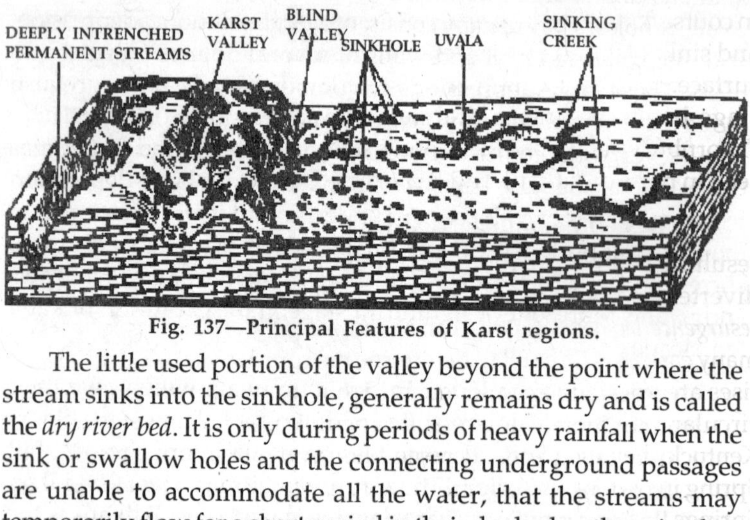

Fig. 137—Principal Features of Karst regions.

The little used portion of the valley beyond the point where the stream sinks into the sinkhole, generally remains dry and is called the *dry river bed*. It is only during periods of heavy rainfall when the sink or swallow holes and the connecting underground passages are unable to accommodate all the water, that the streams may temporarily flow for a short period in their dry beds, otherwise they remain dry for most of the year. Where the stream terminates in a sinkhole and this situation continues for a long time, the stream carves out a fairly deep valley below the level of the karst plain. Under such conditions the river valley ends at the sink or swallow hole and is called a *blind valley*. The dry bed, if any, beyond the sinkhole gradually becomes obliterated through disuse and loses the character of a surface valley. During the rainy season when the increased volume of water is not able to enter the sinkholes, the blind valleys are converted into temporary lakes due to large accumulation of water.

A special kind of blind valley different from the one described above, is also found in the karst regions and is known as either *karst valley or solution valley*. In areas where the upper layer is composed of sandstone or shale and is underlain by a limestone layer, the surface streams developed on the upper layer may in course of time cut down deep enough and start flowing on the limestone bed. In such a situation sandstone ridges may be found on both sides of the valley, but the river flows in a deeply enterched valley on the limestone floor. Here on account of the presence of limestone, the river develops the tendency to sink underground and abandon the surface routes in favour of underground ones. Similarly, the tributary streams which rise from the neighbouring sandstone and shale hills, tend to disappear in sinkholes in the principal limestone valley. In course of time the principal river may go completely underground and sinkholes and other karst features maybe found on the valley surface. Thus "a solution or karst valley represents a transitional stage between surface drainage and underground drainage". (Thornbury, 1969, p. 312). Good examples of such valleys can be seen in parts of Indiana and Kentucky, U.S.A.

The water of a sinking creek frequently returns to the surface, resulting in the formation of springs. The reappearance of water diverted to underground routes on the surface has been called *resurgence or rise*. Some rise can be linked with specific creeks, but in many cases it is difficult to determine the source of water. The larger rises are artesian in character and issue from large and deep sub-circular rise pits. The larger springs of Missouri, Indiana and Kentucky belong to this category. Another famous example is Silver Spring in Florida noted for the clarity and the volume of its waters. Springs that mark the resurgence of sinking creeks vary greatly in the volume and clearness of their waters. Generally speaking, water is clear and less in volume during the dry season but there is increase in the water flow and turbidity during the rainy season. It should, however, be noted that not all the springs in the karst regions are due to the resurgence of the water of sinking creeks, and the smaller springs found along hillsides are mostly simple gravity springs located above an impervious bed.

In limestone areas, after a long period of solutional erosion, some residual hills and hillocks remain which resemble

Fig. 138—Hums or Haystack Hills

monadonocks and bornhardts. These rounded limestone hills rise from the polje surface. They are known as *hums* in the Karst region of Yugoslavia. They are known as *pepino hills or haystack hills* in Puerto Rico and *as mogotes* in Cuba (Fig. 139). They are, however, not usually found in the karst regions of the U.S.A.

Tropical Karst

In recent years the development of karst topography in hot and wet regions has been studied in some detail. These studies which have been particularly conducted in Jamaica, Puerto Rico, Java and South China, show that the tropical karst possesses some features which are not found in the karst of temperate climates. Because of the higher incidence and intensity of rainfall in these hot regions, the sinkholes or dolines are less circular and more irregular in form. Tropical karst has been divided into two types—(a) Kegel karst or Cone karst, and (b) Tower karst or Turm karst. In the *Kegel karst* there are found several cone-like hill features interspersed with irregular depressions known as cockpits. Examples of such landforms are found in Jamaica, East Indies, South China, Vietnam etc. In *Tower karst*, on the other hand, groups of steep-sided hills are found and each hill or group of hills is surrounded by alluvial plain. This is especially so in areas where relatively large streams flow through limestone regions. Examples of tower karst or pinnacle karst abound in Puerto Rico and South China. In South China there has been development of karst topography over an area of nearly 6 lakh sq. kms in the provinces of Yunan, Kwecho and Kwangei. This is probably the most extensive karst area in the world. Its western section is a kegel karst region, while its eastern part (Kwangei) is a tower karst region where the towers are 100 to 200 metres high and have slopes of 80^0 to 90^0.

Our knowledge about topical karst and its development is still very inadequate. This much, however, can be said that in these hot and wet regions both solution and evaporation are fast, resulting in extensive development of stalactites and stalagmites in the underground caverns. Solution is, however, more prominent on the surface than underground and the caverns are therefore smaller in size. Lateral erosion is also important on the surface, and on account of the presence of soil and vegetation, fluvial erosion and corrosion are also relatively important. As a result of these processes, isolated towers are found over relatively flat surfaces which may contain swamps or slowly moving rivers.[1]

Underground Streams and Caverns

Underground streams and caverns are distinctive features of limestone regions. The cavern or cave is a subterranean hollow, formed essentially by the solutional and corrasional activity of underground water. It is not, however, necessary for a stream to be present in all caverns and examples of dry caverns are also found. Dry caverns have usually more than one level. Caverns with two or more levels are referred to as galleried caverns. The influence of joint systems is clearly marked on the cavern pattern and both their horizontal and vertical extensions are largely controlled by joints and bedding planes which have been enlarged by selective solution by water moving along them. This is especially evident in the smaller caverns, but in the more complex caverns this may not be so evident.

Fig. 139—Underground Stream and Cavern.

As the sinking creeks go underground, they flow horizontally along the joints and bedding planes or sometimes fall vertically down as water falls (Fig. 140). By prolonged solution their underground course becomes so wide that a series of underground

1. Sweeting, Marjorie M.-*Karst landforms*, Macmillan, London, 1972,pp.270-296

caverns are formed which are inter-connected by narrow pathways and extend for several miles. At places water may collect and form underground ponds. Thus in course of time below the surface there comes to develop an extensive labyrinth of caverns, interconnecting galleries, rivers, waterfalls and lakes.

Besides solutional features, a variety of depositonal features are formed on the floor, ceiling and walls of the caverns which greatly add to their beauty and interest. *Cave travertine* is a term which is collectively used for these diverse depositional forms. When the underground water drips from the cavern ceiling, it contains carbonate of lime. When the droplets of water are hanging by the ceiling, some water is evaporated and a portion of the carbonate of lime is deposited on the ceiling. This process is repeated with every fresh droplet of water, and the deposit of lime goes on extending from the ceiling downwards. In course of time

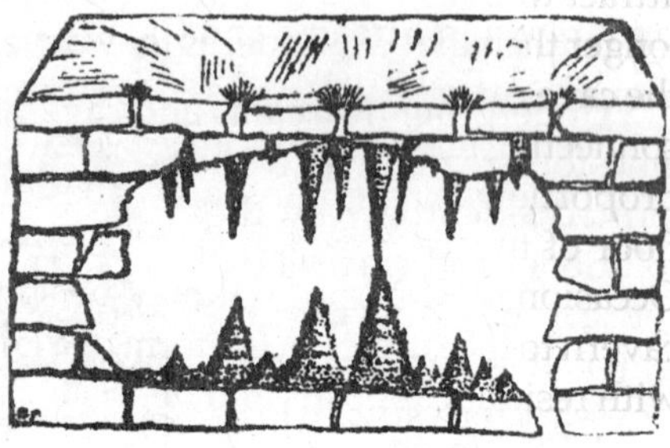

Fig. 140— Stalactites, Stalagmites and Column and pillar in an underground Cavern.

this results in the formation of a long and thin column of carbonate of lime hanging downward from the ceiling. These are known as *stalactities*. They are wide along the ceiling and tapering downwards. The drops of water which fall from the ceiling down to the floor of the cavern also contain carbonate of lime, and with every drop of water falling on the floor there is some deposit of carbonate of lime. Thus from the cavern floor also rises a pillar-like column which is known as *stalagmite*. Stalagmites are usually broader but shorter in length than stalactites. Since the drops of water as they fall down get spattered on the floor, the base of the stalagmite is usually thick and wide and its outline is not so symmetrical, though it tapers towards the top (Fig. 141). Sometimes the stalactites growing downwards and the stalagmites growing upwards meet together and form a continuous column from the floor to the ceiling and are known as columns and pillars (Fig. 141). Where the water drips more or less continuosly along a roof joints, a *fluted curtain or wavy screen* may grow and hang downwards from the ceiling.

In addition to the above depositional features, there is found in some caves another depositional form known as *helictite* which develops where the water entering the cave is not sufficient to give rise to drops that fall but only to keep the surface wet. Under such conditions the helictite deposit does not necessarily extend along vertical lines but may grow in any direction, depending upon the chance orientation of the crystal axes of calcium carbonate. They, therefore, grow in all kinds of shape and size.

In the limestone areas stalagtites, stalagmites and other decorative structures present an extremely weird and exceedingly beautiful sight in the dim light of the underground caverns and attract tourists from distant areas. The caverns are usually much longer than they are wide, as they develop along the joints. Some of the caves may be as long as 60 to 80 kms with narrow passageways connecting large caverns in between, but most are of much smaller proportions. Sometimes large piles of rock debris may lie on the floor of the caverns as a result of roof fall on broken stalactites. Occasionally the roof fall may be so extensiive as to expose the cavern to the surface in the form of a large and irregular depression with residual deposits and debris.

Although chalk is also a carbonate of lime, the development of underground streams and caverns is more characteristic of limestone areas. One important reason appears to be that the joints in the chalk are so closely packed that water instead of entering along the prominent joint planes enters into chalk in small quantities along the minute joints.

Underground caverns are found in India in several limestone areas but very few of them have been studied. These include the Jabalpur region of M.P., the Rohtas plateau of Bihar, the Cherrapunji area of Meghalaya, Kurnool in Andhra, Kumaun Himalayas near Almora, parts of Kashmir and other limestone areas. Several underground rivers and caves exist in the Craven region of West Riding, Yorkshire in the Pennine uplands of England. Here near Ingleborough are found several swallow holes of which the most striking is the Gaping Gill which swallows the water of the Fel Beck river. Below the hole is a cavern which is 500 ft. long, 110 ft high and 90 ft. wide. Connected with this are several passageways and caves for a distance of about 3.5 miles. The water which disappears into the Gaping Gill, reappears on the surface after about a mile.

Gaping Gill cavern has two chambers which are known as Stalactite and Stalagmite chambers.

In the Karst region of Yugoslavia which provides an excellent example of the distinctive limestone topography we find an extensive development of underground streams and caverns. The sinkhole which connects the caverns with the surface is called *ponor* in Yugoslavia. These are usually vertical or stairlike and sufficiently deep, and the caves are located at their bottoms. The caverns usually occur in a series connected by long and narrow passageways. The caverns may also occur at different levels and depths i.e. as galleried caverns.

Hudreds of underground caverns are found in the Indiana Kentucky karst region of the U.S.A. A well known one is the Mammoth cave which is nearly 45 miles long. Another well-known cave is the Carlsbad cavern in New Mexico. It is only 7 miles long, but it has a Big Room which is 4,000 ft. long, 300 ft. high and 600 ft. wide. In the French Pyrenees and the French Alps the underground caverns and their related streams, falls and lakes have been studied up to a depth of 1 km. This region probably contains the world's deepest caverns. World's longest known cave, Holoch Cave is in Muotatal, Switzerland and is about 85 km. long, (53 miles). Good examples of underground caverns also exist in British Honduras, Jamaica and Puerto Rico. There has been more extensive development of caverns in the wetter limestone areas of the world.

Origin of Underground Caverns

There is considerable difference of opinion about the origin of limestone caverns. According to some, they have been formed primarily by solution, but there are others who believe that corrasion has a far more important role than solution. Further, according to some, caverns develop chiefly above the water-table through the action of diverted surface waters, while in the opinion of others among whom Davis is prominent, cavern development takes place below the water-table essentially by phreatic water. To a large extent the difference of opinion regarding the origin of caverns is related to the varying ideas about the nature of water-table in a karst region and the zone of major erosion with respect to it. Some believe that there is no distinct water-table or level of saturation in karst regions. In other words, there is no continuous surface separating the saturated and the non-saturated zones. According to them the entire mass of rock is not filled with water, but underground water is confined to the joints and bedding planes in a discontinuous fashion. But most people are of the view that where limestone is present up to considerable depths, water-table is present just like other areas, though it is true that bulk of the water is held in and moves through the joints and bedding planes.

According to the older view, cavern development takes place largely by mechanical erosion or corrasion by underground stream very similar to the corrasion performed by surface streams. In support of this view it is pointed out that clay, silt, sand and gravels are found in the caves. This shows that when the surface streams enter the sinkholes, they carry with them particles and fragments of rock which enlarge the passageways along the joints by corrasion, and in course of time several holes join together to form a cave. Another evidence which is pointed out in support of the primacy of mechanical erosion is the presence of potholes, grooves and incised meanders on cavern routes. In the face of these direct evidences, it would be incorrect to state that corrasion has absolutely no role in cavern development, but recent studies would seem to indicate that in the formation of most caverns corrasion is only of secondary importance.

Two -cycle Theory of Davis

In 1930 William Morris Davis published a long paper entitled "*Origin of Limestone Caverns*" in which he challenged the existing ideas of cavern formation and put forward his two-cycle theory of cavern development[1]. According to him, although corrasion is not totally absent, caverns are formed essentially by solution. Cavern development. takes place in two stages. In the first stage, the caverns are formed by solution below the water-table by phreatic water. This is followed by a second stage in which following uplift there is lowering of the water-table and a draining of the caverns of their underground water and the entry of vadose water and air in the caverns. It is during the second stage that the depositional features of cave travertine are formed. Davis and his supporter Bretz

---

1. *Bull. Geol. Soc. Am.*, Vol 41, 1930, pp. 475-628. A. Grund (1903) in Europe seems to have anticipated the ideas of Davis but due to lack of familiarity with his writing in German, Davis has been looked upon, especially in America, as the originator of the two cycle theory, Thornbury, W.D.—op. cit, p, 329

have pur forward the following arguments in support of the two-cycle theory :

1. Clear evidences of deep-seated solution below the water-table are available from may karst regions (Florida, Cuba etc.)
2. The caverns present three-dimensional network and sponge like patterns rather than the branching patterns which normally characterize surface streams.
3. It is not possible to explain by stream erosion the blind galleries found in caverns.
4. The general absence of graded longitudinal cavern profiles indicates that the caverns have not been formed by corrasion by underground streams above the watertable.
5. Many wall pockets, ceiling pockets, ceiling tubes and other features unrelated to joints and bedding planes could not possibly have been made by streams that now flow through the caverns.

Water-table Theory of Swinnerton

According to Swinnerton (1929-32) the development of caverns is due to solution by vadose water, and the solution takes place essentially at the level of the water-table by laterally flowing water. The theory has been called the water-table theory because of his emphasis on lateral movement at the level of the water-table. As the surface streams cut down and deepen their valleys, the lateral flow of the water at the top of the water-table is accelerated. This movement of water takes place largely along the joints and bedding planes. Lowering of the water-table with the lowering of valleys is responsible for the formation of caverns at different levels. The surface water which infiltrates through the joints and reaches the water-table is largely responsible for dissolving the cavern passageways. During the rainy season all the joints and holes are filled with water and the water-table rises but this happens only for a short period after which the water-table falls down. During the rains, however, when the caverns are filled with water, network development takes place. It is thus possible to explain the three dimensional network passages of the caverns with the help of the periodic rise and fall of the water-table.

Invasion Theory of Malott[1]

Based on his detailed studies of caverns of Indiana and Kentucky in U.S.A. conducted over several years, C.A. Malott put forward a theory in 1937 which he called Invasion theory of cavern development. According to this theory, the water of the diverted surface streams which moves through underground routes is essentially responsible for cavern formation by the process of solution. He did not agree with Davis that the underground streams flow through caverns which already had been formed below the water-table. Solution could, of course, take place below the water table resulting in formation of poorly integrated passages or networks, but the waters of the diverted surface streams occupied these poorly integrated passageways and developed some selected ones into well-integrated cavern routes at or near the water table.

The views of Malott are thus entirely at variance with those of Davis. According to Davis the caverns are formed by solution below the water-table, and it is only at a later stage that surface water which sinks underground flows through them. On the other hand, according to Malott solution below the water-table could result only in minor extension of the joints and the bedding planes, and cavern development essentially takes place at or near the water-table on account of the subterranean flow of surface waters and the expansion and coalescence of joints by solution by these waters. Malott has cited the example of the Lost River area of Southern Indiana in support of his view (1952). The river consists of three parts— an upper part with unentrenched surface stream, a middle dry bed section where it flows underground, and a lower section with a deeply entrenched surface stream. Though the river extends across limestone in its upper part, karst features are lacking or weakly developed and surface drainage characterizes this stretch. The middle portion extends across a limestone plain with innumerable sinkholes and no surface stream and the river usually flows underground. In the lower portion the river cuts through clastic rocks below which there are thick limetsones, and there are a number of karst valleys which collect the surface drainage from the sandstone ridges. Cavern development has taken place only in its middle part and there is evidence to show that this underground route developed after the entrenchement of the lower course to the

1. Proc. Geol. Soc. Am. for 1937

present base level of erosion which has lowered the water-table and made possible the diversion of surface waters to underground routes with resulting cavern expansion. In the uppar part where stream entrenchment and the resulting lowering of the water-table have not taken place, there is surface drainage and sinkholes are few.

It would appear that there is an element of truth in all these theories, and probably cavern development takes place both below and above the water-table. At the same time it has to be admitted that solution is extremely important below the water-table and we have examples of several caverns which have definitely been formed below the water-table, such as the Holloch cavern in Switzerland, the Carlsbad cavern in New Mexico, and caverns of Florida and Missouri in U.S.A. But this should not lead us to conclude that all caverns have been formed in an identical manner. There is a widespread association between the sinking creeks and caverns in the caverns found in the Indiana-Kentucky region, and there would appear to be a strong case for believing that there is genetic relationship between the two rather than that the sinking creeks have occupied caverns which had already been formed below the water-table.

The Karst Cycle of Erosion

Davis considered the evolution of landscape in the limestone region to be a special and transitory phase marking the mature stage of the normal cycle in regions where the structure was especially favourable to solution. The basis for such a view arises from the fact that even in limestone regions the cycle starts with surface draingae and also ends with surface drainage. In other words, the karst cycle may be said to begin with the transfer of surface drainage to underground routes and end with its re-appearance on the surface. Thus, even while agreeing with Davis, we can conceive of a distinct karst cycle, especially because the concept of the cyclic evolution of landscape is capable of very simple and straight forward application in the case of the karst landscape[1]. Here the structure is simple, because we find only one type of soluble rock. Some complexity has, no doubt, been introduced by faulting in Yugoslavia and folding in the Great Valley of the Appalachian

1. "In fact, there is probably no type of landscape to which the concept of cyclic evolution can be applied with less complication than a karst landscape." Thornbury, W.D.—op. cit, p. 337.

region, but they have influenced the evolution of landforms only in minor details. Further, only one geomorphic process, namely solution is dominant. Thus even while conceding that the karst represents a special phase of the broader fluvial cycle of erosion, it should be possible to use such terms as youth, maturity and old age to the different stages of karst development.

The karst cycle may be initiated in two types of regions (i) in areas where there is a thick deposit of limestoen either on or below the surface, and (ii) in areas where the limestone is overlain by layers of shale, sandstone or other insoluble rocks. In either case, the cycle starts with surface drainage but its translation into underground drainage is effected in different ways. In the first case the karst cycle is initiated rather quickly. As the rivers cut downwards, the water-table is lowered. There is formation of sinkholes and dolines through which the surface drainage sinks below. In the other case, where insoluble rocks are found above the limestone, the rivers have to cut through the insoluble layer to reach the limestone layer and this takes time. But when this happens, the karst valley develops on the limestone floor and sinkholes and dolines are formed.

J.W. Beede (1911) was probably the first to put forward the concept of a karst cycle, dividing it into the three stages of youth, maturity and old age. He called it the cycle of subterranean drainage. But somehow this went unnoticed. In 1918, J. Cvijic published his concept of the karst cycle based on a detailed study of the karst region of Yugoslavia which has received wide publicity and accptance and consequently, often his name is associated with this concept. He recognized four stages youth, maturity, late maturity and old age in the evolution of karst, which are described below :—

(i) Youth may be said to begin with the surface drainage on either the original limestone surface or the one that has been exposed on the surface by erosion. Slowly the surface drainage sinks below and there is progressive development of underground drainage. Sinkholes, lapies and stray dolines develop, but underground drainage is incomplete and there is no formation of large caverns.

(ii) During maturity there is maximum development of the underground drainage. Surface drainage survives only in the form of small streams and sinking creeks which end abruptly in their swallow holes or blind valleys. There is extensive cavern development and cave travertines are formed. In fact, this is the stage when maximum karst development takes place.

(iii) Late maturity marks the beginning of the decline of the karst features. Karst windows are formed and the cavern streams are exposed. These expand to form uvalas, and gradually polje diversified by hums dominate the surface.

(iv) In old age the surface resembles a peneplain from which rise isolated hums as remnants of the original limestone surface. The underground drainage comes to the surface and there is a return to surface drainage.

Just as in an area of river erosion, one section may exhibit the characteristics of youth, another signs of maturity and the third of old age, so also in a karst region it is possible to observe the different stages in different sections. The various stages move progressively into a region following entrenchment of the principal rivers across the terrain. Areas further away from the entrenched streams are likely to be less advanced in the cycle than those which are nearer.

Questions

1. What is meant by water-table ? What is its importance in the formation of springs and the digging of wells?
2. Explain the origin and distribution of underground streams and caverns.
3. What are the characteristic topographical features of a limestone country ? Describe the work of underground water in limestone regions.
4. Describe the distinctive topographical features of karst regions and explain their formation.
5. Discuss the evolution of the characteristic features of karst topography.
6. Discuss the different views regarding the formation of caverns in limestone regions.
7. Discuss critically the concept of the cycle of erosion in karst regions.
8. Discuss the evolution of landforms in the different stages of the karst cycle.
9. Write short notes on :-

   Artesian wells, Sinkholes and Dolines, Polje and Hums, Karst Valleys, Tropical Karst, Cave travertines.

Chapter 23

Wind Action and Aeolian Landforms

Wind as an agent of geomorphic change is much less significant than water. Wind action is particularly important in arid regions where lack of vegetation and presence of extensive bare rock and sandy soil surfaces make wind erosion, transportation and deposition possible. At one time wind action was supposed to be mainly responsible for the shaping of the desert landscape. But recent investigations have shown that even in the deserts running water is a potent geomorphic agent, and aeolian processes are responsible for only relatively small topographical features of the desert. Apart from the arid regions where sand covers extensive areas, wind action plays a part in the shaping of landforms on sandy coasts, downwind of outwash plain in glaciated areas where there is abundance of sand and silt available for wind transport as also in the dry periglacial regions.

Wind Erosion and its Topographic Effects

Wind performs erosion in three different ways—attrition, deflation and abrasion or corrasion. *Attrition* is the mutual wearing down of the rock particles carried along by the wind. *Deflation* is the lifting and removal of loose materials from the earth's surface by wind action. *Abrasion or corrasion* is the cutting of rock surface by moving weathered rock particles or more concretely the sand-blast action of wind-blown sand.

(a) *Attrition*: In the arid regions on account of the great diurnal range of temperature and the action of frost there is rapid mechanical disintegration of rocks. When these rock particles are carried along by the wind, they not only strike against the rocks standing in their way but also strike against one another. Thus by mutual friction and forceful contact they break down further and there is a gradual reduction in their size, and ultimately they are converted into sand and dust particles which cover large parts of the desert. This process of gradual reduction in the size of rock particles by mutual impact is called attrition.

(b) *Deflation*: In the dry regions the wind can easily lift and transport the unconsolidated sand and dust particles. This process is called deflation. The lifting or entrainment of loose particles takes place by a variety of processes such as, shear, lift and ballistic impact. Large scale removal of sand and dust from a desert region may result either in the formation of depressions called *blowouts* or in the exposure of the rocky surface beneath the sandy cover. In semi-arid areas of western U.S.A. where soil has been stripped of its natural vegetation, one witnesses large scale soil erosion caused by deflation. Deflation has possibly been responsible for the formation of *pedestal rocks* which consist of residual masses of weak rock capped with harder rock, such as are found in the intermont basins of south-eastern California where the tabular mesas capped with gypsum have proved more resistant to deflation than the underlying silts and clays which form the prevailing material. But in the sandy deserts *deflation hollows or blowouts* of various sizes are known to form and sometimes large hollows or basin-like depressions may be produced. Several such depressions are known to exist in the desert lying west of the Nile Valley. Of these the most well known is Quattara Depression in western Egyptian Sahara which has been excavated to nearly 134 m below sea level. Similar but relatively shallower depressions are found also in the Kalahari and the West Australian deserts. Where sweet ground water is available at depth of approximately 15 to 30 m, the depressions often form oases, otherwise they are occupied by saline lakes or swamps. The pans of S. Africa and Kalahari and the large basin-like depressions and oases of Egypt and Libya have all been formed in this manner. The Great Plains region of the U.S.A. has thousands of small and big basins or hollows most of which appear to have been caused by deflation. Numerous very large depressions in the Mongolian desert have also been found.

It is difficult to say why deflation begins at a particular point in preference to others. But once the process has started at a particular spot, as a result of the formation of wind eddies the hollow goes on increasing until the ground water-table is reached. The ground water-table is the base level of wind erosion and limits the depth of deflation though the water-table may be lowered as the basins develop. Thus the base level of wind erosion can be much lower than the ultimate base level of the sea.

As the wind removes sand and smaller size particles by deflation, the coarser materials which the wind is unable to lift and carry are left behind. In areas of strong winds the ground may thus be left covered with boulders and pebbles. These concentrations of boulders and pebbles and larger sized particles which are left behind after the smaller sized particles have been removed by wind are called *lag deposits or gravel lags*. Locally they may form striking features in deserts. Such deflation surfaces have been often called *desert pavement*, but they are called by different names in different areas depending upon the size and amount of the lag particles. In Algerian Sahara, where the surface is covered mostly by lag gravel, they are called *regs* and in Libya and Egypt where the surface is mostly covered by pebbles, they are referred to as *serir*. Where the surface has large rocky fragments, the area is called *hamada* in the Sahara. Over extensive areas in the Sahara, deflation by wind has resulted in the formation of hamadas, regs and serirs which provide a striking contrast to the areas of sand accumulation known as *ergs*.

(c) *Abrasion or Corrasion*: Abrasion is the work of winds charged with sand particles. Strong winds carry with them enormous quantities of sand and small angular rock fragments which act as tools of erosion as they strike against rock surfaces. The principal mineral of sand is quartz which is harder than most rock-forming minerals. When wind blown sand particles strike against rocks, they act as powerful sandpaper and homogenous rock surfaces are smoothened and polished. Where there are both hard and soft rocks,

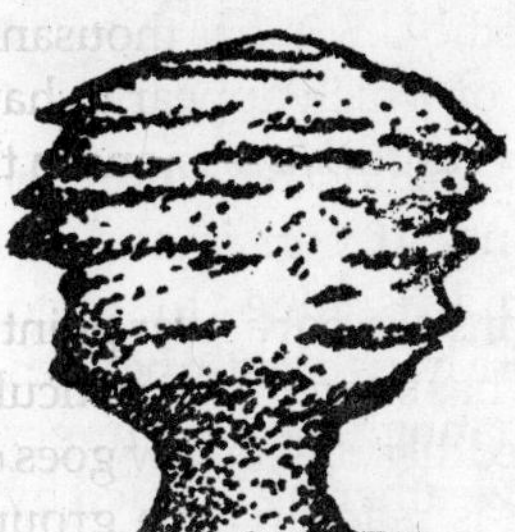

Fig.141— Differential erosion by abrasion

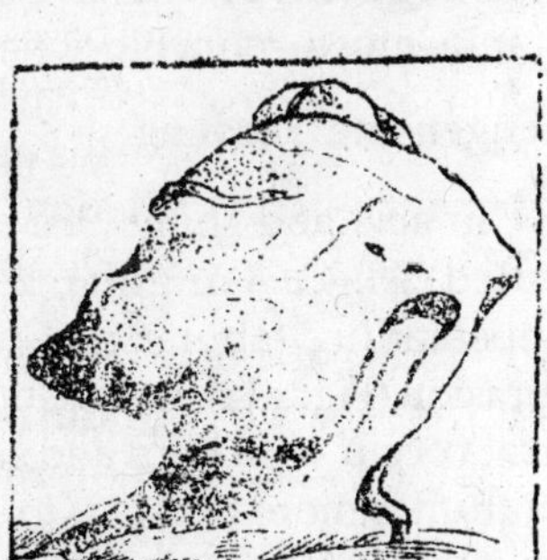

Fig. 142—A granite mushroom block near Jodhpur, Rajasthan.

the soft rocks are more easily cut and the surfaces are grooved and fluted. The Sphinx in Egypt provides a clear example of wind abrasion of this kind (Fig. 142). In conglomerate rocks the relatively soft cementing materials are more easily abraded while the harder gravels and nodules are much less affected with the result that they look like beehives and present honey-combed surfaces, though differential weathering must have also played a part. Pebbles or gravels which are worn and polished by wind abrasion and show one or more polished and faceted surfaces and sharp edges are called *ventifacts or dreikanter*. These are, however, not very common, as they require a combination of strong winds, abundance of sand and absence of vegetation for their development.

Wind abrasion is most effective within a few meters of the earth's surface as the sand is most abundant here, and according to Bagnold, is seldom lifted more than 2 metres above the ground. The maximum abrasion, however, takes place a little above the ground and not close to it, as wind velocity near the ground is impeded by friction with the surface. As the wind cannot lift sand more than a few meters, the effect of abrasion becomes negligible after some height. Undercutting is, therefore, a marked feature of wind abrasion. In this way alcoves and small caverns may be cut at the base of escarpments. But more common are the undercut *mush-room-shaped blocks* called *gara* in Sahara. Where the wind blows from one direction, only the windward face of the rock is under-cut, but where the wind is variable and blows from all directions, the rock is under-cut from all sides. Examples of such mush-room-shaped pedestal blocks can be seen in the western Rajasthan desert where the lower parts of granite blocks have been cut in this manner. An example of granite mush-room block near Jodhpur is illustrated in Fig. 143.

Zeugen and Yardang

In arid and somewhat cooler regions a distinctive landform called *zeugen* (singular zeuge) are found. It consists of an unpstanding tabular rock undercut around its edges by wind abrasion (Fig. 144). It develops in areas where horizontally bedded strata occur with hard and soft layers alternating. The hard upper stratum is more resistant to erosion but its weathering is aided by the accumulation of dews in its joints which are protected from sunlight during the day and freeze at night when the temeprature is particularly low. The freezing causes the rocks to disintegrate near the joints. Wind erosion starts cutting at these vulnerable points in the hard stratum, and once the softer layer is reached, it is cut more rapidly than the more resistant upper layer. In this manner upstanding tabular rock masses with a capping of hard rock and undercut along the sides and looking like ink-pots, develop. These are called *Zeugen* (German). Zeugen are, in fact, a kind of pedestal rocks with a flat surface. Ultimately the basement layer of softer rock may be cut almost entirely and then the tabular block capsizes. Zeugen vary in height from less than a metre to about 30 m. It is obvious that zeugen are formed not entirely be wind abrasion, and differential weathering, dews, rainfall and freezing also play a part.

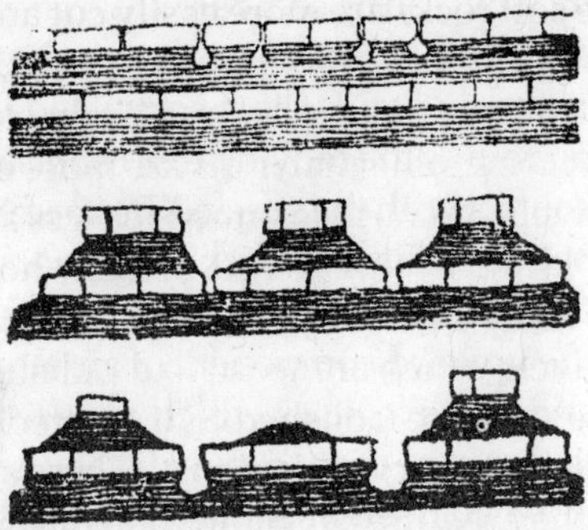

Fig. 143—Formation of Zeugen.

In areas where layers of hard and soft rocks stand vertically in alternate bands, parallel to the direction of the prevailing winds, a distinctive type of land form called *yardang* develops which is almost wholly attributed to wind abrasion. It consists of elongated grooves or furrows which are cut in the softer rocks by wind blown sand in the direction of the prevailing winds and is separated by streamlined, wind-abraded ridges of harder rocks (Fig. 145). These steep-sided linear rock ridges separated by corridor-like depressions occur in groups and are found in Turkestan and Central Asia. Here the rocky vertical ridges have been also undercut by the sand and have an overhanging profile. They are usually less than 2 m high and 2 to 10 m wide, separated by gullylike corridors whose bed is filled with sand. They cover extensive areas and look remarkably similar in their weird appearance.

Fig. 144—Yardangs.

Aeolian Transport and Depositional Landforms

Depositional landforms resulting from wind action are essentially related to its mechanism of transport. Transportation by wind can be compared to transportation by water only to a limited extent. Transportation by wind is not related to slope of the land. In both wind and water, some load is suspended, some bounces at the bottom of the moving fluid and some particles are dragged along in contact with the surface. But because of the great density difference between air and rock fragments, only very fine particles with a diameter of less than 0.2 mm can be maintained in suspension by the upward component of air movement. These are mostly fine silt and clay particles which once lifted fall only slowly back to earth and may be carried long distances by planetary winds before settling down. On satellite imagery clouds of dust have been seen travelling westward in the trade-wind belt from southern Sahara toward the West Indies where they have accumulated on soils. Most loess was transported by this process and ranges in size from 0.02 to 0.07 mm.

In the movement of sands the most important process is *saltation* or movement by intermittent bounces or leaps of varying lengths depending upon the size and weight of the particles and wind velocity. Above a certain velocity the inertia transferred from the wind to the saltating particles is sufficient to maintain a continuous forward transport of saltating sands.

A certain amount of sand is also driven forward over the ground surface without saltating. This process has been called *surface creep* by Bagnold. Surface creep is induced by the saltating sand grains striking particles on the ground and knocking them forward. According to Bagnold (1941)" a high-speed grain in saltation can by impact move a surface grain six times its diameter or more than 200 times its own weight." These creeping particles are large and travel slowly, and thus surface creep contributes to the coarse size-grading of sand deposits, the smaller particles moving by saltation outdistancing the coarser grains that move by surface creep.

The amount of materials transported by the wind depends upon the size and amount of the particles and the velocity of the wind. The materials transported by the wind are also subject to size grading, the smaller silt and clay particles being carried and deposited much further away from the source area than sand, thereby resulting in two distinct types of wind deposits—(a) *Loess*, mostly deposited outside or on the margin of the source area, and (b) *sand deposits* and more particularly *sand-dunes* which are found near the source area.

It has to be noted that wind deposits are not limited to the deserts either in origin or location. In addition to the desert margins, loess covers extensive areas in sub-humid and semi-arid regions. Similarly, sand deposits though most common in the deserts, are also found along some of the coasts, along stream courses in semi-arid regions, in areas of glacial outwash and in areas where the predominant rock is relatively loose sandstone.

Loess

Loess is a wind-blown deposit of fine-grained silt and dust. The two important sources of loess are the weathered materials in deserts (desert loess or hot loess) and the very fine powder from regions of glacial outwash (cold loess). The loess deposits are found away from the source regions and outside the deserts. The deposition of the fine particles takes place either due to reduction in wind velocity or due to rain. The dust particles are so small that they hold together even when dry, and once deposited cannot be easily lifted by the wind. Loess is coherent (bound together) but not cemented and hence is permeable. It is unstratified, and lack of stratification is a clear proof of its aeolian origin.

Extensive deposits of loess are found in many parts of the world. The most extensive deposit is to be found in the region covering much of the North European plain and the Asiatic steppes, extending from France in the west to North China in the east. In France and Germany the area under loess is limited, but as we move eastwards towards Russia, Ukraine and Turkestan, it becomes thicker and larger until the most extensive loess deposits of North China are reached. Other important areas include North Africa, Central United States and Argentina.

The term loess has been derived from the town of Loess in Alsace, France. The loess of North European plain including the extensive deposits of Ukraine has been possibly partly derived from wind-blown silt from the African desert but mainly from Pleistocene glacial flour of the outwash plains whose fine powder has been blown and deposited by wind.

On the other hand, the loess deposits of North China have been essentially derived by deflation from the Gobi desert of Central Asia, possibly supplemented to some extent by the wind-blown fine dust from the flood-plains of the Hwang and the Yangtze rivers. Here the thickness of the loess deposit is about 200ft. (65m), and the high plateaus as well as the low lying valleys are both covered by it. Its colour is yellowish or light gray. It is quite soft and is easily eroded by streams. Deep valleys and deep roads are found in the loess. In the Loess Highlands of North China the Hwangho has cut deep valleys with vertical sides. In Kansu province the smaller streams and rainwater have cut the loess into ravines and badlad topography. But the bigger rivers flow through wide and fertile alluvial plains. Loess forms fertile soil and where the valleys are wide and irrigation is possible, good crops are grown.

The loess deposits of the U.S.A. lie in the Mid-West and along the Mississippi valley and are very similar in origin to the loess deposits of Europe and Asia. In Kansas and Nebraska most of the wind-blown silt is derived from the western semi-arid regions, but in other areas and particularly along the Mississippi-Missouri, Ohio and Wabash valleys it has been mainly derived from Pleistocene glacial outwash deposits and glacial flour which have been blown by the wind and redposited here. The original loess deposits of the Upper Mississippi basin have been eroded and redeposited by the rivers in the flood-plain further down. Thus the loess of U.S.A. is partly of desert and glacial origin but in either case it has been blown and deposited by the wind. The loess deposits which border the Sahara consist of wind-blown silts from the desert, while the extensive loess deposits of the Pampas in Argentina are principally wind-blown glacial powder.

Types of Sand Deposits

The coarser sand particles which are deposited by the wind within the desert itself assume various forms. Bagnold recognizes five principal large scale forms of sand deposits: (a) Sand dunes, (b) Sand shadows or sand drifts, (c) Whalebacks or sand levees, (d) undulations, and (e) Sand sheets. Of these by far the most important are the sand dunes.

Sand Dunes

Sand dunes are large and mobile mounds or heaps of wind-blown sand which are not dependent either on the ground form or a fixed obstruction for their existence and maintenance. They are apparently free to move, divide or grow. They are of different forms—round, crescentic or elongated and narrow, their precise form being dependent essentially on the strength and direction of the dominant winds. They also vary greatly in size, and some dunes may be several hundred metres high and several kilometres long. Groups of dunes are referred to as *dune chains or dune colonies or dune complexes*. Three principal factors influencing the formation of sand dunes are an abundant supply of sand, strong winds and some initial obstruction in the path of the wind for the sand to be deposited.

Sand dunes are typically found in the sandy deserts, but they also occur in those coastal areas where there is an abundant sand supply. Approximately one-fourth of the surface of the deserts in the world is sand covered. Many deserts consist of solid bedrock with some rock debris and pebbles. The deserts where the surface is formed of shale or limestone do not have a cover of sand, but where the predominant rock is sandstone, huge accumulations of sand may develop as a result of the mechanical weathering of sandstone, and in such sandy deserts sand dunes are comnon. Examples of such sandy deserts are the Sahara in Africa, the desert of Utah and Arizona in the U.S.A. the Gobi desert in Central Asia, the desert of Saudi Arabia, the Thar desert in India etc. In addition to the abundant sand supply the other necessary condition for the formation of sand dunes is the existence of strong winds. The hot deserts in the trade wind belt are characterised by strong winds. Strong winds are also found along the sea-coasts. The third condition for the development of sand dunes is the existence of some mechanism or obstruction whereby the wind velocity slackenes and the wind-blown sand is deposited. When the wind charged with sand particles is blowing the presence of dense grass, trees, some rising rock or undulating ground may act as an obstruction and reduce wind speed, leading to deposition of sand. Once the accumulation has started, the accumulated sand itself acts as a barrier and the sand-dune tends to grow. The obstruction, however, is important only in the initial stage of formation, for most dunes attain their maximum development upon relatively flat terrain and do not seem to be dependent upon any obstruction or topographic break for their existence.

Sand dunes have a variety of forms, depending upon wind effectiveness, vegetative cover and sand supply. Some of these have come to be known only recently after aerial photographs of unexplored sandy regions especially in the Sahara and Arabia have become available. Bagnold recognized only two basic dune forms, *the barchan* or the crescent-shaped transverse dune and the *seif* or the longitudinal dune. The *transverse barchan or barkhan dunes* have their long axes at right angles to wind direction. They are associated with large quantities of sand and relatively weak winds and are not very common in the interior of great sandy deserts but form

Fig. 145—Transverse sand dune.

along coastlines. Where the wind is more effective, the transverse dunes assume the classical crescentic form and become barchan dunes (Fig 147.). The barchan dunes including the so called transverse dunes have an asymmetrical profile with a gently inclined windward slope and a steep leeward slope. Since the sand tends to continuously fall from the crest to the leeward side, the leeward side is known as the slip face or the sand-fall slope. Further, while the windward side has a convex slope, the leeward side has

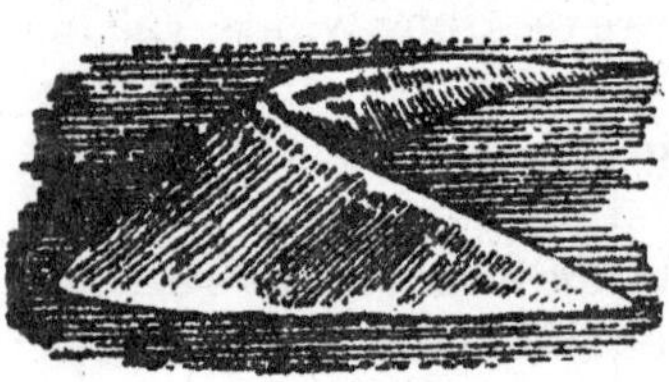

Fig. 146—Barchan.

Fig. 147—Joined Barchans.

a concave slope on account of the sand slip and the eddy action of wind. Because of the persistent action of the strong prevailing winds the two ends of the dune are drawn downwind and project like horns to the leeward making this side concave in plan. In this way the crescentic shaped barchan or barkhan dune is formed.[1]

1. "The slip face of the barchan advances by process of over-steepening and shearing. The flanks of the original mound of sand advance more rapidly than the centre, for the rate of advance is inversely proportional to the height of the slip face. In this way the crescentic barchan form is developed." Sparks, B.W.—op. cit., p. 252.

Barchans do not require any obstruction to develop. Accumulations of sand are enough to help in their formation. Barchans are up to 30-35 metres high, and their width is usually 10-12 times their height. Where the sand supply is more, barkhans are large and dense, but where the sand supply is less, they are smaller and more dispersed. Sometimes barchans are found as isolated mounds, but more commonly they are found in large groups or chains, and sometimes adjoining dunes get joined together (Fig. 148). As the sand blows up the windward side and falls on the leeward, the barchans tend to migrate downwind. Through this process whole groups of barchans may migrate across the desert like irregular waves, maintaining their form as they migrate. They have been found to migrate at a maximum rate of 30-35 metres per year, the larger dunes migrating more slowly than the smaller. The dunes may also occasionally get stabilized by rainfall and by the growth of grass and other vegetation. Good examples of chains of barchan are found in the Sahara (Fig.149). In Turkestan desert there is a seasonal change in the wind direction. The wind blows from the north in some months and from south in the remaining months. The result is that the barchans also change their form and their horns get reversed with a reversal in wind direction.

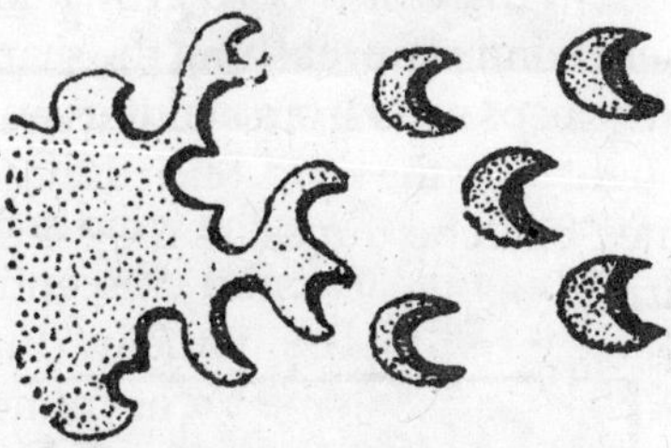

Fig. 148—Barchan chains or complexes in the Sahara desert.

Seif or *longitudinal dunes* are very different from the barchans. Instead of being transverse to the prevailing wind they extend as long and linear ridges parallel to the effective wind direction. These linear ridges of sand may be hundreds of kilometres long, a kilometre or more wide and several hundred metres high. They are much higher than barchans and look like hills with sharp knife-like crests containing peaks and sags and steep slopes on either side. In the Egyptian desert many seifs are about 100 m high but in southern Iran they rise to more than 200 m. According to Bagnold, their width is usually six times their height. Seifs occur in large parallel groups, separated by corridors between adjacent seifs. These corridors are sand-free and are known as *gassi* in the Sahara desert. They are used as passages for the movement of caravans of camels. The linear ridges of the seif with their numerous peaks wear a serrated appearance. Many such parallel seif dunes with intermediate gassi are found in the Libyan desert and extend continuously for 100 km or more. The gassis have been swept clean of sand by the strong wind which by eddy action deposits the sand on either margins.

The highest dune has been recorded in Algeria and is about 400 m high. It follows the line of a scarp which diverts sand flow and allows the dunes to grow in calmer pockets. Some of the highest dunes in the world have grown in zones parallel to escarpments where winds approaching the scarp create a huge standing vortex that sweeps sand from an intervening corridor together with sand brought in by the wind. Such dunes are generally known as 'obstacle dunes' or 'echo dunes' as they reflect the form of a neighbouring scarp.[1]

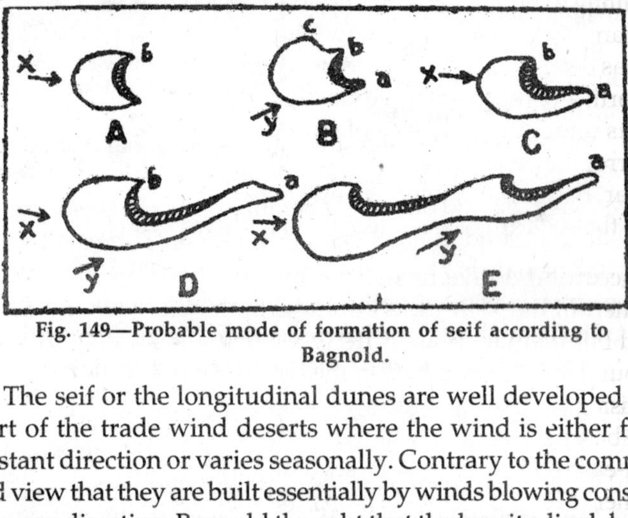

Fig. 149—Probable mode of formation of seif according to Bagnold.

The seif or the longitudinal dunes are well developed in the heart of the trade wind deserts where the wind is either from a constant direction or varies seasonally. Contrary to the commonly held view that they are built essentially by winds blowing constantly from one direction, Bagnold thought that the longitudinal dunes of Sahara grew in height and width largely through the action of cross winds and grew in length downwind during the periods when the prevailing wind was parallel to the trend of the seif chain. He, in fact, maintained that the seif is a modification of the barchan form produced by strong obliquely blowing winds transverse to the prevailing wind direction and the sharp-edged crests represent

1. Warren, A —"Aeolian Processes" in *Process in Geomorphology*, ed. by C. Embleton and J. Thornes, Arnold-Heinemann, Indian Ed., 1980, p. 343

barchan forms superposed upon the sand ridge extending parallel to the prevailing wind.

This is illustrated in Fig. 150 where X represents the direction of the prevailing wind, and Y the direction of the transverse oblique wind. In 150 A the formation of a barchan under the influence of the prevailing wind has been shown. In B, under the influence of the seasonal oblique wind Y, horn *a* is extended and a new horn *c* is born. When the prevailing wind again becomes powerful, the dune is converted into its usual shape with two horns but horn *a* becomes much longer than born *b*. (Fig. C). In Fig. D horn *a* is further lengthened under the influence of wind Y. In the final stage E the dune becomes so long that both the prevailing and the cross winds supply it with adequate quantities of sand and the dune continues to gain in length, breadth and height. The slip face is on the leeward side of the transverse wind Y and not on leeward side of the prevailing wind X. This is an important difference between seif and barchan. Thus according to Bagnold a seif develops from barchan and has been formed in the Sahara by an alternation of winds from the north and north-east. Seifs are found in the interior of the great deserts under the influence of strong prevailing and cross winds and irregular winds are less powerful but sand supply is more regular. Barchans are mobile and keep on changing their position while the seifs are more or less fixed in their position.

According to Bagnold, barchans and seifs are rarely found together. In the Thar desert of India both barchans and seifs are found but their areas are different. Horse-shoe shaped barkhans are found in large numbers in the form of small sandy hillock, but surprisingly while the south-west monsoon is more powerful, many of the barkhans appear to have been formed under the influence of north-east winds. At places several barkhans have been joined together to form a relatively larger hill which has been divided into spurs by intervening valleys. It is possible that the spur represents the windward slope of the original barchan. Examples of barchan can be seen near Barmer in Jodhpur as well as in Bikaner where they are 200 to 1,000 m in length and 180 to 600 m in width.

Longitudinal or self dunes are more characteristic of the western part of the Thar desert where the south-west monsoon winds are more powerful. They have a steeper slope on the leeward side but have a more gradual slope to the crest on the windward side. They extend as parallel linear ridges in the direction of the wind, but at places they have also been joined together by transverse ridges.

Other types of dunes include *parabolic dunes* which are associated with blowouts in semi-vegetated areas with crescentic shapes, and are usually much longer and narrower than barchans. Parabolic dunes have been found in Bikaner and other parts of the Thar desert which are mostly stabilized. Another type recognized in western Rajasthan is the *shrub-coppice dune* which is only about 1 to 1.5 m high.

Very large *stellate and rosette dunes*, several hundred metres high and several km in diameter, with radiating sinuous ridges culminating at a common crest have been found in the Saharan erg near 30°10' N and 8°15'E.

Besides the sand dunes, four other types of sand deposits identified by Bagnold are sand shadow, sand drifts, sand levees or whale-backs and sand sheets.

Fig. 150— Sand shadow

Fig. 151— Sand Drift.

Sand Shadow

Where there is an obstruction such as boulder, bush or cliff in the path of the wind, the wind's velocity is reduced and there is deposition of sand behind and in the shelter of the obstruction. Such a deposit is called sand shadow (Fig. 151). Its height is not higher than the obstruction, for outside the shadow the wind is stronger. Sand continues to accumulate in the lee of the obstruction until its front slope has attained an angle of 34° which is the limiting angle of repose. After this the sand slides down the front of the advancing deposit along the slip-face.

Sand Drift

Where there is a gap between two obstructions, the gap acts like a funnel through which sand is pushed forward and deposited on the leeward side (Fig. 152). This is in fact a variant of sand shadow. Sand drift is a term which is more commonly used for thin deposits of sand but the latter has been called sand sheets by Bagnold. The use of the term sand drift by Bagnold in the above special sense has therefore created some confusion.

Sand Sheets

The term sand sheet has been used by Bagnold for sand areas with extremely flat surface and absence of any topographic relief except ripples which resemble the sea waves. This happens where pebbles are found mixed with the sand and the wind levels the small mounds and is unable to build dunes. Such areas are called "sand seas". A well known example is the Selima sand sheet in the Libyan desert which is nearly flat with only stray line of dunes and covering an area of nearly 8,000 km^2

Sand levees or Whalebacks

Whalebacks are sand ridges with flat tops and aligned in the direction of the prevailing winds. They lack the collapsing fronts found in the seifs and resemble the backs of whales in appearance. They are probably the residue of dune chains that have moved downwind, and the coarser sand has been left behind as a prominent ridge. They are found in the Egyptian sand sea where they are 50 m high, 1-3 km wide and extend upto 300 km. These forms may also represent the unexplained aspect of the effect of climatic change on landforms. Sometimes whalebacks may lack definite form and appear as undultions. In either case they appear to be areas of dead or immobile sand in contrast to the mobile barchans and seifs.

Water Action in the Desert

Most deserts have so little rainfall that surface drainage is practically absent. Whatever little rain that falls is highly irregular. For years on end there is no rainfall, but occasionally in the mountainous areas there may be torrential rainfall. The result is that stream channels which are dry most of the time may experience sudden floods which arrive as suddenly as they depart. These have been called *stream floods* rather than streams by Davis because of their spasmodic and impetuous flow. Besides stream floods, there are sheet floods i.e. broad sheets of water which do not move as definite stream but as system of small enmeshed channels. They result from heavy downpours on barren surface covered with loose detritus and soon become heavily loaded with the detrital materials. Like the stream floods they are short-lived, and on account of the heavy load and strong evaporation they soon dry up and run only for short distances. Most of the detritus carried by the stream floods and sheet floods is deposited in the form of alluvial fans in the mountainous regions and in the foot-hill zone. These fans contain lot of pebbles and stony fragments. Alluvial fans are more typical of the intermontane basins than the plateau deserts. Sometimes several fan-like deposits of pebbles and sand are joined together resulting in the formation of a relatively flat surface called *desert pavement*.

While the desert streams as a rule are short and intermittent, there are some large streams which originate in the rainy mountains outside the desert and contain water throughout the year inspite of their long journey through the desert. Two prominent examples are the Nile in Egypt and the Colorado in the U.S.A. Barring such exceptions, most of the desert streams lose themselves in the desert and the drainage is inland ending up in some low ground within the desert.

In the desert areas of western U.S.A. and Mexico, lowlying ground or basin surrounded by mountains is called *bolson*. In bolson the drainage is centripetal i.e. towards the centre where there is a level plain called playa, marking the site of an existing or extinct lake. Many of the small desert streams end in the mountains, but where the rain water is able to move down through the streams or sheet floods to lower ground temporary lakes are formed. With the evaporation of water, the lake is converted into a swamp and when dry becomes a level plain of fine sediments. Such level alluvial surfaces which either formed a lake in the past or contain remnants of the former lake even now have been called *playas* in the western U.S.A. and the lakes are called *playa lakes*. Most of the playas are dry except after a heavy downpour. Their surface is often covered with white salt, gypsum or other salt precipitated from the water of the temporary lakes. Such salt covered surfaces have been called *alkali flats or salinas* in the U.S.A. and *sholts* in Mexico. Such areas of inland drainage are found in all deserts but are more characteristic of the intermontane basins of Central Asia and N. America, as the rainfall on the mountains encircling the deserts is relatively high. In Central Asia the Tarim Basin is a good example where the centripetal drainage flows into the playa lake of Lop Nor, and in between the central playa lake and the mountains surrounding the basin there is a relatively wide piedmont zone of pebbles. In the mountainous deserts two other important fluvial features are *canyons* and *wadis*. Canyons are extremely narrow and deep valleys mostly found in the plateau deserts. In south Algeria there is a rocky limestone desert in which there is a network of dry gorges. These are called *Shebka or Chebka* which literally means a net. Wadis are river-cut valleys with flat and wide floor and steep and almost vertical sides. The side walls are steep primarily due to two reasons. Firstly, abrasion by wind is most active at the base of the mountains, and secondly, on account of lack of rainfall there is very little corrasion of the valley walls and stream erosion is largely vertical. In the upper courses of rivers, the sudden and short-lived floods carry downwards the rock fragments and debris, but their lower courses are covered with these debris and boulders. The rivers are dry most of the year, with short-lived floods during heavy rains. They are wider near their mouths where the alluvial fans are deposited which gradually slope down to the plain. It has been suggested that the gorges and wadis of the deserts represent relict landforms and were formed when the rainfall was probably higher.

In some regions where there are mountain ranges in the desert, they present a naked and harsh appearance, cut by deep and steep-sided wadis. From above they look serrated and bare. Examples of such landscape are to be found in the Tivesti and Ahgar ranges of Central Sahara and in Baluchistan and Western Arabia.

The occasional downpours of rain in the arid and semi-arid regions have a high intensity. Under their influence, ravines and gullies separated by sharp spurs and ridges are formed especially in areas of relatively soft rocks. Thus the entire surface becomes highly irregular, badly cut with sharp ups and downs. Such dissected lands have been called *bad lands*. In N. America bad lands extend right from Alberta to Arizona.

We thus find that water is a powerful agent of erosion in the desert as well. The influence of rainfall in the shaping of the desert landscape at the present time is clear enough, but it is believed that most of the fluvial features found in the deserts were formed in the

Pleistocene period when these areas had a more moist climate. In other words, some of the features that we find in the deserts today are inherited or fossils landforms that were formed under more humid conditions, and they have retained their forms without much change as wind is a relatively slow and less efficient agent of erosion.

Pediments, Pediplains and Inselbergs

Pediments

In the arid and semi-arid mountainous regions there is usually a sharp break in slope between the mountains and the adjacent plains. The plains have a gentle slope not exceeding 7^0, but the neighbouring mountains have a steep slope which may range from 15^0 to almost vertical. In the arid regions of south-west U.S.A., the area of gentle slope lying between the mountain and the playa is called *piedmont*. The piedmont is normally divided into two distinct parts called the *pediment* and the *bahada* (Fig. 153). *Pediment* is upper section of the piedmont which is situated close to the mountain and has an eroded bed rock surface. It is formed by the degradation of the mountain front. This is thus a plain of degradation and has little or no cover of debris on it. The lower part of the piedmont is called the *bahada or bajada*. Bahada is an aggradational feature and has been formed by the deposition of debris derived from erosion in the hills. The bahada deposit normally

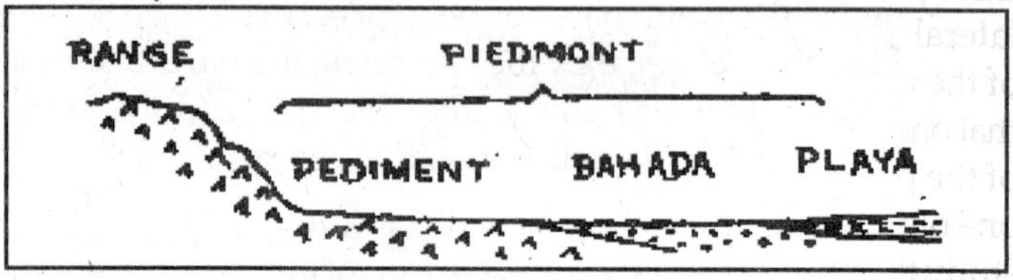

Fig. 152—Piedmont (divided into Pediment and Bahada) and Playa.

occurs on the outer margin of the pediment but occasionally the pediment may also contain a thin cover of debris. On both the pediment and the bahada slope is gentle, but suddenly between the pediment and the hills the slope becomes steep and there is a sharp break in slope where the pediment meets the mountain front. It is evident that deposits of both the bahada and playa have been derived from the erosion of the adjacent hills and have been deposited by the streams coming down from the hills. Consequently, the materials deposited become smaller in size with distance from the mountain. The bahada is formed by coalescing alluvial fans. During the rains the mountain streams bring down huge quantities of debris and deposit them near the foothills in the form of alluvial fans. These deposits have a predominance of pebbles and boulders but there is also some mud and alluvium. The bigger streams form broader alluvial fans with low gradient, but the smaller streams build smaller and steeper alluvial cones. Since these rivers are ephemeral and intermittent, their channels are not fixed and change frequently across the fans, making surface of the bahada undulatory. The alluvial fans may also be built by sheet floods coming down the mountain side. Similarly, the playa deposits are composed of fine alluvium brought by the streams and the sheet floods and deposited further down.

Mode of Formation of Pediment

The mode of deposition in the bahada and playa is relatively simple, and easy to understand. But the origin of the pediment is more complex, and there is considerable difference of opinion about its mode of formation. There are two basic problems in this regard-(1) what is the reason for the sharp break in slope between the pediment and the mountain front? This probably indicates a change in the nature of the process. (2) The second question is as to what processes are responsible for the formation of the pediment. In this connection three possible processes have been suggested—sheet flood erosion, lateral planation by streams, and disintegration and degradation of the mountain front by weathering. Some people have suggested that one of the above three processes may be responsible for formation of the pediments, while others feel that the operation of more than one of the above processes is essential for their formation. We may study the different viewpoints under three broad heads—(i) Lateral planation theory put forward by Page (1912), Blackwelder (1931) and Johnson (1932), (ii) Recession of the mountain front by weathering and the formation of the pediment by sheet floods supported by Lawson (1915) and Davis (1938), and (iii) the composite theory in which all the above processes are involved, put forward by Bryan (1923, 1940), Gilluly (1937), King (1950) and others.

Lateral Planation Theory : According to this theory the pediments owe their origin to lateral erosion by short-lived mountain streams of the desert as they emerge from the mountain front. Both Blackwelder and Johnon considered this to be the dominant process in the formation of pediments. Johnson argued that as the mountain streams debouched upon a piedment slope, they formed rock fans with apex towards the mountain front, and as the river courses shifted laterally, the rock fans were extended and resulted in the formation of pediments. Various objections have been raised to Johnson's theory. Firstly, there is absence of such rock fans in the desert. Secondly, in terms of this theory the mountain front should be indented, but actually we find straight mountain fronts. Thirdly, ordinarily sheet flood erosion is more significant in the desert than stream erosion. Fourthly, as the mountains are cut back the erosive power of the streams should also progressively decline but the process of pedimentation seems to continue. Finally, the presence of inselbergs or residual knobs on the pediment surface is not compatible with lateral planation.

Lawson's Theory : A different theory has been proposed by Lawson which has been supported by Davis. According to this theory the mountain front suffers disintegration by weathering, and the rock debris so received are transported downward by gravity and sheet floods. By this process the mountain front recedes and pediment having eroded bed rock surface is formed, as the sheet floods carry the rock fragments and debris further down but themselves are unable to perform erosion. Only a minor role is assigned to lateral erosion in this theory. Lawson called the bench next to the mountain and not covered with alluvium as subaerial bench while the buried rock bench further down beneath an alluvial till was called by him the sub-alluvial bench. Pediment includes both these sub-divisions, though it is physically exposed only in its upper part.

Composite Theory : According to this theory the pediments owe their origin to a combination of processes rather than to one dominant process. These include back weathering, sheet-flood erosion, rill wash and rain wash and lateral planation. The relative importance of these factors varies from place to place, and different individuals have laid varying emphasis on them. In his study of the pediments of Arizona, Gilluly agreeing with Bryan states that "Pediments are slopes of transportation peculiar to arid regions and formed by co-operation of lateral erosion, rill wash, and weathering with subsequent removal of detritus by rills. One process may dominate in one place and another elsewhere but all are operative."[1]

Rich (1925) believed that lateral corrasion by streams might be a contributory factor in pediment formation but it was not a necessary process. The other factors like back weathering, recession of mountain front and sheet wash were, of course, essential elements. This is close to the views of Lawson and Davis, but lateral erosion may be significant under certain conditions.

An important thing to remember is that the break in slope at the junction of the pediment and the mountain front is indicative of a change in process. The mountain slope is an ungraded gravity slope, but the pediment is a graded slope of transportation down which thin veneer of detritus is in motion which is ultimately deposited in the bahada.

The most suitable rock for the formation of steep slope between the mountain and the pediment is granite, as the rock fragments derived from the disintegration of granite are either boulders which have fractured along joints or are sand particles. The sand particles can be easily transported down a gentle slope. Consequently, the pediment has a gentle gradient, and there is a sudden increase in slope at the junction of the pediment and the mountain. Where other igneous rocks are present their disintegration yields mixed debris of different sizes which are deposited in the piedmont zone and the slope angle is not so marked and sharp.

Pediplains and Inselbergs

Very similar to the pediment areas of arid south-west U.S.A. and adjacent Mexico, the arid and semi-arid areas of Africa and

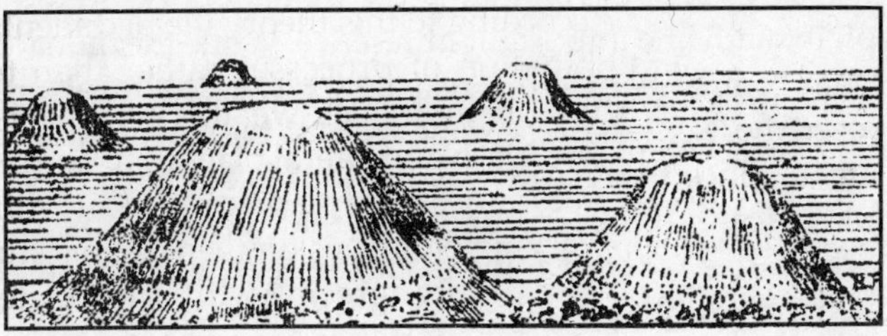

Fig. 153—Dome-shaped Inselbergs in Nigeria.

1. Quoted in Thornbury, W.D.—op. cit., p. 277

more particularly the savana regions of Africa are characterized by a distinctive landscape in which steep-sided, isolated hills of extremely hard rock rise like islands from the neighbouring flat plains. These were studied by German geomorphologists (Bornhardt, 1900, Passage, 1928) early in the 20th century and were called by them *Inselberg-landschaft* or *inselberg landscape*. Unlike the American pediment, this landscape is not related to mountain basins. There are extensive plains which possibly represent the last stage of pediment formation, and have been called pediplain by L.C. King who has made detailed studies of such surfaces in Africa.

The island-like residual hills and knobs found in the Kalahari desert were first called inselberg in German which literally means island mountain. The term, however, subsequently has come to be used for the residual knobs on pediments in arid and semi-arid regions without keeping in mind the distinction in processes between the dry regions and the wet-and-dry savana regions. It is, however, now generally agreed that inselbergs are more truly the result of savana morphogenetic processes and very likely are a relict feature in the desert. Inselbergs are steep-sided residual hills of extremely resistant rocks surrounded by extensive flat surface of the pediplain from where the rocks have been cut and removed by large scale weathering, erosion and deflation by wind. The inselbergs are very similar to the monadonocks found in the peneplain, with one important difference. While the monadoncks represent remnants of more resistant rocks, the inselbergs are composed more or less of the same rocks as the surrounding plains or plateau. Since Bornhardt was the first to study these isolated hills of Africa, it has been suggested that the domelike residual hills which rise without talus from the level erosion surface of pediments or pediplains and which are so common in savana region should be called bornhardts, but this has not been widely accepted.

Fig.154—Conical Inselbergs in Mozambique

The actual shape of the inselbergs is determined by the nature of the rocks. In Nigeria dome-shaped hills of granite rise from the wide rocky plains (Fig. 154). In Algeria these hills are of hard sandstone and rise like steep-sided pillars. In Mozambique the inselbergs are conical in shape (Fig. 155). In southern Rhodesia and in fact throughout the savana region of western and eastern Africa dome-like sugar-loaf hills without talus are quite common. Examples of inselbergs are also found in Western Australia, Arabia and the Rajasthan desert. In most places they are composed of granites or gneisses, as these rocks are highly resistant.

There is considerable difference of opinion amongst scholars who have studied the inselberg landscape of tropical Africa about the process of its development. It was at first thought that it has been formed essentially by wind erosion, and the flat plain dotted with the residual inselbergs was considered to be the last stage of arid erosion cycle. But later on it was suggested that wind was not capable of reducing such extensive areas to low level plains by corrasion and deflation, and very likely water erosion has played an important part in it. Besides the deserts, the inselberg landscape is found extensively in the savana areas of Africa. This was considered to be a proof of the fact that the inselbergs were formed in the Pleistocene Ice age when these areas had a more humid climate and river erosion was more common. Peel (1966) has emphasised that many of the desert landforms appear to be relict features from more humid times. In support of this view it is pointed out that the plains above which the inselbergs rise have much lower gradients than the pediments. Cotton (1942) has suggested that the inselberg landscape of the savana may have been produced largely by lateral erosion by rivers, while the arid and semi-arid areas of Africa may have pedimented landscape in which residual hills similar to the inselbergs rise above coalescing pediments or pediplains.

According to both the new and the old schools of thought two cycles of erosion are considered to be probable for the formation of the bornhardts. According to the old view in the first cycle of erosion there is deep weathering and in the second cycle as a result of the removal of the weathered materials, the more resistant of the weathered rocks remain upstanding as bornhards. This process is not accepted by the new school led by King (1966). It is pointed out the bornhardts are usually several hundred feet high and sometimes may be as high as 1,000-1500 ft. The disintegration of rocks by weathering to such depths is considered improbable. But King agrees with the possibility of the bornhardts being formed in two cycles of erosion. The process of formation as envisaged by him has been shown in Fig. 156. The first requirement is the presence of granites which are not affected by chemical weathering. Other basic or coarse igeneous and metamorphic rocks may also give rise to inselbergs, but in these cases the shape of the inselberg is not quite typical. The other requirement is the presence of a widely spaced system of rectangular joints in the rocks which may determine the drainage pattern.

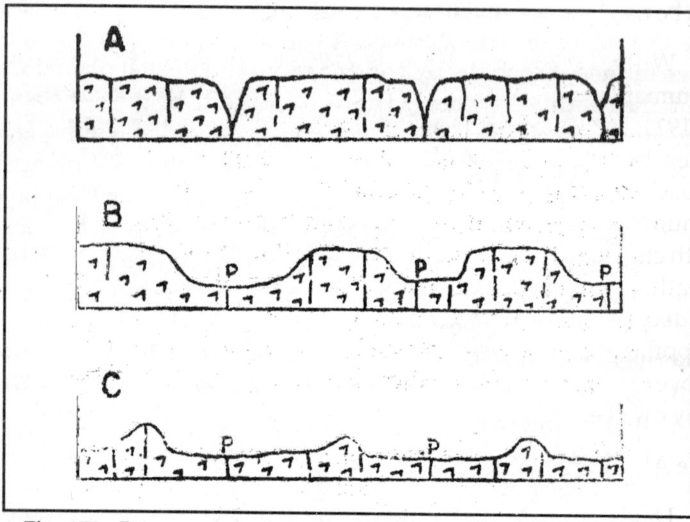

Fig. 155—Process of the formation of bornhardts according to L.C. King.

Bornhardts are formed in the second cycle of erosion of a rejuvenated landscape. By the end of the first cycle the drainage pattern is completely adjusted to the system of rectangular joints. When rejuvenation occurs, these rivers start cutting downwards along the joints (Fig. 156A). As a result of the extension of the valleys, pediments are formed and the rectangular land between the rivers is reduced (Fig. 156 B), and at last these interfluves assume the shape of steep-sided inselbergs (Fig. 156C). Finally, even these bornhardt hillocks are also destroyed and pediplain results by coalescence of the pediments which represents the old age of this cycle.

King has used the pediplain concept on an extensive scale, and in addition to the African plateau, he has tried to interpret the widely separated erosion surfaces in S. America, India, Australia, Europe and N. America as ancient pediplains. Many scientists do not agree with his sweeping generalizations. According to him, the peneplain of Davis is an imaginary landscape which is not to be found any where in nature, and the ultimate feature produced in the cyclic evolution of landscape is pediplain. There is no doubt that the pediplain concept is an important contribution in geomorphological thought and a number of erosion surfaces which were formerly considered to be peneplains, can be better explained by reference to the process of pediplanation. But the concept can not be used indiscriminately to explain all kinds of erosion surfaces.

Willis (1936) and Thomas (1966) have tried to revive the exhumation hypothesis of inselberg formation proposed by Falconer in 1911. In this hypothesis inselbergs are assumed to be domical rises beneath the basal surface of weathering which are subsequently exposed by removal of overlying regolith. The exhumation may be due to uplift and stream rejuvenation, coupled with climatic change from humid to arid which reduces the rate of weathering and vegetative cover and makes mass surface removal of debris more effective. The main difficulty in accepting this hypothesis lies in the great height of some inselberg domes rising above many savana plains, though it is not unlikely that some domes may owe their origin to this mechanism.

The Arid Erosion Cycle

Davis outlined the different stages of the erosion cycle in the arid regions in 1905. For this he has taken the desert of south-western U.S.A. as the initial surface where there are several basins encircled by block mountains formed by block faulting. He has assumed the climate to be semi-arid rather than arid in the initial stages. The progress of the arid erosion cycle as conceived by Davis has been illustrated in Fig. 157. In 157 A, the initial surface has been shown in which there are block-faulted mountains with intermontane basins. On account of the semi-arid climate, there will be active disintegration of the mountains in the young stage and the debris derived from weathering and erosion would be deposited in the basins (157 B), and alluvial fans and playa lakes would be formed. Due to the continued deposition of debris in the basins and the lowering of the intervenning mountains, the deposits of the higher basins would join the deposits in the low lying basins (157 C). The integration in the deposits of the different basins is indicative of the mature stage. In the process of integration of the basins, there may be headward erosion from the lower towards the higher basins, and through this process there is continual lowering of the area. Ultimately due to the lowering of the mountains and the reduction of relief, there will be reduction in rainfall which will slacken river action and increase wind erosion. As a result of increase in aridity and wind erosion, deflation or wind hollows will be formed in the old age (157 D). Thus in the initial young stage of the cycle, the drainage pattern is non-integrated, it becomes integrated in the mature stage, but in the old age it becomes disintegrated on account of deflation. In the old age a rocky plain is formed under the influence of wind erosion the height of which may be lower than the height of the lowest basin of the initial surface. When exactly in the cycle wind action and deflation will become the dominant process will depend on the initial relief, types of rocks and their rate of disintegration into sand and dust as well as the velocity of the prevailing winds. According to Davis graded slope is usually formed by the debris, but in the upper parts this can be rocky (or pediments).

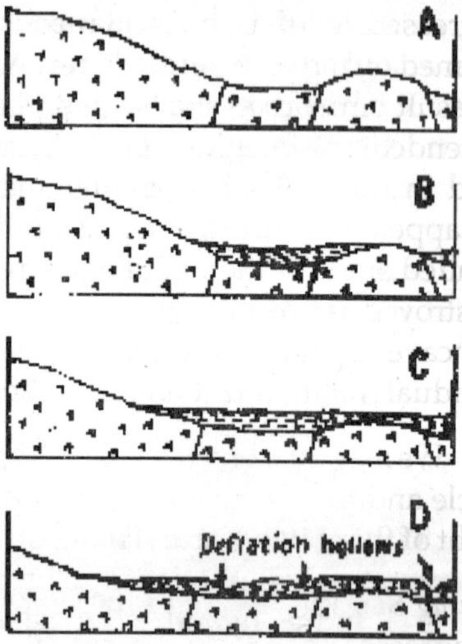

Fig. 156—Stages of the arid erosion cycle according to Davis.

In recent years there has been a tendency to emphasize the formation of pediments and their inward recession in the arid erosion cycle. Davis too has laid stress on the process of pedimentation in arid erosion in his later essays. L.C. King (1953, 1963) is, however, the leading exponent of this view. He believes that pedimentation is the most important process in the evolution of landscape in the arid regions. He has also put forward the concept of a *pedimentation cycle*. According to him, in the young stage there is downcutting by streams and the valleys are formed resulting in increased relief. But simultaneously pediments also begin to be formed on the valley sides in this very stage. In the mature stage as a result of the recession of the mountain front the pediments are extended, the interfluves between rivers are cut by river erosion, and in this way the original surface is nearly destroyed and disappears. During this stage there is also the development of the graded slope of the pediments. In the old age the mountains are destroyed as a result of the coalescenc of the pediments and a multi-concave topographic surface is formed from which rise isolated residual mounds and hills. This surface is the pediplain.

We may now broadly summarize the progress of the arid erosion cycle and the characteristic features in the different stages in the light of the above discussion :—

Young Stage :— In youth the relief is maximum, and as the cycle progresses there is a progressive reduction in relief. There is disintegration of rocks by mechanical and chemical weathering and deposition of rock debris. When there is heavy rainfall, the fast flowing streams carry the rock debris and deposit them in the lowlying basins. The mountain thus begins to be divided by the degrading consequent streams, and simultaneously it suffers reduction in height by weathering and erosion. In other words, the higher parts are eroded while the rivers develop their valleys on the mountain slopes. Thus there is degradation of the higher parts and there is aggradation due to the deposition of debris in the basins. Alluvial fans are built on the mouths of the valleys. The combined effect of all these is a reduction of the initial relief. The rivers are intermittent and shortlived, and every basin has its own drainage system. Drainage is inland, and the debris are deposited separately in each basin. Since the rainfall is uncertain and torrential, there is neither development of master consequent streams nor of subsequent streams. Thus rivers follow different courses at different times depending upon the intensity of rainfall. The mountain is thus cut into several deep valleys which are unrelated to structure and the hardness of rocks.

Mature Stage :—It takes a long time to reach maturity from the young stage, but in the mature stage there is rapid reduction in relief. As in the arid regions the rivers do not reach the sea, local base levels of erosion are more important, and with the accumulation of sediments in the basins, the local base levels also keep on rising. The consequent streams carve out deep valleys on the steep mountain slopes. These valleys are much wider than the river channels that occupy them, as most of the erosion is effected through sheet floods. The rivers build huge alluvial fans at the mouths of their valleys. In the arid regions the large size of the alluvial fans is due to the fact that the entire debris derived from weathering and stream erosion is deposited here and none of it is transported towards the sea[1]. Concurrent with the building up of the extensive alluvial fans the mountain front goes on receding due to erosion, and water divides of the valleys become narrower. The coalescence of several neighbouring alluvial fans results in the formation of the bajada, the upper part of which consists of smooth bedrock surface with little or no veneer of sediment and is called the pediment. It is clear from the deposits of the different basins, that even in the arid cycle the action of water continues to be important till about the middle of maturity, and it is only towards the end of the mature stage that the importance of water action declines as by that time reduction in relief results in reduction in rainfall. From now onwards wind action dominates. The playa lakes start drying up by evaporation, and an extremely flat and hard clay surface called the playa takes their place. The playas are known as *sebchas* in Sahara and *kewire* in Iran. They frequently have a layer of salt on them. By the middle of maturity, the relief is so much reduced that deposits of the different basins coalesce with one another and their drainage also becomes integrated. This landscape is called *bolson*. Towards the end of maturity and in old age deflation by wind becomes important and deflation hollows and other related desert features are built up, and thus maturity passes into old age.

Old Age :— In old age all the upraised parts are degraded and a nearly level surface is formed. Wind action gives rise to different land forms. Wind deflation gives rise to blowouts and wind transportation to sand dunes within the deserts and deposit of

1. "The great volume of material in the fans results from the fact that all the weathering waste of the slopes and all the material removed in making the valleys is contributed to their up-building. In the humid cycle, by contrast the rock debris is continuously in process of transportation at varying rates, down slopes and down valley toward the sea. The alluvial fan, accordingly, has its greatest magnificence of development in desert regions." Von Engeln O.D,-op. cit., p. 409

loess outside deserts. In the deserts the base level of erosion is determined by the water table and is not related to the sea-level. Consequently pediplains or desert plain or hamadas formed at the end of the cycle can be much higher or lower than the sea-level and usually are much more level than peneplain. Occasionally residual inselbergs may rise from the pediplain-surface. The inselbergs represent the last remnants of the initial surface, and are very similar to monadonocks of the peneplains. But inselbergs are formed of the same rocks as the pediplain surface, while the monadonocks represent residuals of more resistant rocks.

Questions

1. Describe the processes of wind erosion under arid conditions and discuss the chief topographical features produced by them.

2. Explain the formation of the different landforms produced by wind transportation and deposition.

3. Write explanatory notes on any two of the following :—

   (a) Sand dunes, (b) Water-action in the desert, and (c) Inselberg landscape.

4. What do you understand by pediments, pediplains and inselbergs ? Give a short account of the different views about their formation.

5. Give a reasoned account of the arid cycle of erosion and the features that characterize its different stages.

6. Discuss the different types of sand deposits and the landforms that they give rise to.

7. Explain with examples the following terms :—

   Seif, Pediments, Barchan, Yardang, Bornhardts, Loess.

Chapter **24**

Types and Characteristics of Glaciers

Glaciers at present cover about 15 million km^2 or 10% of the earth's land area. 96% of the glacier ice, however, occurs in Antarctica and Greenland in the form of thick masses of ice-sheet, average thickness being 2.2 km in Antarctica and 1.6 km in Greenland. The remainder of the earth's present ice cover is mainly found in areas of high altitude where precipitation is mostly in the form of snow. These include the earth's great mountain belts—the Himalayas, the Alps, the Rockies and the Andes as well as Scandinavia, Newzealand and a few other areas.

Though glaciers are relatively unimportant in the present shaping of landscape, glaciers of the Pleistocene epoch covered nearly 30% of the earth's land surface or 45 million km^2 at their maximum extent. They have left their imprint over much of the present landscape of northern North America, north-central Europe and Siberia. In addition, glaciers existed in the mountains where today there are either no glaciers or only small ones and glaciated landscape is found in them 1,000 to 1,500 metres below the present snow-line.

Development of Glaciers

Glaciers develop in regions where snow accumulation exceeds melting. This frequently occurs above the annual snow line either on plateau surfaces or in valley heads. This is the area of permanent snow or the *snow field*. The outer boundary or the lower limit of the snow-field is called the *snow-line*. The height of the snow-line decreases with latitude. It is 15 to 18 thousand ft (4,500-5,400 m) above sea-level in the Himalayas, 9,000 ft (2,700 m) above sea-level in the Alps, and 2,000 ft (600 m) above sea-level in southern Greenland.

Glaciers originate in snow-fields through transformation of snow into glacier ice. This transformation follows a regular sequence. The processes involved are sublimation, recrystallization with the growth of large crystals, melting, refreezing under pressure (regelation) and compaction under weight. All these processes produce an increase in the density of the material. Fresh snow is a loose aggregate of snow-flakes full of entrapped air. It has a density of less than 0.1 to 0.2. Snow flakes readily sublime and lose their frilly margins and become globular. Meltwater often freezes onto snowflake nucleus and thus increases the size of the ice granules. Loose snow changes to granular snow, with a density of 0.3 in a few days or at most, a few months. This 'old snow' has the texture of very coarse sand and is a highly pervious granular aggregate. After the formation of granular snow densification continues by modification of individual grains and by packing which reduces pore spaces and permeability and increases density and compactness. When a density of 0.5 is reached the material is called firn. The semi-crystalline mass of snow that has survived a summer melting season is called firn in German and neve in French. Firn or neve is an intermediate stage in the conversion of snow to glacier ice. As successive annual layers accumulate, the firn is further compacted, the individual ice grains freeze together so that the pore spaces are sealed off and the mass is impermeable to air, and when density of 0.85 has been reached, firn becomes glacier ice. Further densification can take place, so that the density of glacier ice is commonly 0.89 to 0.90. The density of pure ice is 0.917. Air entrapped in glacier ice causes this slight difference.

The rate of transformation of snow to ice varies with temperature and the rate of snow accumulation. Meltwater speeds up the process so that the transformation is more rapid in areas of wet snow than in areas of dry snow. A certain minimum thickness of neve also seems essential for the change to glacier ice. This is believed to be about 100 to 150 ft. Snow-fields that do not attain this critical thickness may not develop into glaciers.[1]

Glaciers have been defined by R.F. Flint as masses of ice and firn formed by recrystallization of snow and meltwater lying on land, which show evidence of either present or past movement. The cause of glacier flow is the accumulation of snow and its conversion to firn and ice in the zone of accumulation, and ice thickness is the

---

1. Thornbury, W, D.—*op. cit.*, p. 348

force causing the flow from the zone of accumulation to the zone of ablation. The ablation zone is that area of the glacier where losses by melting and calving (i.e. spalling of masses of ice from the glacier's leading edge into bodies of water) exceed additions. The balance between additions and losses is the *glacier budget*. During the steady state, the lateral margins and the terminus of the ice mass should be stationary. If accumulation dominates ablation, a positive budget will cause the ice margin to advance. If ablation dominates accumulation a negative budget will cause the glacier to shrink and its margins to retreat[1]. In a very general way small ice masses indicate either very early or very late stages in development, whereas very large ice masses have been in existence for a relatively long period of time.

Types of Glaciers

H.W. Ahlmann (1948) has given a morphological classification of glaciers which is widely used. This is based essentially on the size, shape and position of ice masses. Three major types and a number of sub-types are recognized by Ahlmann :—

1. Glaciers that are continuous sheets from which the ice may move in all directions. These include (a) the continental glaciers or ice-sheets covering large areas, (b) the ice-caps which are smaller than ice-sheet, and (c) the highland glaciers which cover the highest and central portion of a mountain area. For the sake of convenience we may call this group *continental glacier*.

2. Glaciers restricted to more or less confined courses which direct their movement. These include principally (a) cirque glaciers which occupy valley heads or rock basins on mountain sides at high altitudes, (b) valley glaciers of the Alpine type which are long and narrow and confined between the valley walls and are common in areas of high mountains, such as the European Alps. Several cirque glaciers often combine and feed the moving valley glacier. (c) Valley glacier of the outlet type which instead of being fed by a cirque glacier or a series of cirque glaciers is fed by an ice-cap or ice-sheet. (d) Transection glaciers which cover a whole valley system. The mountain may be too deeply dissected to allow an ice-cap to form. In such a situation it is possible for the valley glaciers to become so thick as to overspill across the divides between their confining

1. Ruhe, R.V.—*Geomorphology*, Houghton Mifflin Company, Boston 1975, p.189.

valleys. This entire group having a confined course may be designated *mountain glacier* or *valley glacier*.

3. Glacier which spreads out over level ground at the foot of glaciated areas i.e. *piedmont glacier* formed by the fusion of the lower ends of more than one valley glaciers.

The simplest morphological sub-division of glaciers would thus be into three broad types—(a) *mountain or valley glaciers* that flow between confining rock walls under the influence of gravity, (b) *continental glaciers* or ice-caps and ice-sheets that flow unconfined largely by differential pressures within the ice-mass, in various directions uncontrolled by the underlying topography, and (c) *piedmont glaciers* which form an intermediate and relatively unimportant category.

Valley Glaciers

As stated above valley glaciers are common in areas of high mountains such as the Himalayas, the Alps, the Rockies and the Andes. During a period of falling snow-line, cirque glaciers develop a positive mass balance and extend down to the valley. Several cirque glaciers often combine and feed the main valley glacier. The valley glacier is long and narrow and confined between the valley walls and terminates in a narrow tongue of ice. Some valley glaciers may have steep ice faces at their terminus but others are deeply buried in transported rock debris. Their length may vary from less than 1 km to 80 km and more. The Aletsch glacier, the biggest glacier in the Alps is 16 km long. Most of the Himalayan glaciers are 1 to 5 km long, but there are several Himalayan glaciers which are 50 to 70 km long. It has been estimated that 17% of the Himalayan and 37% of the Karakorm are covered by glacier ice. Some of the longest Himalayan glaciers are found in the Karakorm, where the Siachen glacier is probably the second longest valley glacier in the world, the first being Fedchencko glacier in the Pamirs. Siachen glacier (72 km), Biafo (62 km), Hispar (61 km), Baltoro (58 km), Batura (58 km), Rimo (40 km) and several others are the principal glaciers of the Himalaya and Karakoram. In fact the Himalayas contain most of the world's largest valley glaciers, and most of them terminate as river sources. The Gangotri glacier which is the source of the Ganga (Bhagirathi) is 26 km long. Besides Gangotri, other important glaciers of the Kumaun Himalayas include Milaya (16 kms), Kosa

(11 km), and Kedarnath (15 km). Two notable glaciers in the Nepal Himalayas are Zema (26 km) and Kanchanjunga (16 km). The Himalayan glaciers descend to 3,700 m in eastern Kashmir, 4,000 m in Central Himalaya (Gangotri) and 4,500 m in eastern Nepal and Sikkim. The Khumbu glacier, descending from Mt. Everest terminates at 4, 900 m, while the Kanchenjunga glacier terminates at 4,200 m above sea level.

A valley glacier has a much higher surface gradient than rivers, but in contrast to a river, the long profile of a glacier surface is marked by level areas, steep slopes and even local reversal of slope. On the surface of the valley glaciers in the ablation zone, several ephemeral land forms develop due to melting, collapse of ice tunnels and erosion by streams flowing on or in the glacier. Many of these resemble karst features and "may be collectively called *glacier karst*. Thaw lakes and swallow holes (moulins) mark the ice surface. Segregations of debris-rich ice stand as ice-cored mounds or ridges above the debris-covered surface. Lakes may drain and then abruptly form again when a crevasse or moulin opens at depth and then clay-rich sediments seal it closed again. Ice tunnels within or at the base of the glacier discharge sediments laden water from the terminus[1].

Piedmont Glaciers

Most of the valley glaciers terminate in the mountains before reaching the foot-hill zone. But in very cold climates the valley glaciers sometimes extend beyond the confining valley walls and descend to the foot of the mountains where they coalesce and spread out on the relatively flat and low lying surface, to form a piedmont glacier. Their width is usually more than the sum of the width of each of the tributary glaciers which have combined to form the piedmont glacier. Their movement is also slackened as they spread out and reach the level surface. Sometimes they become almost immobile, and even trees may grow on the rock debris covering their surface. Piedmont glaciers are rather rare and are at present found only in Alaska, Iceland and some other areas. The best known is the Malaspina glacier which is located in the piedmont zone of Mt. Elias in south-eastern Alaska and spreads out as a broad lobe on the Pacific coast lowland and reaches the sea at places. The

1. Bloom, A.L.—op. cit., p. 382

piedmont lobe of the Malaspina glacier spreads out over 2,640 km^2 and the ice is about 600 m thick. It lies wholly in the ablation zone. The outer 8 km are debris-covered and consist of dead ice on which spruce trees 100 years old are growing.

Continental Glaciers

The continent-sized ice-sheets and the smaller but otherwise similar ice-caps are unconfined continuous areas of ice cover. Ice-cap develops at a relatively high altitude on upland plateau surfaces where precipitation conditions are suitable for the development of an accumulation area. The accumulation zone is in the central region of the dome and the ablation zone is on the periphery. The ice in an ice-cap moves in directions largely determined by the slope of the ice surface which is essentially related to the rate of accumulation. Examples of this type of ice-cap are Vatnajokull in Iceland and Jostedalsbreen in Norway, both occupying high plateau areas and having a thickness of 600-760 m in the first case and 300-600 m in the second, with vigorous outlet glaciers descending from the ice-caps.

The only difference between an ice-cap and an ice-sheet is that an ice-sheet is much larger and may cover the underlying topography and develop lines of movement unrelated to the submerged topography. There are only two examples of ice-sheet at present—Greenland ice-sheet and the Antarctic ice-sheet. The two together cover nearly one-tenth of the earth's land surface. The Greenland ice-sheet contains 11% of the world's ice and the Antarctic ice-sheet 85%. It is estimated that the melting of these ice-sheets would raise the sea-level by about 60 metres. Nearly three-fourths of Greenland is covered by a thick ice-sheet which is 1,802,600 km^2 in extent and is nearly 3,000 metres deep in the central part. The Greenland ice sheet rests on a basin-shaped interior plateau surrounded by mountains, and under the weight of the huge mass of ice the plateau surface has at places sunk below sea-level. Along some parts the ice-sheet ends as a vertical ice cliff on land but around much of its margin the ice moves between exposed or partly buried mountain peaks as *outlet glaciers*. The outlet glaciers have many of the properties of valley glaciers. The peaks that rise above the surrounding ice-sheet are called *nunataks*. Some of the outlet glaciers reach the sea with a floating ice tongue, and give rise to ice-bergs by calving and disintegration into separate lobes.

The entire Antarctic continent is buried under an extensive ice-sheet which covers about 12.6 million km^2—an area more than seven times as large as the Greenland ice-sheet. The mean thickness of the ice is about 2,000 metres or possibly more. The surface of the Antarctic ice-sheet is a monotonous dome with little surface gradient, with only a few nunataks rising above the ice surface, mostly near the coast. Much of the perimeter of the Antarctic ice-sheet is a great cliff, but the ice cliff ends not on land but on the *grounding line* in the ocean i.e. the line along which the ice moving seawards starts to float. These form huge floating ice-shelves, very different from the narrow floating glacier tongues found in Greenland. A good example is the Ross Ice Shelf where the floating ice is 300 to 400 m thick and covers an area of 550,000 km^2, an area larger than France. The ice cliffs along the shelf edge rise 40 to 50 m above the sea level. The material of the shelf is partly derived from the outward flow of the inland ice-sheet but it is mainly supplied by the accumulation of snow on the upper shelf surface, as the accumulation is much greater at the edge of the continent. The bulk of the Ross shelf ice is thus formed of firn rather than true glacier ice. This is known as the Great Ross Barrier, after James Clark Ross who first discovered it in 1841 . The calving of the Barrier as the ice shelf is called, produces very large tabular icebergs that occur only in the southern hemisphere, as ice shelves of this type are confined to the Antarctica.

Thermal Classification of Glaciers

H.W. Ahlmann has also suggested a thermal classification of glaciers into two basic types—temperate glaciers and cold or polar glaciers. *Temperate glaciers* are approximately at pressure melting point through out their thickness, except for the uppermost few metres which may become temporarily colder in winter. The temperate glaciers are characterized by the presence of meltwater throughout their thickness which may be found in fissures and channels. Many of the glaciers of the Alps and southern Scandinavia are temperate in character. In this type of glacier the meltwater at the bed helps the ice slide over the wet rock floor and thus the rates of movement are usually greater than in the case of cold glaciers. A *cold glacier* is one which is theoretically below the freezing point throughout the year. Ahlmann sub-divides this type into two—the high polar type and the sub-polar type. In the high polar type, the temperature in the firns remains well below freezing point throughout and even in summer there is no melting on the surface in the accumulation area. On the other hand, in the accumulation area of the sub-polar type, the material consists of crystalline firn down to a depth of 10 to 20 m, and in summer the surface can melt and water can be present. From the geomorphological point of view, the most important property of a cold glacier/ice-sheet is the absence of meltwater at depth in the ice. At the base of the glacier the ice is frozen to the bedrock on which it is resting, and ice masses in this condition cannot achieve as much erosion as the temperate glaciers.

Many complications in the distribution of temperature within ice masses have come to light since Ahlmann first proposed his simple thermal classification. Further, his terminology has also come to be criticized. A. Court (1957) has proposed that terms *permelting, refreezing and non-melting* be used instead of temperate, sub-polar and high polar respectively. These terms are intended to signify that glaciers in the first category are permeated by meltwater throughout, the second type has a layer in which water refreezes below a surface layer of melting, while the third type never experiences melting on the surface. On the whole, however, the terms proposed by Ahlmann continue to be used. It is important to note that the entire ice mass need not necessarily fall into a single category.

The temperature distribution in glaciers and ice-sheets depends on several factors such as, the surface temperature fluctuations seasonally and through a longer period, the thickness of the ice and its conductivity and the geothermal heat flow from below the ice. A glacier ice-sheet need not belong to one type over its whole area and various parts may have different thermal characteristics. In fact even large polar ice sheets are known to have basal areas at pressure melting point. Temperature relationships and changes within a glacier are, however, very important as they influence the rate and mode of movement, the amount of meltwater and the mechanism of erosion.

Under temperate glaciers basal melt-water is present and consequently substantial rates of basal sliding may occur. Under cold glaciers frozen to bed-rock there can be no melt-water present, at least theoretically, and all ice movement must occur by deformation of the ice. As we have seen, many glaciers and ice-sheets do not simply fall into one or the other category. In the accumulation zones of many thick sub-polar glaciers, the upper layers may be cold while the lower layers may be at pressure melting point and basal melting may occur as geothermal heat (and heat from sliding friction) cannot escape upwards from the sole of the glacier. In the ablation zone of the same glacier where the ice is thinner the whole thickness of ice may be cold and frozen to bedrock, It is also possible for a glacier to be completely cold in high altitude accumulation zone, and to flow to much lower levels with a more temperate climate where the glacier may warm up to pressure melting point throughout.

Thermal regime, though extremely important, is by no means the only factor controlling the efficacy of erosion, transport and deposition; lithology, roughness of the bed, ice thickness and some other factors are also important.

Glacial Movement

Actual measurements of glacier motion were first undertaken by G.J. Hugi who built a hut on the Unteraar glacier in the Alps in 1827 and found that in 14 years i.e. by 1840 the hut had shifted down by 1.4 km. Further observations and measurements of ice motion were undertaken by a number of pioneer workers in the mid-19th century, notably by L. Agassiz on the Unteraar glacier and J.D. Forbes on the Mer de Glace. Agassiz fixed several stakes across the glacier in 1835 and by 1840 established that the centre of a glacier flows faster than its sides and that the rate of movement is slower in the source regions and in its terminal parts than in between. In 1845-46 Agassiz also found that the velocity of flow was much greater in summer than in winter. This confirmed Forbes' earlier prediction (1843). Forbes had also found that glaciers often flow faster where their valleys are narrow, and that since the central portion flows faster than the sides it seemed to him likely that the surface layers flow faster than the basal layers. He further suggested that glaciers which may not be frozen to their beds have greater freedom of movement. It is remarkable that these assertions have

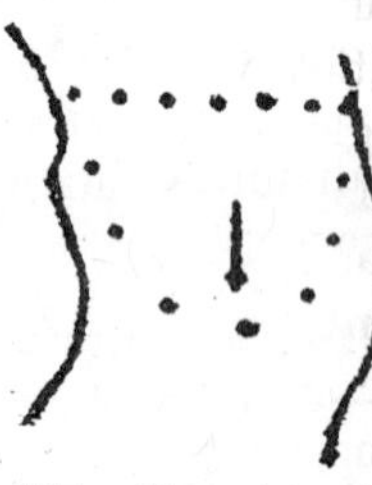

Fig. 157— Experiment by Agassiz showing that the centre of a glacier flows faster than the sides.

been found to be substantially correct by later workers.

Since then not only intensive surveys of glacial motion have been conducted but various theories of ice motion by plastic deformation of ice under stress and differential flow along internal shear planes have been propounded. Still there are considerable gaps in our understanding of this problem.

We know that not only the rates of glaciers motion are different in the different parts of the glacier but they vary enormously from one glacier to another and over periods of time. The majority of valley glaciers move at speeds of less than one metre a day or less than 200 m per year. Most of the Alpine and Himalayan glaciers have a speed of less than 50 m per year, though the Aletsch glacier (Alps) and the Fedchenko glacier (Himalayas) have a speed of about 165 m per year. The rate of movement on the Antarctic ice-sheet is of the order of 30 m per year, but the outlet valley glaciers draining from the ice-sheet move at rates ranging between 300 and 1,400 m per year. The most rapidly moving glaciers are those descending steep slopes and fed by large accumulation areas. Some of the Greenland and Antarctic outflow glaciers have been found to attain velocities of 20 m a day or more. The fact that the lower portions of these glaciers float and calve into the sea helps them to attain these unusually high rates of movement. As indicated earlier, velocity rates across the surface of a valley glacier are usually fastest near the centre line and decrease towards the valley sides. Velocities in temperate valley glaciers also tend to decrease towards the base of the glacier.

Cold glaciers frozen to bed rock, such as we find in Antarctica, do not exhibit any marked seasonal variation in rates of flow, but temperate and sub-polar glaciers show marked seasonal variations. More rapid movement of ice in the ablation zone in summer than in winter is fully established. This is mainly due to the greater availability of meltwater in summer which lubricates the bed of a glacier and thereby enhances the sliding process. Since lubrication by meltwater at the glacier bed is very important, the maximum velocity is found at the middle of the ablation season. There can be also temporary increase in velocity after a period of heavy rain. Summer velocities may be 10 to 20% more than in winter. Monthly average velocities may differ by as much as 40% and the weekly, daily or hourly differences may be much greater, but short-interval variations in flow cannot be explained directly by variation in meltwater lubrication and may be related to weather conditions and other factors. The movements probably proceed in a series of small jerks which are not synchronized at different points.[1]

The movement of glaciers consists of two components—internal plastic deformation or creep and sliding over its bed rock. The relative contribution that each makes varies with surface slope of the ice mass, ice thickness and temperature. The plasticity of ice is clear from the way it moulds itself to the form across which it moves. The glacier ice is a mass of interlocking, roughly equigranular crystals. Ice melts at points of pressure and recrystallizes at points of relief of pressure. Thus plastic deformation in each crystal is achieved by gliding and is accompanied by recrystallization. Thus internal deformation and forward movement go on concurrently. Ice, however, does not flow down channels of constant gradient with uniform rate of discharge. In actual glaciers, longitudinal extension and compression frequently occur. Where the bed slope is reduced, forward velocity decreases and the ice is 'compressed' (not actual reduction of volume but a local thickening of the glacier). Conversely, if the velocity increases down the glacier due to increase in longitudinal gradient, ice is extended. Compression or extension may also be caused by changes in the rates of accumulation or ablation. In both cases thrust planes and shear faults form.

The other component of glacial movement is basal sliding. Cold glaciers as a rule do not exhibit basal sliding since they are frozen to bed rock. But there are some cold glaciers whose base is just at pressure melting point where limited sliding occurs, especially where they are retarded by bed rock obstacles projecting into the cold ice, causing increase in pressure on the up-stream side of the obstacles. In temperate glaciers on relatively steep slopes, basal sliding is the principal mechanism of ice movement and this has important consequences for bed rock erosion. The nature of the form of the bed, the presences or absence of gravel, the permeability of the underlying rock and the distribution and amount of water all

1. Glacier ice consists of a number of irregularly shaped ice crystals in close contact. The apparent flow arises from the fact that the ice melts at point of pressure between the grains, and the resulting water recrystallizes at points of relief of pressure. The mass thus moves on concurrently with the change of shape and transference of material of its grains. Wooldridge and Morgan op. cit., p. 363.

influence the rate of basal sliding. On flat surfaces and up reversed slopes, basal sliding is considerably reduced, and glacial motion is accomplished not by longitudinal pressure but by thickening of the ice-mass until the situation is reached at which the centre of gravity of the ice once again falls downhill sufficiently to permit movement i.e. by the surface slope of the ice-mass.

"In terms of a simple theoretical model, ice velocities are mainly determined by ice thickness and surface slope. Ice tends to flow in the direction of maximum surface slope even if that is up-hill in terms of the slope of the subglacial surface. Velocity is proportional to the fourth power of ice thickness and the third power of surface slope. The product of thickness times surface slope is fairly constant so that where the ice is thin the ice surface tends to be steep and where it is thick the ice surface tends to have a low angle of slope."[1]

It is, however, extremely important to consider temperature since many of the physical properties of ice are dependent on temperature. Temperatures in an ice-mass are affected by two sources of heat—the geothermal heat flux at the base of the ice and the temperature at the ice surface. In the upper 15 m the temperature of the ice varies with changes in air temperature at the surface, and seasonal variations in ice temperature are not detected below this depth. At the same time it has been found that in polar ice masses the temperature increases downwards below a certain depth. The difference can be as large as 11°C to 30° C between the surface ice and the basal ice. The highest temperatures and the greatest shear stresses, therefore, occur at the base of the ice sheets. "In very general terms it can be stated that cold ice will move more slowly than temperate ice and maximum shear will take place near the base of cold ice masses where temperatures are likely to be highest."[2]

Glacier Surges

There are some glaciers in the world that experience short phases of exceptionally rapid advance of ice margins resulting in dramatic increase in forward velocities, known as glacier surges. Glacier surges have been recorded in may parts of the world, but the most complete inventory of surging glacier is that for N. America.

---

1. Price, R.J.—*Gacial and Fluvioglacial Landforms*. Oliver & Boyd, Edinburgh, 1973, pp. 31-32.
2. Ibid. p. 33.

In 1966 a dozen major glaciers in Alaska surged forward, including the Bering glacier, the largest in N. America, which advanced 1,200 m between 1963-66 along a 42 km wide front. Rapid advances of 113 m a day have been reported in the Kutiah glacier of the Karakoram Himalaya in 1953. These surges are usually completed in one to three years and the glacier front may move forward by 10-20 km during the period. Ice velocities during a surge can average 20 m per day but the maximum velocity may be one hundred times greater than its normal velocity. During a surge the upper part of a valley glacier may be lowered by 50-60 m and tributary glaciers may be left hanging, while in the lower part of the glacier the ice may thicken by 50-60 m. Because of the rapid increase in velocity the glacier surface becomes highly crevassed and at times great transverse cracks develop.

Several theories have been put forward to explain glacier surges but basal meltwater lubrication to facilitate the rapid movement is considered one of the most likely mechanisms. In a quiescent phase, the ice in the accumulation area thickens. When it becomes sufficiently thick or the surface slope steep enough, the basal shear stress reaches a critical value and a surge begins. The rapidity of flow suggests a sudden and remarkable decoupling of the glacier from its bed. Once surging begins as a result of accumulation of basal melt-water, the increased velocity of basal sliding will generate further large increase in basal melt-water owing to frictional heat and accelerate the process. Whatever the cause, a glacier surge has important implications not only in terms of the rapid change in the location of ice margins (and sometimes its catastrophic nature) but also the possibility of higher rates of erosion, transportation and deposition.

Crevasses

Crevasses are cracks or fissures found in glacier and are formed by shear stresses set up by differential movement in the ice. In accumulation areas with gentle surface slopes the ice surface tends to be fairly uniform. When the ice begins to move downslope in a valley the drag of the valley-sides creates shear stresses that produce cracks or crevasses at an angle to the glacier margin and pointing up glacier. In valley glaciers longitudinal compression and extension occur. When the velocity increases because the ice is being extended, transverse crevasses will form during extending flow but not during compressive flow. Crevasses may vary in width from hair creaks to gaps many metres wide. But from the geomorphological point of view the depth is of greater significance, as crevasses affect the penetration of melt-water and of debris to the deeper parts of glaciers and in some cases may even reach the rock floor. Measured depths of crevasses rarely exceed 30 m but crevasses exceeding 36 m in depth are known in the Antarctica and Greenland. It is probable that as cold ice can support higher shear stresses, the crevasses are deeper in polar than in temperate glaciers and may be also more sluggish in closing.

Fig. 158—Crevasses.

Several types of crevasses are recognized based on their orientation with respect to the long axis of the glacier :—

(a) *Marginal Crevasses* : They develop along the glacier margins in areas of stress between the relatively fast moving ice in the centre of the glacier and the slow moving ice near the valley walls. As stated earlier they are inclined at an angle to the glacier margin and point up glacier. In areas of compressive flow the crevasses intersect the valley wall at an angle of less than 45^0 but in areas of extending flow the angle of intersection is more than 45^0. When the marginal crevasses are short, nearly equal and grouped in rows, they are referred to as *enechelon crevasses.*

(b) *Transverse Crevasses* : When the surface over which glacier is moving increases in steepness and velocity of the ice increases as the ice is extended, crevasses transverse to the long axis of the glacier are formed. They form a series of arcs curving across the surface of the ice, the convex side facing upglacier. They are formed in areas of extending flow (Fig. 160.B)..

(c) *Longitudinal Crevasses* : are generally straight and parallel to the long axis of the glacier. They develop in areas of compressive flow when the velocity of ice decreases or when the valley widens and the glacier is able to spread out laterally.

(d) *Splaying Crevasses* : are also parallel to the long axis of the glacier but only in their upglacier end and curve towards the margin on their down-glacier end. They form in areas of intense compressive flow (Fig. 160 C).

(e) *Ticktacktoe Crevasses* : are uncommon groupings of crevasses which form criss-cross patterns and probably form over a bedrock knob on the valley floor.

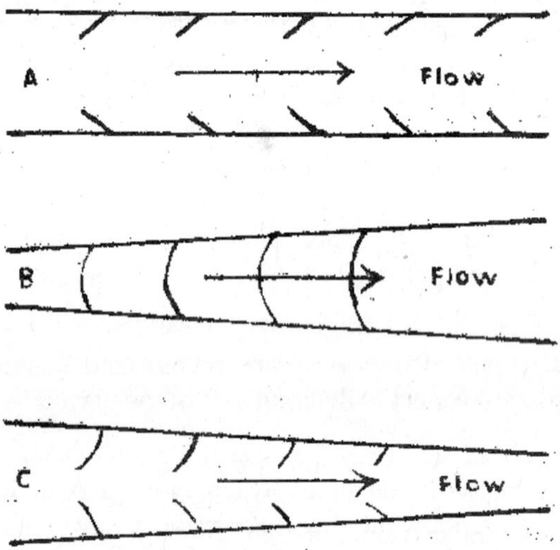

Fig. 159—Development of Crevasses.
A—Effect of sheer stress of valley walls.
B— Effect of shear stress and extending flow.
C—Effect of shear stress and compressive flow.

(f) *Bergschrund Crevasses* : Bergschrund is a special type of crevasse which opens in summer near the top of the neve field where the head of a valley glacier is pulled away from the precipitous walls of a cirque and the ice adhering to the rock wall. Bergschrunds are quite wide and deep, and may be much deeper than 45 metres. It frequently happens that several such fissures may be formed instead of an especially large one.

Sometimes small pools and hills form on the surface of the glacier gathering into streams which mostly fall into the gaping crevasses. Armed with boulders, pebbles and sand this water is able to cut a vertical shaft or moulins (glacier mill) in a glacier down which melt water flows down under gravity, sometimes to the bed rock beneath the glacial ice where it may scour out potholes by grinding action of the boulders, and the water may escape to the snout through a tunnel.

Moraines

Moraine is the name given to the weathered material that is transported by moving ice, and is also used for the resulting landform when this material is deposited. It is a French word and comes from the Western Alps where it means "hill" or "rubble heap" as well as 'rubble' or pebbles. Thus from the beginning the word "moraine' had a sedimentological meaning. In N. America, however, the term moraine is used in its morphological meaning while the term used for the material transported by moving ice is called till.

The glacier ice as it moves along, picks up debris of different size and heterogenous origins along its flanks, over its surface and at its base. A large proportion of the debris is coarse and angular and derived from the rock fragments that fall on to the glacier after frost shattering of the exposed slopes along the sides of the glacier. The sides of a glacier thus become edged with long ribbons of coarse and rocky debris called *lateral or marginal morains*. The Gangotri glacier in the central Himalayas is bordered by a conspicuous cover of lateral moraines. When two glaciers from adjacent valleys meet, the two inner lateral moraines of each unite to form a *medial moraine* on the surface of the united glacier. A trunk glacier which receives the ice of several tributaries thus comes to be ridged with series of medial morains. Barnard glacier in Alaska shows good examples of lateral moraines and their coalescence into a multiple series of medial moraines. In parts of Gangotri glacier the medial moraines are so extensive as to cover the entire surface of the glacier ice with thick morainic debris. Both the lateral and the medial moraines are *surface moraines* comprised of debris falling on the surface of the ice by frost weathering, falling boulders, rock slides, avalanches and other types of mass movement. Sooner or later part of this material becomes covered by ice by the annual snow fall in the catchment area or falls into the crevasses and the surface moraine becomes *englacial moraine*. A certain portion may even reach the sole of the glacier and mix with the material plucked or scoured by the glacier from its rocky bed. This is known as *subglacial moraine or basal till*. The bed rock is worn away by scouring (grinding), scratching and polishing of the rock and by plucking of rock fragments. These materials removed from the bed rock are incorporated into the *ground moraine*. Softerocks are reduced to clay, silt and sand size by continuous motion and load of the ice. More resistant rocks are abraded and become rounded. They also carve striations and grooves into the softer rocks. The boulders develop facets. On the whole, small sized particles predominate and blocks are rare. Often the thickness of the ground moraines is less than a metre, though sometimes the thickness of ground moraines developed by continental ice sheets may be 10 to 20 metres.

All the varied debris carried by the glacier—the surface moraine passively transported on the ice and the ground moraine heavily worked under the ice finally arrives at the terminus of the glacier. This material is a mixture of angular blocks and boulders and finely ground rock flour. When the ice melts, this debris is dumped in a haphazard fashion. Where the ice front remains stationary for some years an arcuate or crescentic ridge is formed called the *terminal or the end moraine* (Fig. 161) and constitutes an important indicator of former glacier distribution. Where the ice retreats with fairly long halts at intervals, several parallel arcuate ridges called *recessional moraines* may be formed. But where the recession of the glacier snout is fast, no piling up of the moraine may be possible and the debris liberated from the retreating front may be deposited as an irregular sheet over the ground moraine.

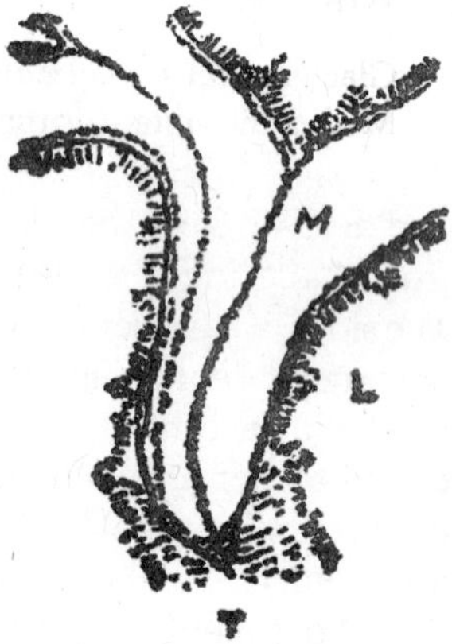

Fig. 160—Moraines, L-Lateral, M—Medial, T—Terminal

The Himalayan glaciers are noted for extensive surface moraines and rock waste which sometimes so cover their upper surface that the ice is not visible for long stretches. Many Kashmir glaciers are used as camping ground by the shepherds for their flock in summer on the moraines overlying the glaciers. The englacial and subglacial moraine is also present in large quantities. As there is a slow retreat of the glacier ends over most of the Himalayas, enormous heaps of terminal and recessional moraines are left behind by the retreating ends of glaciers.

Questions

1. How are glaciers formed ? Discuss the different types of glaciers and their characteristics.

2. Discuss the characteristic features of glacial movement. How is it related to thermal classification of glaciers ?

3. Write an account of the different types of (a) Crevasses, and (b) Moraines.

4. Write short notes on :-

   Glacier surges, Piedmont glaciers, Ground moraine, Cold glaciers, Medial moraine, Terminal and Recessional moraines.

Chapter **25**

Glacial Landforms

Mechanism of Glacial Action

Erosion—Like rivers the glacier does three kinds of work-erosion, transportation and deposition. In the first half of the 19th century there was considerable controversy about the ability of glaciers to erode and modify the landscape. In the opinion of some the ice cover provided protection to the land surface from erosion, while in the opinion of others glacier ice was a powerful agent of erosion. It is true that the ice of a glacier has of itself but little eroding power, and clean ice passing over resistant unweathered and poorly jointed rocks accomplishes very little erosion. But glacier ice is far from clean in its basal layers and the rock over which it moves is often weathered and jointed. Glacial erosion results from the movement of ice over the rock floor, and the main tools of erosion are the rock fragments embedded in the ice at the ice-rock interface. The basal ice is armed with sharp angular rock fragments of various sizes and as the glacier moves under pressure these fragments polish, scratch and cut grooves in the rock surface. This process is known as *abrasion*. The size of the debris incorporated in the basal layer of a glacier largely determines the type of abrasion that takes place. If the debris is rock-flour or sand, the rock surface will become polished. If the debris is of gravel or boulder size, the rock surface will be striated (scratched) or grooved. The striations consist of lines, occasionally a meter or more in length, engraved on suitable surfaces. They have been found to occur in a great variety of situations but are most prolific on gently inclined smooth surfaces which the ice was forced to ascend. They usually occur in large numbers rather than in isolation and thus reflect a relatively high concentration of debris in the basal ice. The larger striations are known as grooves. When a boulder is embedded in the basal ice with a small area projecting, it may be held sufficiently firmly for the projection to scratch and to groove the underlying rock surface. If the block rotates, a new cutting edge may be exposed and a second groove may be cut parallel to but not continuous with the first groove. Some of the grooves are very large and have been possibly gouged out by groups of boulders packed and frozen together.

The other major mechanism of glacial erosion is *plucking or quarrying* or joint-block removal. Abrasion involves erosion at the ice-rock interface without movement of the parent rock and with removal of only small parts of the parent rock. Plucking, on the other hand, involves larger fragments of the parent rock. Plucking is considered to be far more important quantitatively than abrasion. It also provides the tools needed for abrasion. In well-jointed rocks and where blocks of rocks have already been loosened by chemical and frost action, block removal on a vast scale by ice becomes possible, producing a characteristically plucked or quarried surface. Since the strength of the unbroken rock is greater than ice, in most cases existence of fractures either tectonic or developed by pre-glacial frost shattering is essential, though there is evidence that glacial ice can reinforce these fractures by dilatation or by surface drag. In certain situations beneath temperate glaciers, pressure-induced freeze-thaw oscillations may be operative. Well-jointed rocks (quartz-mica schist) are much more susceptible to glacial plucking than massive rocks (granite), and even relatively soft rocks when not well-jointed (unjointed slate) may resist erosion by plucking.

"Plucking gives a characteristically shattered and broken appearance to a landscape (unless it is subsequently till covered), particularly when viewed against the direction of ice flow, and the contrast between surfaces subjected mainly to abrasion and polishing and those affected chiefly by plucking (often adjacent to the former) is a most striking feature of glaciated terrain."[1] This is best seen in the feature known as *roches moutonnees* in which smoothed and plucked surfaces are juxtaposed. The smoothed up-glacier slope represents glacial abrasion and polishing while the shattered down-glacier side represents joint-block removal by freeze-thaw action beneath temperate glaciers. The term was introduced by H.B. de Saussure in 1787 who likened roches moutonnees fields to the rippled appearance of wavy wigs styled moutonnees in those days. An alternative term is *stoss and lee*

1. Embleton, C. and King, C.A.M.—*Glacial Geomorphology*, 2nd ed. 1975, Edward Arnold, London, p. 198

topography. The smoothed stoss side is streamlined while the lee side is broken and shattered. This contrast is used to indicate approximately the direction of glacier motion (Fig. 166). The individual roches moutonnees vary greatly in size, but they usually occur in groups dominating large areas such as along the Norwegian coast. They may occur on valley floors or valley sides or plateau surfaces or on lowland plains.

Transportation—Glaciers are capable of transporting large quantities of debris over considerable distances. Not all the material transported by glacier ice is, however, derived by glacial erosion. Large amounts of material that occur on and within glacier are supplied by mechanical weathering of rocks above the ice, by landslides, debris slide, avalanches, mud flow, and extra-glacial streams that carry material on to the ice surface. Wind may also deposit coarse and fine volcanic materials on the ice surface.

The materials can be transported by glacier ice either on its surface or along its bed or somewhere in between the two i.e. supraglacially, subglacially or englacially. The supraglacial debris is far more extensive on valley glaciers than on ice sheets, as the sources of debris are more extensive both along the side and at the head of valley glaciers, while in the case of ice-sheet most of the rock surface lies beneath the ice cover. The supraglacial debris occurs either in relatively narrow bands as lateral or medial moraines or as extensive spreads in the terminal zone where the material of medial moraines spreads out as a result of differential ablation rates and is also supplemented by subglacial debris brought to the surface along flow lines. Much of the material at the bottom of glacier travels as ground moraine, but some subglacial material may be brought to an englacial position by movement along shear planes and some of the glacial material, in turn, may be brought to the supraglacial position through lowering of the ice surface by melting. Conversely, some of the supraglacial material may fall into crevasses and become englacial. The surface debris are thus carried downwards into the glacier and sub-glacial and englacial debris are carried upward to the surface in the zone of ablation. Thus debris derived by mass wasting or by subglacial erosion may be thoroughly mixed. Rock debris that comprise the lateral and the medial moraines, however, usually stay at a level higher than the glacier.

The immense transporting power of a glacier as well as the

ability of the glacier ice to move uphill is clearly demonstrated by the presence of *erratics* or *erratic blocks*. These blocks which are now found sitting on rocks of entirely different lithology, and which can be traced back to their source outcrop, give an indication of the size of the materials that can be carried by glacier ice and the distance over which they can be transported. Some erratic blocks are up to 800 m long and weigh several hundred thousand tons. The largest is the 'Schollen' of Germany which is 4 km by 2 km by 120 m thick. Apart from the large size of the material carried, glacier ice is also capable of lifting rock fragments high above their original location. Differences of 500 m in altitude between original outcrop and the final resting place of the erratics have been found in Scotland and 1,500 m in the U.S.A. Good examples of erratics are found in New England, U.S.A. where they are found in large numbers on the ground covered by thin drift or boulder mixed soil. Sometimes the erratics are perched precariously on narrow basement made of other rocks and are then called *perched blocks*.

So far as the distance of transport is concerned, erratics have been found several hundred kilometres from their source outcrops. The limiting factors are the distance from the point of entrainment to the terminus of the glacier ice (which is acting as the transporting agent) and the ability of the erratics to survive weathering and erosion during transport.

Deposition—The glacial debris transported by the glacier ice and its associated meltwater is deposited either in the bed or along the sides or in the terminal zone of the glacier. Moraine is an old term which has been used to describe the rock debris transported and deposited by glacier. But originally it was applied to describe the hummocky or the ridged depositional features on the margins of glaciers. The use of the term moraine for both the material and the landform has resulted in some confusion. More recently, moraine is being used as a collective term for depositional landform of direct glacial origin, while glacial and fluvio-glacial deposits i.e. all rock material in transport by glacier ice and all deposit made by it and meltwater, are collectively referred to as *glacial drift*. Glacial drift is subdivided into unstratified drift and stratified drift. The term *till* or *boulder clay* is often used to denote the unstratified drift deposits, though till is a better term than boulder clay as not only does *till* consist of boulders and clay, the particles in a till range in size from clay and sand to large boulders. The unstratified drift or glacial till is further subdivided into lodgement till (ground moraine) and ablation till. When the debris is deposited beneath the ice by basal melting, it is called *lodgement till* or *basal till*. As the ice melts, the enclosed debris is let down from the ice surface on to the ground. This deposit is known as *ablation till* (ablation moraine). While the unstratified drift is laid down directly by the ice, stratified drift is water-laid fluvio-glacial deposit and is differentiated by sorting. Stratified drift has the properties of ordinary alluvium and is transported and deposited by glacier melt-water. This two-fold sub-division of glacial drift on the basis of the medium of deposition is quite useful as there are major differences between unstratified deposits laid down by and from glacier ice with only small assistance from meltwater and stratified drift deposits laid down after transport by meltwater. In reality, however, the medium of deposition is very difficult to establish and the two sub-divisions are part of a continuum and are often found inextricably mixed together in a complex manner.

While the movement of glacier ice is crucial to both erosion and deposition, from the point of view of glacial and glacio-fluvial landforms, erosion and deposition may be considered far more important than transportation. Glacial erosion takes place mostly in the mountains and highlands, while glacial and glacio-fluvial deposition takes place largely in the foothills and lowlands. Glacial action thus gives rise to very different and distinctive landforms in the mountains and lowlands. In the glaciated mountains the erosional features dominate though depositional features are not altogether absent, while in the lowlands, the effects of glacial and glaciofluvial deposition are most strikingly displayed in distinctive depositional landforms. In the mountains the pre-existing topography is modified and altered by erosion, and in the process some distinctive landforms are created, diversified further by morainal deposits. In the lowlands the pre-existing topography is not only covered by a blanket of glacial deposits or drifts but several new and characteristic constructional landforms are built by the combined action of glacier ice and meltwater.

Landforms of Glacial Erosion

We have already referred to the small features produced by glacial erosion—the smooth polish, the striations and grooves and the faceted boulders caused by abrasion and the characteristically shattered and broken appearance of the landscape caused by glacial quarrying or plucking and the formation of roches moutonnees which display a gently sloping striated and polished upstream slope and oversteepened lee slope which has been quarried by the overriding ice.

The ability of glaciers to modify the landscape is most strikingly seen in areas of mountain glaciation, and the highland areas which have been subjected to glaciation display unique topographical features such as the cirques, horns and aretes, the steep-sided troughs marked by a series of glacial steps and rock basins, discordant tributary valleys, truncated spurs and roches moutonnees, all bearing unmistakable evidence of the erosion carried out by the glacier ice.

Cirques and Associated Features

Cirques together with the associated features of *aretes* or *comb-ridges* and the *pyramidal peaks* constitute the most striking landforms in the high glaciated mountains. *Cirques* are semi-circular steep-sided depressions found at the head of valley glaciers where snow accumulated and is compacted to form cirque glacier which flows downslope to feed valley glaciers. These amphitheatre-shaped rocks have the appearance of deep-seated arm chairs (Fig. 162). They are usually filled with neve or glacier ice. In the absence of glacier ice, cirque basins are often occupied by small lakes called *tarns*. Cirques are of wide occurrence in areas of former or present glaciation and are known by numerous local names, such as *corrie* in Scotland, *cwm* in Wales, *kar* in Austria and Germany, *kjedel* in Scandinavia, *botn* in Norway and *nisch* in Sweden. The French term *cirque*, first used by Jean de Charpentiar in his study of the Pyrenees in 1823 is, however, internationally the most recognized and used.

Cirques range in size from small and shallow depressions tens of metres across to the great cirques of Antarctica and the Himalayas which may be several kilometres wide with the back-wall rising to heights of hundreds or thousands of metres. The western cwm (cirque) on Mt. Everest has a maximum width of nearly 4 km and a backwall 2,800 m high. The Walcott cirque on Mount Lister near Mc Murdo Sound in the Antarctica is 16 km wide and has a backwall 3,000 m high. Most of the large cirques in the world are associated with igneous or metamorphic rocks, as the competence of the rock to withstand failure (i.e. the strength of the rock) is important in limiting the heights of the sides and the head walls, though cirques in general are often found cutting across different rock types. Further more, the size of the mountain mass in which cirques are forming together with the duration of glaciation also affects the size to which any one cirque can grow.

Cirques have a strong tendency to form on leeward slope, particularly where shade provides shelter from the afternoon sun. Consequently they tend to develop largely on north-west, north and north-east facing slopes in the middle latitudes of northern hemisphere and south-west, south and south-east facing slopes in the middle latitudes of the southern hemisphere. Factors which affect the location of cirques include the occurrence of shade, the direction of prevailing winds and the general alignment of the high ground. These factors affect the rate and amount of snow accumulation and consequently, the intensity and duration of glacial erosion.[1] Cirques may form in a variety of topographical situations. Some cirques may develop out of preglacial water-eroded features, while others may arise through processes of nivation, deepening any available hollow which is able to collect snow.

The principal morphological features of a well-developed cirque include the following distinctive elements, though it must be remembered that cirque morphology varies considerably and many gradations may be found:—

(a) Firstly, there are the head and side walls, steep and nearly vertical and often shattered.

(b) Secondly, there is the rock basin or hollow of circular outline, a concave floor meeting the headwall in a sharp break of slope and showing evidence of smoothing and polishing.

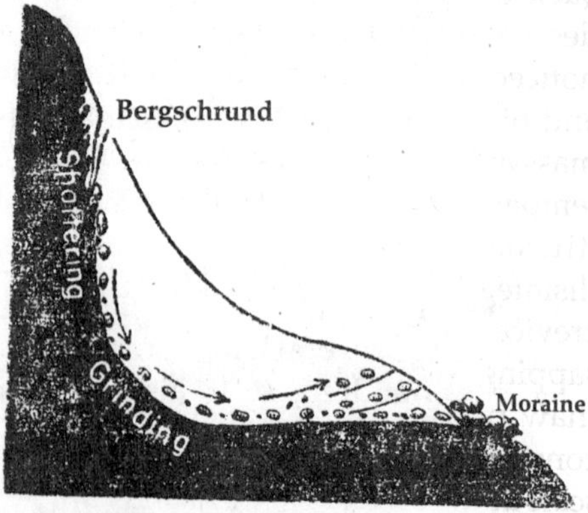

Fig.161—A Cirque with bergschrund.

1. Price, R.J.—*Glacial and Fluvioglacial Landforms.* op. cit., p. 55.

(c) Thirdly, there is the lip or threshold of the basin where the basin floor rises and is sometimes capped with moraine. It is convexly rounded and often shattered on the down-valley side.

(d) Fourthly, the lengths of cirques are usually three times their heights i.e. the length-height ratio is about 3:1.

It is widely accepted that most cirques are features of glacial origin, though the actual mechanics of the erosion processes involved are still a matter of controversy and speculation. The morphological features of cirques are such that they can only be explained in terms of glacial erosion. As we have noted above, the principal features are the rock basin in the cirque floor, the rock lip or threshold at the cirque exit, the steep head and side walls, the smoothed and abraded surface of the cirque floor, and the broken and shattered surfaces on the head and side walls. The steep head wall may be as high as 1,000 metres or more but is free from talus at its base. This suggests that ordinary weathering processes have not played an important role in cirque formation, but rather that some sub-glacial process had operated which permitted the glaciers to remove rapidly the material derived from the headwall.

In 1904 W.D. Johnson proposed his famous theory of cirque formation which has come to be known as the *Bergschrund theory* because of the importance he attached to basal sapping or plucking by freeze-thaw at the base of the bergschrund (the crevasse at the back of a cirque glacier separating it from the backwall). He descended into a 45 m deep bergschrund in Sierra Nevada and noticed marked evidences of frost riving on the cirque headwall and of incorporation of the disintegrated material in the glacier mass at the base of the crevasse. He concluded that the change of temperatures in the bergshrund across freezing point, occurring diurnally in the summer was an effective agent of rock disintegration by repeated freezing and thawing of water in rock crevices, and the cirque essentially owed its origin to this basal sapping. While the importance of head-wall shattering by freeze-thaw processes in cirque development is accepted, it is now considered unlikely that such processes occur on any significant scale at the base of bergschrunds. In 1916 I. Bowman pointed out that bergschrunds were not always present in cirques and that many bergschrunds failed to reach bedrock. He contended that cirque head walls could develop without the existence of bergschrund through plucking and abrasion beneath moving glacial ice.

In 1938-40 W.V. Lewis presented a modification of the bergschrund hypothesis known as the *Meltwater theory* in which he suggested that much of the meltwater in the bergschrunds was of external origin, derived from either summer rain or the melting of winter snow high above the bergschrund and that such meltwater could seep down narrow cracks far below the limits of the deepest bergschrunds, freeze in the rocks and shatter them, thus undermining the headwall and perpetuating its steepness. The loosened rock fragments then become incorporated in the moving ice and abrade deep incisions in the rock floor of the cirque. He also pointed out that the position of the bergschrund on a cirque headwall would vary slightly each season and markedly so during the waxing of a glacial period when the ice surface would rise and fall with changes in the glacier budget.

The work of W.R.B. Battle on temperature measurements in various bergschrunds in Greenland, Norway and Switzerland between 1948-53 showed that within the begschrund there are no large scale or rapid fluctuations of air temperature which might favour alternate freeze-thaw action. He was of the view that freeze-thaw action was much more likely in the upper parts of the open bergschrunds or generally in the zone where rock headwall, glacier surface and freely circulating air meet. More recent work by R. Vivian and G. Bocqulet (1973) on sub-glacial cavities would, however, seem to suggest that the possibility of freeze-thaw action in sub-glacial location can not be entirely dismissed. It is apparent, however, that the most favourable environment for freeze-thaw action would be the exposed rock wall above the glacier or firn limits or gap between the glacier and the rock wall in the upper parts of any headwall. J.E. Fisher (1955, 1963) on the basis of his investigation on some high-level ice and firn areas in the Alps, has suggested that alternatively the process may be associated with particular climatic condition when the boundary between cold ice and ice at the pressure melting point in a glacier migrates up or down part of the headwall, sapping it intensively within this zone. This might tend to undercut the headwall and thereby cause its retreat.

Attack on the headwall and sides by freeze-thaw action (including joint-block removal in areas of well developed joints) will cause the overall enlargement of cirques, but as most cirques have over-deepned floors, another mechanism must be sought to explain the development of the rock basins. This was provided by Lewis in his *Rotational slip hypothesis*. Based on his work in Norway between 1949-53, W.V. Lewis was the first to suggest that abrasion by rotationally moving ice was the principal mechanism for the grinding of the rock basin as well as the perpetuation of the rock threshold. Lewis thought that because of the **high surface slope the glacier would tend to rotate and that the relatively** high velocities of basal ice in a rotational slip were a great advantage in excavating the rock basin. He suggested that "when a section of a glacier is moving in a rotational slip, the bed rock beneath must be moulded into a basin fitted to this rotational movement". J.G. Mc Call's work on Vesl-Skautbreen glacier in Norway (1960) showed convincingly that the rotational movement of the ice favoured the formation of a rock basin. The backwall of the cirque was attacked by sapping, this process being greatly assisted by well-developed joints in the headwall. The material removed from the headwall was incorporated in the basal layers of the glacier which showed a high concentration of debris. This debris acted as the tools for abrasion by the ice as it moved across the cirque floor. As the tendency to abrade the basin floor is related to the slip-line fields in the cirque glacier, maximum abrasion occurs some distance back from the ice front and least abrasion occurs at the ice front. There is very rapid diminution of the rate of basal ice movement and there is little or no ice movement at the ice front, so that not only abrasion is reduced on the cirque threshold but material is piled up beneath the ice, and this tends to accentuate the basin form of the cirques. The rotational slip hypothesis thus provided a simple mechanism for the uplift of ice over the cirque threshold and for the evacuation of debris uphill from the basin floor.

In 1910 W. H. Hobbs tried to relate the development and varying characteristics of cirques to their stage of advancement in a cycle of mountain glaciation. He put forward a scheme outlining hypothetical stages in the destruction of uplands by cirque recession. The "grooved upland' of youth preserves much of the pre-glacial surface with a few nivation hollows. In the second stage (the 'early fretted uplands' stage) the smooth-topped ridges are cut by amphitheatre-like cirques. As the cirques enlarge and recede, the progressive expansion of neighbouring cirques results in the reduction of the unglaciated slopes between them to sharp-edged comb-ridges or aretes which meet in sharp mountain peaks or horns. Thus a series of aretes and horns is created, typical of a 'mature fretted upland'. The stage of old age (monumented upland stage) is reached when the breaking down of aretes creates low smooth cols which allow ice to spill over from one cirque to another and cirque floor may coalesce into a surface of glacial denudation above which stand a few isolated peaks or monuments. It may be pointed out that Hobb's scheme, though attractive and apparently plausible must be regarded as non-proven, as there are many unknown variables, and variation in structure and processes may account for more expanded cirque in some areas than in others.

Fig. 162—Comb-ridges and Pyramidal peaks.

As the cirques enlarge and cut back into pre-glacial relief, the mountain divides become narrower and are turned into steep-sided sharp-edged ridges known as *aretes* in French or *grat* in German. Where three or four cirques meet, the steep-sided ridges disappear and alternating sags or glacial cols produced by intersection of opposed cirques and spectacular pointed peaks or *horns* representing the unreduced portion of the original mountain range are found. Such pyramidal peaks, sharp and very steep-sided, are characteristic of the high glaciated mountains in the Alps, Himalayas and other regions (Fig. 163) This is known as *serrated topography*. The classic example of horn is the Matterborn (4,482 m) in Switzerland, the highest peak of the Pennine Alps, lying on the border between Switzerland and Italy. In the Himalayas a good example of horn is the Shivling peak (6, 545 m) above the Gangotri glacier and close to the Badrinath temple. Other examples are the Machapuchare (6,997 m) in Nepal Himalayas and the Broad peak and Gasherbrum mt group above the Baltoro glacier in Punjab Himalayas. It is to be noted that while headward cirque erosion may proceed actively, the principal erosion above the glacier on the over-steepened slopes results from the loosening of blocks and snow-masses by temperature changes or by expansion and contraction by freezing and thawing plus the general instability due to gravity.

Glacial Troughs or Glaciated Valleys

Below the cirque threshold there is generally an abrupt slope to the floor of the glacial trough or the U-shaped glacial valley. The glacial trough is one of the most distinctive features of glaciated mountains. Glaciers do not carve out valleys for themselves like the rivers do. Most glacial troughs were originally river-cut valleys which were subsequently occupied by glaciers and were so much modified by glacial erosion that they have developed characteristic transverse and longitudinal profiles which are very different from those of river valleys. The importance of the existence of a pre-glacial river valley system and the preparation of the valley floor by deep freezing or deep weathering before the ice advances over it is widely accepted. Another factor assisting erosion of bedrock by a valley glacier is pressure release beneath thick ice. As the glacier bites more deeply into its bed, rock of the density of 2.5 or more is replaced by ice possessing a much lower density of 0.9. The reduction of density causes the development of dilatation joints and overdeepening is achieved by glacial plucking of the loosened joint blocks. An important difference between river and glacial erosion is the size of the channel occupied by water on the one hand and by ice on the other. Whereas the river is in direct contact with only a very small

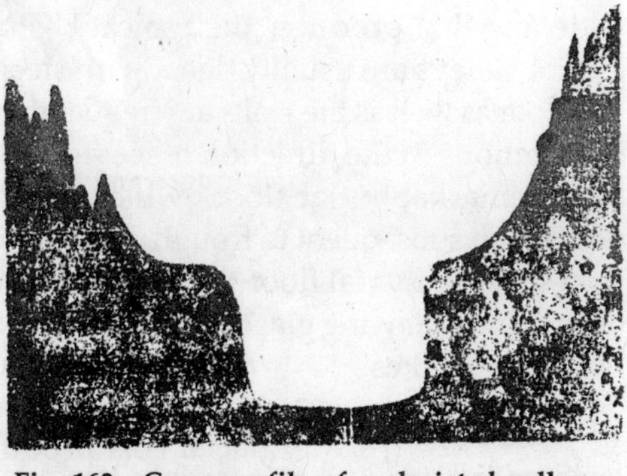

Fig. 163—Cross-profile of a glaciated valley.

proportion of the cross-profile of the valley, the glacier is in direct contact with a very high proportion of the cross-profile.

Glacial erosion is at work both on the valley floor and the valley sides, so that protuberances in both lateral and vertical directions tend to be removed. The typical glacial trough, therefore, tends to be straight, with steep sides, and in cross-profile approximates to a parabola. In fact, the *U shape of the cross profile* is one of the most characteristic features of a well-developed glaciated valley. This parabolic or U-shaped crossprofile can be achieved by the widening of the pre-glacial V-shaped river valley without over-deepening or by both widening and deepening of the preglacial valley by glacial erosion. In order to erode effectively, the ice must exceed a certain minimum thickness, flow fairly fast as a concentrated stream with high velocity of basal sliding, and it must carry considerable number of boulders at the ice-rock interface, for it is these boulders which are used as tools by the glacier to accomplish erosion. Both the ground and the lateral moraines help the glacier in this task. The sides of the glacier are specially "armed" with angular fragments resulting from frost shattering and freeze and thaw along the valley sides, and their frequent occurrence may help to account for the widening that produces the typical U-shaped glacial valley. The glacial valleys are usually deep, with steep sides and a wide floor. The floor as well as the walls are frequently polished and are marked by striations in the direction of ice flow. Portions of glacial valley have remarkably flat floors which commonly are the result of deposition subsequent to trough development. They may represent post-glacial alluvial floor till of the ground moraine and outwash deposition following glacial recession or lacustrine plains produced by filling of lakes.

Another important characteristic of the glacial valley is that its *longitudinal profile* is remarkably irregular and ungraded, unlike the relatively smooth and concave longitudinal profile of a river valley. The glacial valley usually descends in a series of giant steps often with over depeened hollows or basins in between. These are known as *glacial steps* and the profile is known as *glacial stairway*. The glacial steps are generally higher in the upper part of the valley and become less pronounced in the lower section, possibly because the glacier was in existence for a longer period in the upper sections of the valley.

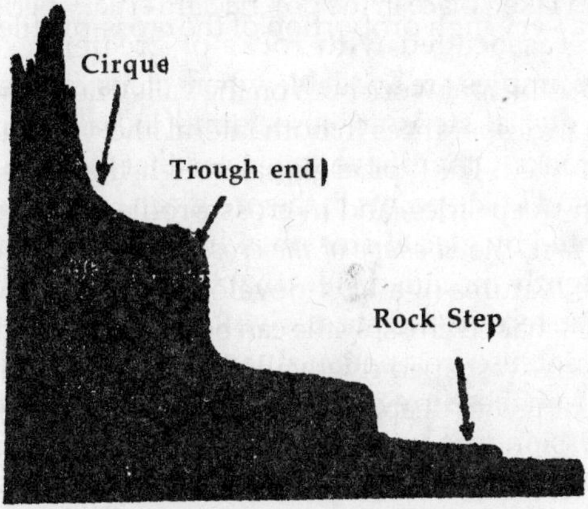

Fig. 164— Longitudinal Profile of a Glaciated Valley and Cirque.

Each glacial step has three parts—*a riser, a tread and a threshold* or *riegel* (an Alpine term). *Riser* is the steep ascent from the step below to step immediately above marking the down-valley end of the latter. A *threshold or riegel* is the rock bar at the crest and the back of a riser. *A tread* is the relatively flat surface of a step. It has a reverse up valley slope which frequently gives it the form of a basin. The rock-basin on the overdeepened tread, sometimes partly dammed by moraines, becomes the site of a lake or chain of lakes. These are called *pater noster lakes* because of their similarity to a string of prayer beads. The basins frequently contain lakes or have been later filled with alluvium, while the steps often show marked signs of glacial scouring and plucking. The steps are rather similar to roches moutonnees, but on a much larger scale.

The steps or the rock bars and intervening basins, which are so common in glaciated valleys are an evidence of the irregularity of glacial erosion. Various theories have been put forward to explain the formation of the steps and the zones of alternating excessive erosion and reduced erosion, but none appear to be of universal application, suggesting that glacial steps are of diverse origin. Some are of the view that they are due to variation in rock structure and hardness. In other words, the risers and the riegels represent hard rock bars in the glacial trough. According to some, where the rocks have more joints and faults the rocks are more weathered in pre-glacial times and are easily removed by plucking and rock basins are formed, as glacial erosion is more effective where deep weathering had taken place in the pre-glacial Tertiary period. Riegels and risers are associated with rocks of greater resistance to weathering. Examples are available where steps can be related to lithology but glacial steps are also found in valleys containing homogenous rocks. Others have suggested that the steps represent irregularities and knick points in the pre-glacial valley which have been accentuated by glacial erosion, while they maintained their position. A slightly modified view that the glacial steps would be produced by the recession of successive knick points, as they would be cut back by intense frost action at the foot of the crevasse which would result from these irregularities, in a manner similar to cirque headwall recession was proposed by Johnson. According to this concept, the steps of a glacial stairway would be essentially migrant features that would shift their position rather rapidly, as the glacial processes are active, regardless of the rock structure.

An ice-fall hypothesis has been advocated by Andrews (1910) according to which the erosive energy of the glacier in its steep ice-fall sectors would be much greater than in the gentle sectors, and thus the steeper sectors would become progressively enlarged and incidentally would migrate headward gradually. A variant of this glacier energy concept explains steps as the result of differential glacial abrasion in wider and narrower sections of the valley. Where the trough is narrow and the passage constricted, the glacier will thicken and its erosive energy would increase and greater deepening will take place than in the wider sections. This theory is applicable to some steps, but many risers do not coincide with constrictions. The theory of extending and compressing glacier flow put forward by J.F. Nye (1952, 1968) throws similar light on the subject. In other words, the form of the bedrock surface over which the ice moves exerts an influence on the ice motion and is in turn modified by the ice movement. In the steep reaches where the flow would be extending, bottom movement would not be so conducive to erosion as in those places where the slope was less or even reversed giving rise to compressive flow and it is in these zones that the ice can erode more effectively.

A possible reason for the formation of steps has been suggested to be the confluence of tributary glaciers with the main glacier. Ice volume and weight will increase below the junction of a tributary glacier and this would presumably result in increased erosion and deepening of the trough below the junction. There are some steps which can be explained in this manner, but not all steps are situated below tributary junctions.

Matthes (1930) in his classic work on the Yosemite valley put forward a rock-competence hypothesis and attributed the steps in the Yosemite valley to the varying erodibility of massive and jointed rocks. Well-jointed or fractured rocks would be more rapidly eroded than rocks that are massive. He pointed out that the inner treads are marked by well-jointed (incompetent) rocks, while the risers and the threshold of the treads are marked by extremely massive and scarcely jointed rocks. Thus there was selective quarrying by the moving ice over rocks of differential competence or strength.

It would appear that the explanation for the steps i.e. alternating zones of excessive erosion and reduced erosion should be sought in a variety of factors — rocks of the valley, form of the valley and the response of the moving ice to this configuration, ice volume owing to incoming tributary glacier, and zones of compressive flow. It is also likely that the ground is prepared in some way, possibly by deep freezing or deep weathering before the ice advances over it.

Truncated Spurs and Roches Moutonnees

Typically the glaciated valleys are in general straighter than the pre-glacial streams valleys which the glacier occupied. The glaciers straighten their valleys by abrasion of spurs projecting from the sides of the original stream valleys and thus produce *truncated spurs*. The truncated spurs may be found in different degrees of truncation so that while some may be partially cut, others may be completely truncated.

Another characteristic feature found in the glaciated valleys is rocky outcrops which are rounded on the upstream side and irregular and shattered by plucking action of ice on the downstream side. These are known as *roches moutonnees* to which reference has been made earlier. Roches moutonnees are, however, not limited to glaciated valleys but are found over large areas where rocky projections have been subjected to the abrasive and plucking action of glacial ice. The glaciated valleys present a rugged and shattered appearance when viewed from below while when viewed from above, the surface of the glaciated valley appears to be smooth and rounded.

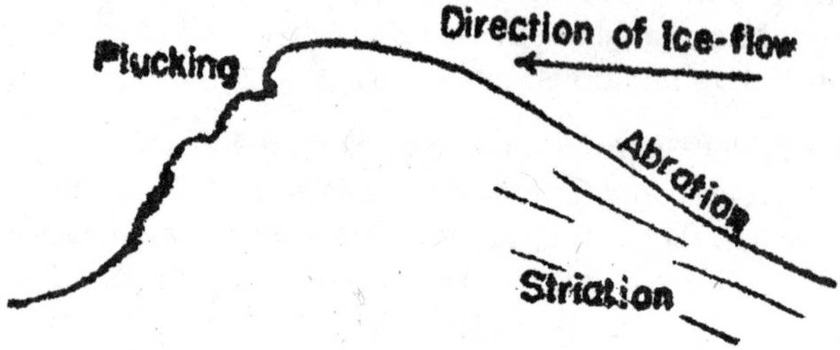

Fig. 165—Roches Moutonnees.

Hanging Valleys

Together with truncated spurs another feature associated with glaciated valleys is the hanging tributaries that occur along the valley sides. In the case of river valleys, the tributary valleys normally join the main valley accordantly. But tributary valleys joining a major glacial trough are quite commonly discordant, i.e. the floors of tributary valleys "hang" above the floor of the main valley. This is so because the glacier occupying the trunk valley is wider and deeper than its tributary glaciers and consequently there is more glacial erosion in the trunk valley than in the tributary valleys. Sometimes the tributary valley may not be glaciated while the main trunk is scoured out by a glacier from further up in the mountain. Such discordant valleys are called *hanging valleys*. After the disappearance of the glaciers when these valleys are occupied

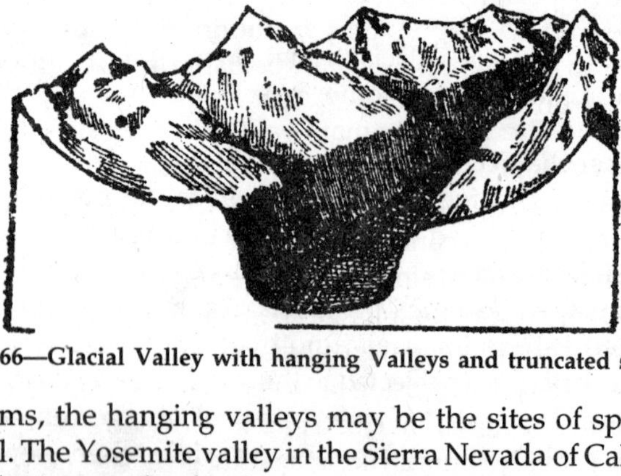

Fig. 166—Glacial Valley with hanging Valleys and truncated spurs.

by streams, the hanging valleys may be the sites of spectacular waterfall. The Yosemite valley in the Sierra Nevada of California is an excellent example of a glaciated valley complete with hanging valleys, waterfalls and cascades, truncated spurs and glacial steps

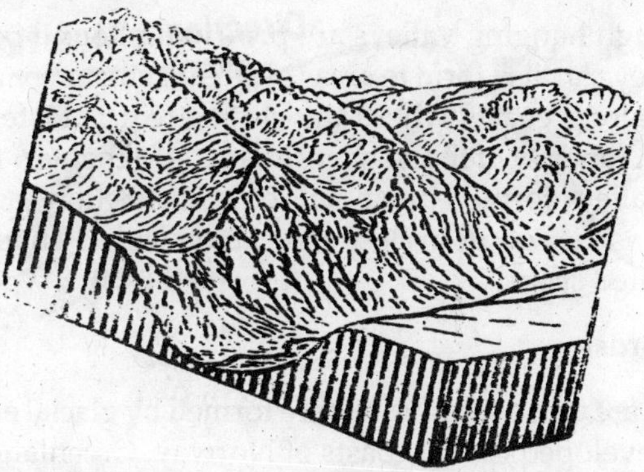

Fig. 167—Landforms of glacial erosion—glacial valley, truncated spur, hanging valley, cirque, arete and horn.

and basins. Another fine example is the Lauterbrunnen valley in Switzerland between the Interlaken and the Jungfran.

There is another view also regarding the formation of hanging valleys which deserves mention. According to those who believe that the ice acts as a protective cover for the rocks, the main valley gets over-deepended by stream erosion in the inter-glacial period. During this period, the ice stays in the tributary valleys for a longer period and there is no stream erosion and the valleys are, therefore, not deepened. As a result the trunk valleys are more deeply eroded and the tributary valleys are left hanging above the trunk valley. An important evidence put forward in support of this view is that all the tributary valleys of a glacial trough do not hang above the main valley in a similar way and some, in fact, do not hang at all. This is so because together with the main valley the ice also disappeared from these tributary valleys and the deepening of the valleys by stream erosion continued simultaneously in both. An interesting fact to be noted in this connection is that most hanging valleys are met with on the leeward side of the slope away from the sun where the ice is likely to stay for a longer period. It seems probable that together with the differential erosion by the trunk and the tributary glacies, the differential erosion between the trunk streams and the smaller tributary glaciers has also a part to play. Sikkim and Kashmir Himalayas provide some striking examples of hanging valleys in which both differential ice and river erosion seem to have played a part.

Although hanging valleys are principally found in glaciated regions, they can also form in areas of stream erosion or along fault scarps or along coasts where the cliff retreat is faster than the adjustment capacity of the smaller streams (see chepter. 15). But the hanging valleys formed as a result of glacial action can be easily identified, as they occur in close association with U-shaped valleys and truncated spurs.

Fjords (Fiords)

Fjords are an impressive feature formed by glacial erosion and are best developed on the coasts of Norway, Greenland, Iceland, southern Alaska, British Columbia, eastern Canada, Southern Chile and the South Island of Newzealand, as also Antarctica where they are mostly buried under ice. They are more common on west-facing than east-facing coasts and occur mostly in higher latitudes, along coasts backed by rugged and dissected highlands which were subjected to severe Pleistocene glaciation. Fjords are very deep and narrow, with long arms of the sea stretching inland between steep and rocky mountainous sides which usually contain hanging tributary valleys. They are essentially glacial troughs which have been partially submerged by sea water. The ice scour has overdeepened some fjords by more than 900 m, and the deepest fjord in the world—Messie channel in Chile is 1,288 m deep, while the deepest in Europe, Sogne fjord in Norway, is 1,210 m deep. They nearly always possess submerged rock basins with the deepest part lying near the coast and a rock bar or threshold at their seaward terminus. The depth of water is much less here than either further seaward or towards the coast. Thus the floor of the fjord resembles a normal rock basin with threshold or sill at its lip. According to many authorities this is the most significant feature of the fjord, distinguishing it from the ria of a submerged non-glaciated coast. It has been suggested that the fjord thresholds are submerged terminal moraines. In some cases this may be true, but in most cases it seems more probable that they are the result of decreased erosion at the terminus where the ice was thin and the glacier's scouring ability was cut down strongly by reduced gradient and more erosion headwards where the ice was thick.

Although most workers attribute the greater part of fjord erosion to ice, some maintain that the ice merely modified the pre-glacial river valleys. According to one view the glacial erosion of the valleys and the formation of the glacial trough had taken place above sea-level and they were subsequently submerged as a result of the rise of the sea level. But the post-glacial rise of sea-level of about 100 m is very small in comparison with most fjord depths, and consequently most workers believe that the fjords are not the result of glacial troughs formed above sea-level and later subjected to submergence, but are glacial troughs eroded by ice below sea-level. Glaciers are capable of exerting normal pressure and eroding their beds even after entering the sea until the floatation actually occurred. The fjord glaciers were, of course, largely guided in their path by well-defined valley systems and had taken advantage of stream-eroded valleys. But the fjord glaciers were capable of eroding bedrock as far down-valley as the point of floatation.

It was once held by some that fjords were essentially of tectonic origin, the result of fracturing under tension, fluvial and glacial action being held responsible for only slight modification and evacuation of the debris. It is, of course, true that some fjords bear a close relationship with lines of structural weakness but most workers now agree that fractures are not necessary for fjord formation. Since fjords pass inland into normal glacial troughs and both possess common characteristics, there is now general agreement that glaciation is responsible for the main features of fjord physiography.

Another similar coastal glacial feature is *fjard* (fee-air-d in Swedish) or *fiard* which is espcially characteristic of the west coast of Sweden and some of the areas of Norway. Fiards are shorter and shallower than fjords. They are glacial scoured, irregular inlets of low relief and lack the high relief and U-shaped cross-section of the fjords. While Norway is a land of fjords, Sweden has few fjords but many fjards.

Erosional Features produced by Ice Caps

Landforms produced by glacial erosion are not limited to mountains or areas of considerable relief. They are also found in low land as well as highland areas of low relief. This is particularly true of features produced by erosion by ice sheets or ice-caps which covered extensive areas, high and low alike. The most distinctive feature of ice-cap erosion is said to be an *ice-scoured plain*, the term plain being used to include tablelands as well. Good examples are the Canadian Shield and parts of Scandinavia and Finland where the surface consists of an assemblage of landforms which are mostly erosional in their origin. These include striated, grooved and polished surfaces, irregular rock knobs and depressions, roche moutonnees with smoothed stoss side and broken, shattered and irregular surface on the lee side, and generally streamlined features with small patches of glacial drift and general absence of the pre-glacial soil.

Landforms Built by Glacial Deposition

The debris directly deposited by ice are not sorted or layered and are marked by heterogeneity of materials and lack of stratification. These deposits, called *till or moraine* are carried beneath, within or along the sides of the glacier and are deposited either by being lodged in place beneath the glacier or by being lowered to the ground as the ice melts. Three types of glacial deposits—ground moraine, end moraine and lateral moraine may be distinguished. A blanket of till forms nearly everywhere beneath a glacier (*ground moraine*) and as a thickened rim at the glacier terminus (*end moraine*). The thickness of the till blanket decreases progressively as we move up from the glacier end. The *lateral moraines* form along the sides of a valley glacier and are found along the sides of glacial troughs, and sometimes small lakes may form between a lateral moraine and the trough wall, above the floor of the trough. The features produced by glacial deposition are more characteristic of the lower slopes of mountains and areas of comparatively low relief which were covered by the continental ice-sheet during the Pleistocene glaciation. But in mountainous areas also depositional features like lateral moraine, crescentic end moraine and some ground moraine mounds are found in the glaciated valleys of the Alps, the Himalayas and the Rockies. Ground moraine, however, is scarce in glaciated valleys and is more abundantly associated with ice caps, as ground moraine which may have been deposited on the valley floor is liable to rapid erosion by post-glacial streams. The ground moraine where present is, therefore, likely to be thin and patchy in the glacial troughs.

Till Plain and Morainic Hills

Glacial deposits result in two types of distinctive morphological features. The first type is characterized by very little local relief and is referred to as a *till-plain* or *till-sheet*. These deposits tend to mask the irregularities in the bedrock surface so that they are thicker along depressions in the bed rock and thinner over ridges or plateaus. The second type has a very irregular surface, consisting of linear morainic ridges or complex of ridges, hollows and mounds and is described as *hummocky topography*. Basically the till-plain is formed by the deposition of the ground moraine and the ridge-like form or the hummocky topography is the product of the end moraine.

The *till plain* is formed as a result of regional deposition of ground moraine by continental glaciers where complete burial of the pre-glacial topography took place. The ground moraine may be deposited as a thin sheet on a relatively flat bedrock surface but a flat till plain has rarely a pre-existing flat bedrock surface. More commonly, the ground moraine has completely filled and masked most of the underlying pre-glacial relief. The term "till" is restricted to those deposits (sediments) which are laid by ice either as basal or lodgement till or as ablation till, but a till plain consists of deposits which are glacial as well as glacio-fluvial. In some places the two types of deposits may be distinguished but in other cases it may not be possible to make a clear-cut distinction between them. Till is

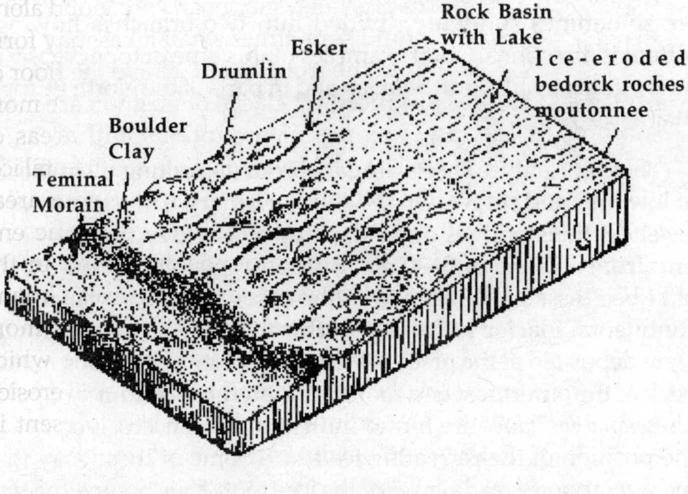

Fig.168—Landforms built by glacial deposition, especially boulder-clay, esker, drumlin, terminal moraine and outwash plain. In the upper part of the diagram rock basins and roches moutonnees produced by erosion can also be seen.

unstratified mixture of sand, silt and clay with modest quantities of pebbles and large boulders. The greater part of the material may be derived from nearby bedrock (1 to 10 km away) but most till contains boulder and stones which differ in character from the bedrock upon which the till has been deposited. These "different" rock fragments are called *erratics* or *erratic blocks*. If the place of origin of an erratic is known, the erratics are called *indicator boulders*. Groups of erratic blocks having a fanlike distribution from their source are called *boulder trains*.

The thickness of the till decreases progressively as one moves further back from the glacier end, and near the glacier's terminus a thickened rim of deposit is found (end moraine). Till plain surface is, however, rarely a completely flat surface but is diversified by small rounded mounds and shallow depressions (sometimes forming lakes) which are irregularly distributed. Streamlined elliptical till hillocks created while the ice is moving, known as drumlins are also sometimes found, though they are rare in Europe. Entire northern Europe was covered by ice-sheet during the Pleistocene ice age and the till deposits are extensively found in the great North European plain. As the surface of the till plain is irregular, the rivers follow tortuous and unexpected courses, and are sometimes suddenly divided into two branches flowing in different directions. Good examples of this type of topography and drainage are available in Sweden and in parts of the North European plain.

Glaciers and ice sheets are capable of depositing till in the form of low hills and ridges. On the southern margin of the continental ice-sheet linear ridges and hills are found extending for several kms from one end of the Great European plain to the other which have been essentially formed by the *end or terminal moraines*. As the continental glacier retreated northwards, the terminal moraines were deposited at the places where the glacier halted for sometime. Back of the outermost end moraine, therefore, a number of parallel *recessional moraines* are found indicating the successive pauses in the position of the retreating ice front. Some of these may in fact mark temporary readavnces of the ice-front. They occur either in the form of linear or irregular ridges interspersed by lake filled depressions. These are all *morainic hills*, which are usually only about 200 metres high above sea-level but some rise to 300 metres or more. The highest hill of the North European plain. Termberg situated near Danzig, is a morainic hill and is about 350 metres high. In western Russia there are several morainic hills, the most prominent being the Valdai hills from where the Volga river rises.

In the course of retreat there are oscillations of the ice front resulting sometimes in a zone of irregularly distributed knolls and basins. Such a topography is described as *knob and basin topography.*

Drumlins

In some till plains, smooth streamlined, oval or elongated mounds or small rounded hills, composed largely of glacial till are found (fig. 169). These are known as drumlins (from Gaelic *druim*, a mound). A typical drumlin is half ellipsoid in shape and looks like an inverted spoon. Though usually having an oval outline and not exceeding 60 m in height, they show great variation in form, height and length, and may sometimes be several km long and nearly a hundred metre high. Commonly length to breadth ratio is 2.5 :1 to 4:1. Most of them are elongated in the direction of ice movement,

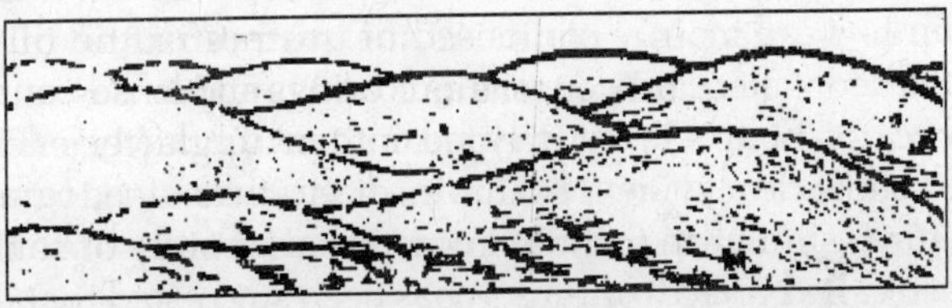

Fig. 169—Basket-of-eggs topography built by drumlins.

and the end facing upglacier is relatively broad and blunt compared with the tapering downstream end or tail. The drumlins rarely occur singly. They commonly occur in groups called swarms or fields. Being distributed in clusters or swarms (*en echelon)* the mounds give rise to what is appropriately described as *'basket-of eggs topography*. (Fig 170) They usually occur a short distance behind the end moraines, and in between the mounds the till covered ground is low with poor and confused drainage and containing ponds, marshes and waterlogged meadows.

Drumlins are absent from most of the areas of Older Drift, but they are widely distributed in those areas that were glaciated during the last major advance (Wirm, Weichsel, Wisconsin). Often, however, they are very localized in any one area, and within any one area, the drumlins tend to be fairly uniform in shape and size and to occur in large numbers. Except for a few areas in Switzerland and Germany drumlins are rare over much of glaciated Europe. In Britain drumlins occur in lagre numbers in some parts of the Newer Drift areas. One of the largest and most densely packed drumlin belt in the world extends across northern Ireland from County Down in the east to Donegal Bay in the west. In England, the high ground of the Lake district and North Pennies is almost entirely surrounded by drumlins. Drumlins also occur in north eastern England, south-west Scotland and the Midland valley of Scotland. In North America there are four major drumlin belts : 1. an area in New England, southern New Hampshire and eastern Massachusetts, 2. a belt in New York and Ontario, south and north of Lake Ontario., 3. Nova Scotia, 4. a belt including parts of Wisconsin, lowa and Minnesota. The New York state is probably the most extensive drumlin belt in the world covering an area 56 km broad and 225 km long between lake Ontario and the Finger Lakes. Generally the drumlins in N. America tend to occur in belts behind morainic barriers on flat till plains.

Drumlins are mainly composed of unstratified till but some have stratified water-laid deposits as coatings and lenses, and some have cores of bedrock, usually near their upglacier ends. All, however, show clear evidence of having been streamlined by moving ice. Features similar in form but composed entirely of solid rock have been called rock drumlins. It has been suggested that it is not desirable to call features of similar form, but composed entirely of solid rock, drumlins, as the main essential of a true drumlin is that it should consist of glacial deposits, shaped to create a streamlined form.[1]

Drumlins thus vary considerably in their size, material, elongation and spacing, but they have one essential feature in common. This is their streamlined form, and it is on the basis of this feature that their origin must be sought. Nearly all agree that they have been formed by moving ice but there the uniformity of opinion ceases. Theories of drumlin formation may be divided into two main groups—those that support an erosional origin for drumlins and those that support a depositional origin. The *erosional theory* explains in terms of a pre-existing drift cover i.e. they have been

1. Embleton G. and King, C.A.M- *op. cit*, p. 403.

formed by two glaciations. The first glaciation provided in irregular till surface and the second scoured this surface leaving the drumlins. The erosional theory can also account for stratified material in drumlins. It is pointed out that there is a range of features composed of pure rock to those composed of till all of which have the same drumlin shape. The solid rock features are clearly erosional and the similarity of form suggests that the same erosive processes produced both the types. One of the problems about the erosional theory, however, is the shape of drumlins. If the dominant process is erosion then they should have a shape similar to the solid rock features i.e. roches moutonnees, with their blunt and steep ends pointing downstream rather than upstream as is common with the majority of drumlins. The roches moutonnees show a reversal of the normal stoss and lee slope of the typical drumlins. The drumlin normally has the steeper slope on the stoss side, while the roche moutonnees has a relatively gentle stoss side and quarried lee slope. This difference suggests that the processes forming drumlins are not similar to those shaping the solid rock. Further, while drumlins usually occur on lower ground, smoothed rock features are characteristic of the steeper slopes and higher areas, though both are essentially streamlined.

By far the most common explanation depends in some way on deposition. It is suggested that "the long low tail is characteristic of the streamline form. It is designed to create the minimum of disturbance in the flow of the moving medium, whether it be air, water or ice. This form is more likely to result from deposition than from erosion because it is this part of the drumlin where the pressure would be least, in the lee of the main body of the feature. Material could therefore accumulate in this position. On a similar argument, the lack of pressure in this position would minimize erosion. It is significant that in roches moutonnees the lee side is steep and rugged, and is eroded by plucking rather than by grinding and smoothing. In so far as drumlins resemble crag and tail features, it seems likely that their shape is partly depositional in character, at least especially the long low tail.

The purely *depostional theory* has been advocated for drumlins formed entirely of till. It has been suggested that they form around a nucleus of frozen till or rock. A few drumlins show concentric layers, suggesting that they have been built up by accretion of layer upon layer. The depositional view of drumlin formation is supported by Hollingworth (1931) and J.K. Charlesworth (1957). The drumlins that are composed entirely of till show no evidence of two stages of development. The distribution of drumlins on the lower ground suggests that they occur in areas that are predominantly depositional. In detail it has been suggested that the drumlins of northern England formed in areas where the pressure of the ice is locally increased by thickening or by increase of velocity. The extra pressure could provide the conditions necessary for drumlin formation in ice in which there is sufficient basal till to form them. The orientation of the stones in the fill of drumlins is a strong argument in favour of their deposition by moving ice, which is also responsible for shaping them into drumlin form. Both the drumlin elongation and the stone orientation have clearly been determined by the same ice flow and, where this can be shown to be true, those particular drumlins can confidently be stated to be depositional in character,"[1] "Some of the observations on drumlins suggest that there is some periodicity in the ice-flow which induces zone of deposition. These cores can then be shaped by the moving ice into drumlin. This would apply particularly to the drumlin fields situated on open areas of low relief. The suggestion that variations in the ice flow itself are a factor in the pattern and occurrence of drumlin is worth close investigation. The approximately equal height of the drumlins in anyone drumlin field probably reflects the thickness of the debris concentrated in the basal ice.

It would seem reasonable to conclude that drumlins can be formed either by the modification of previously existing drift sheets, the readvancing ice causing erosion and deposition in varying proportions or by the direct deposition of till by active ice in areas where till is available and the pressure of the ice is sufficient to produce the streamlining characteristic of drumlins, Dilatancy of the drumlin forming material is an important prerequisite.'[2]

I.J. Smalley and D.J. Unwin (1968) have put forward a new theory of drumlin formation related to the dilatancy of the material forming them. "The basic conditions necessary for till-drumlin formation are said to be: (1) the material at the base must have been continuously deformed by shearing, and (2) the material must

1. Ibid. p. 425.
2. Ibid, p. 427

consist of boulders in a clay-water system, the large particles forming a dilatant system. Dilatancy is a property of granular masses. Smalley suggested in 1966 that dilatancy may be a possible cause of drumlin formation. They form in a layer of boulder clay separating a glacier from certain types of terrain. When flow in this layer is obstructed, part of the layer packs into an obstruction and the rest flows around and moulds it. When a granular mass is disturbed, expansion results, the dilatant materials being then more resistant to shear stress. It is this high resistance that causes drumlins."[1]

The complex internal character of drumlins, however does not suggest a simple depositional mechanism. It seems highly likely that the drumlin with its fascinating streamlined form is polygenetic. Although the term drumlin is usually only applied to streamlined forms consisting of glacial hill, the existence of similar forms consisting of other types of glacial drift and partly or wholly solid rock means that the term has meaning only in a morphological sense and not in a genetic sense.

The Lothian Type of Glaciated Lowland

In the central lowlands of Scotland and more particularly in the Lothian region the glaciated lowlands are covered by a deposit of drift and drumlins. But in this lowland there are patches of igneous rocks where due to plucking by glacial ice one side of the rock masses is found to be rough while the other side is smooth. Where volcanic plugs or any other type of highly resistant rocks are present as protubrances which formed an obstruction in the path of advancing ice, a special type of feature has developed which is known as *crag and tail*. There is evidence of glacial erosion on a crag in the form of striae and fluting on the stoss side and the rock slope is steep and irregular. But in the lee of the crag, the softer rocks are not only protected from erosion but also received a covering of glacial deposition with the result that there is a relatively gentle slope trailing away in the lee of the crag which is known as the tail. The crag and tail topography thus consists of a steep rocky slope on one side (the crag) and a relatively long gentle slope on the other side (the tail). An excellent example is found in Edinburgh where the Edinburgh castle is located on the crag composed of a volcanic plug, while the high street follows the gentle slope of the tail. The

1. Ibid, p. 426

main interest of the Lothian type of glaciated topography lies in the fact that it provides an example of both glacial erosion and glacial deposition at one and the same place.

Fluvioglacial Landforms

The features which are produced by the combined action of glaciers and meltwater are called fluvioglacial or glacio-fluviatile landforms. The huge amount of debris acquired and transported by the glacier is ultimately got hold of, transported and deposited by the meltwater streams that flow on, in, under and beyond the glacier ice. The meltwater released by the ice is capable of a great deal of erosion, transportation and deposition, and thus produces landforms that are different from either purely glacial or purely fluvial landforms. The fluvio-glacial or the glacio-fluvial deposits retain some of the characteristics of glacial deposits but show a certain degree of assortment and stratification on account of the action of water. The longer the distance they are transported by the streams before being deposited, the greater the degree of assortment and stratification that they display. Fluvio-glacial deposits not only tend to be stratified but also consist of rounded or sub-rounded particles as they are the result of melt water transportation and deposition. The most common fluvio-glacial features found in the areas of mountain glaciation include esker, kame, outwash fans and deltas and others. These are essentially depositional features, but in some areas fluvio- glacial erosion may supersede fluvio-glacial deposition, as meltwater flowing on, at the margin of, within and beneath the ice is known to be capable of eroding deep channels when it comes into contact with solid rock or drift.

Types of Fluvio-glacial Deposits

Fluvio-glacial deposits can be sub-divided into two broad categories—proglacial deposits and ice-contact deposits. *Proglacial fluvioglacial deposits* result from meltwater transport and subsequent deposition in locations some distance from, and undistrubed by, the actual presence of glacier ice. These are deposits made by meltwater beyond the limit of the glacier ice and show undisturbed sedimentary characteristics and forms. On the other hand, *ice-contact deposits* result from deposition of sediments by meltwater in close association with the ice mass and are laid down against it, beneath it, within it or on top of the ice surface. Such deposits often show disturbed sedimentary structures. The position, extent and form of the deposits are determined by the ice mass or by its subsequent slumping resulting from loss of support of ice walls or blocks or ice buried under the debris.

Ice-Contact Fluvio-glacial Deposits

The two principal features produced by ice-contact fluvioglacial deposits are *eskers* and *kames*. The term *esker* is used for features that run at right angles to the ice edge, while the term kame is used for features lying parallel to the ice front. Both are ice-contact landforms. The majority of the ice-contact landforms result from deposition by meltwater streams and are characterized by channel deposits. But some of these landforms result from accumulation of sediments in lakes which may form in lateral, frontal, subglacial, englacial or superglacial positions and thus exhibit lacustrine rather than channel flow characteristics.

Eskers

The term 'esker' originates from either the Irish word *eiseir* or Welsh word *escair* meaning crooked or winding. An esker is essentially a long narrow winding and low ridge of coarse sand and gravel (see Fig. 169). The course of an esker is similar to that of a stream, being at times sinuous and at times straight. The low elongated ridge is generally aligned at right angles to the ice front and provides an indication of the direction of ice flow. It is formed of stratified fluvioglacial deposits—principally well sorted sands and gravels—laid down by meltwater streams as channel (tunnel) deposits in or beneath the slowly moving or stagnant ice. Eskers range in length from a few metres to 400 km. The longer ridges are usually higher and wider than the shorter ridges. Broadly speaking, eskers between 50 and 200 m in length range between 3 and 10 m in height, while eskers which are over one km long range in height between 10 and 100 m. Even the larger eskers are, however, rarely higher than 40-50 m and wider than 400-700 m. Many of the smaller ones which are 200-300 m long are 40-50 m wide and only 10-20 m high. Eskers are usually flat-topped and steep-sided, but their cross-profiles show great variation in form depending upon the environment in which they were deposited. They may occur as single continuous ridges of uniform dimensions or ridges with varying height and width or ridges consisting of a series of mounds linked by low ridges. They can also occur in the form of complex systems with a series of tributaries or distributaries. Examples of eskers are to be found in north-eastern United States, eastern Canada, Ireland, southern Finland and Sweden. In parts of Sweden and Finland, because of the extensive presence of lakes and marshes, eskers provide the only high ground for the construction of roads and railway. A noteworthy characteristic of eskers is that their courses are independent of topography. They cross valleys, climb slopes and hills, rising and falling as they are traced across moraine landscapes.

The most favoured method of esker formation is the hypothesis which suggests that eskers are produced by the deposition of sediment in the channel of subglacial or englacial streams. Meltwater descends down the ice surface down crevasses and moulins, and tunnels develop at the base of the ice to allow movement of meltwater towards the ice margin. Channel floor deposits accumulate in these sub-glacial tunnels between walls of ice. The actual cross-sectional shape of the tunnel determines in large measure the cross-sectional shape of the esker, as the deposition takes place encased between the tunnel walls which are finally removed as a result of melting. They are slumped and sharp crested where the tunnel is high and narrow and flat topped where the tunnel is low and wide.

The meltwater streams can develop considerable hydrostatic pressure because the meltwater largely originates on the ice surface and descends to the base. Pressure can also be built up by constrictions developing in the tunnel. Water flowing under hydrostatic pressure can flow up-hill and this has been used to explain the fact that some eskers run up-hill in places. A simpler explanation of this lack of conformity between the trend of eskers and the underlying topography was however suggested as early as 1902 by Cros by. According to him the eskers first formed in englacial tunnels and they were later let down on to the subglacial surface and superimposed on the underlying irregular surface as the ice melted. The suprimposition hypothesis of Crosby was not accepted for long, as it was found difficult to believe that ridges of sand and gravel would survive the process of being let down to the subglacial surface without obliteration of the distinctive form of an esker. It was also pointed out that for long esker ridges to develop on the ice surface or in tunnels in the ice, it was necessary for large areas of the ice mass to be without crevasses, other wise the suparglacial or englacial streams required to produce the esker deposits would become subglacial. Recent studies by various workers especially by

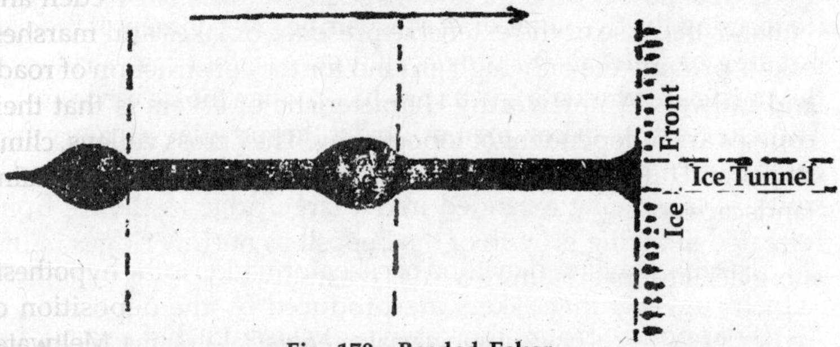

Fig. 170—Beaded Esker.

Price between 1964-69 in areas of existing esker formation in Norway, Alaska and Iceland, however, indicate that eskers can be let down from the surface of a glacier on to the sub-glacial surface without being destroyed. It is, of course, possible that some parts of an esker ridge were deposited subglacially while other parts were subsequently let down by the melting of ice which once formed the floors and sides of englacial tunnels or supraglacial channels.

Eskers are found in the country behind the terminal moraines. They were deposited in the ice tunnels as the glacier margin retreated upwards. Occasionally an esker consists of elongated thin ridges joined at intervals by thick mounds of sand and gravel resembling beads on string. The term *beaded esker* has been used to describe such features (Fig. 171). The mounds or swellings mark pauses or still-stands in the retreat of glacier when they probably formed under lacustrine conditions against or within the walls of stagnant ice, while the elongated ridges accumulated when the glacier was retreating. Eskers which are very close to the terminal moraine have a fan-like appearance and are referred to as *esker fans*.

According to another view of esker formation the esker deposits were laid down by sub-glacial or englacial meltwaters as they emerge from the glacier in lakes or ponded water formed by the ice blocking the normal draingae channels. It has been suggested that *eskers* will form in certain conditions by the annual addition of small mound of sediment as the icefront retreated. This leads to the formation of beaded esker type. The summer period of rapid water flow and adundant material adds each bead, while the winter is represented by the intervening gap. Eskers of this type are rather rare. The deposition takes place either within the channel with the ice forming the two sides or in a lacustrine environment if there is ponding of water in the proglacial area. In either case a ridge of fluvioglacial sediments will be produced when the ice melts away. The concept of esker formation in the proglacial marginal environment was suggested by DeGeer (1897) when he envisaged a series of fans being extended into a large ridge as the ice front retreated allowing growth of the deposit as narrow ridges in the sub-glacial tunnels leading to the lakes.

It appears from the above discussion that eskers are polygenetic in origin and may be produced in a variety of ways by one or more of the following mechanisms :- (a) subglacial deposition in tunnels, (b) deposition in englacial tunnels or supraglacial channels and subsequent lowering on to the sub-ice surface, (c) deposition at the ice margin to produce a fan or a delta in standing water which becomes elongated as the icemargin retreats, (d) deposition in ice-walled re-entrants at the ice margin, or (e) ridges of fluvioglacial deposits may be produced by the wastage of buried ice from beneath a spread of sand and gravel.[1]

The necessity for the ice to be stagnant in order that true eskers may be formed has restricted their occurrence mainly to areas covered by large ice-caps rather than by valley glaciers, especially by temperate ice-caps where the ice was at the pressure melting point and much meltwater was available.

Kames

The term kame (*cam or kaim*) is of Scottish origin and means either crooked and winding or alternatively a steepsided mound. Both kames and eskers have a common relationship to stagnant ice and both are normaly composed of sand and gravel, but a kame differs from esker in being a mound rather than a ridge. A kame can be defined as (after C.D. Holmes, 1947) "a mound composed chiefly of sand and gravel whose form has resulted from original deposition modified by any slumping incident to later melting of glacial ice

1. Price, R.J.— *Glacial and Fluvioglacial Landforms*, Oliver & Boyd, Edinburgh 1973, pp. 159-60.

against or upon which the deposit accumulated. " Normally kames show a disorganized pattern and are often found as isolated mounds. Though freqeuently associated with terminal moraines, they may be found scattered through the ground moraine area in areas of moderate relief and irregular topography. When numerous mounds are found as a closely associated group, they are referred to as a *kame-complex* or *kame-field* or *kame group*. Another term which has been used to refer to a numebr of mounds separated by depressions in fluvioglacial deposits is *kame and kettle topography*, the depressions or kettle holes representing the location of glacier ice upon or against which the sediments accumulated.

Various hypotheses have been put forward to explain the formation of kames. According to one view kames are formed where a meltwater stream, carrying a heavy load of debris, debouches into a proglacial lake. The material is depositied in a deltaic form, and the side of the feature close to the ice later collapses to form a steep ice-contact slope on the proximal side. Where several sub-glacial streams enter a lake close together the deltaic deposit may be large or fan-shaped aligned parallel to the ice front. Where a delta-form rather than a more generalised mound is produced the term *delta-kame* is applied. Locally such deposits may be so extensive that a terminal moraine may consist entirely of kames in which case it is known as *kame moraine*. But acually moraines differ from kames in that while the moraine consists of unassorted rock debris, kame is formed of sorted sand and gravel deposited by meltwater.

To explain the occurrence of kames as isolated mounds J.H. Cook (1946) put forward what is known as the perforation hypothesis. He thought that the masses of sand and gravel forming the kame collected in pools or cavities on the surface of stagnant ice. As the pool warmed, it gradually melted its way down to the ground beneath. The debris that had collected in the pool then became the kame mound as the surrounding ice slowly melted away. Such mounds could be produced by accumulation in cavities at the ice-margin. Some of these deposits would have lacustrine characteristics if they accumulated in pools of water in the cavities under, in or on the ice.

Another hypothesis put forward by Holmes (1947) suggests that kames form where heavily laden meltstreams flow into small ponds on or within a stagnant ice-mass. At the point of inflow the stream would deposit some of its debris to form a small deltaic deposit. This would form an isolated mound when the surrounding ice melted away. Glacial meltwater streams often develop moulins, and as these deepen, the debris accumulating in them will be let down on to the sub-surface ice.

Apart from the kames formed at or near the snout of an ice-mass, another type of kame is formed along the valley-side margin of glacier. This is known as *kame terrace*. It is formed by deposition from meltwater flowing between the ice-mass and a valley wall. Kame-terraces are especially well developed in regions of valley glaciers but they are not unknown in regions of continental glaciation. Since deposition occurs against the ice margin, when the ice wastes away slumping occurs and a steep irregular topped slope develops to form the scarp face of the terrace. These terraces slope gently down-valley and often have a slight tranverse slope away from the hillside.

Both kames and eskers have a rather similar geographical distribution and are found in areas where the ice-sheets stagnated, melting away slowly and in such a way that much fairly coarse debris was available. Kames and eskers are found in large numbers in central Ireland, Scotland, lower parts of Sweden and Finland and in similar areas relative to the ice sheet in parts of Canada.

Proglacial Fluvioglacial Deposits (Valley-train and Sandur)

The large volume of meltwater that is dischargeed by the melting glaciers and ice sheets from their snouts is usually heavily laden with coarse debris as well as fine silt. These sediments are transported and deposited by the meltwater streams beyond the limit of the ice cover and the end moraines. There are two major proglacial landform types that result from this deposition, depending on the nature of local relief. If the glacier is confined between valley walls, the deposited material is similarly confined in a relatively narrow zone and the resulting deposit is generally known as a *valley train or valley-sandur*. Valley trains develop between valley walls and usually have only one main channel that branches out to create a network of smaller channels. The slopes of the valley-sandur can be very steep or gentle. Where, however, there is a wide ice margin terminating on a broad lowland area the melt streams can spread widely and build up an extensive surface. This is called an *outwash plain* or by the Icelandic name of *Sandur* (plural sandar). Sometimes the term plain-sandar is used to distinguish them from the valley sandar. Valley sandar or valley trains are associated essentially with valley glacier and the outwash plains or plain sandar with ice sheets. Valley trains owe their characteristics to the glaical meltwater streams that form them and the fact that streams are confined by valley walls. Most valleys (e.g. the great Alpine valleys) that have been glaciated in the past show some evidence of valley train development. On the other hand, plain sandar have few constraints on their lateral extent and develop as a result of aggradation of a network of braided streams that produce fan-shaped features which may join together to produce an extensive plain. The debris provided by the ice is very poorly sorted and consists of a wide range of size of material. Normally the coarser material is deposited near the ice margin and the finer sediment is carried further down. It is this material that gives rise to the Icelandic name sandur which means sandy area. Plain sandar accumulate both as channel deposits and over-bank flow deposits. Because of their large load, the meltwater streams aggrade their valleys and follow braided courses that are continually changing in pattern. Another characteristic of the outwash streams is their variable flow. The seasonal variation is particularly important. In summer because of ablation the flow increases markedly for a few months and becomes low during the winter. Some glacial rivers are also liable to more violent and short-lived variation in the discharge rate. These result from glacier bursts or *Jokulhlaups* as they are called in Iceland. These sudden outbursts are caused by the rapid drainage of lakes dammed up by the ice or of englacial or subglacial waterbodies. The largest outwash plain in Iceland, Skeioarar sandur is particularly liable to experience *Jakulhlaups* owing to the periodic drainage of the subglacial lake of Grimsvotn. In addition to the large volume of water suddenly discharged on to the sandur, a great deal of debris is also carried down by the water. When these vigorous flood streams emerge, they at first erode deep, wide channels. Very soon their heavy load causes the rivers to spread out into the typical braid pattern, the accompanying deposition causing rapid aggradation of the sandur. Due to the rapid fall in the discharge the river suddenly loses its transporting power and much of its load is then deposited. The sudden and wide variations in discharge thus help to explain the aggradation that goes on the sandur.

The level of the sandur or outwash plain is commonly linked with the position of the end moraine, to which the outwash is graded. The aggradation of outwash material, therefore, is likely to take place most effectively when the ice front is advancing and then remains stationary for a terminal moraine to form. Most of the fluvioglacial outwash probably accumulates during the advancing stages when the erosion capacity of the glacier is also greater so that the meltwater streams carry a larger load. The relationship between terminal moraines and outwash spreads is clearly seen in the valleys on the eastern side of the southern Alps of Newzealand. Another important area where outwash fans are associated with terminal moraines is in the northern Alpine foothills.

Outwash plains or plain sandar are very common in Iceland and occur as fossil forms in Newzealand, Poland, North Germany, the United States and Canada. There are extensive outwash plains in North Germany where they form wide sandy heaths. The material gradually changes from gravel near the moraines to sand and then to finer clay deposits southward. Large and thick sandar are also found to the east of the Baltic. There are very extensive outwash deposits in N. America also in the region extending from New England in the east to the Dakotas and Manitoba in the west.

Outwash plains sometimes have a hummocky surface resulting from the development of kettle holes. Near the glacier snout, lumps of ice may be broken off and left stranded on the sandur as the flood subsides, eventually melting out to form kettle holes. The kettle holes are often found within the large channels of powerful streams and result from the melting of ice buried under fluvioglacial deposits. This modified sandur surface with kettle holes has been called *kettled sanduar or pitted outwash or pitted sandar*. In fact, kettled sandar are both proglacial and ice-contact deposits. They are proglacial in that they are proximal parts of sandar that are essentially products of melt-water stream deposition, and they are ice-contact deposits in that the fluvioglacial deposits are either laid down on top of ice or contain large blocks of ice which subsequently melt. There has been extensive development of proglacial deposition including the kettled sandar in Iceland in the form of great plains of sand and gravel, and consequently there has been wide acceptance of the Icelandic terms for these features.

Pleistocene Ice Age and its Effects

There are evidences of major ice ages in the pre-Cambrian, Permo-Carbonifoerous and Cenozoic periods of the earth's geological history. The evidence about the pre-Cambrian glaciations is very fragmentary but the approximate world extent of the permo-carboniferous glaciation is known. On the whole, however, our knowledge about these two earlier glaciations is extremely limited, and we do not know how long glaciations lasted and whether there was a series of glacial and interglacial episodes. The permo-carboniferous glaciation was followed by a long warm period, through the Mesozoic era, when the temperature was somewhat higher than today. In the Cenozoic era (a term which embraces both the Tertiary and Quaternary periods) great continental ice sheets and glaciers covered large parts of the earth during the Pleistocene epoch. This is the most recent and the largest glaciation and is referred to as the Pleistocene Ice Age or the Great Ice Age. It is supposed to have begun about 1.8 million years ago and has continued till about 10,000 years ago. Recent investigations have, however shown that substantial regions experienced glacial morphogenetic conditions well before the Pleistocene, and we can no longer think of a sudden onset of the Pleistocene Ice Age. The Pleistocene ice at its maximum extent coverd an area of nearly 45 million km^2, more than three times the present ice cover of the world and included large parts of northern and central N. America, Greenland, northern and central Europe, Siberia and the mountain heart of Asia, southern part of S. America and the Antarctica i.e. roughtly the areas lying north of 40^0 N and south of 40^0 S. It was not one long and continuous period of glaciation but during this period there were repeated advance and retreat of the ice sheet following climatic fluctuations.

The period during which there is an advance of the ice-cover is called the glacial period or *glacial stage*. When there is retreat of the ice-cover as a result of increase of temperature between two glacial stages, the intervening temperate period is called the interglacial period or the *interglacial stage*. The glacial stage itself may consist of short periods of ice advance, called *stadials* separated by periods of retreat but colder than an interglacial, called *interstadials*. In the opinion of some, the present period which started some 10,000 years ago, is an interglacial. It is called the Holocene or Recent or post-glacial. Pleistocene and the Holocene together constitute the Quaternary period. Consequently the Pleistocene ice age together with the Holocene interglacial is sometimes referred to as Quaternary Ice Age. It can not, however, be stated with any certainty whether the present period should be considered to be an interglacial stage. In fact it is almost impossible to say wheter the climate will become warmer or colder in the coming years. In either case the consequences for man would appear to be disastrous. At the present time all the ice-sheets and glaciers are in a stage of retreat. If this be taken as an indication of warming climates, then the melting of all the ice will cause the sea level to rise by at least 35 metres even after isostatic adjustment, resulting in the worldwide submergence of the great towns and ports along the coast. If, on the other hand, the climate cools further and there is advance of the ice-sheet, many areas which are now densely populated will become unsuitable for human habitation.

The concept of the sequence of glacial and interglacial stages in the Pleistocene Ice Age was first put forward in a systematic and forceful manner by the famours Scottish geologist James Geikie in the 1894 edition of his book, "*The Great Ice Age*". His view did not evoke much interest but when Penck and Bruckner published in 1909 the resutls of their ivestigations in the Alps in this regard, they provided a powerful confirmation. Penck and Bruckner presented evidence for four periods of advance and four periods of retreat of the ice cover. The four glacial periods are called Gunz, Mindel, Riss and Wurm. In between these four glacial periods there are three interglacial stages called G/M, M/R and R/W repsectively. This glacial sequence is broadly confirmed by the studies of terminal moraines and other glacial deposits in Britain, Northern Europe and North America and leads to the conclusion that the glacial and interglacial stages were broadly contemporaneous in the high latitudes throughout the world. In North America the four glacial stages are called Nebraskan, Kansan, Illinoian and Wisconsin repectively, and the three interglacial between them are known as Attonian, Yarmouth and Sangamon respectively. Later on, in the Alps a glacial stage earlier than Gunz has also been found. This is known as Donau or Danube and goes back to early Pleistocene.

Several of the interglacial stages were probably longer than the glacial periods, but the duration of the Mindel-Riss interglacial stage is thought to have been the longest. It lasted at least 300,000 years. It is, therefore, commonly referred to as the Great Interglacial dividing the Ice Age into two major divisions (Fig. 172) — the period of older glaciations and the period of new glaciations. In each of these two major glacial periods, there were interglacial periods when the ice-cover retreated but did not disappear. But during the

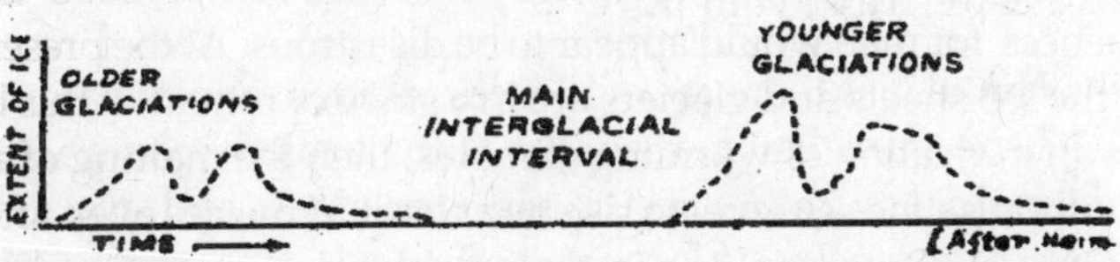

Fig. 171—The Great Interglacial Interval dividing the Pleistocene Ice Age into two major divisions—the period of older glaciations and the period of new glaciations.

M/R interglacial, the ice had disappeared and the temperature was 2^0 C higher than at present. Evidence for this is found in both Europe and N. America, and in the latter the duration of the Yarmouth interglacial is very long. The division of the Ice Age into older and younger is equally valid for both N. America and Europe. In both the areas there is a marked difference between the older drift and the new drift.

While Donau and Gunz glaciations are lower Pleistocene in age, Mindel is middle Pleistocene, and Riss and Wurm belong to the upper Pleistocene. It appears that the maximum spread of the ice took place towards the end ot the middle Pleistocene and the beginning of the upper Pleistocene, and Riss (or Saale) glaciation in Europe and the Illinoian in N. America were the most extensive of all galciations. It began about 0.5 million years ago and consisted of several advances separated by local retreats. There are numerous evidences to show that the last interglacial, called the Sangamon in N. America and Riss/Wurm or Eemian in Europe was somewhat warmer than the present. This interval (140,000 to 120,000 years ago) was marked by extension of deciduous forests into what is now conifer zone, a sea level a few metres higher than present and the widespread development of coral reefs, and less glacier volume than today. Around 120,000 years ago the earth's climate cooled sharply and the last major glaciation began. It is called Wisconsin in N. America and the Wurm or Weichsel in Europe and lasted till 10,000 years ago. It consisted of short periods of ice advance or stadials, separated by periods of retreat but colder than an interglacial, called interstadials. Each succeeding stadial was more severe than the earlier, and the maximum advance of the ice took place only 15,000 to 18,000 years ago. Since the ice of the last glaciation retreated only about 10,000 years ago, the erosional as well as depositional features produced by it are well preserved and are still quite fresh and youthful.

In the Himalayas there are evidences of extensive glaciation in the Pleistocene age. In all parts of the Himalayas we come across heaps of terminal mornines, ice-transported blocks and smoothed and striated hummocky surfaces and other evidences of ice action at elevations thousands of metres below the present level of descent of the glaciers The upper Siwalik abounds in conglomerates the top strata being of glacial origin. This indicates that an ice age prevailed in this region. It is difficult to estimate the extent of land actually covered by ice, but studies of flora and fauna indicate that the ice age affected the climate of large areas and the climate of South India and Sri Lanka became temperate enough to account for the migration of the Himalayan fauna and flora. It has been suggested that man must have come into existence before the Ice Age since primitve stone implements have been found in the glacial moraines of both upper Siwalik and upper Karewa formations.

In the Himalayas too, four glacial periods with intervening interglacial periods were initially identified from the evidence available in the Karewa lake deposits of the Kashmir basin and the well developed Pleistocene terraces of the Kargil Basin, 120 km north-east of the Kashmir basin. But further studies have cast doubt on the existence of the oldest of the four glaciation periods as the evidence for it has been found to be weak and inconclusive. The glacial influence is very pronounced in the upper Karewas but rather uncertain in the lower Karewas. There was a rapid and remarkable uplift of more than 2,000 m after the deposition of the Lower Karewas during the post-Siwalik orogeny of the late Pleistocene period. Prior to this uplift the Himalayas probably had not reached the necessary height for an earlier glaciation. The main glacial stage came soon after this uplift and is followed by a clear interglacial period. Subsequent to this interglacial, two more glacial stages are generally reported, interrupted by an interglacial or possibly only by an interstadial, as the third and fourth glacial periods were rather small. The direct evidence for these earlier glaciations consists of old morainic ridges both lateral and terminal, U-shaped valleys and disproportionately small streams flowing through wide and gently inclined glacial valleys. In the Gangotri area the lateral moraines of the different glacial phases are found along the valley of the Bhagirathi river, the oldest lateral moraine being the highest and standing more than 200 m above the present valley bottom and considerbly dissected. In the Everest area prominent terminal moraine heaps occur much below the terminus of Khumbu glacier, and lateral moraines also occur at various levels along the old glacial valley now occupied by the Imja Khola stream. In the Liddar valley of Kashmir too three phases of earlier glaciation are identifiable, each being marked by glacial deposits, striations and epigenetic gorges.

Physiographic Effects of the Pleistocene Ice Age

The Pleistocene Ice Age had far-reaching geomorphological consequences from several points of view. These effects were by no means limited to the lands covered by ice and the adjacent periglacial zone but even distant lands were afected in several ways by radical climatic changes which affected the earth's water balance. The sea-level rose and fell several times with the waning and waxing of the great ice sheets. The discharge of rivers fluctuated according to change in precipitation and temperature, and lakes expanded and shrank according to the changing hydrological conditions. There were thus world-wide changes in land and sea levels as well as in fluvial activity.

We may divide the lands actually coverd by ice-sheet and valley glaciers into three parts. (1) There was, firstly, the area which was right in the centre of the ice-sheet, from where the ice moved out in different directions. Here the effect of the ice cover was largely proctective and consequently we do not come across evidence of either glacial erosion or glacial deposition. (2) Beyond this lay a middle zone where the landforms carved by glacial erosion are dominant, although occasionally depositional features are also found. (3) Next was the outermost zone of the ice sheet where we find predominance of glacial deposits. Besides the above three zones, and close to but outside the zone occupied by icesheet and glaciers, lay the periglacial zone where the climate resembled the present tundra and physical weathering by frost action and solifluction were extremely important. This was covered by glacio-fluvial and glacio-lacustrine deposits.

The most distinctive among the features produced by ice-sheet erosion is the *ice-scoured plain*. The Canadian shield and parts of Scandinavia and Finalnd are good examples of this. The term plain has been used here in a very comprehensive sense. The ice-scoured plains are, in fact, an assemblage of different landforms which owe their origin essentially to erosion by ice. They have a typically smooth, grooved and striated surface, diversified by rock basins, roches moutonnees and glacial drift. Where the ice-sheet covered a mountainous area, such as the eastern U.S.A., we do not find ice-scoured plains, instead the old surface has been smoothened and streamlined by ice erosion. In both types of landscapes., however, the old soil cover has been swept away by ice and extensive areas are found to be unfit for cultivation.

In areas of glacial deposition the most characteristic feature is the *till plain* diversified by parallel linear ridges or mounds built of terminal morainic deposits, left behind by the retreating ice sheet. Sometimes lakes and marshes have been formed in between the parallel ridges. In addition to the terminal morainal ridges, drumlins, eskers, kames and outwash plains are found in extensive areas. In fact glacial deposits cover about one-third of Europe and nearly one-quarter of North America. In the marginal areas of the ice-sheets in Europe and North America glaciation has brought about significant changes in the drainage, and several lakes have been formed. The Great Lakes of N. America owe their origin to the repeated advance and retreat of the Pleistocence ice-sheet. In fact in the glaciated areas of northern U.S.A. (Minnesota, Wisconsin, Michigan), Canada, Sweden and Finalnd innumerable lakes are found which were formed under varying conditions.

In the coastal areas, when the sea-level was low, the glaciers had cut deep valleys which were later submerged when the sea-level rose. This has resulted in the formation of fiord coasts in Norway, Scotland and other areas.

In the mountainous areas of Europe and N. America which were influenced by valley glaciation, cirques, pyramidal peaks, U-shaped valleys, hanging valleys and other erosional features present extremely attractive natural scenery. In the Himalayas where glaciers were of the mountain type, U-shaped valleys formed by the earlier glaciation together with lateral and terminal moraine ridges are quite prominent especially in the middle Himalayas which contain no glaciers today. Since the retreat of ice has taken place only about ten thousand years ago, the landforms produced by glacial erosion and deposition wear an extremely fresh and youthful appearance. In fact, it is impossible to understand the present landscape of the earth without reference to the geomorphic effects of the Pleistocene glaciation.

Apart from the direct geomorphological effects, there are three other kinds of influences which must be borne in mimd as these are of major worldwide geomorphic significance:

(1) In the Ice Age there was equatorward migration of the world's major climatic zones. This had two effects. Firstly, very probably the climatic zones lying between the equator and the ice-covered areas shrank latitudinally and some climatic types were altogether eliminated in the process. Secondly, large parts of the present day hot deserts experienced not only lower temperatures but increased rainfall and reduced evaporation which is shown by the extensive presence of features produced by fluvial action and of pluvial lakes in the hot deserts. In fact, the climatic changes during the Great Ice Age were so radical and world-wide that almost all the areas of the earth must have experienced morphogenetic changes as the river adjusted their discharge and load in response to the changing climatic conditions and the changing vegetation cover.

(2) Secondly, as a result of the extension of ice sheets during the glacial period and the liberation of meltwater during the interglacial period, there were several world-wide glacio-eustatic sea level changes. These not only brought about changes in the coastal features, but over large parts of the earth's surface the erosion rates and processes were profoundly affected, because sea-level is the real base level of erosion[1].

(3) Thirdly, the enormous burden of ice that accumulated in some areas led to subsidence of the crust, and later when deglaciation occurred, the crust was unloaded and gradually recovered its

1. The term 'eustatic' was introduced by Edward Suess in the last 19th centruy to refer to world-wide sea-level changes, caused by several possible factors, not only those related to ice ages.

former level. This is known as glacio-isostatic crustal deformation. It is important to remember that the ice masses were not simply loads added to or removed from particular areas but simultaneously equivalent masses of water were taken from or returned to the oceans, resulting in the transfer of sub-crustal matter from glaciated areas to extensive oceanic areas and vice-versa.

Fluctuations in Sea Level

During the glacial period large quantities of water are withdrawn from the ocean and accumulate on land-masses in the form of snow and ice. This results in a drop in sea-level. Conversely, during the interglacial period when there is increase in world temperatures, the ice melts and most of the meltwater finds its way to the sea and causes a world-wide rise in sea-level. As there were four glacial periods of unequal intensity in the Pleistocene Ice Age, the fall in sea-level in the different glacial periods was not the same. During the great interglacial, the climate was warmer than today, and in the present time we still find the existence of ice-sheets in some ares. Consequently, it is highly probable that the present sea-level is lower than the sea level during the great interglacial.

Several estimates have been made of the difference in the low sea-level of the glacial periods and the high sea-level of the interglacial periods, and the several lines of evidence now suggest a lowering of 100 to 140 m in each main glacial period. The evidence comes from sub-merged marine terraces and wave-cut platforms, submerged littoral or terrestrial sediments, and buried channels of coastal rivers. The marine terraces and the wave-cut platforms were formed when the sea level was high in the interglacial period. Later as a result of fall in sea-level these features are now found about 100 metres above the present sea level. When the sea-level was low during the glacial period, the rivers met the sea near the coast, but during the interglacials when the sea-level rose, the river valleys along the coast were sumerged in the sea.. Such drowned river valleys when found in dissected highlands are called *rias* and when occuring in relatively lowlying areas are called estuaries. In support of low sea-level in the last glaciation, it is pointed out that buried channels of river in western Europe often grade down to levels approaching -100 m, and many submarine terraces fringe the coasts at about this level.

In the warmer seas the development of coral reefs and atolls was made possible by fluctuations in sea-level. In 1910 R.A. Daly expounded his theory of glacial control in the evolution of coral reefs, involving the upgrowth of corals from platforms cut during low sea-levels. Many oceanic atolls possess basal platforms at about — 100 m

Several islands in the world owe their origin to the rise in sea level in the post-glacial period. The separation of Great Britain from the mainland of Europe and the formation of the Isle of Wight have happened in this manner.

Apart from the above facts, special significance attaches to fluctuatuions in sea level on account of the fact that sea-level is the ultimate base level of erosion, and the fluctuation in sea-level means fluctuation in the base-level of erosion. Since it is not always possible to determine whether the sea has risen or the land has subsided, when the land rises relative to the sea-level (i.e. the sea-level drops) it is referred to as negative movement of base level. On the contrary, when the land falls relative to the sea level (i.e. the sea level rises), it is called positive movement of base level. From the point of view of its geomorphic effects on the land surface, negative movement is far more significant than the positive movement of base level., because when the sea level rises (posititve movement), the earlier coastal features are submerged, river valleys are drowned and there is increase in aggradation in the lower reaches of streams. But the effects of a fall in sea-level (negative movement) are far more profound and far-reaching. It results in the formation of raised beaches or raised strand lines and marine terraces, and more importantly it causes rejuvenation of rivers which start eroding with renewed vigour, resulting in the formation of knick-points and incised meanders along the longitudinal profile of rivers and formation of river terraces along the cross-profile.

Subsidence and Uplift of Land in Glaciated Areas

The idea that isostatic deformation of the earth's crust may result from the weight of thick ice sheets was first advanced by T.F. Jamieson in 1863. This has been found to be correct by later workers as evidences of subsidence and later uplift have been found in Scandinavia, Finland and N.America. It is also known that glacial loading today keeps nearly one-third of Antarctica below sea-level, and in one area where the ice is 1500 m thick above sea-level, the bed rock surface is 2,000 m below sea level. Similarly in Greenland the bed-rock surface in the centre has been depressed under the weight of ice and has become basin-shaped.

After deglaciation, adjustment to unloading is slow and the land rises very slowly. In the marginal areas of ice sheets where the subsidence is least, uplift takes place first. In the central parts of the ice-sheets the uplift or "crustal rebound" as it is called, takes a much longer time, as here the ice sheet is present for a much longer time and the subsidence is also greater. It seems that the crustal uplift is complete over a span of 15,000 to 20, 000 years. The uplift is dome-shaped i.e. more in the centre and less on the periphery. Evidences of such isostatic depression and recoverry are found in the elevated strand lines and the deformed shorelines. Studies of deformed shorelines have made it possible to calculate rates of uplift consequent upon ice unloading, and to delineate the areas affected by glacio-isostatic deformation. It has been found that the boundaries of these areas are roughtly parallel to the limits of the last glaciation. The pattern of uplift is shown by *isobases* or lines joining points of equal uplift on the affected strand lines. Examples of such uplift are to be found in eastern Canada, the Great Lakes area as well as in the Baltic sea region of Scandinavia.

The mechanics of subsidence and uplift are, however, incompletely understood. It has been suggested that the maximum subsidence beneath a large ice-sheet should be 25-30% of the maximum ice thickness, but correspondingly less beneath small ice-sheet assuming that the density of sub-crustal matter (3.0 to 3.3) is about 3 or 4 times that of ice. Andrews (1970) has identified three stages in the process of glacio-isostatic revovery. The first is the phase of restrained rebound begining at the time when ice thinning begins, the second stage is that of post-glacial uplift, extending from the time when the ice had almost disappeared to the present, and the third is the residual component, comprising the amount of recovery still to take place before final equilibrium is established. At present. uplift of 9 mm a year is in progress in the north of the Gulf of Bothnia to 4 mm a year near Stockholm. The maximum rate of uplift in eastern N. America at present is about 13 mm a year over southern Hudson Bay. In the Great Lakes, water is rising on the southern shores and receding from the northern ones.

Questions

1. Describe the processes (mechanics) of glacial erosion, transportation and deposition.

2. Discuss the landforms resulting from *either* glacial erosion *or* glacial deposition.

3. Describe and account for the typical landforms of *either* glaciated mountains *or* glaciated lowlands. Give examples.

4. "In glaciated lowlands the most important features concern the glacial deposits or "drifts" which not only obscure the pre-existing topography but themselves build characteristic landforms". Discuss.

5. Give a reasoned account of the Pleistocene Ice Age and its geomorphic effects.

6. Write an explanatory note on *one* of the following :—

   (a) Glacio-fluvial deposits, (b) Longitudinal profile of glaciated valleys, (c) Fjords.

7. Discuss the formation of *any two* of the following :-

   (a) Sandur, (b) Drumlins, (c) Hanging Valleys, (d) Cirque, (d) Roches Moutonnees.

8. Discuss the formation and characteristics of Eskers and Kames.

Chapter **26**

Periglacial Landforms

Periglacial Environment

Periglacial literally means near-glacial or peripheral to glacial. The term periglacial was first used to refer to the region in Europe that was peripheral to the margins of the Pleistocene glaciers where there was no ice cover but frost action was particularly active. The periglacial environment today is to be found principally in the Tundra regions including the cold, dry continental interiors of Asia (including Russia) and North America and the maritime boreal coasts of Iceland, Scandinavia, Canada, and Alaska as well as on the low-latitude mountain tops above the timber-line where the climate is very similar to the tundra. In the Pleistocene Ice Age the tundra climate was much more extensive than today. Consequently the periglacial landforms are not limited to the regions having tundra climate but are also found outside this climatic zone in the periglacial regions of pleistocene times. Ideally, the periglacial environment is a zone of intense freezing and seasonal thawing so that while the sub-soil is perennially frozen, the top soil experiences seasonal thawing.

The tundra climate has the following characteristics which present particularly favourable conditions for the periglacial processes:—

1. The winter is long and severely cold and the temperature rarely exceeds 32° F. There is frost almost everyday for ten months in the year, but the snow cover is thin and discontinuous. Consequently, frost penetrates the soil to a depth of 2 to 15 ft.

2. The summer is short, and though the surface temperature may rise to 75° F (24° C) or more, the temperature in the sub-soil 4 inches below the surface is usually below 40° F. At such a low temperature no chemical weathering or hydrolysis is possible beyond the

solution of carbonates. Because of the short summer, the winter frost does not thaw completely, and most of the sub-soil remains permanently frozen and is referred to as *permafrost*. Above the permanently frozen sub-soil is a thin layer of soil in which the frost melts in summer, this is called the *seasonal thaw zone*. The depth of the seasonal thaw zone varies from 1 to 2 inches in swampy soils to about 15 ft. in sandy soils. Below the thaw zone the thickness of the permafrost layer may be anything from 10 ft. to thousands of feet.

3. On account of the permanently frozen sub-soil the downward movement of water in the soil above is impeded in summer. Consequently the water-table is quite high and the soil remains water-logged, and swamps and lakes are found in large numbers. The roots of plants and trees can not penetrate deep into the soil.

4. On account of the deep frost mechanical weathering is active in the sub-soil, and the rocks on the surface are also liable to continual disintegration as a result of frost action. Early in winter when the freezing begins, the thaw zone is subjected to pressures of various kinds which result in the mixing, churning and heaving of the regolith, and through these processes some distinctive landforms are built. These processes are described at a subsequent stage in this chapter.

Permafrost and the Active Layer

Although the literal meaning of permafrost is permanently frozen land, but in its extended sense the permafrost also includes the seasonal thaw zone just above it which melts in summer and starts freezing again with the onset of autumn. This upper layer of seasonal thawing is called the *active layer* because as a result of the seasonal changes in temperature various processes are active in this layer (Fig. 173). When the active layer which has thawed

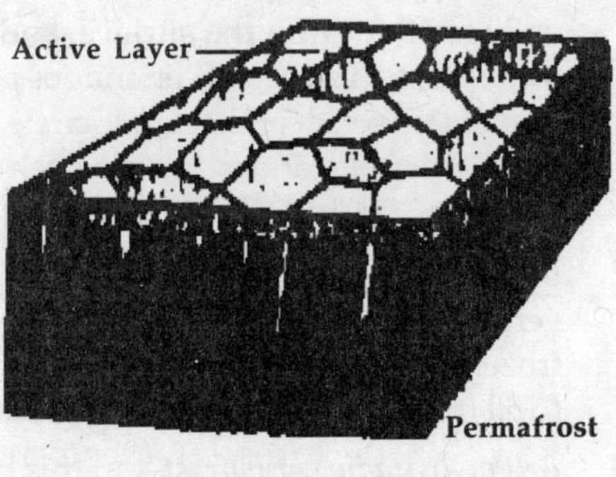

Fig. 172—Permafrost and the Active Layer.

in summer, begins to freeze in autumn, the process of freezing starts from the top and extends downwards with the fall of temperature until it reaches the level of the permafrost below. Similarly the process of summer thawing in the active layer also begins from the top and extends downwards. In other words, in the beginning of autumn, in between the top frozen part of the active layer and the permafrost is a thawed layer which goes on contracting with the fall in temperature. There is considerable difference in the depth of the active layer in different places, and it may vary from a few inches to as much as 15 ft. This depth depends upon temperature, the amount of insolation received, conductivity of the soil and insulation of heat by snow cover or vegetation. The active layer is extremely important in the formation of periglacial landforms, as all the landform buiding processes which owe their origin to temperature changes, are essentially confined to this layer. In the permafrost layer the subsoil as well as the rocks are found in a perennially frozen state.

The regions of the world in which permafrost is found have been shown in Fig. 174. This map also shows the limit of permafrost in the Pleistocene times. In its extended sense, we may divide the permafrost regions of the world into three classes:-

1. Belt of continuous permafrost,
2. Belt of discontinuous permafrost,
3. Belt of sporadic permafrost.

1. *Belt of Continuous Permafrost*— In this belt the sub-soil never thaws. The temperature remains lower than 23^0 F up to depth of 30 to 50 ft. from the surface, and may be as low as 10^0 F in some places. The belt of continuous permafrost is found in North Canada till south of the Arctic circle (about 1,500 miles south of the North pole) and in Siberia even further south to a distance of about 2,000 miles from the North pole.
2. *Belt of Discontinuous permafrost*— In this belt the sub-soil, though generally frozen also contains parts which are not permanently frozen. The temperature ranges from 30^0 to 23^0 F to a depth of 30 to 50 ft. This belt covers a large area in Canada and Siberia.
3. *Belt of Sporadic permafrost*—In this belt a great part of the sub-soil remains frost-free, but permanently frozen sub-soil is found in isolated patches. The temperature is only slightly below

freezing point up to a depth of 30 to 50 ft. This belt is quite extensive in central Asia and Canada.

The conditions favourable for permafrost are long and severe winter, short and relatively mild summer, and low rainfall. In other words, the maximum development of permafrost takes place in severely cold and dry climates. But the ground should be free from snow cover. There is no permafrost in regions which are covered by ice or ice-caps. The most favourable conditions are found in regions where the average temperature ranges between 24^0 to 30^0 F. The permafrost is receding from some areas and advancing in some other areas at the present time. But broadly speaking, it would be correct to say that the permafrost limit is shrinking, as the winter is becoming warmer in the middle latitudes during the last 70-80 years. In the opinion of Soviet scientists, the southern limit of the permafrost in Siberia has shifted towards the north in the last 100 years. In the

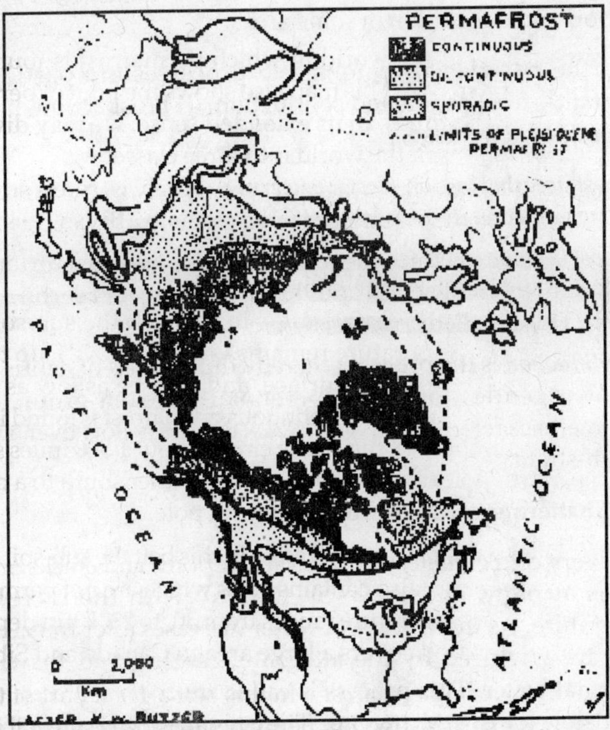

Fig. 173—The Permafrost Regions of the World.

glacial periods of the Great Ice Age the southern limit of the permafrost extended down to much lower latitudes.

Permafrost now underlies about 20% of the world's land area. The maximum extent of permafrost is to be found today in Russia including Siberia and Canada. It is estimated that about 50 percent of Russia and Canada is covered by permafrost, its area in the former exceeding 10 million sq. km and in the latter approximating 5.7 million sq. km. In addition to these countries, permafrost is also found in about 85% of Alaska and Greenland and parts of Scandinavia, China and Mongolia.

Periglacial Processes

The periglacial processes are dominated by the intensity and frequency of freezing and thawing and mass movement of the active layer over the permanently frozen subsoil. But fluvial erosion is not entirely absent and wind becomes a relatively important agent of transport and deposition in some areas.

In the periglacial regions two processes are particularly important—frost action and solifluction. (i) Frost action includes two kinds of activities :-

(a) *frost weathering or frost shattering* which is referred to as *congelifraction* in the periglacial terminology (Bryan, 1946), and

(b) *Frost heaving* in which the rock fragments are lifted up (and not broken) by frost action and there is mixing and sorting of the soil. This is called *cryoturbation*.

(ii) *Solifluction* is the process whereby fluid sheets of debris move down gentle slope over perennially frozen ground. It is sometimes referred to as *gelifluction* to specify flow over a frozen sub-strata.

Frost Shattering or Congelifraction

In very cold climates the water in the pores and cracks of rocks freezes at night and thaws in the day with the increase in temperature. As the volume of water increases after freezing, the rocks are fractured by the alternate freezing and thawing of interstitial water. This process is called *freeze-thaw process* and is responsible for frost shattering. The presence of cracks or joints is necessary for the process to be effective. The process is by no means limited to permafrost areas but is particularly active and vigorous in such areas. Where well-jointed rocks are found on a relatively flat or gently sloping surface, rocks disintegrate into big blocks or fragments by this process and the ground is covered with large, angular blocks. Such areas are called *block-fields* and are commonly found in the periglacial environment. But the ultimate product of frost shattering is the rock waste in which the size of the particles is very small. In fact, silt and angular rock fragments or rubble are found mixed together. By the process of alternate freezing and thawing great quantities of rock-debris or masses of rubble called *congelifractate* are formed through sheer mechanical weathering. Mechanical weathering by alternate freezing and thawing is an extremely important periglacial process in the rocky regions of the tundra.

Frost Heaving or Cryoturbation

As indicated earlier, by this process there is mixing and sorting of the rock debris and soil as a result of alternate freezing and thawing in the active layer, and the rock fragments are separated from the soil and lifted up to the surface. When the saturated layer

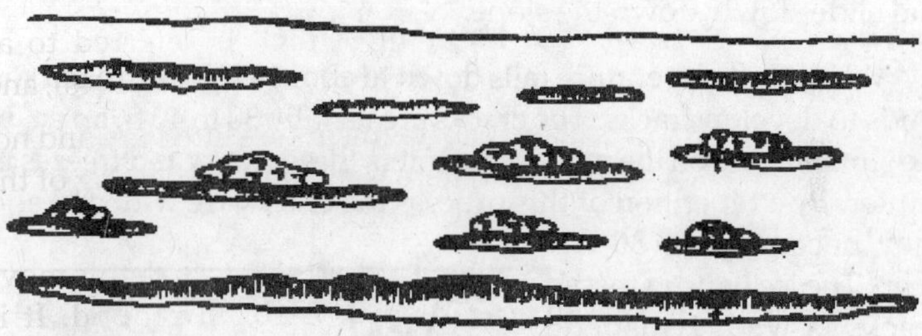

Fig.174—Segregation of ice in the soil by formation of ice lenses and layers. The ice lenses are especially formed below the rock fragments.

freezes, ice nuclei form in the soil, water present in the soil migrates towards these nuclei which grow as veins or lenses of ice and the adjacent sediments are dewatered. Thus there is segregation of ice in the soil with the freezing of the saturated layer. The actual distribution of ice within the layer is dependent upon the rate of freezing. When the freezing is rapid, the entire soil is full of small ice crystals. When the freezing proceeds slowly, ice is segregated in the form of lenses and layers, and in some places a continuous horizontal layer of ice is formed. It is on account of formation of the ice lenses or layers within the soil that rock fragments and stones are pushed upwards towards the surface. The ice nuclei first form below the stones because the thermal conductivity of rocks is more than that of soil. Consequently while the soil freezes slowly, there is segregation of ice below the rock fragments and the ice pushes the rock fragments upwards. Further, where the rock fragments are embedded vertically in the soil, the upper part freezes earlier than the lower part. So when the frozen soil expands, the rock fragment in which there is less expansion and which is embedded in the soil is subjected to lateral pressure and is pushed up. In summer when a sufficiently thick layer of the soil melts, the rock settles down again but simultaneously some sediments from the adjacent soil also migrates underneath the rock. Thus with every freeze and thaw the rock fragments are raised up and up. Because of the frost and thaw there is an up and down movement of the rock fragments in the soil. In a fluid medium of this kind the rock fragments tend to stand vertically. This explains why the long axis of the rock fragments which rise to the surface, is vertical. On reaching the surface the rock fragments fall down from their vertical position and slide slowly down the slope.

When the temperature falls down to about - 5^0 F the frozen soil tends to develop cracks. The cracks are usually 30 to 40 ft. apart. In the summer season the melt water enters these cracks and freezes in winter. By a repetition of this process the cracks are widened and they become 3 to 30 ft. deep. The ice lying in these cracks is wedge-shaped and tapers downwards. These are therefore called *ice-wedges* (Fig. 176). The fossil forms of ice-wedges have been found in Pleistocene gravels and the gravel beds are bent downwards in the old ice-wedges. The ground is raised up on both sides of the ice-wedge, and gravels and sand from the sides slide down and settle in the cracks when the ice melts in summer.

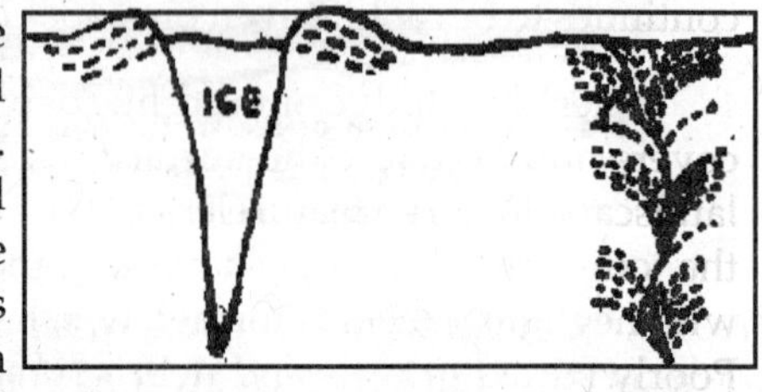

Fig.175—Ice Wedges. Fossil form of the ice-wedge has been shown on the right hand side.

Sometimes as a result of the formation of ice-lenses in the soil below the surface, there is swelling of the surface and small rounded mounds may be formed ranging in diameter from 60 to 70 ft. and varying in height from 2 to 20 ft. When the ice below melts in the summer, the ground slumps and the soil layers on the surface become bent downwards. This explains why in Britain the Pleistocene gravels which have passed through periglacial conditions, are found contorted in their upper layers.

Solifluction or Solifluxion

Solifluction is an extremely important process in the periglacial environment which is responsible for the formation of several distinctive landforms. When the active layer melts in summer, the melt water is not able to infiltrate downwards because of the permafrost layer lying underneath, and the debris in the upper layer becomes saturated and mobile. There is absence or dearth of natural vegetation in the tundra regions on account of the water-logged soil in summer coupled with the low summer temperature and the extremely cold winter. The roots of natural vegetation which may bind and hold together the saturated and mobile layer in summer are absent. Consequently in summer this water saturated debris tends to slide down easily even on comparatively gentle slopes. It is this process of mass movement of the water-mixed debris and soils as a result of gravity which is called *solifluction*. This term has been later modified to *gelifluction* to specify flow over a frozen substrata. This debris is so fluid that even a slope of 2^0 is enough to make it mobile. Where solifluction starts on a steep slope the debris continues to be mobile even on slopes of less than 2^0.

When this sludge or mud-like debris moves down the slope it covers and blankets the surface irregularities. This is why the landscape looks so featureless in the arctic regions. As a result of the joint action of frost shattering and solifluction, a landscape with negligible relief is formed which normally looks almost flat. Poorly sorted gravels and debris transported by solifluction are found in several parts of Britain and Europe and have been called by various names, such as "head" or 'coombe rock' and so on, but probably the most appropriate name for them is solifluction gravels. Their nature changes with the nature of the rocks but generally they are found in unbedded and unsorted condition and angular debris are mixed with silt and clay.

Other Erosion and Transport Processes

In addition to the above two fundamental processes, nivation, fluvial processes and the action of wind are also significant in the formation of some periglacial landforms. These are discussed at the appropriate place in the sequel in relation to the features produced by them. We may, however, note here that *nivation* refers to the erosive effect of immobile patches of snow that persist into the summer in the periglacial environment. These snow patches that persist in sheltered position have been found to produce nivation hollows after a few seasons. The snowpatch provides a localized source of water around its edges. Some of this water percolates through the downhill edge of the snowpatch and carries away particles of rock and soil from underneath. Freezing and thawing may also be intensified at the edges of the snowpatch. In this way a hollow starts building. Nivation hollows tend to be self-enlarging, as the hollow once formed catches more snow and shelters a larger snow patch in successive years. These hollows are usually shallow and wide as permafrost encourages widening rather than deepening of the hollows. Periglacial hill slopes are commonly lined with shallow nivation hollows especially on the shaded and downwind side of the ridge tops where snow accumulates and is sheltered from sunlight.

Periglacial Landforms

We may divide periglacial landforms into two broad classes—landforms produced by frost action and landforms produced by the joint action of frost and solifluction. (i) Frost shattering leads to the formation of block fields, talus slopes and frost soils. *Blockfields* develop in areas of well-jointed rocks over a relatively flat or gently sloping surface. Masses of large angular blocks and fragments or rocks lie strewn over the area and some of the larger blocks may even be found standing vertically. In *frost- soils* silt and rocky debris are found mixed together and as a result of frost heaving the coarse debris are sorted and lifted up to the surface of the soil and form a variety of patterns. (ii) The great majority of the periglacial landforms owe their origin to the joint action of frost and gelifluction (solifluction). The intense frost action, permafrost, lack of vegetation and brief and ineffective river action combine to produce a very distinctive landscape. The periglacial plains are waste lands of patterned ground, blister-like domes, seasonal bogs and shallow lakes, chocked river valleys, gelifluction lobes and sheets and rubble strewn hillslope. In mountainous regions block-fields, rock glaciers, massive taluses and a variety of terraced slopes shaped by nivation and gelifluction add other striking features to the periglacial landscape. Many of these features are depositional rather than erosional because of the intense frost weathering and mass wasting and ineffecient stream erosion.

Features associated with Ground Ice

A number of features, in themselves quite small and constituting what may be considered minor landforms are associated with the action of ground ice. On the plains, 'permafrost overlain by the active layer contorts, erupts, or collapses the land surface into a maze of polygonal patterns, bogs or shallow lakes, ice-cored domes, and pits'.[1] These features are described below.

Periglacial Involutions — Involution refers to contortions of bedding which are sometimes found in unconsolidated stratified deposits whereby one strata appears to have entered into the other strata. Because such features may be produced by a variety of reasons and are found even outside the periglacial regions, the involutions met with in the periglacial regions have been called *periglacial involutions*. Sometimes on account of involution the materials of one layer are so much injected into the materials of the other layers that their horizontal bedding is completely deformed. According to A. John (1956) there are three kinds of periglacial involutions— fold, pillar and amorphous structures. In fold there is downward movement of fine materials into underlying beds, while in pillar there is upward migration of fine material into overlying beds. In the third type, the original stratification becomes so much deformed that it cannot be distinguished.

Although the involutions associated with fossil ice-wedges are clearly periglacial in origin there is difference of opinion as to their origin. According to R.P. Sharp (1942) who has made detailed study of involutions of this kind in the region south-west of Chicago, there is expansion in the silt and clay layers in the process of ice segregation and they enter the neighbouring layers of unfrozen materials. But according to Washburn (1956), as freezing extends

1. Bloom, A.L. *Geomorphology*—A systematic analysis of late cenozoic landforms, Prentice-Hall of India, New Delhi, 1979, p. 362

from the upper to the lower layer, there is contraction and contortion of the soft and wet layer lying between the upper frozen layer and the permafrost and in this way there is both horizontal and vertical deformation in the stratification. This latter view has been more widely accepted. Examples of such structures are found in Central Europe which were first described by Troll in 1944. More recently R.B. Williams (1969) has found several examples of Pleistocene periglacial involutions in South England.

Earth Hummocks and Turf Hummocks— In the periglacial plain areas small mounds or hummocks are found in large numbers. They generally rise half a metre above the swampy or grass covered level surface, and their diameter is about 1.5 metres. They are generally clothed with tundra vegetation. Sometimes they are found close to each other and sometimes they are far apart, and are generally made of clay, silt or sand. These mounds are called *earth hummocks or turf hummocks or thufur*. According to Sharp (1942) who has studied this feature in Yukon, on account of expansion of the frozen layer, there is squeezing in the adjacent areas resulting in the formation of small knobs or mounds. According to Taber (1952) these mounds are the result of frost heaving resulting from segregation of ice-lenses. But Sigafuz and Hopkins (1954) agree with Sharp and have pointed out that the ground in between two mounds freezes first and there are definite evidences of lateral pressure so that tongues of silt layer enter into the mounds from the adjacent areas. Whatever the process of formation of the hummocks, they contain enough evidence of involutions, and in areas of extensive involutions they are found over large areas which are called *hummock fields*.

Palsa— Hummocks of a special kind, known as palsa are met with in the arctic and sub-arctic regions. The height of palsa is up to 10 metres and its diameter 10 to 20 metres. These mounds are found in groups, generally in swampy areas and are made of peat with thin layers of ice inside. They are formed by frost heaving resulting from ice segregation. Some times they contain some vegetation. Palsa starts disintegrating after the rupture or fracture of its upper surface. Sometimes these mounds are also destroyed by the rise of water-table in the adjacent swamps. Palsa is mainly found in areas of sporadic or discontinous permafrost.

Pingos— Pingo is an Eskimo word literally meaning isolated dome-shaped hill or mound. Under certain hydrologic conditions, laccolithic intrusions of ice arch the overlying surface into a fractured dome which is called a pingo. The term was first used by Porsild in 1938 for ice-cored domes of this kind in the Mackenzie delta of Canada. Pingo constitutes a distinctive feature in the periglacial areas of Alaska, Canada, Greenland and Siberia. It is called *Bulgunnyakh* in Russia. The height of pingo varies from 1 to 60 metres and its diameter is up to 600 metres. It varies greatly in form and size, but a common feature is the presence of a core of massive ice or icy sediments 1 to 10 metres below the surface. The upper part of small pingos is rounded, but the upper part of bigger pingos is usually fractured resulting in the formation of a crater which may contain a small lake on account of the melting of ice. This may be considered to be the initial stage of the destruction of pingos, because as a result of the melting of ice below, the dome-shaped sedimentary layer breaks and falls down and finally only a shallow depression with raised sides is left. Although permafrost is absolutely essential for the formation of pingo, and pingo is a striking landform in periglacial regions, especially in areas of low relief, it is not found in all periglacial areas. According to F. Muller (1968) pingo is found in areas lying between 65^0 to 75^0 N. latitudes where the permafrost is either discontinous or continuous but thin. But pingos have been found in areas of thick permafrost as well, and as our knowledge about them increases it appears that it would be incorrect to place any such definite limit about their occurrence.

F. Muller (1962) has given a description of the Ibyuk pingo in the Mackenzie delta of Canada which is a good example of pingo. It rises about 41 metres above the neighbouring level swamp, and the circumferenee of its base is 900 metres. It has a crater at its top which contains a small lake. It contains a 40 metre thick ice core which is overlain by a 15 metre thick layer of frozen sediments, and the lake has been formed as a result of the fracture of the upper part of the frozen sediments. In fact the Mackenzie delta is full of such pingos and probably in no other part of the world are pingos found in such large numbers.

Pingos are of two kinds. One of the types is that which is characteristic of the Mackenzie delta described above and has been called the Mackenzie type. The other type of pingo is found in East Greenland, middle Alaska and some other areas and has been called the East Greenland type. F. Muller (1959) has suggested that the Mackenzie type may be called *closed system pingos* and the East Greenland type may be called *open system pingos*.

(a) *Mackenzie Type or Closed System Pingos*— The closed system pingos are found in areas of continuous permafrost and especially on flat alluvial surfaces. Normally this type does not occur in groups but is found in isolated form. The finest examples of this type are found in the Mackenzie delta of Canada. In this region every pingo is found either in a shallow lake or an old lake basin. Pingos in the Mackenzie delta have been studied extensively by J.R. Mackay (1962, 1973). According to him, closed system pingos are formed when the permafrost extends towards a shallow lake basin which is either drained or filled up. As a result of the freezing of the surface layers a closed system of highly saturated and unfrozen soil is formed beneath the lake. As a freezing plane advances in the unfrozen zone, the pore water is expelled in front of the freezing plane as a result of pressure. "This expelled pore water provides the continuing water supply and hydrostatic pressure necessary for pingo growth. The growth of the ice then leads to the updoming of the surface sediments and the formation of the pingo"[1]

In the opinion of Mackay the formation of pingo is not entirely due to ice-lense heaving, as the pingo contains about 95% fine and medium sand which is not frost-susceptible, and fine-grained materials like silt and clay which are more susceptible to freezing are found only in small quantities in the pingo. Further, in sand there is rapid movement of pore water. Fig. 177 explains the theory of pingo growth as put forward by Mackay. Initially there is a big lake which is surrounded by permafrost. Underneath the middle of the lake lies unfrozen land. The upper surface of the lake is frozen, but some water is present in the bed of the lake. Slowly the entire lake including the sediments present in the bed freezes and in place of the lake we find permafrost and a thick layer of ice. In this way the unfrozen land underneath is surrounded by permafrost and ice layer and a closed system is produced from which water cannot come out (Fig. 177 B). This closed system of unfrozen soil is subjected to pressure of the advancing permafrost from all sides. Because of the continuous increase in pressure, pore water is expelled up from

1. French, H.M.—*The Periglacial Environment*, Longman, London, 1976. p.98.

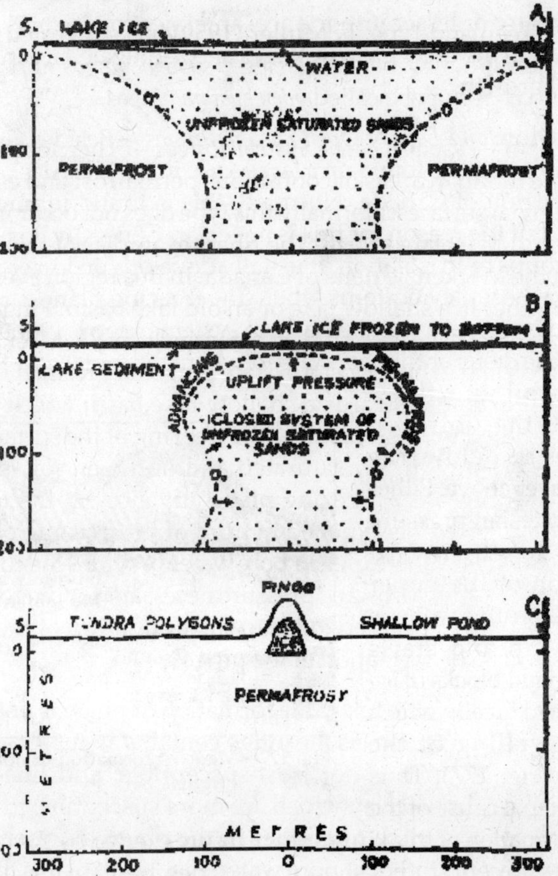

Fig. 176—The mode of formation of closed system pingos according to Mackay.

the unfrozen sand. Because of the increasing pressure there is updoming of the surface and just below this arched-up surface the entire expelled water freezes into pure ice. This explains why underneath every pingo there is found accumulation of almost pure ice.

More recent studies by Mackay, however, indicate that the majority of pingo growth is through the process of ice segregation and not by the freezing of a pool of water. It is, of course, likely that at different stages of pingo growth different types of ice formation

1. Ibid., p. 100.

are of greater or lesser importance, and injection ice may be more important in the early growth stages and segregated ice in the later stages of growth.[1]

(b) *East Greenland Type or Open System Pingos*— Besides Greenland this type of pingo is found on a large scale in middle Alaska and the Yukon region of Canada. Normally this is found in unglaciated regions, but the reason for this is not clear. Generally it is found in small groups or in isolation. It usually develops on lower slopes of valley or the valley bottoms where the freezing plane is regularly supplied by surface water. The open system pingo is found in areas of discontinuous or thin permafrost where the surface water enters the subsoil and flows through the unfrozen sediments below the frozen layer above. Either due to artesian pressure or due to downward extension of the upper frozen layer this water is pushed up and starts freezing and blocks of ice are formed locally which cause swelling of the surface (Fig. 177). Thus the principal cause of the pingo formation of this kind is the pressure exerted by the growth of ice lenses. Recent studies show that besides injection ice segregated ice has also a hand in open system pingo growth. Some people are of the opinion that bigger pingos which have bigger accumulation of ice core have been formed by injection ice resulting from rising water, and smaller pingos where the ice is present in the form of frozen sediments, segregated ice is responsible for pingo growth.[1]

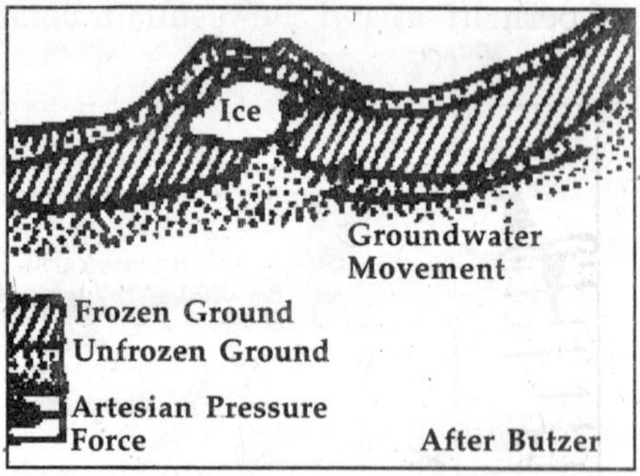

Fig. 177—Mode of formation of open system Pingos.

In addition to the two types of pingo described above, some pingo like mounds are found which do not quite fit into this classification and which appear to have been formed in other ways but our knowledge about them is extremely limited.

1. Ibid., pp. 88 and 98.

Hydrolaccoliths and Frost Mounds — In the present day periglacial areas small mounds containing ice lense or ice core are found in large numbers. These are usually formed temporarily every year and then are destroyed. These are not included under pingos. They are essentially limited to the active layer and are diverse in form. These are usually formed as a result of local ice injection or ice segregation below silt or peat. Oval-shaped mounds are found in Siberia which are 5 to 10 metres high, 100-500 metres long and 50-80 metres wide, and are referred to as *Bugor*. Similar low and rounded mounds are also found in the Arctic regions of N. America.

Thermokarst — In 1932 the Russian scientist Yarmolev used the term thermokarst to describe the features produced by the melting of permafrost and the resultant collapse of the ground over the

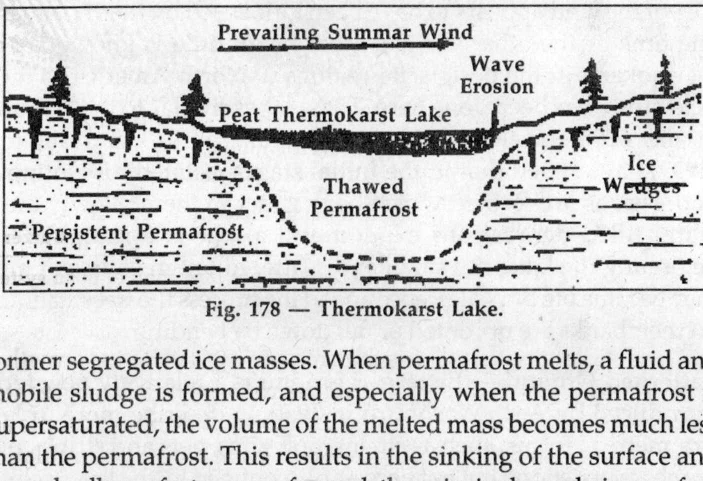

Fig. 178 — Thermokarst Lake.

former segregated ice masses. When permafrost melts, a fluid and mobile sludge is formed, and especially when the permafrost is supersaturated, the volume of the melted mass becomes much less than the permafrost. This results in the sinking of the surface and several collapse features are formed, the principal ones being surface pits and basins, funnel-shaped sinks, dry valleys and ravines and thermokarst or thaw lakes. Before the actual sinking of the surface, underground caverns or hollows may sometimes be formed as a result of melting of underground ice. Where several sinks join together, a long continuous depression with steep sides and uneven floor develops. These are called *dujodas* in Yakutia (Russia.) When the dujodas are more extensive and developed, round or oval shaped depressions are formed with vertical sides and level floor. These are called *alas*, and frequently contain thermokarst or thaw lakes. For the best development of thermokarst, it is necessary to have several thick blocks of frozen ice in the upper part of the permafrost in order that the overlying surface may collapse when they melt. There are obvious similarities between the thermokarst features and the karst forms developed in limestone regions. But thermokarst is not a type of karst. The chief process in the development of karst is solution which is a chemical process. But the chief process in the formation of thermokarst is ground ice melting which is a physical process and is limited to areas where permafrost occurs below the surface.

Examples of thermokarst are met with in several areas in Russia where they have been extensively studied. Here thermokarst is found in a large area extending from the arctic and sub-arctic regions of North Siberia to the mountainous region of middle Tienshan in the south. In parts of central Yakutia and eastern Siberia nearly 40% of the land appears to have been influenced by thermokarst at one time or the other. On the other hand, little is known about thermokarst in the periglacial regions of North America as little field study has been done here. Thaw lakes similar to those found in 'alas' of Siberia also exist in Alaska. Wallace (1948) and Hopkins (1949) have stated that in the initial stages isolated discontinous depressions are formed which have water in them. Slowly these water-filled depressions expand and a lake is formed. In the beginning the lakes are small and scattered but in course of time they become big by coalescence and in the process the trees standing on their banks are uprooted or fall down by bending.

Patterned Ground — Patterned ground is a relatively new term introduced by A. L. Washburn in 1950, to describe more or less symmetrical forms, such as circles, polygons, nets and stripes that are characteristic of, but not necessarily confined to, the layer subject to intense frost action. Patterned ground forms are characteristic of the micro-relief of periglacial zone. In the periglacial regions the land surface is divided into a variety of polygonal patterns. The patterns may consist of sorted or unsorted debris. In level areas where the upper melted layer consists of mud, the surface is divided into *mud polygons* (Fig. 180). Their formation is clearly related to contraction and expansion, but our knowledge about the exact mechanism of their formation is incomplete. One reason for this is that contraction takes place not only when the ice is melting but also when the temperature of ice falls sufficiently below freezing point. The density of ice at the temperature of 0^0 C is 0.92 and at -22^0 C it increases to 1.03. In other words, its volume decreases by 11% when its temperature falls from 0^0 C to -22^0 C. The other reason for uncertainty is the fact that it cannot be said with any definiteness whether the polygons were formed under conditions similar to those existing today.

Fig. 179 — Mud polygons and Ice-wedge polygons.

The most prominent polygonal patterns are found in areas of continuous permafrost where the mean annual temperatures are extremely low and average at least -6^0 C to -8^0 C (18^0 F to 22^0 F). In such regions networks of *ice-wedge polygons* cover large areas. Each polygon is about 3 to 30 metres in diameter and is outlined by shallow troughs, and beneath each polygon is a wedge-shaped mass of ice, a metre or more broad at the top and extending downwards from 3 to 7 metres or more. Most authorities accept the contraction theory of formation of the ice-wedge polygons. The cracks surrounding the polygons have developed by contraction resulting from fall of temperature of the frozen ground during winter. In these cracks, ice, water and sediments blown by wind are deposited. Each winter these cracks open and new layer of ice is introduced, and in this way ice wedges are formed which gradually become wider and deeper. The spacing and intersections of these ice wedges outline the ice-wedge polygons. The edges of the active ice-wedges are at least a metre high and there are dome-like hummocks within each polygon because of the expansion of the ice wedges. These are called *ice-wedge polygons*, as they have been formed by the ice wedges. Generally the ice-wedge polygons are made of unsorted sediments and there is compressional deformation of the frozen sediments adjacent to the cracks around the polygon rather than by sorting and translocation of rock fragments within polygonal prism of soil.

Where the active layer is a mixture of fine debris and stones the situation becomes more complicated. There is sorting and rearrangement of rock fragmentes and the coarser materials are lifted to surface by frost heaving (cryoturbation). Initially probably the polygons are formed by thermal contraction and the central portion of the polygons is raised in the form of hummocks[1]. Because of the raised centre the rock fragments are thrust to the edge of the polygons. The edges of the polygons are therefore, made of stones

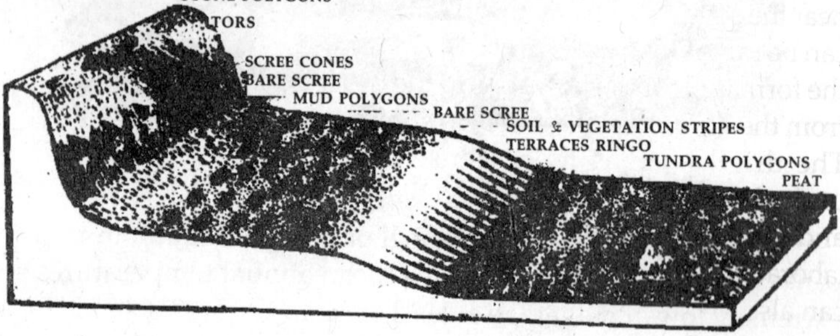

Fig. 180 — A cross-section of Periglacial landforms, showing different types of polygons.

or rock fragments, and the polygon has raised rims. These sorted polygons are known as *stone polygons*. In patterned ground of this type the diameter of each polygon is usually less than a metre, the edges are marked by stones or rock fragments, while the central upraised portion consists of soil or/and fine-grained sediments.

1. "Of the suggested explanations for polygonal ground,that favoured by Richmond (1949) seems the most likely. He points out that freezing waste has been known to crack suddenly and loudly, and that contraction-cracks often divide frozen mud into polygonal blocks. Although water expands slightly as it cools towards freezing-point, ice contracts slightly as its temperature falls from freezing point downwards. Hence Richmond infers that the formation of stone nets begins with the freezing, further cooling and cracking of the waste-mantle' Dury, G.H.-*The Face of the Earth*, Penguin Books, 1959, p. 176.

Fully developed polygons are characterized by well-sorted stone borders and a central area which is free from coarser material. The stone borders may be deep or shallow. The central area of sorted polygons made of finer materials tends to be convex in developing form and concave in declining forms.

Some people have tried to explain the formation of stone polygons with reference to convection currents. In summer when the upper layer of the soil thaws, then at the junction of the thawed layer above and the frozen layer below the temperature of water is 39^0 F. Since the density of water is highest at 39^0 F, the water from above sinks below and the lighter and colder water from below rises up. In this way convection currents are set up. It has been found in laboratory experiments that under these conditions, there is development of polygonal patterns where the current rises upwards in the centre of the polygon and the currents come down near the edges of the polygon. But it seems doubtful if rock fragments can be raised up by this process. Some other people have attributed the formation of the polygons to the differences in density resulting from the drying up of the soil. In summer the upper layer dries first. The dried layer is more dense and tends to sink. The layer underneath is wet with thawed water and is relatively less dense and tends to rise up. An attempt has been made with the help of laboratory experiments to show that this kind of density differences can also produce polygonal patterns.

Whatever the mode of their origin stone polygons constitute a distinctive landform in the tundra or periglacial regions, especially where the climate is relatively humid. Where they are found in fossil form, they provide proof of the existence of periglacial conditions and intense freeze-thaw activity in the past.

The form of the stone polygons changes with slope. (Fig. 182). On relatively level land where the gradient is 2^0 or less, the polygons are regular and the rock fragments surround the fine sediments in the middle in the form of a ring. These are called *stone rings*. Where the slope is between 3^0 to 7^0 the polygons tend to be elongated by soil creep and they look like garlands. They are known as *stone garlands*. When the slope is still steeper (8^0 to 26^0), the water saturated mud in the middle of the polygon breaks through the front of the stone garland, and the polygon is converted into long stripes in which the coarse and fine sediments are segregated in parallel bands trending down the hillslope. These are the called *stone stripes*. They tend to remain stable at usually steep angles (say 30^0) on account of the excellent drainage provided by the stone stripes.

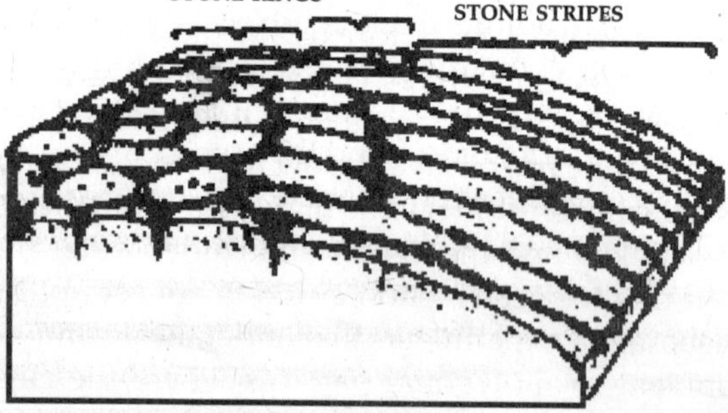

Fig. 181—Diagram showing changes in the form of stone polygons with increase in slope.

A.L. Washburn (1956) has classified the landforms found on patterned ground into five groups based on their form namely, circles, nets, polygons, steps and stripes. Further based on the amount of sorting of materials, they may be sorted or unsorted. In the sorted forms the edges are marked by stone fragments, while in the unsorted forms the stone margins are absent. Their form is primarily dependent upon the slope of the ground. Circles, nets and polygons are found on level or near level surfaces, while steps and stripes are formed on land where the slope is usually between 6^0 to 15^0-30^0. On slopes between 2^0 to 6^0 the forms are intermediate (i.e. steps) between circle or polygon and stripes. Where the land is steeper than 30^0, conditions are not favourable for the development of patterned ground.

The shape of these features slowly changes with the change in slope gradient. Elongated steps or circles are called garlands, and when elongated still further they change into stripes. Circles are formed in areas where the active processes are divided into different centres, but where two or more active centres come in close contact, nets or polygons tend to be formed. In fact, the irregular form of the nets tends to develop in situations midway between the circles and polygons. Earth hummocks made of unsorted materials are an example of the net pattern.

Washburn (1970) has proposed a detailed genetic classification of patterned ground. The chief characteristic of his classification is the division of the processes responsible for the formation of patterned ground into two major divisions—(a) processes in which cracking is essential, and (b) processes in which cracking is not essential. Washburn admits that the subject is extremely complex, and our knowledge about the processes underlying the formation of these forms is very limited and incomplete. Even so the classification helps us to understand the process of their formation. According to his classification cracking is an important process in the creation of polygonal forms. The circular forms owe their origin to primary sorting, differential heaving and mass displacement. For example, non-sorted circles or mud circles are formed as a result of the rising up of the sub-surface materials to the surface. In the formation of stripes we see the influence of mass movement on circular and polygonal forms. In both sorted and non-sorted stripes the movement is maximum in the middle, and the movement decreases towards the margins and also with depth. In non-sorted stripes the movement is confined to sections without vegetation cover, and where the stripes have a cover of vegetation, the movement is almost absent.

Periglacial Mass Movement and Slope Deposits

Gelifluction

In the periglacial regions the conditions are highly favourable for the mass movement of water saturated debris and soils and this process has been called solifluction. The name was first proposed by J.G. Andersson in 1906, but subsequently the word solifluction has come to have a wider connotation. K. Bryan (1946) has called all kinds of mass movements including solifluction by the name of congeliturbation. J. Dylik (1951) has used the term congelifluction for solifluction in periglacial areas, but this does not include frost creep, rapid mud flows and sheet wash. H. Baulig (1956) has suggested the use of the term gelifluction in place of solifluction to specify flow over a frozen substrata and indicating a specifically periglacial process. This has met with wide acceptance and Washburn (1973) too has accepted it. *Gelifluction* means the mass movement of water saturated debris in periglacial regions, but it does not include downslope creep of debris resulting from alternate frost and thaw. The term *frost creep* is used for the latter process.

Periglacial Slope Deposits

For periglacial slope deposits a number of terms are used in England and Europe, such as head, coombe rock or trail. Gelifluction deposits are mostly related to the local rocks, because it is but rarely that debris are transported to a distance of more than one or two km by this process. Erratics are, therefore, rarely found. Because of their short journey the rock fragments are not rounded but remain angular. The thickness of the deposits shows great variation. The deposits are particularly thick in the depressions and valleys. The longer axis of the stones is in the direction of flow of the debris, but where there is some obstruction in flow such as, on the edge of a lobe or terrace, the rock fragments tend to move at right angles. These deposits have also a very loose texture and are poorly sorted. These characteristics distinguish these deposits from glacial drifts.[1] But sometimes it is difficult to distinguish the head deposits from glacial drift, especially when the source of the periglacial slope deposits is glacial drift.

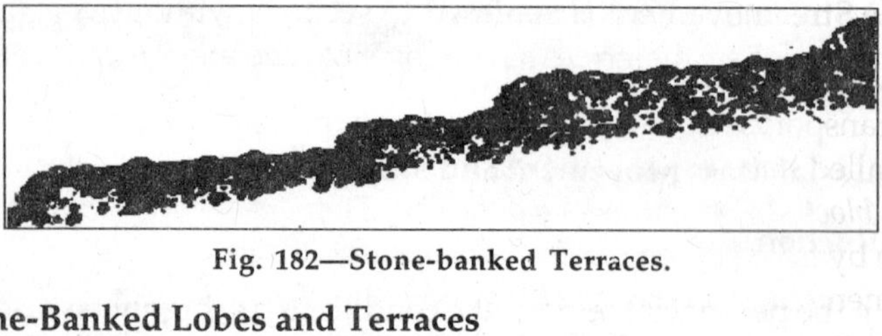

Fig. 182—Stone-banked Terraces.

Stone-Banked Lobes and Terraces

Stone-banked lobes and terraces are gelifluction deposits which are found on intermediate slopes with gradients of 10 to 25^0. They have raised stony edges on the sides and in the front. The terraces slope down gently in the direction of the general slope but the lobes have their greater extension in the direction of the slope and have less extension along the contour. It appears that the stone fragments are thrown up to the surface by frost heaving and tend to move downslope faster than fine debris but it is not quite clear as to how these features come to be formed.

1. "Periglacial slope deposits may sometimes be distinguishable by the downslope orientation of stones, the more angular nature of the stones, they absence of erratics and the generally looser texture". Embleton, C. and King C.A.M.—*Periglacial Geomorphology*, London, 1975, p. 107

Talus or Scree

Where the slope is steep, talus or scree is formed in the foothill zone. In the hilly areas wherever mechanical weathering by frost is active, talus is likely to develop. But the conditions are particularly favourable for its development in the periglacial regions. Here large accumulations of big angular rock fragments which have slid downslope by gravity are found along the foot and up the slope for some distance. Sometimes they look like block-fields, but actually they are very different, as block fields are found on the relatively flat or gently sloping upper parts of the mountains. Blockfields are extensively found in the arctic regions having shales and sandstones, as these rocks are particularly susceptible to disintegration along joints and bedding planes by frost action. In talus rock fragments are not only comparatively smaller in size but also they occur locally and in relatively smaller accumulations. Further, in talus downslope movement of the rock fragments is absolutely essential which is not true of block-fields.

Stone Streams or Rock Streams

When masses of rock debris stream down mountain sides or are transported down in the form of a river through the valleys, they are called *stone-streams* or *rock streams*. Sometimes they are referred to as *block streams* or *block slopes* also. There is sorting of debris in them by frost heaving and other causes, with the result that bigger fragments are found on the surface and finer materials further below. Good examples of Pleistocene rock streams are found in the chalk country of Wiltshire and Dorset in Britain. Relict rock streams are also found in Appalachian mts. of U.S.A., consisting of large angular boulders and extending 1,000 metres downhill from a cliff or ledge, and were almost certainly formed by periglacial mass wasting.

Rock Glaciers

Rock glaciers are frozen rubble with glacier-like motion. There is no ice on the surface of a rock glacier. Its surface consists of coarse and angular rock fragments extending up the hill slope to the cirque. But within one metre of the surface the pore spaces between the rock fragments are packed with ice. In the lower layers of the rock glaciers relatively fine materials such as, sand, silt and mud-mixed stony fragments are present. These also contain interstitial ice. In some rock glacier the proportion of debris by volume has been found to be not more than 10 to 12 percent, but normally it is much more. In fact the difference between a rock glacier and a glacier is that the former is mostly rock debris with interstitial ice, whereas the latter is mostly ice with some included rock fragments. Examples of rock glaciers are found in Alaska, Switzerland, and Colorado in the Rocky Mts. Larger rock glaciers in Alaska are ice-cemented rock fragments. Rock glaciers move forward by internal plastic deformation of the ice and debris-slide at their termini. The movement on their surface is about 1 to 2 metres per year, that is, much slower than glacier flow. This is but natural in view of the higher proportion of debris which imparts them greater rigidity. Active rock glaciers are either tongue-shaped along valley axes or lobes on valley walls, and have steep frontal slopes.

The surface of the rock glaciers is marked by micro-relief. The melting of ice may produce at places longitudinal furrows and conical pits and the whole surface wears a wrinkled look on account of the downhill movement.

Altiplanation or Cryoplanation Terraces

In periglacial regions while the summit areas may be flattened, the mountain slopes are marked by bench-like terraces. The terraces have gentle slopes but are separated by steep scarps. These benches have been called *altiplanation or Cryoplanation terraces*. They contain

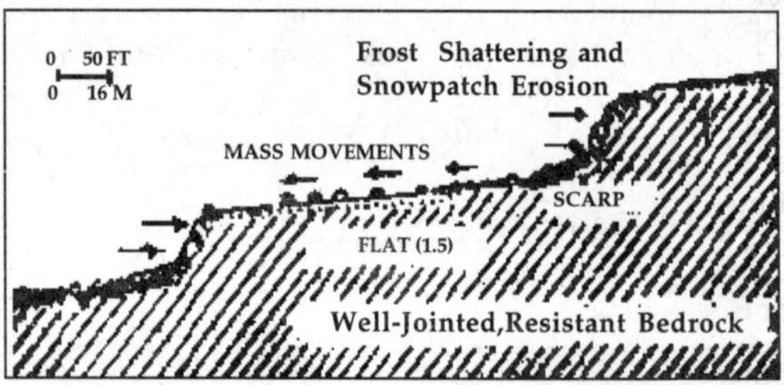

Fig. 183—Altiplanation or Cryoplanation Terraces.

a thin veneer of debris but they are not constructional features. In fact, they are erosion surfaces formed by the combined action of nivation, frost shattering and mass movement. R.S. Waters (1962) has explained their origin by his bed-rock erosion theory. According to him, the formation of the terrace starts with the accumulation of snow in small depressions along the hillslope. The rocks break by the processes of nivation and congelifraction and the rock debris are removed by downhill mass movement. By repeated action of frost sapping and mass movement of debris a surface of low local relief or bench-like surface is produced which is separated from its lower neighbour by a steep faced scarp. The debris continue to move from the higher benches to the lower benches. Thus while the lower benches are covered with debris, the higher benches are relatively bare. Frost shattering of the steep scarp face remains active. The scarp face is, therefore, rugged and tends to recede, resulting in extension of the benches. J. Demek (1969) has confirmed the views of Waters through his own studies (see Fig. 186)

Such cryoplanation erosion terraces are found in several places in Europe, Siberia and the northern regions of North America. They are called *golcovaya terrasa or goletz terrace* in Russia. These terraces vary greatly in size. The smaller terraces may not be more than 50 metres long or wide, but at several places they have been found to be 400 to 600 metres long and 100 to 200 metres wide. In Siberia some of the terraces are more than a kilometre wide and several kilometres long. Similarly the frost shattered scarps separating these terraces also vary greatly in height, depending on their gradient and the hardness of the rocks. Where the gradient is comparatively gentle and the rocks are unconsolidated, the scarps may not be more than 1 to 2 metres high. But where the gradient is steeper and the rocks are relatively harder, the scarps are 10 to 20 metres high. The slope of the terrace benches is usually between 1^0 to 12^0 and on fully developed benches the slope is commonly around 7^0. Usually the larger benches have less slope and the smaller benches have steeper slope. The benches are usually covered with debris, and sometimes there may be even development of patterned ground and rock streams on the larger benches.

Tors or Rockstacks

Sometimes big rock fragments or blocks rise vertically from the general surface like stacks or pillars. These are called *tors*. Tors are found in several climatic regions and are not directly related to periglacial conditions. They are usually remnants of extremely hard rocks and are found together with talus deposits, but they are the product of erosion and not deposition. At the present time examples of their formation under periglacial conditions are found in northern

Canada and the Antarctica. There is a great deal of controversy about their formation, as they are formed under different climatic conditions. Demek (1964) has suggested two possible processes for the development of tors in the Bohemian upland of Czechoslovakia. (1) According to the first process they have been formed in two stages: (a) Possibly during Tertiary times there was development of thick regolith on account of weathering under hot and humid conditions, and the rocks which had distant joints broke into huge rock fragments due to weathering. (b) During the periglacial stage of Pleistocene the rock debris were removed by frost action and solifluction and the bigger rock fragments took the form of tors. Demek has tried to explain the formation of several tors in Western Bohemia in this fashion. (2) In the other process suggested by Demek the mechanical disintegration of rocks as well as the removal of debris have both taken place in the periglacial environment. According to Linton (1964) the tors in south-west England have been formed in the manner of the first proces, that is the rock fragments were formed in hot and humid climate and the loose debris have been removed later by periglacial mass movements, leaving the bigger fragments upstanding in the form of tors.

Fluvial Action and Features

In the periglacial regions we find two kinds of rivers (1) Firstly, there are the big rivers, like the Mackenzie in North America and the Ob, Yenesi and the Lena in Siberia which have their sources outside the periglacial region and which enter the Arctic ocean after covering a distance of thousands of kilometres. In these rivers the characteristics of water discharge and sediments are not influenced to any great extent by the periglacial environment. (2) In addition, there are several smaller streams which originate within the periglacial region and whose water discharge and sediment characteristics are fully influenced by the periglacial environment.

The Arctic rivers remain frozen for the greater part of the year on account of below freezing temperatures and their flow is confined to a short period in the summer. It is, therefore, generally believed that running water has a negligible role in the formation of the periglacial landscape. This, however, is not entirely true. Recent studies show that even in the periglacial regions running water is an important agent of transport and erosion. In spring the ice melts quickly and within two to three weeks the situation is one of rapid discharge and flooding in the rivers. Thus even in regions where the rainfall is small and is distributed in the different months, 25 to 75 percent of the total run - off is concentrated within a few days. Further, in the arid regions most of the precipitation comes in the form of rain in summer. On account of the lack of vegetation and the presence of permafrost below the surface the rainwater is not able to infiltrate into the ground and flows on the surface. In some areas water is also derived from a neighbouring snowfield or glacier as a result of summer melting. Thus for a short period these rivers experience a rapid and swollen flow, and by the end of summer the smaller streams freeze solid and completely cease to flow.

The rivers have also an unusually heavy sediment load. In addition to sand and gravels the river bed is full of rock fragments and boulders. The river valleys are usually shallow and the river channels are braided. But on account of the heavy load and the rapid flow the rivers erode fast their bed and banks. Lateral erosion of the banks is particularly important. A special kind of erosion which the rivers perform in the periglacial regions is *thermal erosion* of the banks. The flowing water of the river has the power to melt the permafrost. Thus during the period of flooding the rivers not only erode the banks by lateral corrasion, but also melt the permafrost along the banks by thermal erosion. As a result of melting of the permafrost in the lower portion of the bank the uncosolidated sediments are removed at the flood water level, and a long linear hollow called the *thermal-erosional niche* is created. This niche may be sometimes as high as 10 metres or more. On account of the formation of linear caves in the lower part of the banks, the frozen sediments above tend to fall down at weaker spots. This process has been observed in several rivers of Alaska and Siberia. This process is also responsible for erosion of the sea-coast in the region of permafrost. As a result of thermal erosion lateral erosion becomes very active, the river valleys are widened, and the rivers receive large quantities of debris from bank erosion which results in shoaling and formation of bars and the braiding of river channels. In the periglacial region, the conditions are especially favourable for the development of braiding in rivers, and the existence of braided or multiple channel form is a common feature of periglacial rivers. Braiding is related to both heavy sediment load and rapid

1. French, H.M.—op. cit., p. 174

and large variations in run-off, any decrease in competence of the stream resulting in bar development and multiple channels.[1]

The cross-profile of river valleys in periglacial regions is quite often asymmetrical, with one bank being much steeper than the other. In Siberia, Greenland, Alaska and northern Canada the north-facing slopes have been found to be steeper than the south facing slopes, and this would appear to be broadly true for high latitudes till about 70^0 N. in the northern hemispehere. Where the valleys trend north-south the asymmetry disappears. Till about 70^0 N. latitude the south and south-west facing slopes receive more insolation and consequently the active layer is deeper on these slopes. In other words, higher insolation leads to increased thaw, and more bank debris fall into the river by solifluction.

One important reason for this phenomenon is the fact that the north-facing slope remains shaded and therefore cold and frozen for a much longer period than the south facing slope. It is less prone to gelifluction and the original steepness of the slope continues. On the south facing slope the longer period of sunshine and more insolation lead to increased thawing and gelifluction and downslope sliding of debris is more intense. Permafrost holds unconsolidated debris together even on steep slopes by the cementing action of ice, but after the melting of ice the debris cannot stay on a steep slope and collapse down and are deposited at the bottom.

Secondly, on account of the accumulation of large quantities of debris at the foot of the south-facing slope, the river flow is pushed towards the north-facing slope, where there is increased erosion of the bank base by lateral corrasion and thermal erosion and the upper portion of the bank slumps vertically below making the slope steeper.

Besides the above factors, local conditions also influence the face of the steep slope. Among these, wind direction and differences in vegetation cover are particularly important. In the latitudes north of 70^0 N the differences in the receipt of insolation between north-facing and south-facing slopes become less, as during the long polar days there is not much variation in the sun's inclination. Consequently in Spitsbergen (78^0 N latitude) the steep slope has been found to be facing all directions.

Influence of Wind

The periglacial regions are either arid or semi-arid. Because of the low percipitation, freeze-drying of exposed sediments and the presence of strong winds, wind becomes a relatively important agent of transportation and deposition in the periglacial environment. Clouds of sand and dust rise from the river valleys after the spring flood subsides. Dunes are formed on the downwind banks of rivers. Sandsheets which are less sorted than dunes and are sometimes stratified are widespread in Northern Europe. They are believed to be wind blown sediments which have settled on snow. Large areas of present and former periglaciation in the plains of Central Europe and Central Asia are covered with wind-blown loess deposits.

Periglacial cycle of erosion

In 1950 L. C. Peltier pur forward the concept of a periglacial cycle of erosion.[1] This is similar to the Davisian concept of the

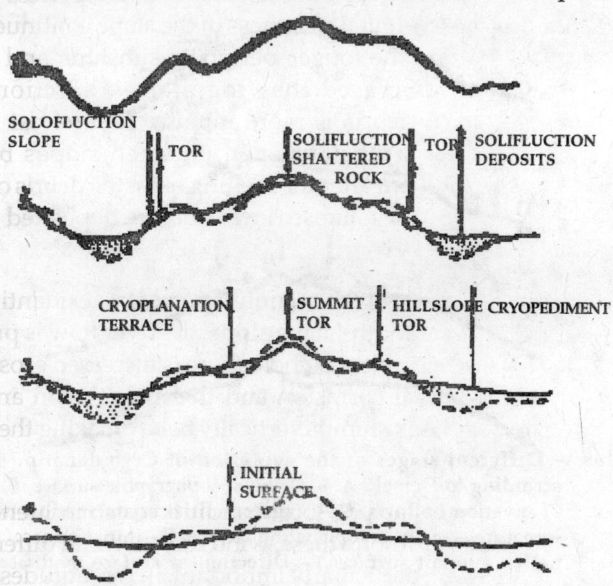

Fig. 184—Periglacial erosion cycle and its different stages according to Peltier. 1. Initial surface, 2. Early or young stage, 3. Mature stage. 4. Old stage of Cryoplanation.

---

1. Peltier, L.C.—The Geomorphic Cycle in Periglacial Regions as it is related to climatic geomorphology, *Annals, Association of American Geographers*, Vol. 40, 1950, pp. 214-236.

normal cycle of erosion, and the periglacial landscape attains old age after passing through the stages of youth and maturity. During this entire period there is erosion of the higher parts and deposition in the lowlying parts, resulting in overall reduction in relief, and the slope profile gradually becomes smooth and flat. The process by which there is gradual flattening of the surface in the periglacial regions is called *cryoplanation*. In this process there is parallel retreat of the scarp face by frost shattering and the development and gradual

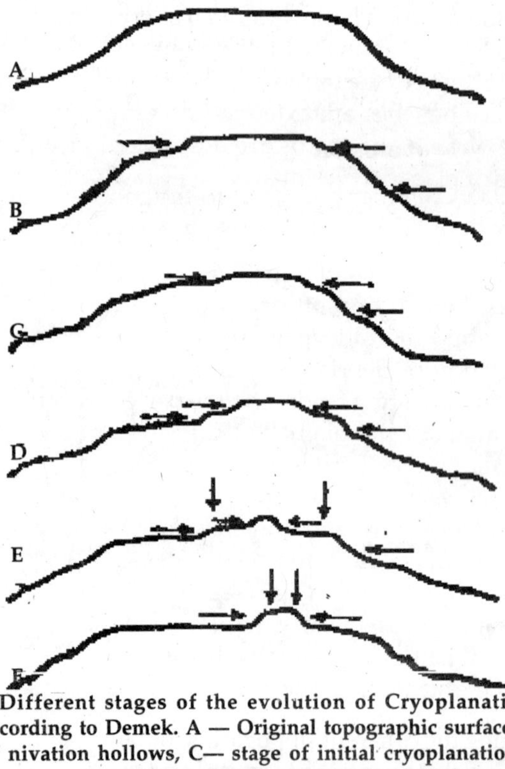

Fig.185 — Different stages of the evolution of Cryoplanation terraces according to Demek. A — Original topographic surface, B — stage of nivation hollows, C— stage of initial cryoplanation terrace, D —mature stage of cryoplanation terrace, E—Initial stage of cryoplanation summit surfaces.F—Direction of surface modification.

extension of gentle slope at its base by deposition and transport of debris. Cryoplanation is thus achieved primarily by the processes of intense frost action or congelifraction and solifluction or congelitubation. The penultimate landform is a surface of low local relief, not controlled by any base level, and has been called a *cryoplain* or an *altiplain*.

The evolution of the periglacial eorsion cycle and its different stages have been shown in Fig. 185. In 185 (1) the initial surface has been shown. In diagram 185 (2) the cycle begins with the onset of periglacial conditions. In this initial stage congelifraction is especially active as a result of which there is shattering of the bare rocks on the upper slopes, and block-fields are formed by the accumulation of angular rock fragments. The rock fragments move downhill along the slope by gravity, and talus or congeliturbate mantled slope is formed at the base of the slope (Fig. 185, 2). During this period nivation hollows as well as cryoplanation terraces also begin to be formed on the higher slopes. Till the end of the young stage, the scarp of the terraces starts retreating by frost action, and the terrace benches are extended by scarp retreat. With further passage of time, the scarps are destroyed by continual recession and in their place only residual or tor features are left on the summits or slopes. This is the beginning of maturity when the original landscape is lost and the valleys and lower slopes are covered with a mantle of frost shattered and soliflucted materials (185, 3). Towards the end of maturity and in old age the higher land is further reduced and levelled on account of solifluction, and the accumulation of debris goes on increasing in the neighbouring valleys and lowlying lands, and the periglacial surface is coverted into an almost level plain. The deris become fine on account of continual frost weathering and solifluction (gelifluction) by further disintegration and abrasion. The finer debris are transported by the wind, and loess and sand dunes are formed at places, and areas of ventifacts and lag depsotits develop by the blowing away of finer materials by the wind. The rivers which were filled with coarse debris in the young or early mature stage, are now in a position to transport the fine debris relatively easily, and the further wearing down of the surface may be controlled by the sea-level.

Taking the different slope forms found in periglacial regions as the basis, Peltier has tried to synthesize them and arrage them in a sequence. His hypothesis thus provides an overall framework for the understanding of the slope evolution in periglacial regions, but it does not provide an analysis of how exactly slope form is influenced by frost shattering and solifluction. Further, adequate attention has not been paid to other processes, espcially the influence of running water. But the concept of scarp recession by frost shattering and the resultant extension of the terraces is by itself an important and useful concept which helps us to understand the formation of cryoplanation terraces, summit tors, nivation hollows and cryopediments. J. Demek (1969) has made a detailed study of the evolution of cryoplanation terraces. The different stages in the evolutionof cryoplanation terraces as conceived by Demek have been shown in Fig. 186 which broadly supports Peltier's hypothesis. While cryoplanation terraces are a characteristic feature of periglacial landscape and cryoplanation is recognized as a distinct process, the implied penultimate landform of cryoplain or altiplain has not been seriously accepted as a climatic analogue of peneplain or pediplain.

Questions

1. Discuss the principal characteristics of the tundra climate which provide favourable conditions for periglacial processes.

2. What do you understand by permafrost and the active layer ? Give the distribution of permafrost in the world and explain the significance of the active layer in the formation of periglacial landforms.

3. Explain the main processes in the evolution of periglacial landscape.

4. Acount for the salient features of a periglacial landscape.

5. Discurss critically the different views regarding the formation of pingos.

6. Describe and account for the formation of patterned ground in periglacial regions.

7. Write short notes on any *three* of the following :-

   Thermokarst, Patterned Ground, Rock Streams and Rock glaciers, Pingos, Cryoplanation terraces, Block-fields and Talus.

8. Explain what you understand by Gelifluction, Cryoturbation, and Permafrost.

9. Give an account of the periglacial cycle of erosion as proposed by Peltier.

10. Give a reasoned account of the fluvial and aeolian features found in periglacial regions.

Chapter 27

Types of Lakes

Lakes are inland bodies of standing water. A lake has been defined as a mass of still water situated in a depression in the ground without direct communication with the sea. Lakes vary greatly in area, depth, location and other characteristics. They may be very small or quite big. Their water may be sweet or saline, though most of the lakes are fresh water lakes. Most of the lakes are situated on the land surface above sea level, but there are some, such as the coastal lagoons which are at sea level and other like the Dead Sea and the Caspian Sea which are below sea level.

Distribution of Lakes

Of all the landforms, lakes are probably the most universal in their geographical distribution and are found in all latitudes and climates. All the different landscapes which have been described in the earlier chapters contain some kind of lakes. Some of the lakes owe their origin to the work of rivers and their changing channels, while other lakes are found in regions where the landscape has been principally formed by the action of wind, glaciers and waves, or volcanic and tectonic activity. Lakes, however, are most abundant in glaciated regions, and this explains why the largest number of lakes are found in higher latitudes and in the high mountainous regions. Lakes abound in the higher latitude of North America, Western Europe and South America which have been recently glaciated. In the humid regions lakes are common alongside rivers with low gradients and wide flood plains, where they are connected with changing river channels. Lowlying areas bordering the sea may also have lakes. In the arid and semi-arid lands the lakes are usually shallow and their area contracts or expands markedly with the season. In the Eastern Highlands of Africa there is an extensive chain of lakes which are connected with tectonic movement and rift valleys and include the lakes of Albert, Edward, Tanganyika, Nyasa and Victoria. In the U.S.A. and Canada there are the five Great

Lakes— Superior, Huron, Michigan, Erie and Ontario which constitute the largest chain of lakes in the world and which have been formed essentially by glaciation. Besides these are numerous smaller chains of lakes in Canada which are similar in origin.

There are important lakes in every continent, but the following deserve special mention. The Caspian Sea in the biggest lake in the world. It has an area of 1,70,000 sq. miles (436,400 sq. km) and is a saltwater lake. To the east of it is the Aral sea in the Soviet Union which has an area of 25,000 sq. miles. The Baikal in Siberia is the deepest lake in the world and has a depth of 5,600 ft. The lowest lake in the world is the Dead Sea in Israel whose surface is 1,268 ft below sea level. It has salt water. The lake floors of L. Baikal, Dead Sea and the Caspian Sea are far below sea level. Lake Titicaca located in Peru and Bolivia is usually considered to be the highest lake in the world. It is 12,500 ft above the sea level. There are, however, some lakes in the Himalayas which are at a higher altitude, such as the Devatal lake in Garhwal Himalayas, Mansarowar, Gaurikund lake etc. But these are relatively small. The Chad lake in Sudan is only about 8 to 20 ft deep but no other lake in the world is marked by such great seasonal variation in area. In the dry season it is only 10,000 sq. miles but swells to 50,000 sq. miles in area in the rainy season.

The Origin of Lakes

Two things are essentially for the formation of a lake. Firstly, there should be a depression or basin where the water may collect, and secondly, there has to be supply of water to fill the basin. Most of the fresh water lakes have an inlet through which a river enters the lake, and also an outlet through which the river again drains out. The bottom of the basin is at a lower level than the outlet, because then only it is possible for the basin to hold some water. Normally the salt lakes have no natural outlet, and on account of evaporation the water level never rises so high as to fill the basin completely or to overflow it.

The lakes receive their water supply principally from rivers which drain into them. Some lakes receive directly the melt water from glaciers, and in some the principal source of water supply is either direct rainfall or/and run-off from the neighbouring areas. In still others water may be obtained from underground sources. Some lakes get water from springs, while is others the water seeps from the neighbouring higher land and collects in the basin. It is thus clear that it is not necessary for precipitation to be in excess of evaporation over the lake area for a permanent lake to be formed, as direct rainfall over the lake surface supplies only a very small part of its water supply. This explains why permanent lakes are found in areas where evaporation exceeds rainfall.

In the hot and dry regions where the rainfall is scanty and evaporation is rapid, even the larger depressions remain dry. But where the surface drainage of a large area and more especially that from a neighbouring mountain collects in a basin, the water does not stay the year round on account of the intense evaporation and the lake either dries up completely or is converted into a swamp in the dry season. Where permanent lakes do exist, the basin is rarely full and the water never overflows or drains out. These lakes, therefore, have no outlet.

The first requirement for the formation of a lake is the existence of a basin which may be surrounded on all sides by relatively high ground. If the water supply is adequate, the basin will be filled up to the brim and the excess water, if any, will go out through the outlet. The depression or the basin may be formed either by deposition or erosion or by earth movement. Consequently lakes may be formed in any one of these ways. These depressions are referred to as lake basins.

Types of Lakes

Lake basins may be classified according to the agencies which were responsible for their formation and the same may be used for the classification of lakes. Accordingly following Hutchinson. (1957) we may classify lakes into ten principal types based on the mode of origin of the lake basins.[1]

1. Tectonic lakes,
2. Lakes associated with volcanic activity,
3. Lakes formed by landslides,
4. Lakes formed by glacial activity,
5. Solution lakes,

1. Hutchinson, G.E.— *A Treatise on Limnology, Vol I, Geography, Physics and Chemistry*, Wiley, New York, 1957, Chapter I, pp. 1-163

6. Lakes due to fluvial action,
7. Lakes formed by wind action,
8. Lakes produced by marine processes,
9. Lakes formed by organic or animal activity,
10. Lakes produced by meteorite impact.

W.M. Davis (1882) has classified lake basins into three broad classes on the basis of groups of processes responsible for their formation namely, *constructive, destructive and obstructive.* In the constructive process the rim surrounding the basin is built up by a positive process, in the destructive process the lake basin is formed by subsidence or excavation and in the obstructive process there is damming up of a pre-existing valley on account of some obstruction. This classification is to a great extent artificial and does not help us in understanding the regional distribution of lakes or the exact mode of their origin. For example, in a volcanic region, a basin may be formed either by the destructive process of explosion or subsidence or the rim of the crater may be built up by the constructive process of accumulation or a valley may be blocked by flowing lava. Similarly, in glaciated regions hollows may be scooped out by the destructive process and result in the formation of cirque lakes or lakes may be formed by obstruction created by morainic deposits in a valley. But glacial lakes, whether they are the product of constructive, destructive or obstructive processes, are found in regions which have been subjected to glaciation. Thus the threefold classification suggested by Davis is not helpful in explaining the regional geographical distribution of lakes. Its usefulness is limited to explaining the various processes that are responsible for the formation of lakes of glacial or volcanic origin in a particular region. Hutchinson's classification based on the agency responsible for the creation of lake basins would appear to be more useful and scientific because it enables us to easily understand and interpret the origin and geographical and regional distribution of lakes over the globe.

Tectonic lakes

The lake basins which owe their origin to earth movements are called tectonic lake basins. Lake basins may be formed by earth movements in several ways. Epeirogenic earth movement may isolate portions of the sea which may take the form of inland seas surrounded on all sides by land. Such relict seas are called *relict lakes* and have saline water. Caspian sea and the Aral sea are good examples. In the early Tertiary period they formed part of the Tethys sea which extended up to the Atlantic Ocean in the west but by the late Miocene period due to the uplift in this area, it was separated from the Atlantic and the big lakes came into existence. The Caspian Sea is the biggest lake in the world.

Sometimes when continental shelves are uplifted to form land the irregularities on its surface take the form of lakes. Such lakes are called *newland lakes.* In Florida, U.S.A. there are several shallow lakes of this type, the most prominent being lake Okeechobee.

Tectonic uplift across the course of a river may impound the natural drainage or gentle tilting may lead to the reversal of drainage, In such cases the upper portion of the stream may be converted into a lake. Lake Eyre in the Great Basin of South Australia, and Lakes Kioga and Kwania situated in the upper part of the Kafu river in middle Africa are examples of lakes formed in this manner. Local subsidence caused by earthquakes sometimes results in the formation of depressions which when filled with water become lakes. Reelfoot Lake in Tennessee in the U.S.A. has been formed in this way.

Lake basin may also be formed by folding. The synclinal basins formed by folding are the sites of Lake Rudolf in Middle Africa and of Lake Fahlensee in Switzerland. Such lakes are called *synclinal lakes.*

Among tectonic basins the most important are those formed by faulting. Such lake basins are of two kinds—(1) lakes associated with a single fault, and (2) lakes associated with rift valley or graben. They are called *fault lakes and rift-valley or graben lakes* respectively. In fault lakes the margin on the side of the fault scarp is steep. Abert Lake in Oregon, U.S.A. is a good example of fault lake. In fact there are several examples of fault lakes in the Sierra Nevada and Great Basin region of western U.S.A.

Graben or rift valley lakes are long and deep with vertical and steep sides. There are several lakes of this type in the Great Rift Valley of East Africa, the more important ones being Nyasa (Malawi), Tanganyika, Edward, Victoria, Albert (Mobutu) and Rudolf. The most important of these is Lake Tanganyika which has a maximum depth of 1,435 metres. Only Lake Baikal in Siberia is deeper than this. Lake Tanganyika is 650 km long and covers an area of 31,900 sq km. The Dead Sea is also a rift valley lake whose floor is 792 m below sea-level and the lake level is about 400 m below sea-level. There are several examples of graben lakes in Central Asia, Lakes Baikal and Balkhash being the most prominent. They are long, narrow and deep. Lake Baikal is the deepest lake in the world. It is 674 km long and has a maximum width of 74 km. The basin of lake Baikal was formed essentially in the Pliocene age. Other examples of graben lakes are Lake Matana in the Celebes and Lake Torrens in South Australia.

Most of the Himalayan lakes are of tectonic origin having been formed by the Tibetan rivers encountering the rising Himalayas, though some are considered to have been formed due to obstruction caused by glacial morains in river valleys. The tectonic lakes are of cause much older than the glacial lakes. Tibetan lakes, Lakes Baikal and Tanganyika are at least 20 million years old.

Lakes associated with Volcanic Activity

Volcanic activity gives rise to lake basins in various ways. Leaving aside the arid areas, lakes are found in large numbers in volcanic regions. Different kinds of volcanic lakes are to be found in the volcanic regions of Iceland, Eifel District of Germany, Auvergne region of France, Newzealand, north-western U.S.A., Central America and the Andes.

Lakes may be formed by craters of extinct volcanoes. Craters are formed either by explosive eruption or by collapse of the overlying rock after considerable lava has been drained out from the volcanic vent. Caldera is an extended form of crater. Although a distinction is sometimes made between a crater lake and a caldera lake, both are commonly referred to as *Crater lakes*. The Lonar lake in Maharashtra occupies a circular caldera (diameter 1800 m) surrounded on all sides by 500 ft (168 m) high steep hills. It is considered be a collapse structure. In East Africa, north of Nayasaland right up to Ethiopia several crater lakes are found in the zone of the Great Rift valley. The Crater Lake in Oregon, U.S.A. has been formed in a caldera. It is nearly circular, has an area of 20 sq miles and is surrounded on all sides by steep sided hills which rise 500 to 2000 ft above the lake surface. The biggest caldera in the word lies in Sumatra which is partially occupied by Lake Toba.

Here both volcanic activity and faulting have taken place. Examples of crater lakes are also found in Japan.

Sometimes the pyroclastic material ejected with the lava accumulate as a circular ring round the vent. The crater wall in such cases consists primarily of ash, lapili and rock fragments with very little lava. In such craters relatively small but deep lakes are formed and are called *maars* in Germany. Besides the Eifel District in Germany, such lakes are found in several volcanic regions in Iceland, Italy, Java and New Zealand. The Pulver maar in Germany, Lake Viti in Iceland and Lake Tikitapu in the North Island of Newzealand are good examples of this type. Quite often maars are found on the parasitic craters along the flanks of volcanic hills. Lake Grati situated on the lower flanks of Tennger hill in Java is a good example of parasitic maar. A large group of parasitic maars is found on the slopes of Gunung Lamongan volcano in eastern Java.

Some lakes have been formed by the combined influence of volcanic and tectonic activity. These are called *volcano-tectonic basins.* Examples of lakes formed in this manner are to be found in Eastern Middle Africa, North Island of Newzealand and Sumatra. As already stated, Lake Toba in Sumatra occupies a volcano-tectonic basin.

When a valley is blocked by a barrier of lava, a lake is formed above the barrier. This is possible in two ways. Either volcanic hills may be formed in the river valley or the valley may be dammed by a lava barrier. In either case the natural drainage is blocked and a lake is formed. Lake Texcoco in Mexico is an example of a lake created by the formation of a volcanic hill in a river valley. Probably even a more beautiful example is provided by Lake Kivu in East Africa. As a result of the formation of a volcanic hill which is even now an active volcano, the upper course of the river Ruchuru, a tributary of the Nile, has been blocked and has been converted into a lake. Examples of lakes formed by the damming up of river valleys by lava flow are found in several volcanic regions in Japan, New zealand, Philippines and France. Lake Lanao in Mindanoa, Philippines and Lake Omapere in the Waitangi valley in Newzealand are examples.

Lakes formed by Landslides

In hilly areas with steep slopes there is sometimes sudden mass movement of rock fragments under the influence of weathering and heavy rainfall. This is particularly true of shale, clay or decomposed igneous rocks. Landslides are also common on steep as well as mild slopes in areas affected by earthquakes. Sometimes as a result of heavy landslide alongside a river valley, the course of the river may be blocked and the portion above the landslide dam converted into a lake. Lakes of this type are usually extremely short-lived and temporary, as the water level above the dam keeps on rising and soon overtops it or the dam ruptures after some time under the impact of water. When the dam breaks, there is heavy flooding in the valley below. Several examples of such short-lived lakes are available from the Himalayas. In 1840 a lake 64 km long and 300 metre deep was formed in the Nangaparbat area as a result of the damming of the Indus river by a landslide caused probably by an earthquake. The barrier lasted for nearly six months. It broke in June 1841 and the lake was drained of its water within 24 hours. In September 1893 in one of the head waters of the Ganga, a 250 high dam was built as a result of landslide at a place called Gohana and the lake above it lasted for abut 8 months.

Where the stream is small and the landslide big, there is a possibility that the landslide dam may become permanent, and then the lake may not be short-lived. There are several small lakes of this type in the Warner range of north-eastern California such as, Clear Lake, Blue Lake, Peat Lake and Lost Lake etc. Lake Sarez in the Murgab Valley in the Pamirs is a beautiful example of a lake held by a landslide and is quite stable. Examples of lakes formed by landslides are also found in France, Eastern Alps, Japan and Newzealand.

Glacial Lakes

Lakes formed by glacial activity exceed in number lakes formed by all other agencies put together. Glacial lakes are found in large numbers in the regions which were affected by the Pleistocene glaciation in North America, Europe and Asia. In the glaciated regions lakes have been formed both by erosion and deposition.

1. *Lakes formed by glacial erosion are broadly of three different kinds—ice-scour lakes, valley rock-basin lakes and cirque lakes.*

(a) Small ice-scoured depressions are formed by ice-sheet erosion, especially along zones of fracture or weakness. Lakes formed in these basins are small but usually occur in large numbers. Good examples of such lakes can be seen in Norway, southern Sweden and Finland. Extensive lakes may be formed where the basins produced by continental ice-sheet scouring are on an enormous scale, of which the Great Lakes of North America are an excellent example.

(b) The valley rock basins formed below the snowline by glacial erosion and plucking form valley rock-basin lakes of which many of the lakes in Alpine valleys, the larger lakes of the English Lake District and the Fjord lakes of Norway are examples. Sometimes the valley rock basin lakes are long and linear and resemble the fingers of human hands. They are, therefore, called *finger lakes*.

As we move down the glacial valley, we come across a series of steps which are called glacial stairways. At the mouth of every stair there is usually a rock bar riser or riegel behind which lies a depression. These depressions in the glaciated valleys form a chain of lakes at different levels. These are called *paternoster lakes*. Examples of this type are found in Alaska and the Western Cordillera of North America.

(c) Cirque, corrie or cwm lakes develop in high mountains at the head of glaciated valleys after the glacier has melted. Good examples of cirque lakes are found in the cordilleras of N. America, Scandinavia, English Lake District, eastern Alps, Tibet, New-zealand etc. While the ice-scoured lakes are generally irregular in shape, the cirque lakes are usually sub-circular.

2. In addition to the lakes formed by glacial erosion there are *lakes formed essentially by glcial deposition*. On the disappearance of continental glaciers the ground moraine is deposited unevenly on an extensive area in the form of boulder clay or till. The ground has an uneven surface with many shallow depressions scattered irregularly. These may be converted into shallow lakes which are almost completely surrounded by glacial débris. Even more important than these are the lakes which are formed by terminal moraines and recessional moraines. As the glacier retreats, the recessional moraines are deposited behind the terminal moraine. The terminal moraine acts as a natural barrier, and the water held up in the depressions behind the barrier forms a lake. Both the above types of lakes formed by glacial deposition are found in the North European plain and more especially Germany. Both valley

glaciers and continental glaciers give rise to lakes of glacial deposition. But in the valley glaciers it is only rarely that the terminal moraine constitutes a complete barrier and more commonly it is cut through by a stream issuing from the glacier. Further, the terminal moraine is unconsolidated and includes big rock fragments through which the water tends to escape. Generally the lakes formed by the terminal moraines are not very big or deep, but several small lakes owe their origin to this process.

In the outwash plains sometimes blocks of ice get trapped underneath the outwash deposits of glcial drift, and when the blocks of ice melt the overlying deposits sink and form depression called kettles. Lakes filling these depressions are called *kettle lakes*. They are usually small and shallow but are extremely common in the glaciated regions of North America and Europe.

Under certain conditions the glaciers itself acts as a barrier and gives rise to ice-dammed lakes. Many glaciers come down sufficiently below the snow line. When the main valley is filled with glacier up to a sufficiently low level while the rivers flow in the tributary valleys, the glacier in the main valley impounds the water in the tributary valleys, giving rise to lakes. A good example is lake Marjelensee in Switzerland situated in the lateral tributary valley of Marjelen. The main valley is occupied by the Aletsch glacier. Sometimes the barrier breaks, the then the water flows down in the form of a flood.

Sometimes the lateral glacier may impound the water of the river flowing in the main valley. A good example is the Gapshan or Shyok ice lake in the Muztag-Karakoram mountains. The Shyok is a tributary of the Indus. In the head-waters of the Shyok river three glaciers come down from the west and meet, and usually the flow in the Shyok is held back by one of the three glaciers. But sometimes the ice barrier breaks and them there is a heavy flood down below. The lake, therefore, is not permanently present.

Though the lakes formed by glacial erosion and glacial deposition have been considered separately, there are several glacial lakes which owe their origin to the combined action of both. In the Andes region of Patagonia in South America there are several lakes which have been formed by the deep scouring of the valley by glacier together with the blockage of the valley by the terminal moraine. The lakes of the English Lake District have been mostly formed by glacial excavation but there are also evidences of morainic damming The Great Lakes of North America which are the largest group of fresh water lakes in the world have been formed by the repeated advance and retreat of the Pleistocene icesheets, involving both glacial erosion and deposition, with isostatic rebound and perhaps earth movements. There was lowlying land here which was further excavated by the glacier, and excluding Lake Erie the bottom of these lakes is very much below sea-level and is also quite irregular, indicating local glacial scouring.

Fluvial Lakes

Various kinds of lakes may be formed by fluvial erosion and deposition, but such lakes are usually small, temporary and short-lived. The big waterfalls create a plunge pool below them by their corrasive power. When for some reason the river changes its course, a lake is formed in the plunge pool depression created by the falls. They are called *Plunge pool lakes*. Examples of such lakes are found in the Grand Coulee, Washington state, U.S.A. which are located in the old course of the Columbia river e.g. Falls Lake, Castle Lake, Lenore Lake etc.

In wide intermontane valleys when lateral mountain streams join the main valley, alluvial fans are built in the main valley by lateral deposition, and sometimes the entire valley may be blocked by alluvial fan deposits. In such cases a lake is formed above the barrier. This is especially likely to happen in semi-arid regions in valleys of tectonic origin which are comparatively wide but the river itself contains little water and is unable to cut through the alluvial barrier. Lake Tulare and the Owens Lake in California are examples of lakes formed in this manner.

A special kind of fluvial lakes is *lateral lakes* which are formed when the course of small tributary streams is obstructed by high aggradation along the natural levee as well as bed of the main river. The high rate of silt deposit in the main valley not only blocks the course of the tributary stream but during flood the water of the main river starts flowing into the tributaries causing delta-like deposits at the mouth of the tributary stream and further strengthening the barrier of silt deposits. In this way lateral lakes are formed along the tributary stream. Good examples of such lakes can be seen in the tributaries of the Red river and the Yangtzekiang in China. In the Yangtzekiang valley Lake Tung-ting is a relatively large lake of this kind. Other examples are lateral lakes along the northern bank of the Lower Danube and in the valley of the Darling river in New South Wales, Australia.

Basins formed by abandoned channels in mature floodplains of rivers form distinctive types of lakes called *ox-bow or cut-off lakes.* Lateral erosion of the meanders, their increasing curvature and finally separation from the main channel followed by deposition along the meander mouth are the processes involved in the formation of these lakes. Such lakes are found in the flood-plains of the Ganga and its tributaries, the Mississippi, the Murray-Darling and other rivers.

In the deltaic stage the river is divided into several channels on account of the heavy deposition of sediments. The river meets the sea in U-shaped channels with the mouth of U opening towards the sea. But slowly the mouth of U-shaped banks is closed by deposition resulting in the formation of lakes. Lakes also form in the depressions between the levees of the distributaries in a delta. As these lakes are close to the sea and in contact with sea water, their water is usually saline. Such lakes exist in the deltas of all major rivers like the Mississippi, the Danube, the Rhone, the Nile, the Yangtzekiang, the Ganga and the Indus. In the Ganga delta these lakes are known as *bhils*. Possibly some of them are ox-bows which very such resemble the deltaic levee lakes in shape and size.

Sometimes lakes are formed in the depression lying between two contiguous deltas. The Kolar lake lying between the deltas of the Godavari and the Krishna rivers on the eastern coast of India is an example.

Solution Lakes

We have seen that in the karst regions solution of limestone by water gives rise to roughly circular depressions such as dolines, uvalas and polje. They may be filled by water to form lakes which are collectively called *karst lakes.* They are also given individual names after the nature of the depressions, such as doline lakes, uvala lakes or polje lakes. The most favourable condition for water filling up these depressions is when there is a rise in water-table either on account of change in climate or change in the sea-level. The solution lakes of Florida, U.S.A. have been formed by this process. The dolines were formed in the Pleistocene Ice Age when the sea-level and water-table were low, but when the sea level and water table rose after the Ice Age, the dolines were filled with water. There are several examples of solution lakes in the limestone region of Florida. The Deep Lake is a beautiful example of a circular doline lake. The Long Lake is a polje lake formed by solution in a tectonic basin, in which several sinks are joined together in a fault scarp. Solution lakes are found at high altitudes in the limestones of the Alps, but they also show evidence of glacial action. The Muttesee lake in Switzerland is a good example of lake formed in an uvala. The poljes in the karst region of the Balkan Peninsula are now mostly dry, but there are evidences of their being occupied by lakes. In Croatia there are the Plitvic lakes which are located in a complex combination of polje and uvala.

Examples of solution lakes are also available from other limestone regions of the world. Lakes Otjikoto and Guina in S.W. Africa are doline lakes. The Nainital lake in Kumaon Himalayas is a solution lake according to some authorities, though faulting has also played a part.

Lakes formed by Wind Action

Wind is not an important agent in the formation of lake basins. In the arid and semi-arid regions the wind is capable of building depressions both by erosion and deposition, but on account of the scarcity of rainfall, these depressions rarely get filled up with water permanently.

Deflation basins or hollows are formed by wind action in sandy deserts. Such deflation basins are found in large numbers in Texas, New Mexico and Nebraska in the semi-arid high plains to the east of the Rocky Mts in U.S.A. Where the drainage of the neighbouring mountains or highland collects in the large basins, temporary saline lakes called *Playa lakes* are formed. Examples of lakes in this type of basin occur in the dry lands of S. Africa, Central Asia and Australia, though some of them occupy rock basins.

Long, narrow and shallow lakes may be formed in the depression between dunes especially where the depressions are underlain by impermeable rock. Such lakes are found in the Tarim Basin of Central Asia, and in Cherry country, Nebraska, and in Indiana south of Lake Michigan in U.S.A.

Sometimes the small desert stream may be dammed by wind blown sand dunes, giving rise to temporary lakes. Examples of this type occur in the Namib desert of S.W. Africa. Some water is able to flow out of the sand dunes, and, therefore, the water of the lakes does not become saline. In N.E. Colorado and western Nebraska there has been a constant struggle between the small streams and the wind blown sand. The streams are weak and seasonal, while the wind is more powerful. As a result several small valleys have been blocked by mass of wind-blown sand above which temporary lakes are found.

Coastal Lakes

The formation of coastal lakes and lagoons is related to the formation of bars along the coast. The process of bar formation has been described in the next chapter. When the bar is deposited by longshore drift at the mouth of the bay, then on account of the bay mouth bar the water situated between the coast and the bar gets separated from the sea and becomes a *lagoon*. A lagoon is an elongated body of shallow water laying parallel to the coast and separated from the open sea by sandy or shingle barriers built by the waves. Lagoons are most common along low coasts, usually athwart stream estuaries, where large quantities of river-borne sediments are available for the formation of barrier islands. Lagoon water is commonly saline, as the sea water enters the lagoon at high tide. But when the lagoon receives adequate water supply from the rivers, there is decrease in salinity.

On the western coast of India there are several lagoons associated with river mouths. They are locally known as *Kayals*. Most of them are parallel to the coastline and are small and elongated. Of the western coast lagoons the largest is the Vembanad lake in Kerala which is 75 km long and 8 to 15 km wide. The Pambiar river drains into the lake. The eastern coast of India is comparatively lowlying and contains several long and narrow lagoons. Largest of these are the Pulicat lake on the Madras coast and the Chilka lake along the Orissa coast. The Pulicat lake is 55 km long and 5 to 16 km wide. The salinity of the Pulicat lake and other lagoons on the eastern coast increases during the period of the north-east monsoon. The Chilka lake is approximately 60 km long from NE to SW along the Orissa coast and its width varies from 7 to 20 km. Hills occur on its north-western and south-western margins and there are a few rocky islands in the lagoon. In reality, it is a baylet which has been made shallow by the deposits of the distributaries of the Mahanadi, and subsequently it has been separated from the sea by the growth of barrier islands at its mouth by longshore drift.

The most favourable conditions for the formation of lagoons exist (a) when on account of a rise in sea-level the river estuaries are drowned, followed by a fall in sea-level which stabilizes the bars, (b) when there are extensive sandy hills and sand-dunes along the coast which obstruct the course of the rivers under the influence of winds.

Apart from the formation of bars near the river estuary and bay mouth, it is possible for lagoons to be formed in other ways. Where some island situated near the coast is joined to the coast by two tombolos, the water between the two tombolos becomes isolated and a lagoon is formed. Examples of such saline lakes or lagoons are found on the Italian coast. Sometime a cuspate spit or hook may bend and be joined to the coast again, leading to the formation of a lagoon.

Coastal lakes may be formed also by accumulation of organic matter. Among these those formed by coral reefs and atolls are by far the most important. The most suitable conditions for the growth of corals are to be found in the tropical seas where the water temperature is at least 20^0 C (68^0 F) and the water is clear, mobile and not more than 7 metres deep. Consequently the distribution of *coral lagoons* is limited to tropical seas between 25^0 N and S. latitudes. Coral lagoons are found in the western Pacific Ocean, parts of the Indian Ocean and the Caribbean sea. The water in these lagoons is calm, and they provide good harbours for the inhabitants of the Pacific islands, although the passage to the lagoon through the coral reefs may sometimes be quite treacherous and dangerous. Coral lagoons are irregular in shape, but those formed by atolls are circular. While the water in the lagoon behind bars near river estuaries is muddy, the water in the coral lagoon is clear. Most of the coral lagoons are less than 20 metres deep, although there are some which are more than 60 metres deep. The lagoon in the Mayotte Island in the Indian ocean which lies behind a barrier reef has a maximum depth of 92 metres, but its average depth is much less. It has a maximum width of 16 km.

Lakes formed by Meteoric Impact

When large meteorites strike against the surface of the earth, the impact leads to the formation of a circular hollow. Because of the explosion (impact) the hollow is larger than the meteorite and the meteorite is also thrown out of the hollow. Depressions formed in this manner are mostly dry, as possibly the preservation of these craters (*called astroblems*, meaning star-wounds) is easier in the arid lands than in the humid ones. The Great Arizona crater which is at present dry was occupied by a lake in the Pleistocene and its bottom contains above 30 m thick lacustrine deposits.

There are a few lakes in the world which are believed to be meteorite crater lakes without any doubt. These include Kaalijaru lake in Osel island in the Baltic, Laguna Negra lake in Granchaco of Argentina, and Chubb or Ungava lake in Quebec. The crater of Chubb lake is one of the biggest meteorite craters in the world and has a diameter of about 3 km and a maximum depth of 251 metres. The largest known is the Ashanti crater occupied by lake Bosumtwi in Ghana with a diameter of over 9.5 km.

The Disappearance of Lakes

From the geological point of view lakes are transitory features with a very limited life-span. It is inevitable for lakes, large or small, to disappear at one time or the other. The origin of lakes is usually related to catastrophic and revolutionary events of diastrophism, volcanic activity or ice age, but gradually and imperceptibly they attain maturity and old age, and then die quietly. Some of the lakes are in fact extremely short-lived.

Except for the lakes found in the flood-plains and deltas of rivers, almost all other lakes are in their young age and are indicative of a relatively young topography. One important reason for this is that the majority of the lakes are of glacial origin and were formed either during or after the Pleistocene age. Since they are quite recent geologically speaking, not enough time has elapsed for their basins to be filled with sediments or vegetation or for rivers to drain them out completely.

Whether lakes will take a long or short time to disappear depends primarily on their own characteristics and the related conditions. The disappearance of lakes is possible in three ways-by the lowering of outlet below the bottom level of the lakes, by the filling up of the lake basins, and by the tectonic uplift of the lake bottom. The rivers going out of the lake erode the outlet vertically downwards and when the outlet becomes sufficiently low the lake water drains out rapidly. By deposition of sediments also the bottom of the lake is raised and the lake not only becomes shallower but its water drains out more easily while the outlet is cut deeper and deeper. In a way, the lowering of the outlet and the filling of the lake basins are complementary processes.

Rivers which enter a lake bring sediments with them which are deposited in the basin. Some rivers which pass through several lakes deposit most of their sediments in the upper lakes while the lakes further below receive little sediments. In the semi-arid regions the rivers carry a heavy load of sediments. The Colorado river in North America which flows through a semi-arid region, deposits so much sediment in Lake Mead that it has been estimated that the lake will be filled up completely in about 200 years. On the sea coast also the lakes are liable to be filled up by deposition of debris brought by waves. In the arid regions the wind transports and deposits large quantities of sand and silt in the lakes. In some places volcanic ash has been responsible for the filling up of lake basins. Similarly the lake basins may be filled up by the extension of vegetation, landslides, rock-slide by frost action or by morainic deposits.

The third way in which lakes may be destroyed is by the vertical uplift of their bottom by earth movements. This has proved quite effective in some of the older lakes. In Kashmir old lacustrine deposits called Karewas prove the existence of lakes in tectonic synclinal troughs in the Pleistocene times.

Lakes occupy only about 1% of the earth's land surface, but they are very useful in reconstructing its past environment. Lakes are areas of net accumulation of sediments, just as mountains are areas of net erosion. The lacustrine sediments preserve a record of past ecological conditions and geomorphic processes which can be studied by modern scientific techniques. They thus provide a valuable record of regional climatic change in the geological past and act as a natural archive for the surrounding landscape.[1]

1. Roberts, Neil - Climatic Change and Lake Levels in *The Oxford Companion to the Earth,* Oxford Univ. Press, 2000, p. 125

Questions

1. Classify lakes on the basis of the mode of their origin and give examples.
2. What are the conditions necessary for the formation of lakes ? Why are some lakes saline and other fresh water? Give examples.
3. Give with examples an account of the formation of lakes by tectonic and volcanic activity.
4. Explain the mode of formation of the different types of glacial lakes and give examples.
5. Write short notes on any *three* of the following:-

   Lagoons, Solution lakes, Fluvial lakes, Graben lakes, the disappearance of lakes.

Chapter **28**

Marine Erosion and Coastal Landforms

Coastal Terminology

It is convenient to start with the definition of some of the terms which are used to describe coastal and shoreline features. The *shoreline* is defined as the line where land and water meet, and the shore is the zone between mean low tide line and the line of the highest tide or the inner edge of wave transported sand. The *shore* is thus a zone of periodic submergence, and remains sometime above

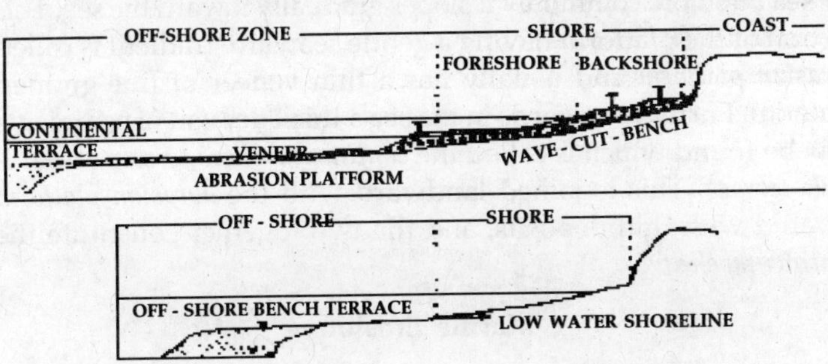

Fig. 186—Coast, shore and the off-shore zone.

the sometime covered with water. The water-line periodically migrates over this zone. The *shoreline* is thus not a fixed line, and several shorelines may be recognized, such as low water and high water shorelines or spring and neap tide shorelines. The coast is the broad zone directly landward from the shore. It includes the sea-cliffs and elevated terraces as well as the lowlands inside the shore. The distinction between the coastline and shoreline is not very clear but *coastline* is synonymous with the cliff-line and usually refers to the maximum high-water shoreline.

The shore is divided into two parts—the foreshore and the back shore. The *foreshore* is the outer sloping part of the shore zone which lies below the bermedge[1] and extends down to the low water mark. The *backshore* is the nearer section of the shore zone which lies between the foreshore and the coastline. The breaker zone directly seaward of the low tide line is called the *offshore zone* (Fig. 187).

In the zones described above we find distinctive erosional and depositional landforms. The section of the shore zone which is covered with unconsolidated debris extending landward from the mean low water line to the effective limit of ordinary storm waves where there is change in material or physiographic form, is known as *beach*. This deposition takes place on the wave-cut bench and is also the result of wave action. When this bench extends seaward into the off-shore zone, this is known as *offshore bench or beach terrace*, while the upper section of the bench which is covered by beach deposit is called the *back-shore terrace*. The wave-cut bench or platform may sometimes end abruptly with a sharp break in slope towards the sea but more commonly it slopes gradually toward the sea. This section of the platform having a gentle seaward gradient is called *abrasion platform* and usually has a thin veneer of fine-grained material. Further outwards in the sea a thick deposit of sediments may be found which is called the *continental shore terrace or wave-built terrace*. This is joined landward with the *abrasion platform* covered with thin deposits, and the two together constitute the *continental shelf*.

Marine Erosion

The principal agent of marine erosion is breaking waves armed with rock fragments. Earlier it was believed that ocean currents and tidal currents also have an important role in marine erosion. But recent studies have shown that these currents have little influence on the formation of coastal features, and their principal work is to transport the fine debris produced by wave attack.

The Origin and Nature of Waves

In order to understand the erosional and depositional work of waves, it is necessary to know the origin and the nature of waves. Waves are caused by friction on the ocean water surface by winds.

1. Bermedge is the seaward limit of the nearly horizontal part of a beach.

Under the influence of winds the level surface of the ocean water develops ups and downs and becomes undulating. In waves some part of the water rises up while some other part is depressed. The highest part of the wave is called the *crest*, and the lowest or the depressed part of the wave is called its *trough*. The horizontal distance between two consecutive crests is called the *length of the*

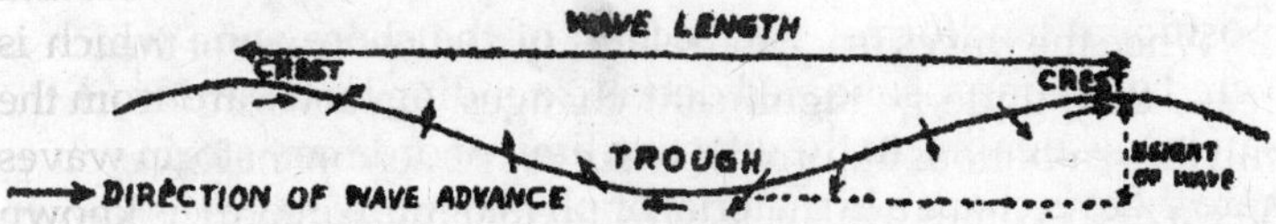

Fig. 187—The length and height of waves.

wave. The vertical distance between the crest and the lowest point of the trough is called the *height of the wave* (Fig. 188). Waves may be of two kinds—free waves and forced waves. When friction by blowing winds causes waves to develop on the ocean water surface, the waves may be transmitted to long distances from their place of origin independently of the wind. Such waves are called *free waves*. But when the wind is strong, it presses against the side of the trough as a result of which the height of the wave increases. Such waves are called *forced waves*. Normally the height of the waves does not exceed 20 to 30 feet but the height of the forced waves may go upto 50 ft. In the open seas the length of the waves ordinarily does not exceed 300 ft. but in storms this has been found to reach 2,500 ft.

An important thing to remember is that in the open seas the water does not move forward with the wave; it is only the wave form that advances. Just as the sound waves pass through the atmosphere and the earthquake waves pass through rocks in the earth's interior but this does not result in the forward movement of either air or rocks, so the water also acts merely as a transmitting medium for the waves, and in the open seas it is only the waves that advance and not the water. The speed of a 300 ft long wave is about 30 miles per hour.

In deep water inside the waves the water particles have a circular motion near the surface of the ocean water. On the crest of the wave they have a forward motion, and in the trough they have a backward motion. With increase in depth the circular orbit or cirumference decreases and the circular movement also decreases, and at a depth equal to the length of the wave the motion stops altogether. Broadly speaking, it would be correct to say that on sea coasts, erosion by waves become negligible at a depth equal to one-half of the length of the waves. If the average length of the wave for most places may be taken as 200 ft, then it can be said that erosion by waves is limited to the zone between high-water mark and low-water mark.

When the waves reach the relatively shallow water of the coastal areas., they undergo significant changes. On account of friction with the sea-bottom the orbit of water movement in the lower part of the waves becomes oval instead of circular and ultimately becomes linear. Friction also reduces the rate of movement, resulting in increase in the height and length of the waves. Because of the increase in height the waves become unbalanced and the crest breaks and is thrown forward. This is known as *plunge* (Fig. 189). Normally the waves break when the depth of water below the trough becomes equal to the length of the waves. When the waves break, water rushes forward towards the coast with considerable speed. The uprush of water from the breaking waves is known as *swash* or *breaker*. After striking the coast with considerable force this water again returns to the sea. This backward movement of wave water is called *backwash or undertow*. The backwash is essentially a bottom flow and carries along with it fine sand and other particles towards the sea. There are some who are of the opinion that the backwash does not return in the form of bottom flow but moves seaward principally in localized lanes or streaks of water called *rip currents*, in contrast to the non-localized flow of undertow. There is no doubt that some water returns to the sea in the form of rip currents, as marks of shallow channels cut by the backwash are found on the sea-floor near the coast.

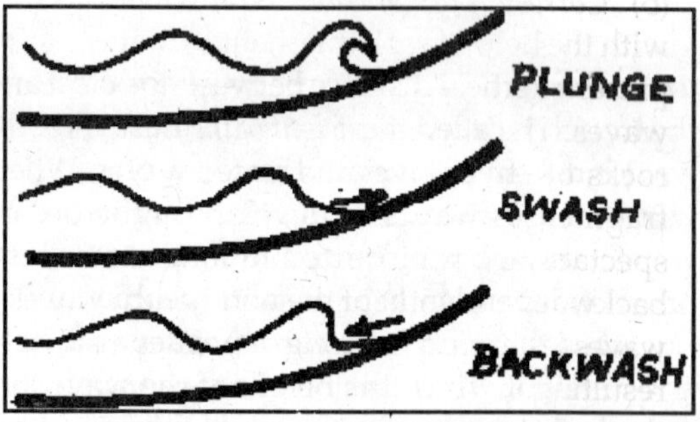

Fig. 188—Plunge, Swash and Backwash.

Processes of Marine Erosion

Five distinct processes may be said to operate in marine erosion by waves. They are corrosion, corrasion or abrasion, attrition, hydraulic action, and the shock pressure of breaking waves.

(a) *Corrosion* : The waves dissolve the soluble rocks or the soluble parts of other rocks along the coast. This is a chemical process and is called corrosion. This process is not so significant as its influence is limited to limestone and other similar soluble rocks. Further, on account of the relatively high percentage of calcium carbonate in the ocean water, solution is slow.

(b) *Corrasion or abrasion* : The waves accomplish erosion mainly with the help of rock fragments. Coarse sand, pebbles and boulders present in the sea-water near the coast act as tools of erosion by the waves. The waves armed with these tools strike against the solid rocks of the coast with the force of an artillery, break them into fragments and transport the fragments towards the sea. Less spectacular but more common than this action is the forward and backward movement of sand, pebbles and small boulders by the waves (especially large waves) against the cliff and over the platfrom, resulting in the corrasion or abrasion of he floor. In fact corrasion is the most important process of marine erosion, and is performed with the assistance of rock fragments which act as tools of erosion. It may also smoothen the surface of the shore platform as abrasive materials are moved backward and forward across it.

Abrasion implies actual physical break-up of cohesive rock. Some what similar process called quarrying is the pulling away by waves of particles of materials which have been prepared by some other agency specially blocks provided by jointing or cleavage. Closely jointed rocks like some basalts or those with pronounced cleavage (slates and schist) are especially favourable to quarrying process. Where the sea is cutting into unconsolidated material and directly freeing pebbles and sand particles, it is also quarrying. In this process the sea disaggregates rather than disintegrates. Quarrying is more effective in high energy coasts of temperate latitudes where big storm waves are more common. Greater quarrying means more tools for the abrading waves.

(c) *Attrition* : In the process of corrasion the fragments of rock also strike against one another and keep on continually breaking into smaller pieces. This breaking up of rocks by mutual impact and their conversion into finer fragments is known as attrition. As a result of this process the rock fragments ultimately become so small that under-tow and rip currents can easily transport them towards the sea.

(d) *Hydraulic action* : Even when the waves do not contain rock debris, they are able to erode the coastal rocks by their movement and forceful impact. When the breaking waves strike against the coastal rocks like a big hammer, this impact is sufficiently powerful to break and disintegrate the rocks.

(e) *Shock pressure of breaking waves* : When the breaking waves strike against the coastal rock, they exert an enormous amount of pressure, especially on the rock joints and cracks. In these joints and cracks, the air becomes compressed on account of the enormous pressure of the waves, and it appears as if suddenly wedges have been thrust into the cracks. Then suddenly as the water returns with the waves, the pressure falls sharply and the air expands and the rock starts exploding and breaking along the joints and cracks. When water enters these cracks, not only there is increase in mechanical stress, but solution activity also increases in soluble rocks.

Factors Affecting Marine Erosion

We may divide the factors affecting marine erosion into two broad groups : (a) firstly, the influence of the waves themselves including their shape and size and the manner of their approach to the coast, and (b) secondly, the characteristics of the coast such as, height of the cliff, rock structure, alignment of the coast, amount and nature of debris on the beach etc.

Effect of Waves

The effect of the waves depends mainly on the intensity of impact of the striking waves and the presence and nature of rock fragments along the coast.

Where the sea is deep right up to the coast, the waves strike against the coast without breaking and return after being refracted. The returning refracted waves act as impediment in the path of new waves. The result is that the waves become stationary near the coast and their crests appear to be fixed and motionless. The shaking caused by these waves has been found to damage structural joints of buildings, but their influence on jointed rocks is not exactly known. It would be, however, true to say that where the sea water is deep near the coast and the coast is steep the impact of the waves is not very forceful or active. On the other hand, where the coastal water is shallow, the waves impinge on the coast with greater force and erosion is more effective.

The breaking waves exert a great amount of pressure on the coastal rock, but the intensity of this pressure depends on the distance between the cliff and the breaking of waves. Sometimes an air pocket may get trapped between the cliff and the breaking waves, and in such cases the air exerts an enormous pressure on the cliff.

Apart from the manner of breaking of waves, their size has an important bearing on the erosion of the coastal rocks. The size of waves depends on wind velocity and the extent of sea water. The stretch of water on which the wind blowing from a particular direction gives rise to waves, is called the *fetch*. Wind of moderate velocity blowing on a large expanse of water can give rise to bigger waves than strong wind blowing over a smaller fetch. Where the fetch is large, the size of the waves is dependent on how long they have been moving under the influence of the wind. This will depend on the duration of the wind as well as on the width of the fetch.

Earthquakes and volcanic activity may sometimes give rise to very big waves in the sea called *tsunamis*. They may be 100 miles long and have velocity of more than 400 miles per hour. In the open seas their height may not be more than 1 to 2 ft but by the time they reach the coast, their height increases enormously. Along the coasts of Japan and Chile their height has been found to the nearly 100 ft or more. The erosion done by tsunamis during a relatively short period may exceed several years of erosion by ordinary waves. But such waves are comparatively rare.

Another important element in the erosion of cliffed coasts composed of hard rocks is the amount of boulders and rock fragments brought by the waves. The effects of these rock fragments are different from the pressure of the breaking waves. The breaking waves break the coastal rocks along joints, but when the boulders and rock fragments strike against the coastal rocks with the waves, they break the rough and angular projections of rock and in the process they themselves also become smooth and rounded. This is why the cliff rocks below the high tide level are found to be generally smooth and the rock fragments on the beach are usually rounded.

Where the coastal rocks are relatively soft or incoherent, the pressure exerted by the waves is not so important, and most of the erosion is performed by the stones and boulders striking against the coast with the waves and breaking the fine rock particles with comparative ease. The base of the cliff is thus easily cut. As a consequence the overhanging upper part of the cliff becomes unstable and falls down. On such coasts extensive shingle beaches are usually found, and the shingles have been derived from erosion of the cliff. But subsequently these shingle deposits tend to protect the cliff from further erosion.

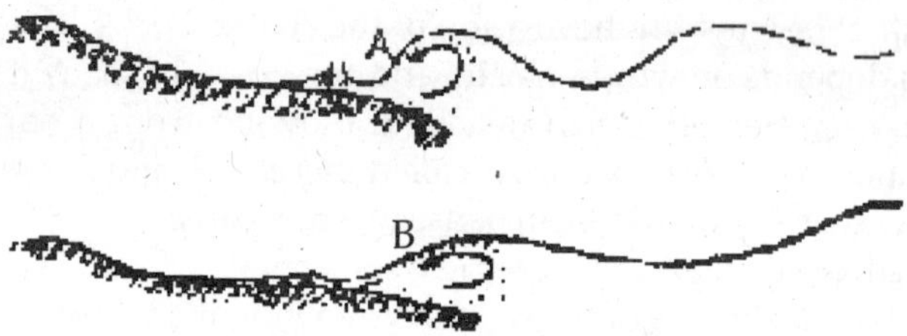

Fig. 189—Destructive (A) and (B) Constructive waves.

Where there are shingle deposits along the coast, erosion by waves depends upon whether the waves are constructive or destructive. High frequency waves (13-15 per minute) are destructive and low frequency waves (6-8 per minute) are constructive (Fig. 190). This difference is due to the fact that the direction of fall of water after the breaking of waves is different in the two types. In high frequency waves the motion of water is nearly circular. Consequently when the waves break, the direction of fall of water on the beach is down ward. Therefore the velocity of swash i.e. of the water moving from beach to the coast is weak. Further, on account of the high frequency of the waves, when the second wave arrives, the backwash of the first wave is flowing underneath in the opposite direction. This further weakens the already weak upward movement of water. Such waves succeed in dislodging the deposits on the beach and transport them towards the sea.

In low frequency waves the motion of water is oval or elliptical. The result is that the swash or the upward surge of water is quite powerful and this carries along with it upward the shingle deposits of the beach. The backwash which has to return through these shingles, is sluggish and in spite of favourable slope is unable to transport down the shingles carried upward by the swash.

High frequency or destructive waves are usually related to onshore winds. When the on-shore winds are strong, the level of water rises along the coast, and on the sea-bed there are counter currents flowing from the coast towards the sea. Although these counter-currents are quite weak, they are still helpful in transporting down the material loosened by the waves. Low frequency waves may originate by winds hundreds of kilometres away from the coast and advance as free waves towards the coast. At the time of their arrival near the coast, light breeze may be blowing from any direction. Thus together with the wave which is itself a constructive wave, light offshore breeze may blow. When this happens, then on the sea-surface there may be seaward movement of water and on the sea-bed water may be moving towards the coast, and in this way the constructive activity of these waves may to some extent be further strengthened.

But the difference between the constructive and the destructive waves is neither so simple nor so clear, and big destructive waves are quite capable of transporting and depositing the beach shingles above high water mark and thus perform constructive work. Big waves, whatever be their frequency, erode the beach, and on the other hand, small waves, whether of high or low frequency, assist in building the beach.

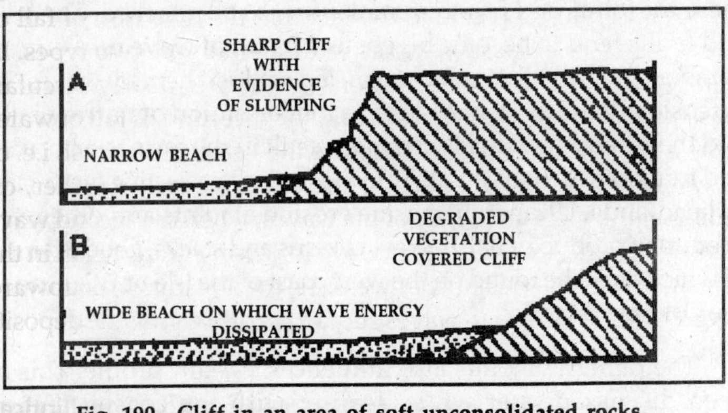

Fig. 190—Cliff in an area of soft unconsolidated rocks. A—At the time of active erosion, B—After erosion has ceased.

Effect of the Nature of the Coast on Erosion

Under the effects of the nature of the coast, we may consider the effects of the following :- (a) the nature of the rocks i.e. their hardness, the influence of chemical weathering on them and the presence of joints and cracks in them, (b) the height of the cliff, and (c) orientation and indentation of the coast, and nature of the shore away from the coast.

Two important things relating to the nature of rocks are their hardness and the existence of joints in them. The harder the rooks the fewer will be the joints in them and lower will be the rate of coastal erosion. Maximum erosion will take place in uncosolidated or incoherent rocks. Where soft unconsolidated rocks are present, the rate of erosion may be from 5-10 ft to 100 ft. per year. For erosion to continue at a rapid rate, it is necessary for the sea to remove the eroded rock debris. Otherwise they will be deposited on the beach and form depositional features on which the energy of the waves is wasted and which protect the base of the cliff from wave attack. On coasts where erosion is active, the base of the cliff shows signs of scouring and the unconsolidated rocks above it either slump or fall down. On the contrary, on coasts where erosion has stopped, the slope is relatively gentle and low and is usually covered with vegetation. In cliffs of soft rocks erosion is very active in the beginning, but later on this becomes very sluggish on account of the deposition of debris derived from coastal erosion.

The process of erosion in the case of hard rocks is different. In these the joints and fissures are the first to be attacked. The joints and faults tend to become bigger as a result of wave pressure and abrasion by the rock fragments hurled against them by waves. The extension of the joints may lead to the formation of narrow inlets, and the extension of the faults may result in the formation of caves and tunnels which may enter into the narrow projecting landmasses or headlands. Ultimately marine erosion of joints and other weak spots may produce small narrow caverns and stacks, good examples of which are to be found on the west coast of the lsle of Wight and in the Orkneys islands.

The pattern of joints also influences the cliff profile. This has been discussed later while dealing with the coastal features produced by erosion.

Though in most areas the sea water is saturated with calcium carbonate, the influence of chemical action on limestone and other soluble rocks cannot be ignored. In the opinion of some, the solution on the coast is performed not so much by sea water as by rainwater flowing from land to sea. But this view does not appear to be wholly correct, as landforms produced by solution are found even in such dry areas as the Red Sea. One reason given for solution by sea water is that with decrease in temperature the solubility of carbon-dioxide increases. Consequently, when the temperature of sea water falls at night, solution activity increases as a result of increase in acidity. But a more plausible explanation seems to be the difference in amount of carbon dioxide between day and night due to the activities of algae. During the day the algal absorbs carbon-dioxide from water in the process of photosynthesis, and due to decrease in carbon-dioxide the sea water starts precipitating minute particles at calcium carbonate. The waves carry these precipitated particles When the algae releases carbon dioxide at night, the acidity of the sea water increases again resulting in not only solution of the precipitated calcium carbonate particles but also in chemical attack on the rocks.

There are some non-calcareous rocks which too are chemically affected by sea water. Some rocks and minerals which weather more rapidly in sea water than in rain water include basalt, obsidian, orthoclase felspar and horneblend.

The height of the cliff also exerts an important influence on its erosion or rate of recession. It is clear from Fig. 192 that for a given amount of marine erosion the amount of debris increases in the same ratio as the height of the cliff. In Fig. 192 A the erosion of the coast provides debris from the area a b c d, while in Fig 192 B similar amount of erosion results in the accumulation of debris derived from the area a b e f. This debris collects at the base of the

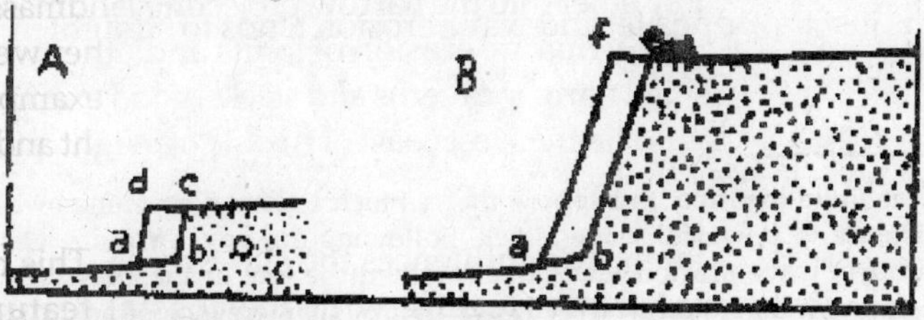

Fig.191—Influence of cliff height on its recession

cliff and protects it from further erosion by the sea. So long as this debris is not broken into fine particles and removed from the cliff base, it is difficult for the sea to perform further erosion. Consequently, the rate of recession of a high cliff is much slower than that of a low cliff.

The portions of the coast which project into the sea as headlands are the first to receive the attack of the waves. It is here also that the wave attack is at its maximum. When the waves advance towards the coast, the front of the wave is bent in sympathy with the coastline and is refracted. Water in front of the headlands is usually shallow, and when the waves reach shallow water, their flow is retarded,

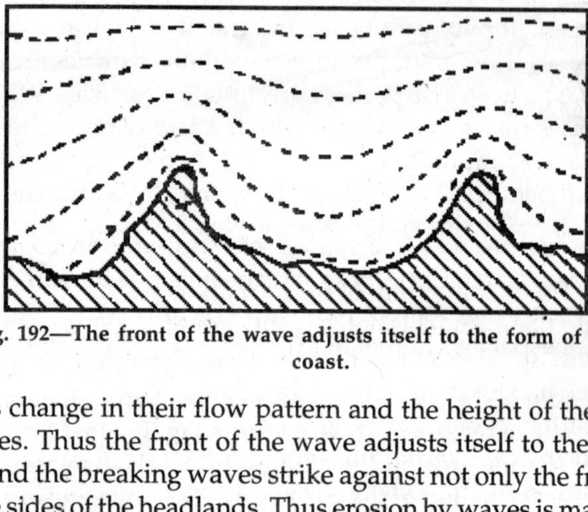

Fig. 192—The front of the wave adjusts itself to the form of the coast.

there is change in their flow pattern and the height of the waves increases. Thus the front of the wave adjusts itself to the coastal form, and the breaking waves strike against not only the front but also the sides of the headlands. Thus erosion by waves is maximum on the headlands and the bays are relatively protected. In this way over a period of time the indentation of the coast-line becomes straight. It is incorrect to suppose that the coast-line becomes indented under the influence of powerful wave attack Actually its effect is just the opposite, and wave erosion tends to straighten the coastline.[1]

1. "It is quite incorrect to suppose that a much indented coastline owes its existence to powerful wave-attack, hollowing out bays or gulfs. On the contrary, wave erosion tends to straighten the line of the coast. A high degree of indentation nearly always inplies submergence of an irregular land surface. Wooldridge and Morgan—op. cit., p. 324.

Landforms resulting from Marine Erosion

Two major landforms produced by marine erosion are the sea-cliffs and the wave-cut benches or platforms. In addition, there are some smaller but striking features related to these, such as notch, cave, arch, stack etc. which deserve consideration.

Sea Cliffs

In quite a few places the seacliff formed by marine erosion is a distinctive feature. When the waves first start their work of erosion on a sloping land mass (Fig. 193) they cut a small nick or notch at

Fig. 193—Development of sea-cliff and the wave-cut platform.

place a. This is the g point of the formation of the sea-cliff and the wave- cut platform. This notch is cut at the base of the slope where the waves are able to reach at the time of high tide. The debris produced by erosion gets transported towards the sea. The height of the cliff goes on increasing with the continued erosion of the coast, and below it the width of the wave-cut platfrom also goes on increasing simultan-eously. Sea-cliff formed in this fashion by marine erosion has a steep slope and at the base of the cliff there is a notch cut by wave attack. When the notch becomes big, the overhanging part of the cliff becomes foundation less and unstable and starts falling down. By this process the cliff continually recedes towards the land. The notch occurs near high-

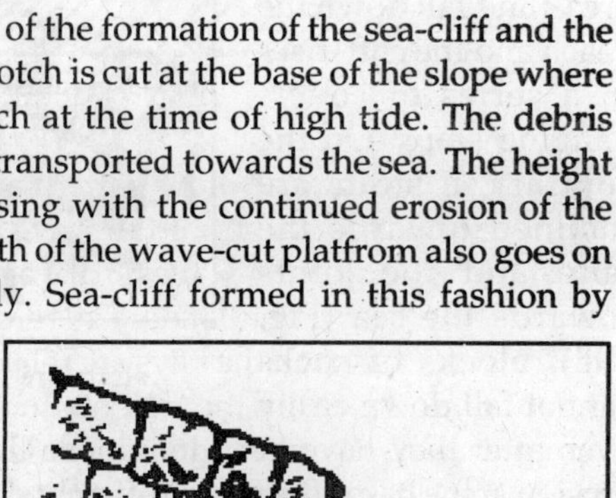

Fig. 194—Erosion of the sea-cliff where the rock strata are inclined towards the sea.

water mark, because it is here that the waves strike against the coastal rocks with maximum intensity. Inside the notch the rocks become smooth and rounded by abrasion, but the rocks above the notch are rough and angular on account of weathering and disintegration by falling. The portion below the notch is either covered with rock debris or beach pebbles or is joined to the beach-platform by a gentle slope on account of erosion. Although there is no straight-forward relationship between the height of the notch and the sea level, it is possible to get an idea of past sea level with the help of notches. At the present time we come across some examples of notches which are above mean sea level but below high water level. But in a general way it may be stated that in seas where the tidal effect is nominal, the notches will be found at the height of the mean sea-level.

The form of the sea cliff depends primarily on the type, nature and structure of rocks. Joints and bedding planes particularly influence the cliff form, as the rocks break most easily along the joints and the bedding planes. Where the rock strata are inclined towards the sea and the joints are inclined toward land (Fig. 195), big blocks of rocks easily break and fall down the beach and the cliff rises in a series of over-hanging steps. On the contrary, if strata are inclined towards the land and the joints towards the sea (Fig. 196), blocks of rocks cannot fall down easily even after they have been loosened. Under such conditions the erosion at the base affects the cliff only slowly and the cliff usually slopes towards the sea. If the strata of rocks are inclined steeply towards the sea (Fig. 197), blocks of stone easily slide down along the bedding planes, and the front of the cliff is smooth and cannot be easily eroded by the waves.

Fig. 195—Erosion of the sea-cliff where rock strata are inclined towards land.

Similarly, the form of the granite cliff is different from that of basalt cliff. Where the rocks are hard sandstone or granite, and especially when they are massive and lacking in joints and bedding planes, the cutting and recession of the cliff are very slow. But where

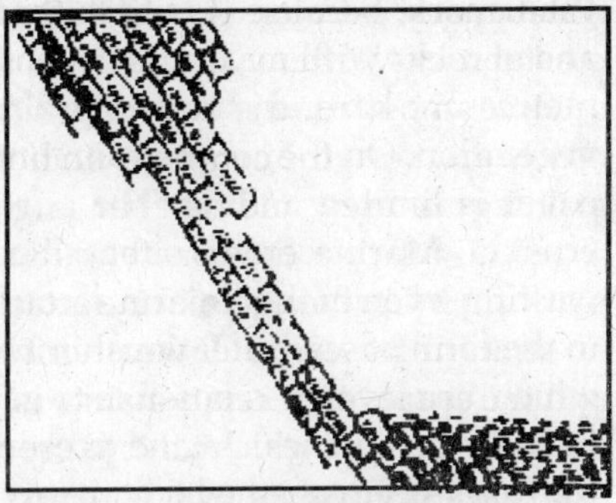

Fig. 196—Erosion of the sea-cliff where the rock strata are inclined steeply towards the sea.

the rocks are weak shale or glacial till, the recession of the cliff is very fast and there is rapid slumping and land-sliding above the notch. On the east coast of England where costal rock is weak Pleistocene sediment or sand, the coast is receding at a fast rate.

When the coastline is receding towards land on account of wave attack, it is called *retrograding coastline*, and the process is called the recession or retrogradation of the sea cliff. Apart from the intensity of the wave attack, the rate of recession of the sea-cliff depends also on the hardness of the rocks. As a result of differential marine erosion, the waves cut the softer coastal rocks more rapidly, and consequently, at such places the sea water enters into the land, while the hard rock coastal areas project into the sea as headlands or buttresses. In the soft rock coastal areas we find semi-circular or semi-oval coves, bays or bights which have been cut by wave erosion. At the same time it is doubtful if the waves are capable of producing significant indentation along the coast by differential marine erosion and it appears that they are successful largely in producing only small and secondary indentations by this process.

The close relationship of cliff form with rock type and the presence of features inherited from past denudation make generalization about the geographical distribution of cliff morphology extremely difficult. J.A. Steers (1953) has shown the extreme variety that may exist within a limited area such as the British Isles. [1] Based on the relative efficiency of masswasting and wave attack in different parts of the world, it is possible, however, to make some broad generalizations. Tricart and Cailleux (1962) have suggested that there is significant defference between arid and humid coasts in the tropics.[2] On hot wet shores, such as the

1. Steers, J.A.—*The Sea Coast*, Collins, London, 1953
2. Sea Davies, J.L.—*Geographical variation in Coastal Development*, Oliver & Boyd, Edinburgh, 1972, pp.92-93

Pacific coast of Colombia and the coast of Liberia, cliffs recede slowly and at relatively low angles by mass wasting. Only the lowest few metres are bare, the rest being covered by almost continuous vegetation. On the contrary, on hot, dry shores plant and regolith cover is limited and steeper angles are encouraged by marine erosion. Marine erosion tends to be more effective than mass wasting, even though marine attack is relatively weak and largely in the form of sea water weathering and solution. The main factor which upsets this relationship is the occurrence of limestones, including corals which tend to produce arid type cliffs throughout the tropics because they do not generate substantial waste mantles of weathered material.

As we move polewards from the tropics, quarrying by big waves becomes proportionately more effective. The temperate regions are high wave energy latitudes where marine attack tends to dominate on all rock types and to evolve steeper and more spectacular cliffs. Further poleward, however, along Arctic and Antarctic shores, wave energy decreases once more and periglacial mass movement processes maintain relatively low angle slopes.

In temperate latitudes the presence of two-cycle slope- over-wall cliffs has been attributed to strong degradation of older cliffs by Pleistocene periglacial mass movement and subsequent steepening of their lower section by postglacial sea (Cotton 1952, Bird, 1968). Slope-over-wall cliffs occur in may parts of the world and they probably reflect the intervention of a low sea level phase in which marine erosion was absent and subaerial processes held sway. Where both marine and subaerial processes were in turn particularly effective, such as in Western Europe, where periglacial solifluction extended below present sea level during glacials and wave energy was high during inter-glacials, the slope- over-wall form is especially well-marked.

On cliffs where marine processes dominate, a notch may form at the base. But it is now widely recognized that a notch is best developed in low energy environments where weathering, solution and biological processes are more effective in cliff retreat than wave quarrying and abrasion. This is because the big waves tend to destroy the upper limb of the notch and the varying wave height gives rise to a broad, relatively shallow indentation. Maximum notch development is found on tropical limestone coasts on account of various solution processes and relatively weak wave attack within a limited vertical range. A low tidal range is supposed to be of great importance in the formation of notches.

Distinctive Erosional Features of Cliff Coasts

The waves striking against the coastal rocks produce a number of small but quite distinctive erosional features. Where the coastal rocks have well-developed joints or weak dykes, the waves cut them to form *sea caves*. The sea-caves can be formed both by mechanical abrasion or by chemical corrosion. The size of the caves goes on slowly increasing. The interiors of the caves extend a little beyond the reach of waves. This is so because the waves are preceded by compressed air which enters these cracks under great pressure. When the waves return, and the mouth of the cave opens, there is decrease in pressure and the compressed air expands with explosive intensity and breaks the rocks. By this process not only are the

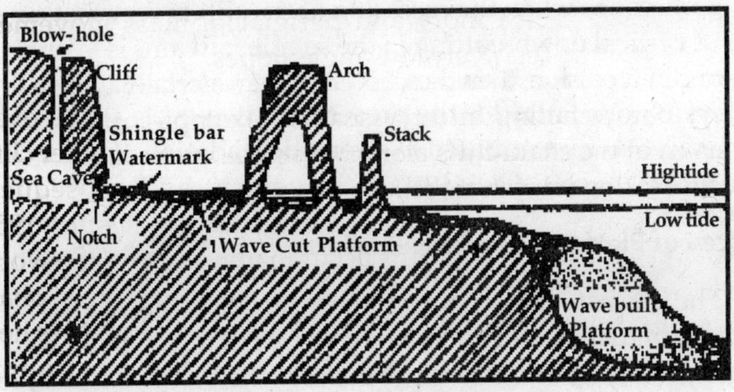

Fig.197—Some distinctive erosional landforms of Cliff coasts

caves enlarged, but sometimes the roof of the caves cracks along joints right up to the rock surface above through which the air escapes above with a hissing sound. Such vertical holes are called blow-holes (Fig. 198).

The narrow arms of land which project into the sea as headlands or buttresses may develop caves on either side which ultimately may join together from side to side while the upper portion of the buttress may stand like an arch. These are known as *sea-arches*. In course of time when the arches fall down or the projecting portion of the headlands is separated from the cliff, vertical pieces of rocks may be found standing in the sea separated from the cliff. Such features are called *stacks or sea stacks*. They have also been referred to as chimney or skerry. Beautiful examples of sea arches and stacks can be seen on the Normandy coast of France. Here the cliff is composed of chalk. Good examples of sea arches and stacks are also found in the Capri Islands of Italy. Arches, stacks, blow holes etc. are relatively rare in nature and are largely confined to high wave energy coasts where structure and lithology are suitable for their development. Within tropics minor examples occur on limestone coasts and in association with corals but otherwise they are uncommon.

It is important to remember that caves, arches, buttresses or stacks are secondary features which are formed in different stages in the process of cliff recession. By this process the cliff continually keeps on receding towards the land and the beach platform gets extended.[1]

Sometimes when the recession of the cliff is very fast, and the rate of vertical down-cutting in the smaller streams is less than the rate of cliff recession, then these rivers form waterfalls along hanging valleys before falling into the sea. An example of this may be observed in the chalk cliffs along both sides of the strait of Dover. (See the section on waterfalls).

Wave-cut Platform or Beach-Platform

There is usually a bench or platform extending seaward from the base of the sea-cliff and having a slope of less than 5^0. Its width may extend to several hundred metres. The bench is composed of bare rock or it may contain a temporary deposit of rock debris, pebbles or sand. It is formed by wave erosion, and as the cliff receds under the impact of marine erosion, so does the width of the wave-cut platform go on increasing. Thus the platform gets extended towards the land by erosion of the coast. The debris derived from coastal erosion become smaller by attrition and the undertow transports them towards the sea and deposits them after some distance. Consequently, while the upper portion of the platform is wave-cut or abrasional platform, it gradually becomes a wave-built

---

1. "In the general view and also in the long view, caves, arches, buttresses and stacks are no more than details in the form of a receding line of cliffs." Dury, G.H.—*The Face fo the Earth,* Penguin Books, 1960, p.100

or depositional platform towards the sea. In areas of hard rocks the surface of the wave-cut platform consists of bare rock and is relatively flat, but where there is deposit of shingles and sand, the slope is concave.

There is some controversy about the formation of the wave-cut platform. There is a sudden change of slope at the junction of the platform and the cliff. This point is called the *nip*. One controversy relates to the relation between the location of the nip and the high tide level. This relationship is important, because it is possible to explain the ancient high level marine terraces with its help. It is clear that there is no direct relationship between the height of nip and the high tide level. Nonetheless there is a close relationship between the two, although the high tide level is not a fixed level. Further, erosion is possible above the high tide level because the splash and spray of breaking waves which rise above this level are also capable of marine erosion. Consequently, it is accepted by all that the inner edge of the platform should be slightly higher than the high tide level. In areas where the stratification of rocks is suitable, just above the base of the cliff a subsidiary bench or stepped wave-cut platform is found which is known as *high-water platform.* Weathering and rainwash also play a part in its formation. But it becomes difficult to ascertain if these benches have been formed by splash and spray from the existing sea level or they were formed at a time when the sea-level was higher and the present day splash and spayr have merely modified their form.

The other controversy relates to the width of the wave-cut platform. In the opinion of British scholars it is possible for a sufficiently wide wave-cut platform to be built by marine erosion, but other European and American scholars have expressed doubt

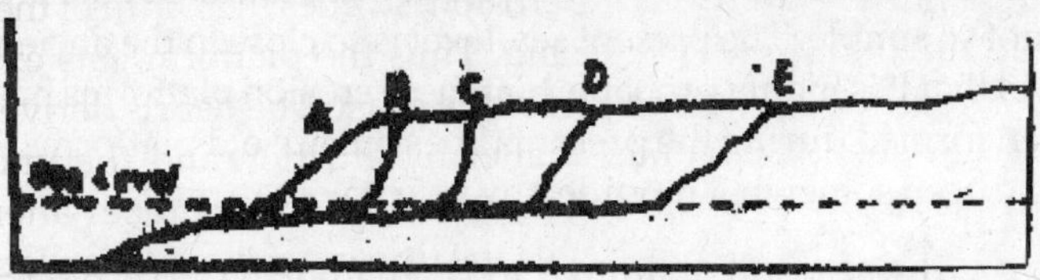

Fig.198—Stages of cliff development.

in the ability of marine erosion to cut very wide platforms. In the opinion of the latter this is possible only if the land is being progressively submerged or where the sea may have merely modified the surface built by subaerial erosion. Further, in the opinion of some, it is only in the early stages of cliff development that marine erosion is active (Fig. 199, A to C), and in the later stages the rocky platform becomes so wide that the strength of the wave is lost on the platform and erosion fails to be effective near the base of the cliff, and the slope of the cliff is reduced by weathering and other agents of erosion (Fig. 199 D & E). On the contrary, there are others who hold that wave-cut platform is built at such a level that it does not impede erosion by waves at the foot of the cliff, and marine planation continues for a long time. It looks as if the truth lies some where in between these two extreme views.

Plains of Marine Denudation

There was a time, especially between 1845 and 1875, when the sea was considered to be an extremely important agent of erosion. But since then there has been a significant change in this view point. In the 20th century the greatest supporter of marine planation has been Douglas Johnson. He put forward forceful arguments in his book "*Shore Processes and Shoreline Development*" (1919) in support of the view that it is possible for extensive plains of marine denudation to develop. But most other people do not agree with this view.

For plains of marine denudation to develop it is necessary for the sea-level to remain stand-still for a long time, but we know that the sea-level has not remained constant during the past several lakh years. Some people have cited the examples of existence of wide platforms in areas of hard rock which appear to be related to the present sea level, and from this they have tried to prove that marine planation is active during the present times. But the problem is not so simple. The present sea-level is so close to the ancient sea-level that it is highly probable that these erosion platforms have not been formed during the present times but are old platforms which have been somewhat modified by recent wave action.

Marine erosion is limited to such a narrow part of the land surface that the formation of extensive marine plains by this process is likely to take a much longer time than erosion by wind or water.

1. Thornbury, W.D.—op.cit. p. 432.

Whereas marine erosion cuts a small area like a saw, wind and water cut extensive areas and bring them down. If there is even slight uplift of land, the marine erosion cycle starts again, but on the contrary uplift causes fluvial erosion to increase. If it appears improbable for the land to remain stand still for such a long time that peneplain stage may be reached in the fluvial cycle, then it seems even less likely for the sea level to stand still for such a long time that extensive marine plains may ever be cut.[1]

Landforms Built by Marine Deposition

Longshore Drift

Debris derived from erosion of the cliff as well as those brought by rivers are deposited near the sea coast. These debris are carved into different landforms by the const-ructive action of waves. The coarser debris are deposited on the beach near the coast, while the finer debris are transported towards the sea. But apart from the seaward transport of debris from the coast, the debris are also transported along the coast. This kind of transport is called *longshore drift or shore drift*. This is accomplished in two ways—by beach drifting and by longshore or littoral currents. The breaking waves usually strike against the coast obliquely, and together with them the debris are also transported towards the coast obliquely. But on the return journey the debris slide down towards the sea on a straight path with the backwash. The second wave again carries the debris to the coast on an oblique path, and it again returns straight down to the sea with the backwash. The result is that as the particles move up the beach with the breaking waves and then roll down in the direction of the steepest beach gradient, they move along the beach with a type of zigzag motion.[1] This kind of lateral movement is known as *beach drifting* (Fig. 200). By this process large quantities of debris and sediments are moved

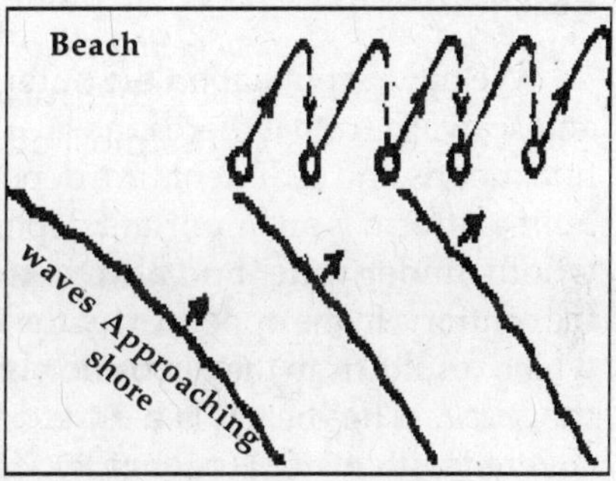

Fig.199—Beach Drifting.

---

1. Sparks, B.W.—op. cit., p. 184

along the coast, and in the course of transport various kinds of depositional features are formed on the beach.

Lateral movement of debris along the coast takes place in another way also. When the breaking waves strike against the coast, the level of water rises near the coast, and under suitable conditions and wind, this water begins to flow in the form of current parallel to the coast. Such currents are called *longshore currents*. These particularly develop at the time of high tide.

Both beach drifting and longshore currents move simultaneously and in the same direction. But longshore drift is mainly due to beach drifting caused by waves, while the work of currents consists essentially of transporting sand and smaller particles. The direction of drift is essentially in the direction of the prevailing wind. For example, on the southern coast of England where the winds are westerly, the movement of debris is towards the east, while on the eastern coast of England where the winds are from the north-east, the debris are transported towards the south.

Beaches

We have seen that on the outer edge of the wave-cut platform and seaward from it there is a wave-built platform or terrace where finer debris and sediments are deposited on the sea-bed. The wave-built platform is an important depositional feature, but it is formed wholly under water and always remains covered with water. On the contrary, in the upper part of the wave-cut platform, the landform which results from the deposition of debris along the coast, is called the *beach*. The beach is a zone of periodic submergence and is covered with water at high tide. This deposit is usually temporary as there is a constant movement of debris by the waves. When the waves are strong at the time of storms, the beach is cut and eroded, and at the time of ordinary waves there is increase in the deposition of debris. The seismic tsunami waves sometime either completely wash away the beaches or push them towards the land. Where the coast is receding, the beach is especially short-lived, and where the coast is prograding or advancing towards the sea, the beaches are semi-permanent.

The beach may extend along the coast for several kilometres, just as we find on the south-east coast of the U.S.A. or they may be of limited length. Where the coast is broken into headlands and bays, crescent beach is formed in the relatively quiet waters of the bays where the debris and sand derived from erosion of the headlands by waves drift and are deposited. Such small beaches are known as *pocket beaches*. Examples of pocket beaches can be seen on the western coast of India near Ratnagiri and Goa.

Most of the debris that goes into the formation of the beach is derived from the land to which rivers, landslides, weathering of the cliff, marine erosion of the coastal rocks all contribute. But in addition some debris are also derived from the sea which are brought by the waves. The formation of the beach starts in a narrow zone below the cliff or crescents in the bays and them advances progressively towards the sea.

The beach may be formed both on the backshore and the foreshore. Usually the upper portion of the beach is in the backshore and the lower portion of the beach is on the foreshore area. The nearly horizontal part of a beach inside the sloping foreshore is called *berm or beach berm*, while the seaward limit of a berm is known as the *berm crest or berm edge* where the slope becomes more steep. The sloping section of the beach below the berm normally exposed to the wave uprush is called the *beach face*. The deposition of particles on the beach may consist of either rounded rock fragments, pebbles, shingles, or sand or a mixture of all these. This depends on the source of the debris and the action of the waves. Some rocks such as basalts and slates form pebbles readily while granites are relatively rare as pebbles. But since lithology varies rapidly along most coasts, it fails to explain the broad zonal distribution pattern of pebbles, as beach materials which show strong concentration in temperate and colder latitudes and appear to be related to the ability of waves to quarry and to morainic and outwash materials brought by glaciation and deposited during glacial sea levels. On hot wet coasts pebbles are rare or absent. Usually coarser particles such as, rock fragments and pebbles together called shingles are found in the upper section of the beach. In the foreshore i.e. in the lower section of the beach the deposit is usually sand. Good examples of this can be seen on the eastern and southern coasts of England where shingles occur in the upper part and sand in the lower part of the beaches. There are two reasons for this. One is that swash pushes the shingles towards the coast but the backwash is unable to transport them down towards the sea. Secondly, when the waves reach the shallow waters of the coast, there is an increase both in the height and length of the waves and long troughs alternate with high crests. On the crests there is rapid forward movement, but when the flat troughs pass above those areas, there is no movement. In other words, the debris are pushed forward with force with the crest but they remain practically unaffected by the troughs. The shingles which are transported to the upper portion of the beach by both the above methods get deposited there like a mound. These *beach ridges* are low and long ridges of beach materials piled up landward of the berms. They are particularly piled up by strong stormy waves. They usually consist of coarse sand, gravel or shells. They occur singly or in a series of more or less parallel ridges. The shingle ridges have steep slope both towards land and sea. Shingle beaches generally develop along quiet protected places. Consequently, pocket beaches formed in small quiet bays usually consist of shingles.

Where large amounts of sand are received from rivers or the coastal rocks consist of sandy sediments, a long and wide stretch of sandy beach may be found along the coast. On the east coast of India where the coast is low, extensive sandy beaches are found between the mouths of the Mahanadi and Godavari rivers. The beach east of the town of Puri is particularly wide where the width of sand and shingles is nearly 4.5 kms. Crystalline and laterite hills are present near the Mahanadi delta and the Mahanadi brings large quantities of sand by their erosion. The beach near Madras (Chennai) is also sandy, and the entire coast is covered with sandy deposits.

The slope of the beach face is the result of a dynamic equilibrium between swash (run-up) of water up the beach face and the return flow (backwash) of water down the face. Although primarily related to the beach permeability, the slope shows a strong correlation with the grain size of the beach sediment, because the latter determines the permeability. While sand beaches have a beach face slope of 1^0 to 7^0, beaches having pebbles and cobbles have slopes from 12^0 to 24^0. The loss of run-up because the discharge into the beach is ten times greater for beach grains 4 mm in diameter (granules and small pebbles) than for 1 mm (very coarse sand), and the discharge through sand less than 1 mm in diameter is almost negligible for times as short as a wave period.[1] Thus there is almost as much water in the

1. Shepard, F.P.—*Submarine Geology*, 3rd ed., Harper & Row, N.Y.&London, 1973, p.127.

backwash of a fine sand beach as in the run up, and the resultant beach face cannot stand at a steep angle except where a scarp has been cut recently. All coarse sand beaches are soft, making them poor for walking or traffic and the offshore commonly has bars and troughs. The average slope of beach face in the coarse and very coarse sand is 7^0 to 9^0.

Fine sand beaches have very gentle foreshore slope (1^0 to 3^0). The sand is generally packed hard on the foreshore and is firm enough for walking or to support an automobile

Beaches also differ somewhat according to the amount of tide and their exposure to wave attack. In tideless or nearly tideless seas, beaches are often bordered by a series of longshore bars and troughs whereas seas with large tidal ranges are likely to have broad terraces, each with at most one large bar exposed at low tide level.

The beaches of the high latitudes in the Arctic and the Antarctic are very different from those of warmer climates, because of intense frost action, the almost continuous sea along the coast, and the glacial tongues that enter the sea. The pack ice along the shore allows wave action only for short periods and even when the sea is open, the off flying ice pack prevents heavy wave action, and therefore high storm beach ridges are rare. However, low ridges of gravel are very common because of the greater importance of the physical than of chemical weathering in high latitudes. Some beaches may have covering of silt coming from solifluction flows during summer thaws. Sometimes the beach may have chaotic piles of boulders and gravel left by melting ice.

In the Antarctic, many gravel beaches rest on ice. When ice melts, it leaves very irregular surfaces with many collapse depressions in the gravel. The beaches terminate abruptly against snow drift. The beach gravel is generally more angular than elsewhere as it has not been much acted upon by the sea.

Elevated beaches are a common feature in northern and eastern Canada where the land depressed during the glaciations has been recovering its former elevation since the ice melted. These beaches consist mostly of low concentric gravel ridges and follow the contours of the elevated areas.

Cusped Beaches or Beach Cusps

On the seaward margin of many beaches boulders, pebbles and sand are deposited by waves in such a manner that angular ridges are formed roughly at equal distances. The ridges project towards the sea and in between the ridges there are shallow arcuate embayments. These angular projecting ridges are called cusps. They

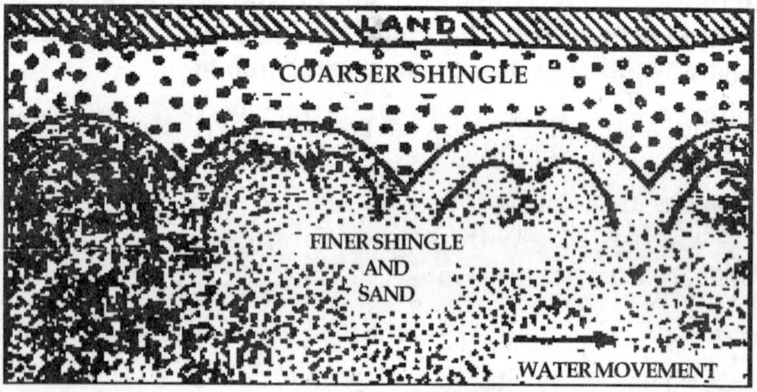

Fig.200—Beach Cusps.

are chiefly formed on shingle beaches. The height of the cusps is usually less than one foot, but some cusps have been found to be 11 ft high. On shingle beaches the distance between two cusps is about 15 ft, but on sandy beaches this may be upto 200 ft. Where the beach is composed of a mixture of shingles and sand, the shingles are deposited in the form of cusps and the smaller particles of sand are deposited in the embayments

The exact reason for the formation of the cusps is not known. But according to Johnson the waves by their swash and back-wash action tend to increase the minor irregularities present on the seaward slope of the beach with the result that small embayments are formed. This process can be understood with reference to Fig. 201.

Bars

In the extended sense the term bar is used for all embankments or ridges of sand and shingles which are deposited on the seabed by waves and currents. These embankments may be found either above or below sea-level. In fact, bars are of several kinds and are given different names according to their location and shape. The most common type of bar is called spit. Other forms of bar are connecting bars or bay bars, tombolo, hook, loop etc. In the formation of all these bars longshore drift plays an important role.

Spits

According to O.F. Evans (1942) a spit is a ridge or embankment of sediment attached to the land at one end and terminating in open water at the other. It is formed essentially by the sediments brought by longshore drift, and its alignment is principally determined by the direction of the longshore drift and the prevailing wind. On the North Sircar coast of India where south-west monsoon winds blow parallel to the coast, the south-western end of the spit is attached to the land and the north-eastern end terminates in open water. Good examples of bars and spits are found also on the Kerala coast in S.W. India where lagoons are usually found behind the spits. Spits form a variety of features which vary in their location and form, such as double spits, winged headlands, hook, loop etc.

Landforms produced by spits

Sometimes long and narrow embankments of sediment are deposited on both sides of the headland by longshore drift and wing-shaped spits fan out on both sides of the projecting headland. Such headlands are called *winged headland*. When the bar is deposited across the mouth of the bay or embayments, this is known as *bay mouth bar or bay bar*. The bay mouth bar separates the water lying between the coast and the bar from the sea and converts it into a lagoon. When the mouth of the bay is totally closed by the bar, the lagoon is completely detached from the sea, but when one end of the bar remains open the lagoon continues to connect with the sea. Sometimes the bay mouth bar may be broken in the middle by tidal or stormy waves and

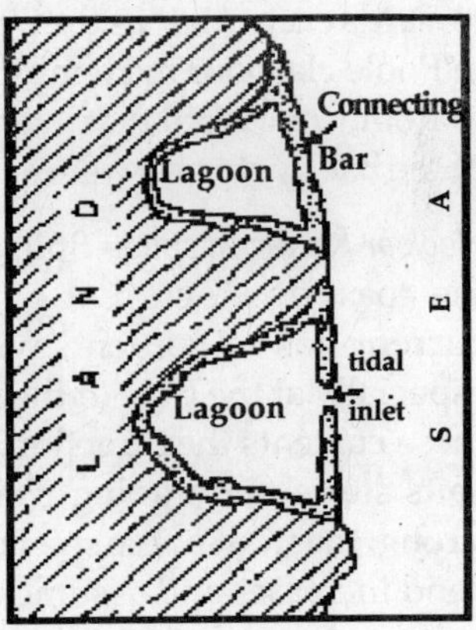

Fig.201—Connecting bar or Bay mouth bar, Double spit and Tidal inlet.

1. Ahmad, E.—*Coastal Geomorphology of India*, Orient Longman, New Delhi 1972, pp. 94-96.

divided into two parts separated by an inlet in between which the lagoon connects with the sea. This feature is called *double spits*. Several good examples of double spits can be seen on the western coast of India, between Cananore and Mangalore. Here there are spits both to the north and south of Kasaragod Harbour and the harbour inlet lies in the middle. Towards the north, where the Shiriya river enters into the sea, double spits occur both on the north and south of the river mouth. It would appear that apart from the tidal or stormy waves, the heavy discharge of water during the monsoon from the rivers descending down the Western Ghats has also contributed to the formation of the inlet.[1] Connecting bay bars which completely cut off the water of the bay from the sea are probably not found along the Indian coast, but on a small scale examples can be found between Nileshwar and Cananore on the western coast.

Sometimes double spits maybe formed not by the breaking of the connecting bar in the middle but in a different way. The spits projecting from two neighbouring headlands continually advance towards each other as they grow, until only a narrow inlet may remain between the two. This kind of deposit is possible where the direction of the longshore drift changes with the seasons, and under such conditions spits in opposite directions start forming from the neighbouring headlands and the spits so formed tend to advance towards each other as they grow. An example of this is to be found in Poole Harbour in Hempshire of England, although it is the view of Robinson (1955) that even this has been formed by the breaking of the bar by stormy waves.[1]

Hook or Recurved spit— Spits are usually formed roughly parallel to the coast down-current of headlands. But the principal littoral current has to contend with some other conflicting currents, especially at the time of the stormy waves. Under the influence of these currents the direction of the deposit at the outer edge of the spits starts undergoing change. When the curving currents are strong, the front of the spit is bent towards the coast. Gradually this bend increases and assumes the form of a sickle. Such spits which are bent towards the coast are called *hook or recurved spit or cuspate spits*. But the spit also continues to build in the forward direction away from coast and a second hook may result by repetition of this

1. Sparks, B.W.—op. cit., pp. 202-204

process. When a spit has several hooks, it is called *compound hook or compound recurved spit*. Examples of hook and compound hook can be seen on the western coast of India in the double spits formed near the mouth of Shiriya river. The formation of the hook has been made possible by the fact that the direction of the longshore drift is from south to north, but strong on-shore winds blow from the west which are successful in bending the front of the spits towards the coast. It is possible that compound hooks have resulted from the formation of different hooks in the spits at different periods as the coast has receded.

Loop—In special circumstances when the other conflicting currents become more powerful than the longshore currents, the hook may be so much curved towards the coast that it may ultimately touch the latter after forming a semi-circle. Thus a loop may be formed in which both ends of the spit touch the coast and inside which the water is ultimately completely separated from the sea. This is a more developed form of cuspate spit. As a result of the successive formation of cuspate spits and loops the coastline tends to move forward. Examples of such land formations are Cape Kennedy in Florida and Cape Henry in Virginia, U.S.A.

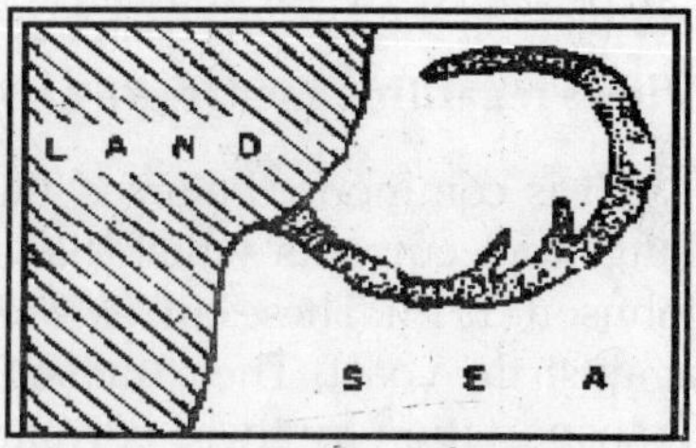

Fig.202—Formation of the Loop.

Connecting Bars and Tombolo—When the spit is so much extended that it connects two headlands or two islands or an island with the coast, then such a bar is called a connecting bar (Fig. 202). A connecting bar may have different locations and is consequently known by different names. As a rule the term connecting bar is used for a bar that joins two neighbouring headlands. When a bar joins an island with

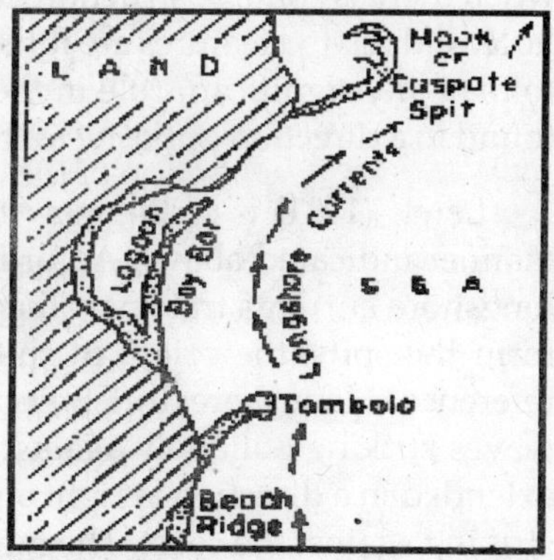

Fig.203—Formation of Hook, Connecting Bar, Bay Bar and Tombolo

the coast, it is commonly known by the Italian term "tombolo" (Fig. 204). Tombolos are principally made of shingles and the most favourable conditions for their development are to be found in areas of submerged drumlins where there may be several small islands and abundant debris are present for deposition of bars by the waves. Good examples are to be found on the Nova Scotia coast, near Halifax. The islands which are joined by the tombolo with the coast are called *tied islands*. Examples of tombolo are found also along the Indian coast. The tombolo near Tuticorin joins Hare island with the coast and there is a lighthouse on this island. On the western coast of India several examples of tied islands and tombolo can be seen between Malwan and Ratnagiri.

Views regarding the Origin of Spits and Bars

It is commonly believed that spits are formed by action of longshore currents which have been called wave currents by Johnson (1919). These currents result as the waves strike obliquely against the coast. They transport the debris and deposit them. Transportation by these currents has been called shore drift by Johnson.

According to one view spits are formed when the currents do not follow the irregularities of the coast but flow in front of the headlands in a straight course. Their speed becomes less in deep water, and some of the sediments being transported are deposited. In course of time these deposits become sufficiently long, wide and high. Their length is aligned in the direction of the currents, and when there is some obstruction or change in the direction of flow, hook and related features are formed. Tidal currents do not appear to play any significant role in the formation of spits, as spits are found in a direction opposite to the direction of tidal currents.

Lewis (1931) is of the view that spits are not formed in the manner indicated above. According to him, while it is true that longshore currents transport the sediments which on deposition form the spits, the origin of spits cannot be explained only by reference to longshore drift. Spits are essentially formed by stormy waves striking obliquely against the coast, and spits or bars are extended in a direction at right angle to the direction of the waves striking against the coast. The most effective are the waves which have travelled over an extensive fetch. In other words, the formation of spits is essentially due to the influence of the waves and not of the currents. The swash is weak in the case of high frequency waves, as the length of the waves is less and the movement of water particles is circular. But the debris move downward without any impediment with the backwash. On the contrary, in the case of low frequency waves, the wave length is more, and there is rapid horizontal movement in swash which therefore is able to carry and deposit shingles on the beach. Consequently, stormy waves which have high frequency and low height essentially determine the alignment of the spits.[1]

According to Steers (1948) the most important element in spit formation is beach drifting. It is by beach drifting that the beach debris are transported and the spits are formed. Beach drifting transports the debris to the terminus of the spits and deposits them there, and then these deposits are turned into ridges by wave action. The outer edge of the spit continues to advance in the direction of the principal beach drifting.

Offshore Bars or Barrier Beaches

Offshore bars are long ridges of shingle and sand which lie above high water level and are formed some kilometres away from the coast. They are not connected with the coast but are separated from the mainland by a lagoon or a marsh. They extend roughly parallel to the cost. Where rivers fall into the sea, the bigger bars are usually broken at places forming tidal inlets.

The offshore bars are formed by the breaking waves in the offshore zone. For the waves to break in the offshore zone it is necessary for the sea-bed to have gentle slope. Johnson (1919) has made a detailed study of offshore bars, and the stages of their formation as outlined by him have been shown in Fig. 205. In the initial stage (Fig. 205 A) the breaking waves scoop out some debris from the ocean bed and give rise to a submarine bar which gradually increases in height with the deposition of debris and finally rises above sea-level (Fig. 205 B). In the shallow lagoon situated between the bar and the coast marsh vegetation starts growing and gradually

1. "While longshore drift provides the material for building, the building itself results from spasmodic progradation, by "frontal" wave accretion, in part, at least at the time of major storms. Wooldridge and Morgan—op. cit, p. 341

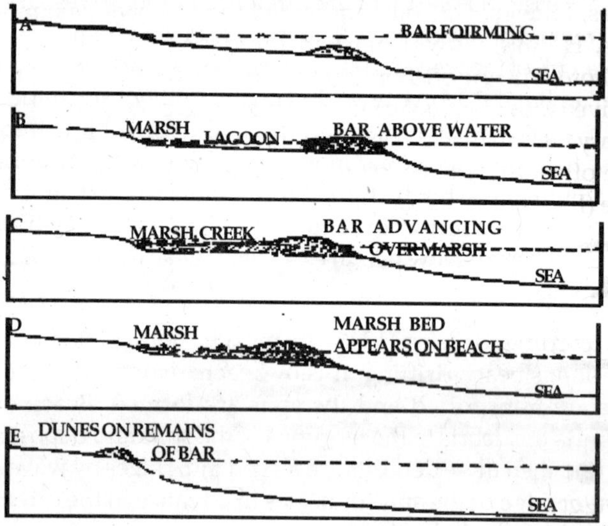

Fig.204—Mode of Formation of Off-shore Bars according to Johnson

the lagoon gets filled up with sediments and the remains of vegetation (Fig. 205 C). Simultaneously the bar is pushed towards the coast by the waves and advances over the marsh, as during storms the waves cut the sediments from the front of the bars and throw them across the bars towards the marsh. The result is that the marsh bed appears on the beach (Fig. 205 D) on the seaward margin of the bars. In course of time the bar is pushed so much inwards that the marsh disappears altogether, and remains of the bar and the associated sand ridges and dunes come right on the coast (Fig. 205 E). Examples of the last stage depicted by Johnson can be seen on the east coast of India between the Mahanadi and the Godavari rivers, where the lagoon and the marsh are absent and sand ridges and dunes are present on a large scale on the coast.

There is some difference of opinion about the source of the materials which constitute the offshore bars. According to Gilbert (1890) the debris for the formation of offshore bars are derived essentially from longshore drift and they are merely worked by the breaking waves. But in the opinion of De Beaumont (1848) and Davis (1909) the debris for the formation of the bars are derived from the ocean bed in front of the bars by wave attack. After detailed studies, Johnson (1919) has accepted the views of Beaumont and Davis. He has shown that had the debris been derived from longshore drift, the offshore bar would be in the nature of merely an additive feature on a surface of uniform gentle slope. But on the contrary we find that the slope becomes steep on the seaward margin of the offshore bars while on the lagoon side the seabed slope is gentle (Fig. 206).

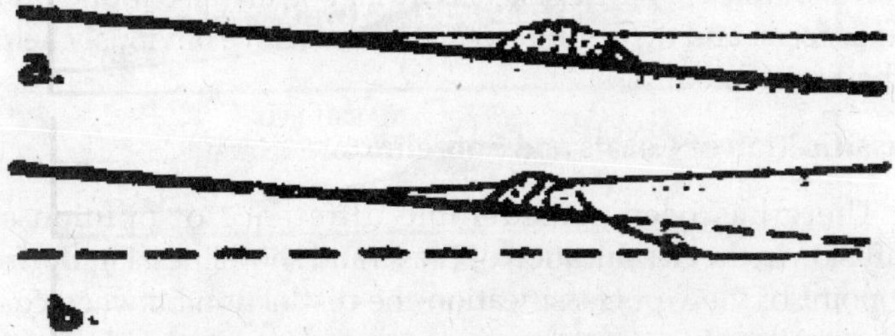

Fig.205—a.Uniform gentle slope and additive feature. b. Steep seabed slope on the outer edge of offshore bar.

The concept and terminology of off-shore bars have undergone considerable change since the time of Johnson. According to Shepard (1952, 1973) the bar is an elongate slightly submerged sand ridge which may bare at low tide. It does not therefore seem proper to use the term off-shore bar for oblong sandy islands away from the coast. Such features can be more than 32 km long, 16 km wide and more than 30 meters high, and it does not seem appropriate to call them bars. According to Shepard a more appropriate term for these features would be *barrier*, a term which has been also used by some other scholars. Where the barrier is built only of sand deposits and has no sand dunes or marsh, it may be designated *barrier beach or barrier island*. Where a chain of such features is found over long distances, it may be called a *barrier chain*.

Shepard is of the view that debris brought by lognshore drift and the shelf sands brought by waves both contribute to the sediments required for their formation, and the action of both longshore drift and waves is important in their formation. In addition, the rise of the seabed has also an important role in their formation, for it is because of this that the deposition of debris is made possible for some height on the barrier island and on both sides of the ridges.[1]

1. Shepard, F.P.—*Submarine Geology*, 1967, pp. 188-190.

Good examples of barrier are found on the eastern coast of U.S.A. A long and almost continuous barrier exists on the east coast of India from the Krishna delta in the north to Point Calimere in the south. Behind it lies a more or less continuous line of tidal lagoons or creeks surrounded by marshes.[1] Where the rivers discharge into the sea, the barrier is broken. The barrier varies in width from 3 to 7 kms, and the sandy deposit rises about 10 metres above sea level. Good examples of barrier backed by lagoon are also found off the Kerala coast, and the strong onshore winds have obviously helped in their formation.

Classification of Coasts and Shorelines

There has been considerable difference of opinion and confusion in the classification of coasts and shorelines. Firstly, from the point of view of classification the distinction between coasts and shorelines is not only unnecessary but impracticable. Even so some people have attempted separate classification of coasts and shorelines. The other reason is the complex nature of most coasts, as in the Pleistocene and Recent times there has been rise and fall of the sea-level several times, and in all coasts and shores there are evidences of this rise and fall in the form of submergence and emergence of the coast. Three factors are extremely important in the classification of coasts and shorelines : (a) the general character and structure of the land which comes in contact with the sea, (b) dominant recent relative movement of land and sea, and (c) the changes brought about by marine erosion in the initial relief.

The broad features of coasts depend to a large extent on the structure and configuration of the landmass, and it is possible to easily identify the similarities and differences produced by them, though it is possible that even where the broad feature of the coasts are similar, the form and development of shorelines may differ widely on account of the nature and stage of marine erosion. For understanding the classification of shorelines, it is necessary to have an idea of the classification of coasts based on their major features, because the type of the coast has an important bearing on the initial form presented to wave-attack.[2]

---

1. Ahmed, E., —op.cit. p. 87
2. Wooldridge and Morgan—op. cit. p. 352.

The most recent classification of coasts by Inman and Nordstrom (1971) based on major features makes use of plate tectonics,[1] and divides coasts into three main types : (a) Collision coasts, (b) trailing-edge coasts and (c) marginal seacoasts. *Collision coasts* are the coasts which are found on the collision or subduction side of continents and island arcs. These include principally the young mountain coasts of the world which are almost entirely located along the collision and subduction coasts. They are characterized by high seismicity and volcanism and narrow continental shelves. The collision or subduction coasts can be subdivided into two types— the continental and the island arc types. Western South America is a collision coast of *continental type.* where two plates are coming into contact and the crust is being consumed. The *island arc type* of collision coast is exemplified in the Western Pacific.

The *trailing edge coasts* are those which lie on the sides of continents and islands that are moving away from the rising and spreading mid-oceanic ridges. An example of a trailing margin is the coast of western Europe and Africa along which the rift of the Atlantic ocean has spread. A more recent example is to be found in the Red Sea and Gulf of Aden. In fact, the trailing edge coasts may be subdivided as *Neo-type, Afro-type and Amero-type,* the Red sea exemplifying the Neo-type, the west coasts of Africa and Europe and the east coast of Africa the Afro-type, and the eastern coast of North and South America the Amero-type. The *neo-trailing-edge coasts* where new rifts have recently opened and only a limited amount of separation has taken place, are precipitous with few or no continental shelves. The *Afro-type* has a predominance of plateaus alternating with plain and has relatively narrow continental shelves. It is defined by both coasts being trailing on either side of the ocean, hence erosion is limited. In the *Amero-type* the opposite coast is a collision coast and therefore mountainous, so that erosion is greater. The Amero-type has broad coastal plains and broad continental shelves. Both the Afro and Amero types have low volcanic and seismic activity. The *marginal sea coasts* are those which lie protected behind island arcs, such as the mainland coast of eastern Asia.

---

1. Inman, D.L. and Nordstrom, C.E.— "On the tectonic and morphologic classification of coasts" *Jour. Geol.*, Vol. 79 (1971), pp. 1-21. For a summary see Shepard, F.P.—*Submarine Geology*, 3rd ed., 1973, pp. 104-106.

Suess (1888) was the first to draw our attention to the two fundamental types of coastline— the Pacific type and the Atlantic type of coastline. In the Pacific type the coastline is parallel to the folded mountain ranges and is, in fact, formed by them. The Pacific ocean exemplifies this type of coastline. In the Atlantic type the coastline is transverse to the structure and alignment of mountains and is met with on both the coasts of the Atlantic ocean. Supan and then later Butcher (1939) have suggested that in order to bring out the essential difference between these two types they may be called concordant (Pacific) and discordant (Atlantic) coasts.

Von Richthofen has divided coasts into five types:

(a) Longitudinal coasts,
(b) Transverse coasts,
(c) Coasts forming part of the margins of foundered basins e.g. around parts of western Mediterranean basin.
(d) Block coasts, bordered by plateaus such as in south and central Africa and eastern South America. This is principally found in areas which have remained unaffected by recent earth movements.
(e) Regional alluvial coasts which are primarily composed of new sedimentary rocks.

In the above classification (a) and (c) together are equivalent to the Pacific type of Suess, while (b) and (d) together represent the Atlantic type of coast.

The concept that longitudinal coast is characteristic of the Pacific basin does not apply wholly to the mainland coast of eastern Asia. J.W. Gregory has, therefore, proposed that from the structural point of view the real margin of the Pacific basin may be taken as a line starting from Kamachatka and passing through Japan, the Philippine islands and New Guinea to New Zealand, and this should be considered as the primary Pacific type of coast. Further west towards the mainland, there is a series, of collapsed basins resulting in the formation of a secondary Pacific type of coast on the mainland which is really type (c) in von Richthofen's classification.

Ria, Dalmatian and Fjord Coasts

The influence of the regional structure of land on the coast and shoreline is particularly marked where the landmass near the coast is submerged as a result of rise in sea-level. This influence shows itself through the influence of regional structure on the relief of the submerged land. Good examples of this can be seen in the ria and dalmatian types of coasts. Where the coast is tranverse i.e. the structural grain (alignment) of the land is tranverse to the coast and there are river valleys in relatively softer rocks parallel to the mountain ranges, the river valleys are drowned with the rise in sea-level, and long arms of the sea enter into the land along the drowned valleys which are separated from one another by high lines of mountains prejecting into the sea. Such drowned river valleys which widen towards the sea are called ria, and such indented coast is called a *ria coast*. South-west coast of Ireland and the north-west coast of Spain provide good examples of ria coast (Fig. 207). In the ria coast the influence of the initial transverse coast is quite clear. On the contrary, where the initial coast is longitudinal and the structure is parallel to the coast, submergence causes the mountain ranges to form long chains of islands parallel to the coast and in between them long narrow arms of the sea enter the land through the old valleys, parallel to the coast. Such a coast is called *dalmation coast*, as its best example is found on the coast of Dalmatia in Yugoslavia (Fig 208). Another good example is the coast of southern Chile. On such coasts the drowned river valleys are found somewhat inland but parallel to the coast, and they communicate with the sea by narrow valleys between the ridges. The influence of the longitudinal structure of land is well marked on such coasts.

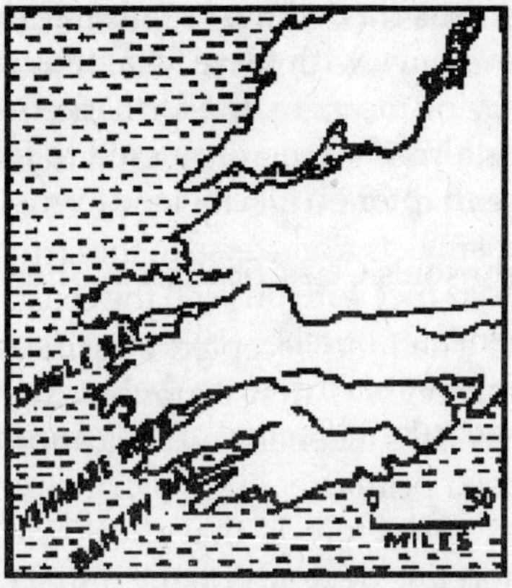

Fig.206—Ria coast of South-west Ireland

Fjord coast is really a modified form of ria coast where the river valleys were converted into glacial troughs by glacial erosion prior to their submergence in the sea. The coast is marked by long, narrow and deep indentations of the sea into the land. The walls of the fjords are high, vertical and relatively straight, in contrast to the sinuous rias. Almost all glaciated coasts have bays with these characteristics in varying measure. The formation of fjords has already been discussed in detail in the chapter on glacial landforms.

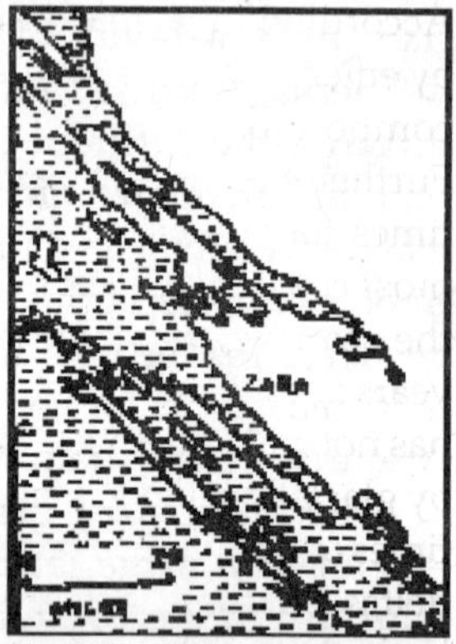

Fig.207—Dalmatian coast of Yugoslavia. The sea has been shown by broken lines.

Johnson's Classification of Shorelines

The long accepted classification of Johnson was first proposed by him in 1919. It is a genetic classification and is based primarily on two principles. The first principle is that the form of the shoreline depends on the initial form of the land which is in contact with the sea. The second principal is that it is important to keep it mind the recent history of the coasts in relation to the relative rise and fall of land and sea in understanding the present form of shorelines. Prior to Johnson, Davis (1902) had classified coasts into coasts of submergence and coasts of emergence based entirely on the latter principle. Johnson added two more divisions. Thus according to Johnson there are four types of shorelines :

(a) Shorelines of submergence,
(b) Shorelines of emergence,
(c) Neutral shorelines,
(d) Compound shorelines.

The form of neutral shoreline is related neither to submergence nor to emergence but is essentially determined by the deposition of debris or by tectonic reasons, for example, by delta, alluvial plain, outwash plain, coral reefs, volcanoes and faults.

The compound shorelines owe their origin to the combined influence of any two of the three processes stated above and particularly of both emergence and submergence.

Johnson's classification has several merits and J.B. Lucke (1938) has particularly drawn our attention to them. Besides being genetic, it is simple, comprehensive and practical. But Johnson's classification has been criticized by Shepard (1937, 1963, 1973) and the latter has given an alternative and more detailed classification. The principal objections of Shepard are : (a) According to Shepard most shorelines have been influenced both by emergence and submergence and should therefore be classed as compound shorelines in terms of Johnson's classification. (b) Further, the post-glacial rise of sea-level continued into such recent times that all coasts are potentially coasts of submergence, and most coast may even be considered as coasts of emergence, since the sea level was higher than at present some 80,000 to 150,000 years age during the interglacial episode. In Shepard's view Johnson has not paid adequate attention to the sea-level changes produced by glaciation and glacial melting during the Pleistocene and recent times (c) The chief examples used by Johnson for neutral coasts are deltas, but there is clear evidence that instead of being neutral most large deltas are actually subsiding. (d) Straightness of coasts and the presence of offshore bars (called barrier islands by Shepard) have been taken as important indicators for shorelines of emergence by Johnson but this is actually not true. We now know that straight coasts have been principally straightened either by wave erosion or wave deposition after the virtual termination of rise of sea level during deglaciation. Also drilling in barrier islands has shown that they have been growing upward as they submerged along various coasts of the world.[1]

In his defence Johnson has pointed out that he has fully kept in view the change in sea-level resulting from glacial control, and in fact this is the principal basis of his classification. So far as the question of offshore bars is concerned they may be characteristic of coasts of various kinds, but characteristics and basis of classification are two different things.

Valentin's Classification

Valentin (1952) has developed a classification which has of late drawn much attention. He divides coasts into two major types: Coasts that have advanced and coasts that have retreated. The advancing coasts may be formed either by uplift or emergence of sea-bottom or by deposition, while receding coasts may be formed either by submergence or by wave erosion. His classification is given below where both the advancing and the receding coasts have been subdivided into two sub-types each:

1. Shepard, F.P. — *Submarine Geology*, 1973 p 106.

I. *Coasts that have advanced*

(A) Emergent coasts

1. Sea bottom coasts, where sea bottom has been exposed by uplift

(B) Outbuilding (Prograded) coasts

1. Organically upbuilt or prograded (a) by corals, (b) by mangroves.
2. Inorganically prograded : (a) by marine deposition, (b) by fluvial deposition (deltas)

II. *Coasts that have retreated*

(A) Submergent coasts

1. Drowned glacial relief, (a) Erosional topography (fjord), (b) Depositional topography.
2. Drowned fluvial topography (Drowned river valleys) (a) Young folded, (b) Old folded, (c) Flat-folded

(B) Retrograded or Eroded Coasts, (a) Cliff-bordered.

Shepard considers Valentin's classification much more realistic than that of Johnson but asserts that some of the objections to Johnson's classification could also be applied to that of Valentin. For example, he shows few areas where he has called the coasts emerging (emergent coasts). Then many of his faulted coasts (sub-division of submerged coasts) could be included under emerged coasts. It is, however, obvious that although Valentin's classification is genetic, it lays greater emphasis on recent changes rather than on the initial form.

Shepard's Classification of Coasts and Shorelines

Shepard first presented his classification of coasts in 1937. Lucke (1938) criticized this classification as he thought that being limited to coasts only, it was incomplete and failed to bring out the evolutionary changes. In 1948 and again in 1963 Shepard presented a revised classification in which he included coasts as well as shorelines. This was further modified in 1973 in the light of later studies.[1] This classification has two major sub-divisions—primary and secondary. The configuration of the primary coasts has been determined mainly by non-marine agencies. (1) The *primary coasts and shorelines* include four types : (a) Land erosion coasts, shaped

1. Shepard, F.P., Ibid, pp. 111-122.

by subaerial erosion and partly drowned by postglacial rise of sea level, (b) Subaerial deposition coasts, (c) Coasts formed by volcanic action or volcanic coasts, (d) Coasts shaped by diastrophic movements. (2) The *secondary coasts and shorelines* of Shepard are those which have been shaped primarily by marine agencies or organisms. They may or may not have been primary coasts before being shaped by the sea. (2) *The secondary coast and shorelines* are of three types : (a) Coasts and shorelines shaped by marine erosion, (b) Coasts and shorelines shaped by marine deposition, and (c) Coasts and shorelines shaped by growth of corals or marine organisms. The four types of primary coasts and the three types of secondary coasts are further sub-divided into a number of sub-types by Shepard. For example, *land erosion coasts* partly drowned by rise of sea-level include, (i) ria coasts which are drowned river valleys, (ii) fjord coasts which are drowned glacial erosion coasts, and (iii) the rather uncommon type of drowned karst topography (embayments with ovalshaped depressions). *Subaerial deposition coasts* include (i) river deposition coasts such as deltaic coasts and alluvial fan coasts, (ii) glacial deposition coasts, such as partially submerged moraines and drumlins, (iii) wind deposition coasts such as dune coasts or dune prograded coasts or fossil dune coasts, (iv) landslide coasts. The *volcanic coasts* include (i) lava-flow coasts with conical slopes continuing from land out under water, or (ii) tephra coasts where volcanic products are fragmental and quickly modified by wave erosion, or (iii) volcanic collapse or explosion coasts. *Coasts shaped by diastrophism* may be either (i) fault coasts, (ii) fold coasts, or (iii) sedimentary extrusions such as salt domes.

Among the three types of secondary coasts the following sub-types have been identified by Shepard. *Wave erosion coasts* may be either (i) wave straightened cliffs bordered by gently inclined sea floor in contrast to steep incline of fault coasts, or (ii) made irregular by wave erosion. *Marine deposition coasts* i.e. coasts prograded by waves and currents include (i) Barrier coasts, with barrier beaches (shingles ridges) or barrier spits (connected to land or bay barrier i.e. sand spits have blocked bays or lagoonward extension of barriers due to storm surges, (ii) Cuspate forelands, (iii) Beach (sand) plains with no lagoons inside, (iv) Mudflats or salt marshes. *Coasts built by organisms* include principally the various types of (i) coral reef coasts including elevated reef coasts, (ii) mangrove coasts, and (iii) marsh grass coasts which could also be classed as mudflats or salt marshes.

It is obvious that Shepard's classification is quite detailed and comprehensive, but he does not give any place to emergent coasts. The subdivision of primary and secondary coasts has been made with reference to the specific agent, terrestrial or marine, which has had the greatest influence on coastal development. It is recognized that numerous gradational types exist which are difficult to classify in one single category. But most coasts appear to have been shaped only by one dominant influence which may be responsible for their major characteristics.

The above discussion shows that the classification of coasts and shorelines presents a complex problem. It is difficult to think of a classification which may be simple and practical and yet comprehensive. Further, it is not possible to make a clear-cut distinction between coasts and shorelines while classifying them as this is bound to lead to unnecessary overlaps. Coasts and shorelines may best be treated together from the point of view of their classification.

Shoreline Development and Cycle of Erosion

The theoretical cycle of marine erosion suggests that shoreline development follows a cyclic order and passes through different stages with characteristic landforms until the retreating coastal zone takes the form of a plain of marine denudation. There is probably, however, a limit to which erosion can proceed with a static sea level. In the initial stage most coasts are irregular, but the normal tendency of marine erosion is to make the coasts regular in course of time. Where, however, there are different kinds of rocks on the coast, there is a tendency in the young stage for increase in irregularity in the coast on account of the differential erosion of hard and soft rocks.

Johnson (1919) has made a detailed study of the cycle of shoreline development. According to him the sequence of development is different in shorelines of submergence from that in shorelines of emergence. Consequently, he has presented the development of the cycle along submergent and emergent coasts separately. This is broadly true, but it is important to bear in mind that what has been presented is an idealized picture of shoreline development, as in nature there is hardly a coast in the world where the development would take such a simple course.

Cycle of shoreline development along Submergent coasts

When a coast is submerged either by rise in sea-level or by sinking of land, the initial form of the shoreline is highly irregular. The drowned river valleys form ria coast and the drowned glacial troughs form fjord coasts. The ridges in between the valleys form headlands and project into the sea like peninsulas with bays in between them. Where submergence is marked, a number of small islands appear along the coast (Fig. 210 A). Marine erosion begins its work on this initial irregular coast. Differential erosion by waves further accentuates this irregularity, and the headlands and islands are particularly subjected to erosion by wave attack. In the youthful stage the sea-cliff and the related erosional features are formed. Initially the shore profile is ungraded, but after the formation of wave-cut bench or abrasion platform and the wave-built terrace on its front, graded profile tends to develop towards the close of the youthful stage (Fig. 209). Wave erosion results in the formation of caves, arches, stacks and related features. The coast tends to be straightened by the erosion of headlands and islands. Simultaneously, availability of abundant sediments and debris leads to the formation of several depositional features. In this way various kinds of bars and spits, connecting bars, hook, tombolo and beaches

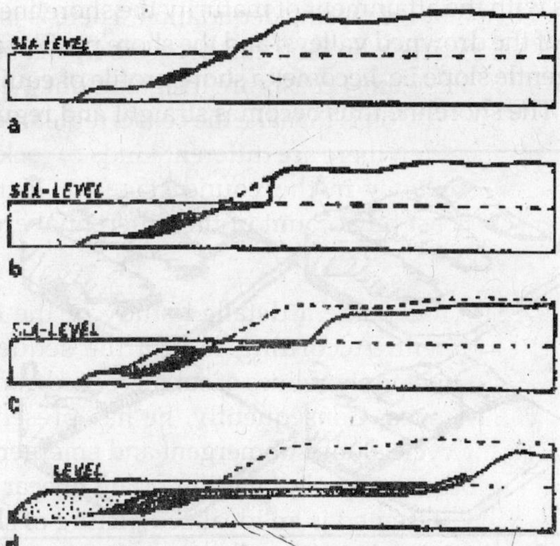

Fig.208—Stages of development of shore profile along submerged coast. (a) Initial stage, (b) & (c) Young stage, (d) Mature stage.

are built on a large scale. Apart from the extensive beaches along the coast, headland beaches, bay head beaches and bayside beaches are also formed (Fig. 210 B).

Because of the erosion of the headlands in the youthful stage and the formation of spits and bars along the bay mouth the sea coast starts becoming straight. In the early stage of bar development there are, in fact, two shorelines one in front of the bars and the other along the coast on the inner side of the bars. But when the bay mouth is partially or wholly closed by the bars, the inner coastline becomes ineffective and the portion behind the bars is converted into a lagoon.

This stage when the bay mouth is closed by bars and the headlands are retrograded, markes the end of the youthful stage and the beginning of maturity (Fig 210 C.) . In the mature stage, most of the landforms built during youth start getting destroyed. The lagoon starts filling up as a result of accretion of sediments and debris from land and is converted into salt marshes. Various depositional features formed earlier also start getting eroded, and are completely destroyed by the end of maturity (Fig. 210 D). The coast continually receds backwards, and the remains of bays and headlands are obliterated, and the soreline reaches the original bay heads. Thus with the attainment of maturity the shoreline recedes to the head of the drowned valleys, and the shore profile shows an extremely gentle slope i.e. becomes a shore profile of equilibrium. (Fig 209 d). The shoreline thus becomes straight and regular, and

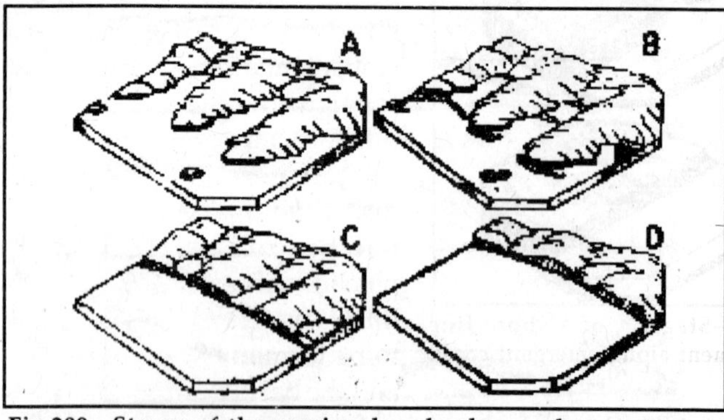

Fig.209—Stages of the erosional cycle along submergent coast A—Initial stage, B—Young stage, C—Early mature stage, D—Mature stage (*After Johnson*)

the entire coast assumes the form of a continuous cliffed coast.

In principle, in old age continued marine erosion results in low coast and shoreline, and adjacent land also is eroded to form a lowlying plain. But as has been pointed out earlier this is in reality only an abstractlon, as the possibility of the sea level remaining constant for such a long time is so remote that it seems highly improbable for plains of marine erosion to be formed.

Cycle of Shoreline development along Emergent coasts

An emergent coast may develop either by uplift of land or by fall in the sea-level. Under such conditions, the form of the coast is simple and straight, and there is a gradual slope from the coast towards the sea (Fig. 211). This is so because on account of emergence the continental shelf rises above sea level and this zone usually has a gentle off-shore slope. Under these conditions the wave attack on the coast becomes less powerful as the waves start breaking in the offshore zone. As a result, in the offshore zone where the waves break, sub-marine embankments begin to build by deposition. Gradually these bars or embankments rise above the sea-level and emerge as linear offshore bars or barrier islands. The offshore bars are usually discontinuous with intervening patches of sea. These open patches or breaks are called tidal inlets (fig 211 B). Lagoon comes to be formed between the coast and the barrier islands. Breaks or tidal inlets in the line of barrier islands are found more particularly in those places where rivers discharge into the sea. The tides rise and fall through these inlets, and in shallow and gently sloping

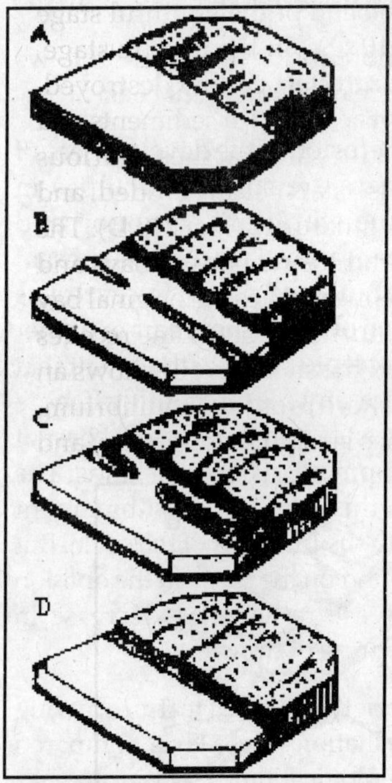

Fig.210—Stages of shore-line development along emergent coast.

rivers the influence of the tides is felt upstream for quite some distance. Consequently at the time of low tide extensive tidal flats may be seen behind the barrier islands.

With the onset of maturity the lagoon starts getting filled up by the sediments received from the rivers. Wind also transports and deposits sand in the lagoon. Gradually the lagoon is converted into a marsh and is occupied by marshy vegetation. The most significant change that takes place during the mature stage is the coastward migration of the barrier islands under the impact of wave attack. By the end of maturity these sandy islands have been pushed right up to the coast, and both the lagoon and the marsh disappear. The coast is covered with deposits of sand ridges and dunes, and the seaward slope along the coast becomes more marked than formerly.

In the old age the sequence of shoreline development follows broadly the same pattern as in the case of submergent coasts.

Although the concept of cycle of erosion in the development of shorelines has certain validity, there are some difficulties in accepting it. Most shorelines of the world are so young and immature, having been formed since sea-level reached its present height about 5,000 years ago, that only a small part of the cycle has been completed. Further, the correlation of the offshore bars or barrier islands with emergent coasts can be accepted only to the extent that the offshore slope is very gentle. If for any reason gentle offshore profile is found along submergent coasts (as in the Norfolk coast of Britain), then also we find the development of barrier islands. The third difficulty is that according to this concept, on submergent coasts the coast becomes straight with the progress of the cycle. But where both hard and soft rocks are found on the coast, in the opinion of some there is possibility of increase rather than decrease in irregularity of the coast with the progress of the cycle.

At present there is an unusual variety of coastal features, owing to the geologically rapid and large oscillations in sea level, compared with those operating over most of geological times. These changes have influenced not only the coastal zone, but also large areas of the continental shelf, which is still out of adjustment with current processes.[1]

1. King C.A.M.—*Introduction to Marine Geology and Geomorphology*, Edward Arnold, London, 1969, p. 95.

Questions

1. Examine critically the factors affecting marine erosion.
2. Give a reasoned account of the topographic features resulting from either marine erosion *or* marine deposition.
3. Give a critical account of the classification of coasts and shorelines as proposed by Johnson, Valentin and Shepard.
4. Discuss the different stages in the development of shore profile following
   (a) Submergence, and (b) Emergence.
5. Explain the formation of wave-cut platforms, wave-built platforms and beaches.
6. Explain the following :-
   (a) Longitudinal and tranverse coasts,
   (b) Off-shore bars.
   (c) Retrogradation and progradation of coastline.
7. Give a concise account of the significant features of coastal topography.
8. Write short notes on :-
   Longshore drift, Undertow, Ria coasts, Tombolo, Origin of spits and bars, Arches and Stacks.
9. Distinguish between (a) Ria, (b), Dalmation and (c) Fjord Coasts.
10. Discuss the formation of spits and the different land forms produced by them.

Chapter 29

Geomorphology of the Ocean Floor

During the last four decades there has been a revolutionary change in our knowledge of the geomorphology of the ocean floor. Use of new sophisticated instruments and techniques has provided us with far more accurate profiles of the topography of the ocean floor and has given us new insight into the physical properties of the marine sediments, the existence of linear magnetic anamoly belts and many other significant geophysical facts. The new knowledge has provided the basis for the new concepts of sea-floor spreading and plate tectonics which have completely changed our ideas of the nature and formation of the ocean basins. In fact, it has become necessary to have a knowledge of the geomorphology of the ocean floor in order to understand the global tectonics and the geomorphic characteristics of the continents, as it provides the key to the understanding of the structure and configuration of the earth and throws new light on several geo-physical, geological and geomorphic phenomena. In a sense, this is but natural, as 71% of the earth's surface is covered by the oceans, comprising 60% of the northern hemisphere and 80% of the southern hemisphere.

Based essentially on depth and the underlying geological structure, the submarine topography can be subdivided into two basic and fundamental units—the *continental margins* and the *deep sea floor* or the *ocean basin floor*. The continental margin is the submerged portion of the continental crust while truly oceanic crust characterises the ocean basin floor.

The Continental Margins

The continental margins constitute the zone linking the continents with the deep ocean basins. They cover a relatively small area in comparison to the much larger area of the deep-sea floor. Though the zone of continental margins is covered by the ocean, structurally it is not a part of the oceans but of the continents. About 15% of the continental crust is submerged to form the continental margins. Geomorphologically it has three distinct components or parts—the continental shelf, the continental slope and the continental rise[1]. The continental margins cover approximately 20.6% of the total ocean area. The zone is divided into 27 million km^2 of continental shelves, 28 million km^2 of continental slope and 19 million km^2 of continental rise—a total of 74 million km^2. The *continental shelf* is a shallow and gently sloping platform extending out from the adjoining continental land masses into the sea. At the seaward margin of the continental shelf known as the *shelf break,* there is an abrupt increase in gradient. This zone of sharp slope and deeper water adjoining the shelf is called the *continental slope.* There are deep and narrow trenches on the continental slope which are known as submarine canyons. These canyons act as the route along which the sediments from the continental shelf move down to the ocean basins. These sediments are deposited on the ocean floor at the base of the continental slope and form features resembling alluvial fans and piedmont plains. This zone of sediment deposition at the foot of the continental slope is known as *continental rise.* While the continental shelf lies on the continental block and the continental slope forms the side of the continental block, the continental rise is either downfaulted or downwarped edge of the continental block or the upturned or depressed margins of the ocean basin.

Classification of Continental Margins

The continental margins can be divided into two principal types- (a) stable or passive continental margins, and (b) active continental margins. In the *passive continental margins* there is absence of tectonic activity. The margins of the Atlantic ocean belong to this type, and the type therefore is referred to as the Atlantic type, though it is found also along the east coast of Africa and India. On the stable or passive continental margins all the three elements—continental shelf, continental slope and continental rise are present. The *active continental margins* are characterized by strong seismic activity and vigorous vulcanicity. These are particularly found all

---

1. According to some the coastal zone is also considered to be a sub-division of the continental margins, but it appears more appropriate to consider coastal scenery to be part of the continental subaerial landscape. See King. C.A.M.—*Introduction to Marine Geology and Geomorphology*, Edward Arnold, London, 1975 p. 68.

around the Pacific Ocean and are referred to as the Pacific type. They lack the continental rise, and quite often are bordered by a trench. In fact, the active continental margins may be sub-divided into two types—the *Andean type* and the *island arc type*. In the Andean type the continental shelf is narrow and below the continental slope there are oceanic trenches. In the island arc type, there is a volcanic island arc which is separated from the continent by a small stretch of relatively shallow ocean basin and the trenches are found beyond the island arc. Some of the island arcs are about 2,000 km away from the continents and these are not included in the continental margins.

Continental Shelf

Continental shelf is the shallowest part of the ocean and surrounds the continents as a shallow platform which is terminated seaward by a sharp break in slope, called the *shelf-edge* or *shelf break*. The depth at which this change of gradient occurs is variable and so is the shelf width. The continental shelf has a slope which varies from less than 1^0 to 3^0. The average gradient or mean slope is 0.07^0 or less than 1^0, being somewhat steeper in the inner than in the outer half. The continental shelves altogether cover about 10 million sq. miles or 27 million sq. km which is equal to 7.5% of the total area under the oceans or 18% of the area of the continents. Though traditionally the maximum depth of the continental shelf is taken as 100 fathoms or 200 metres, the average depth at which the greatest change of slope occurs at the shelf margin is 130 m. The maximum depth actually shows great variation in different parts of the world and has a range of approximately 20 to 550 metres. The continental shelf has an average width of 75 km but actually the width varies widely in the different parts of the world. Along most of the western coast of S. America and all around the African coast the continental shelf is very narrow and is only a few km wide. But along the southern and eastern coast of U.S.A. it is nearly 200 km wide. Along west European coast the width varies from about 50 km in the south to 180 km off the French coast and about 160 km off Norway but it widens to 300-600 km off the English channel and the Irish Sea. Most of the northern coast of Europe and Russia has very wide shelf extending to about 550 km off north-eastern Siberia. The 600 km wide continental shelf along the Alaskan coast of the Bering sea is one of the widest in the world. Broad shallow areas occur along the China coast off the old mouth of the Hwang Ho and the present

Yangtzekiang river. Another of the world's widest continental shelves is in the Gulf of Siam which curves to the east and maintains a width of 400 km or more through the South China and Java seas. Along the east coast of the Indian penisula the shelf is narrow and usually 40 km wide, becoming only 20 km wide around Sri Lanka, but the east-west shelf skirting the Ganga-Brahmputra delta at the head of the Bay of Bengal is much wider. Along the western coast of India the shelf is 110 km wide in the south widening to about 350 km off the Gulf of Cambay.

The widest shelves in the world are found inside the island arcs along the East Asiatic coast and in the Arctic. Inside the barriers of the island arcs, the marginal basins from the Bering sea to North Australia have been filled with the sediments deposited by large rivers inside the barrier. The broad Arctic shelves north of Siberia are also related to deposition of large rivers. Although the sea remains frozen during most of the year, some of the rivers discharge sediment underneath the ice pack, as is the case with the north flowing rivers of Alaska and the Mackenzie in northern Canada.

Origin of the Continental Shelf

Four possibilities have been suggested for the formation of the continental shelves. First is that they could be constructional features, built up by deposition from the land. Second, they could result from the marine erosion of the offshore islands while the intervening basins became filled with fine sediment. A third possibility is that they are submerged deltas. A fourth possibility is that they are submerged and down-warped wave-cut terraces. Shepard (1963) is of the view that the continental shelves in the different parts of the world have probably originated in different ways and cannot be explained by reference to any one theory. All geological studies show that the shelf "is merely the slightly submerged continuation of the continent from which it may be distinguished only by the thin layer of water above it and by the sedimentary features caused by recent marine activity[1]." Several scholars are of the view that the continents really extend up to the outer margin of the shelf. Nearly 70% of the sediment cover of the continental shelves of the world are relict land deposits. These sediments were deposited at times of lower sea-level, probably much

1. Boillot. G.—*Geology of the Continental Margins*, (translated by A. Scarth), Longman, London, 1981, p.2.

of it during the last glacial maximum about 15,000 years ago. In the Pleistocene ice age the sea level was about 100 to 150 m lower than the present sea level and the zone now comprising the continental shelf was dry land where the rivers deposited their sediments in the coastal plain. Several modern estuaries can be traced as submarine valleys on the shelf right to the shelf break. In most well-surveyed areas, the shelf-break lies between 130 and 150 m depth and this is considered to be a good approximation of sea-level as its lowest point at which the sea stood for an appreciable period. When the sea level rose as a result of the melting of ice sheets, these coastal areas of fluvial, aeolian and estuarine deposits were submerged by the sea and have since been affected and reworked to some extent by Holocene submarine processes.

As indicated earlier, deposition of sediments is an important process in the formation of the shelf. The land surface is being continuously worn down by rain and rivers and the coastal areas are subject to erosion by waves. The debris derived from all these sources are deposited near the coast. These debris are further reworked and redeposited away from the coast by waves and currents. The effect of the littoral currents is felt up to a depth of about 200 m which closely approximates to the shelf margin.

Fig. 211—Formation of the continental shelf by deposition of sediments.

Other processes have possibly contributed to the formation of the shelf. The sea waves by process of marine erosion give rise to extensive platforms which have a cover of sediments on them. Further, as a result of coastal erosion, there is recession of the coastline and in the process the erosional platforms also get extended towards the land. In this way the sea slowly covers the land and the process helps in the formation of continental shelves. The view that the shelves may be submerged and down-warped wave-cut terraces has been

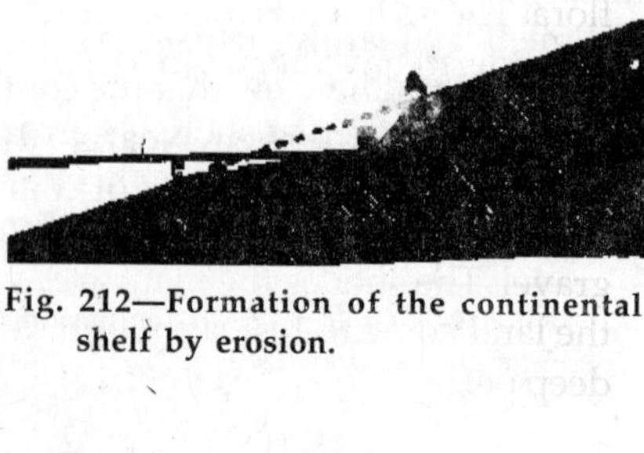

Fig. 212—Formation of the continental shelf by erosion.

particularly advocated by Bourcart (1950) in his theory of continental flexure. He has suggested that the continents may flex about the position of the present sea-level, the inland area tending to rise, while the offshore zone tends to sink. He discusses the character of the coast off Morocco where he finds evidence for the down-warping of a succession of continental shelves. This type of warping is probably applicable to limited areas and may have played a part in the making of continental shelf in some areas.

In has been suggested that the shelves which have been formed by wave erosion or by glacial erosion with little or no sediment accumulation are youthful, while the shelves which have a thick prism of sediments on igneous and metamorphic rocks are mature and indicate long periods of accumulation, probably accompanied by subsidence. The sediments have been held in place by tectonic dams as along the west coast of U.S.A. and in the Yellow sea or by reef dams as in eastern Gulf of Mexico. There are considerable stretches where there are no dams and the sediment lies at the angle of rest. This applies to large stretches of the east coast of Africa and India, much of eastern S. America and eastern U.S.A., western and southern Australia and east Asia and parts of north Russia.

The details of the sediment on the top of the shelves have been greatly affected by sea-level changes of the Pleistocene. There is no even grading of the sediments, outward gradation of sediments is exceptional, and coarse sediments are common on the outer shelf, and tracts of sand border many areas covered by mud on the inner shelves. Sea-level 15,000 years ago, when it was at its lowest, lay at the seaward margin of the shelf. Since then the sea has risen across the shelf and influenced its present superficial character. Submerged barrier beaches are found on some shelves, while others have drowned glacial deposits. Many of the sediments are not yet in equilibrium with present conditions. There are not only relict sediments but relict features and also evidences of relict fauna and flora. The sediments that show mixed characteristics (relict as well as modern) have been called 'palimpsest' sediments.

The continental shelf has great economic potential. 90% of the marine food comes from this zone. It is becoming increasingly important for natural gas and petroleum as well as for sand and gravel. The shelf is the main area where the erosional remnants of the land come to rest, although some of it may find its way to the deep oceans.

The shelf and the slope are closely interrelated and Shepard in 1973 (3rd) edition of his book '*Submarine Geology*' treats them together under the term *Continental Terrace*. He identifies eight principal types of continental terraces which are described below:

Types A and B are terraces with faulted margins and probably are very common around the slopes of the world, although the fault scarps have generally been greatly modified by slumping and erosion and by deposition in the depressions left by slumping or in the depressed zones between step faults. In *A* the shelf has been eroded, developing a type of wave-cut terrace related only in part to present sea-level. The material eroded from the shelf could accumulate on the slope or if the slope was sufficiently steep, to cause sediments to slump to the base, and the fault scarp might be preserved. In *B* the shelf is built upward by deposition as the underlying crust sank. Probably there would be extensive slumping around the shelf margin so that the edge could grow.

Type C is a forward-built slope, usually called a wave-built terrace. Prograding has been very important and has definitely widened many shelves, notably along the Atlantic coasts. In most places, prograding has simply added to the outer margin. Around many great deltas, however, the margin has advanced over the deep ocean for many kms (the Niger and the Nile deltas). Very likely the forward growth here has been as foreset beds of the delta rather than as an actual wave-built terrace.

Type D—a combination of deposition on a subsiding basement with prograding of the slope beyond the slope, is relatively common.

Type E—Dam along the outer edge of continental shelves is the primary cause for the shelves. Deposition has occurred on the inside of the hard rock barrier (San Francisco area)—such situations are referred to as tectonic dams or as basement ridges.

Type F— Another type of shelf-margin dam is built by coral growth (the slope beyond may be tectonic). Coral reef dams are found off the east coast of Florida.

Type G—Having a dam as the result of a diapiric intrusion (North-west Gulf of Maxico). Here the diapirs are believed to be mostly salt domes.

Type H— Volcanic dams (Pribilof islands, along the southern margin of the Aleutian shelf)

Each of the narrow seas along the east and south-east Asiatic coast has a framework of islands that is allowing sedimentation to create a shelf on the inside. Wave erosion may bevel the islands to produce a submerged dam.

Geomorphic Features on the Continental Shelf

The data available from the measurements of depths on the shelf indicate that the shelf is characterized by highly irregular relief, and the typical shelf is found to include many terraces, ridges, hills and depressions. *Sand ridges and hills*, usually rising only a few metres above their surroundings are particularly common in the shallow water near the coast but are found also on the middle and outer shelves. The nearshore ridges that are parallel to the coast and those on the outer shelf parallel to the shelf margin appear to be drowned barrier islands. Some of the transverse sand ridges of the outershelf are also the result of tidal currents which were directed towards bay mouths during periods of lower sea level. In tropical regions away from large rivers, the profuse growth of corals on the shelf accounts for numerous banks that rise to or nearly to the surface over wide areas. But even more important than the influence of wave and current erosion and the growth of corals would appear to be the influence of shelf glaciation and the changing sea-levels. The outer shelves have many banks that rise above the level of the inner shelves. These are a combination of morainic ridges dumped at the glacier margins and outwash fans built along the glacial terminus.

Even more striking than the ridges and banks is the existence of a variety of sea valleys or *shelf-channels* that diversify the surface of the shelf. Kuenen (1950) has classified the shelf channels into three types—drowned river valleys, tidal channels and drowned glacial troughs.

Drowned River Valleys—We come across river valleys on the shelf which were formed on land by subaerial erosion and portions of which have been subsequently submerged as a result of rise in the sea-level. The lowering of the sea-level during glacial stages allowed rivers to flow out over the shelves and cut shallow channels into the exposed surfaces and most of these were submerged as the sea rose again. A good example is the valley of the Hudson river in the U.S.A. which extends on the shelf for nearly 200 km off the present coast. There is a drowned river valley in the North sea which represents the old course of the Rhine when this region was above sea-level and the Thames was a tributary of the Rhine. Similarly the submerged valleys of the Elbe and the Weser can be traced as far north as the latitude of Edinburgh. Other good examples are the drowned valley of the Ganga in Bay of Bengal, the drowned valley of the Sunda river in the shelf between Borneo and Sumatra and the submerged valleys of the South China Sea.

Tidal Channels—In the Sunda Sea, off the south-eastern coast of North Sea and along the east coast of U.S.A. there are found submerged channels between islands which are not drowned river valleys but channels which have been cut by tidal scour. They are usually small in length and it is probable that some of them represent former river valleys cut further by tidal scour.

Drowned Glacial Troughs—Fjords which are found on many shelves in high latitude are admittedly submerged glacial troughs. But apart from these there exist certain trough-like depressions about whose origin there is difference of opinion. According to Shepard and others they are glacial troughs which have been cut below sea-level by glaciers or which were cut in the shelf when the shelf was dry land and have been later submerged with the rise of sea level. Many of these drowned glacial troughs connect with the fjords but they are wider and shallower than the fjords. Their longitudinal profile is not graded but frequently shows basining and reverse slopes. They are frequently joined by hanging valleys from the sides and their courses are not sinuous like those of the drowned river valleys. All these characteristics favour their glacial origin. Shepard (1931) is of the view that from the mouth of the Saguenay river out to the edge of the continental shelf, the 1200 km long trough in the Gulf of St. Lawrence is essentially a product of glacial erosion. On the other hand, Johnson (1925) thought that the Gulf of St. Lawrence is essentially a submerged river valley modified only slightly by glacial erosion.

Continental Slope

The continental slope is the relatively steep descent from the shelf break to the deep-sea floor. From the shelf break which is usually less than 200 m deep, the continental slope plunges down at least 1 km and usually 2 or 3 km., to the top of the continental rise. The average gradient of the continental slope is 4^0 17' to a depth of

1800 m. The continental slope is usually a very irregular area, being diversified by valleys or basins. In general, the gradient decreases at greater depth, but in the Pacific most slopes increase with depth. On the whole, the continental slopes are steeper in the Pacific at 5^0 20', than in the Atlantic **(average 3^0 05') and the Indian ocean (average 2^0 55'). Its width is, however, only** 16 to 32 km and the continental slope covers only about 8.5% of the total area of the oceans.

The depths of the base of the continental slopes vary extensively. They are generally deeper (-5,000 to -10,000 m) on active margins than on stable margins (-200 to -3,000 or- 4,000 m). They are consequently deeper in the Pacific, averaging greater than - 4,000 m and as much as -10,000 m where the slopes terminate in the deepest trenches in the western Pacific. In the Atlantic many of the continental slopes terminate at less than - 3,000 m. The same is true of the inland seas and large embayments such as the Bay of Bengal and the Arabian sea. The typical sediments of the continental slopes are finer than those of the shelves and appear to be different from land sediments, though originally most of them must have been derived from deposits brought by rivers. Roughly, about 60% is mud, 25% sand and the remainder rock, gravel, shells and ooze.

The continental slope is a major relief feature which lies at the junction of two fundamentally different structure types, with two very different levels—the continental platforms and the ocean basins. The steep, escarpment-like character of the continental slope has been shown to be the result of various kinds of structural contol. Some are fault scarps, and others seem to result from a variety of 'dams' i.e. ridges that have trapped shelf sediments behind them and have themselves been covered by sediments to a varying degree. Some of the kinds of dams that have been suggested include horsts and titled fault blocks, coral reefs, volcanic islands, salt domes and anticlines. Shepard suggests that most continental slopes represent a zone of recurrent faulting where the continental and oceanic crusts meet. It seems reasonable that very appreciable slopes should separate the light continents and the heavy ocean basins. According to Shepard (1973) the continental slopes on the two margins of the Atlantic were initiated during the Jurassic or Triassic when the east and west hemispheres were separated and heavy sima rose in the crack that developed between the two masses. The deep-sea floor sank as the continents drifted apart and thus increased the height of the escarpment separating continents and ocean floor. Most of the Pacific continental slopes had a different origin according to the sea-floor spreading plate tectonics hypothesis. They are largely associated with collision zones and down bending of the great ocean trenches, along with the subduction of the margin of oceanic crust under the continents. The base of these slopes is commonly steeper than the upper portions because of the sinking of the trenches. The upper slope appears to represent a series of step faults that have been partly covered by deposition from the land.

"If most continental slopes have developed as the result of sea-floor spreading, then the slopes formed by the separation of the two sides of the Atlantic should have originated in the Mesozoic and only slightly steepened because of (a) very gradual sinking of the ocean as it moved further from the active rift of the mid-Atlantic Ridge, and (b) isostatic adjustment due to accretion of the continental rises. However, the continental slopes along most of the Pacific margin should be in a development stage, being constantly renewed by collision of the continental and oceanic plates and the subduction of the ocean crust beneath the continents or island arcs. The seismic profiles certainly show a great contrast between these two types, especially in the general absence of continental rises along the Pacific margins and their good development in the Atlantic. One can imagine that if the major rifts were shifted to new centres of activity, like the Red sea and the rifts valleys of Africa, the subduction coasts would stabilize, and sediments would begin to fill the moribund Pacific trenches and then build continental rises. Apparently this happened in the remote past along the margins of the Atlantic."[1]

Submarine Canyons

The surface of the continental slope is smooth and regular but is incised by deep and narrow valleys and canyons. These distinctive sea valleys found on the continental slope are called submarine canyons. The submarine canyons are very similar to river-cut land canyons. Like the river canyons, they have narrow V-shaped cross profiles, winding courses and dendritic tributaries. They are best developed in the slope zone, though some extend back into the continental shelf. Many submarine canyons are directly off river valleys, but the river valleys have much wider floors and much gentler gradients than the adjoining submarine canyons. Some of

1. Shepard, F.P.—*Submarine Geology*, 3rd ed., 1973, p.300.

them can be connected via shelf channels with the estuaries of drowned river valleys, such as the large submarine canyon at the mouth of the Congo river in Africa or the canyon adjoining the mouth of the Hudson river in the New York Bay. But there are many submarine canyons in which no such relationship with a river valley exists.

The cross profiles of typical submarine canyons are V-shaped with several tributaries and resemble those on land but their logitudinal profiles are much steeper. The longitudinal profiles are also steeper near their heads than in the lower portion. They follow a sinuous or winding course and are usually about 120 km long, but some have been found to be 200-300 km long. The Congo canyon is 200 km and the Hudson canyon is 288 km long. Most canyons extend to the base of the continental slope where they either end or change into leveed fan valleys cut in the sediments of the continental rise which forms an apron below the relatively steep rocky continental slopes. The depth of the submarine canyons varies usually between 300 to 1,000 m but some have been found to be as deep as 4,000 m below the surface of the slope. They are thus easily comparable in depth and steepness of their walls to the world famous Grand Canyon of Colorado. The walls and floors of many of the canyons are rocky and consist of shales as well as hard rocks like granite and even quartzite. Large amounts of sediments of sand and clay, however, move through them, and some of the canyon floors are covered with a thin layer of sand, gravel and mud and a few seem to be undergoing filling. Heads of active canyons have access to rapid influx of sediment and to have extremely unstable sediment fill, which is carried seaward periodically by flows, slides or unusual currents. Current ripple marks are found at all depths on the canyon floors. The organisms contained in sand layers, both on the canyon floors and on the adjoining fan valleys below, include many shallow-water forms and land plants that have been transported down-valley.

The submarine canyons are world-wide in their distribution and are found on continental slopes of all the continents except Antarctica. In fact our knowledge about the Antarctica coast is so insufficient that we have no definite idea about their non-existence or existence. Canyon distribution does not seem to be related to the varying geological histories of the coasts. They are found off stable as well as unstable coasts, in areas of uplift as well as of subsidence, off deltas and associated with estuaries. Many of the largest canyons, however, appear to be related to shelves that have a long history of submergence and upbuilding of sediments such as the shelves of the Bering sea, the Bahamas and the west coast of Europe.

The continental slope along the east side of the wide Bering Sea shelf contains probably the largest and longest submarine canyon in the world. Bering canyon, the longest in the world, extends for at least 400 km and Zhemchung canyon is thought to have the largest volume in the world with 8,500 km^3. The Great Bahamas canyon, found incised between two of the low islands of the Bahamat, has a maximum wall height of 4,347 m which is highest in the world and is perhaps greater than any land canyon. It is also one of the longest with a total length of 278 km.

The canyons off the Californian coast have been most intensively studied. The most impressive among these is the Monterey canyon characterized by high walls and a branching dendritic pattern, deeply incised into the edge of the shelf and the slope and resembling in profile the Grand canyon of Colorado. At a depth of 1,900 m, the narrow floor of the canyon broadens to 5 km and at 2,700 m to 6 km where it is bordered by levees. Another canyon of the California coast is San Lucas canyon whose walls consist of intrusive crystalline rocks but the falling rock and sand flows testify to active erosion and where coarse sand and gravel are as common at canyon mouth as in its upper portion. La Jolla and Scripps canyons and the La Jolla Fan valley built into the San Diego trough off La Jolla are probably the best studied of any of the marine valleys. The Scripps canyon is a narrow gorge with precipitous rock walls and overhanging cliffs but the floor is covered with a thin layer of sand with some grass and plant material. Compared with Scripps canyon the head of La Jolla canyon has less vertical walls and a broader floor and changes to a fan- valley at its lower end with well-developed marginal levees. Many of the canyons of the California coast spread out as large fans at their lower ends, and in may instances the volume of the fan is greater than that of the canyon above it, which shows that the canyons are used as routes for the transport of sediments from shallower water by powerful turbidity currents.

Delta-front Troughs

Delta-front troughs are submarine valleys which are very

similar to submarine canyons and are found associated with deltas. They are trough-shaped with floors sloping continuously seaward down the continental slope and beyond, and rarely have tributaries or hard rock outcrops. They have U-shaped tranverse sections and comparatively straight courses. They occur off the large deltas of the Ganga, Indus, Mississippi and the Niger. The delta front troughs are almost as long as the longer canyons, have floor depths hundreds of metres below their surroundings and are important corridors for sediment transportation. Transverse seismic profiles suggest that the troughs were cut through shelf deposits but do not prove it. Shepard has suggested that they may be simply non-depositional, being maintianed by currents as the shelves were built up. Like some of the canyons, they have indications of thick deposits on their floors, and may probably have undergone similar history.[1]

Origin of Submarine Canyons

Various theories have been proposed to explain the origin of submarine canyons and it is possible that they have been formed in different ways in the different regions. Some may be related to diastrophic movements and faulting (Gregory), some may have been formed by vertical downcutting by rivers when the sea level was low and have been subsequently submerged (Hall and Shepard), while many may owe their origin to turbidty currents and mass movement (Daly). Many submarine canyons are cut into solid rock and are clearly erosional features. There are two theories about submarine canyons that deserve special mention. One is the view first proposed by Daly that they have been cut by turbidity currents. The other is the hypothesis of Shepard that they were cut subaerially during the Pleistocene low sea-level and later submerged and since then have been maintained and enlarged by various marine processes. It will be perhaps true to say that most geologists now agree that turbidity currents have played an important role in the formation of submarine canyons.

Daly's Theory regarding Origin of Submarine Canyons

R.A. Daly was the first to propose in 1936 that the submarine canyons may owe their origin to turbidity currents. About a hundred years ago a Swiss engineer had suggested that the canyon on the floor of Lake Leman has been cut by silt laden water flowing from the Rhone river, but this had remained unnoticed. In 1936 Daly of

1. Shepard, F.P.—*Submarine Geology, 1973, pp. 336-337.*

Harvard University was the first geologist to recognise the importance of turbidity currents in the ocean as a means of transporting sediments along the continental slope and as an agency of erosion of the slope. He put forward the hypothesis that it is possible for the turbidity currents to have cut the canyons below sea level. For some time people were not prepared to accept this idea, and, in fact, there was doubt about the existence of such currents. Now the existence of turbidity currents in the oceanic areas of the continental margins is well established. These currents are so powerful that they can cut deep calleys on the continental slope by erosion. Turbidity currents operate on the principle that water with a load of sediment in it is denser than clear water and therefore descends a slope displacing clean water. As it flows down, it tends to gather more sediment, its density increases further and its effect becomes cumulative. Once the currents are initiated they have the ability to continue flowing idenfinitely down slopes by stirring up more sediments along their path. An increasing number of observations show that some such process is active on the continental slopes. The character of the sediments found on the floor and beyond the mouth of the submarine canyons shows that sediments can be carried from shallow water, along the submarine valleys, to accumulate at their mouth in the form of fans.

The evidence cited in support of the fact that turbidity currents do attain substantial velocities which make them capable of eroding the slope of canyons, is the successive cable breaks on the continental slope off eastern Canada that followed the Grand Banks earthquake of November 18, 1929. In a study made by Heezen and Ewing in 1952 it was shown that all the cables within a radius of 72 km of the epicentre had broken instantly but during the succeeding 13 hours a number of other cables had broken in an ordelry sequence in keeping with their distance from the epicentre. From this it was concluded that the earthquake resulted in an avalanche flow of slope sediments, and as this flow went downslope, the cables snapped at intervals one after the other. From these succesive breaks velocity of 50 knots (92 km/hr) was calculated. The existence of turbidity currents was also confirmed by the fact that a thick layer of silt was found deposited below the lowest broken cable.[1] Shepard and Dill (1966) have doubted the validity of this conclusion and have suggested a current velocity of 15 knots (28 km/hr). It has to be, however, admitted that even at this velocity a turbidity current is capable of much significant erosion. When tthe specific gravity of turbidity currents is higher than the neighbouring waters, they can continue to flow down slopes for very long distances and can transport heavy load of sediemtns and rock fragments. Sediment slide even on a slope of 2^0 can initiate turbidity currents beneath the sea which may flow down the slope with great speed.

Turbidity curents are not likely to be formed by river water entering the sea as the river water is usually lighter than sea water and mostly spreads out over the sea surface beyond the river mouth. But the earthquakes may cause landslides and the huge deposit of silt at the river mouth may slide down as turbixdity currents. Other causes for turbidity currents include sea waves that stir up sediments from the bottom in shallow water and also slide along the sloping sea floor which may carry considerable sediments in suspension. In the Ice Age large quantities of sand and gravel flowed down towards the sea from the melting ice-mass. Under the influence of gravity these debris may flow as turbidity currents on the outer edge of the shelf and on the slope. Through repetition of such processes over long periods of time the turbidity curents may cut these canyons on the continental slope.

The Views of Shepard

Shepard disagrees with the turbidity current hypothesis of Daly. For quite some time he held the view that the submarine canyons were originally cut by river by subaerial erosion when the sea level was low during the Pleistocene Ice Age and were subsequently submerged as the sea level rose. A number of objections have been pointed out against the subaerial erosion origin for the deep submarine canyons. Firstly, during the Pleistocene the sea-level never went down so low as to expose the continental slope to subaerial erosion. Secondly, if subaerial erosion was the principal cause, submarine canyons should be well-developed where the coasts have undergone large submergence and rare or non-existent along coasts which have a history of emergence. But canyons are common off rising as well as sinking coasts. Thirdly, if subaerial erosion were important, we should expect distinct breaks in the

---

1. Heezen B.C.—"The Origin of Submarine Canyons," Scientific American, offprint 807, *Readings in Earth Sciences* Vol. 1, Indian Reprint, 1975. pp. 62-64.

continuity of the canyons at the level where subaerial erosion ceased. But no such breaks exist, and the canyons continue without modification past the 100-200 m level of the maximum glacial stage down to the base of the continental slope with little change. Fourthly, if submarine canyons were produced largely by submergence of land canyons, they should show a more striking resemblance than they do to the adjacent land canyons.

In the face of these formidable objections Shepard modified his views in 1966 and favoured a combination of processes or what might be called a theory of compound origin for the origin of the submarine canyons. He admits that turbidity currents are an active contributory cause of submarine canyons and are particularly important in the formation of the lower parts of the canyons. But in his view, in addition to turbidity curretns, other types of current are also important. Mass movements, rock falls, gravity induced slides and slumps and various types of currents not yet understood, may all have contributed to the development of canyons. But, according to him, there are some canyons or at least their upper parts which have been cut by subaeiral erosion. Although there have been considerable modification and change in the views of Shepard, he is not prepared to accept that turbidity currents are the sole or the primary cause of submarine canyons. He emphasizes that considerable differences are found in the canyons of different regions, and it is unlikely that they all have been formed in the same way. His latest view appears to be that the (1) "canyons were initiated by drowning of land valleys and have been maintained and enlarged by various marine processes, such as mass movement and turbidity currents, and (2) that the principal cause of the canyons have been the marine processes, but that great heights of the walls are due to upward growth by sedimentation on the surrounding marginal areas.[1] In his opinion, "the advantage of the combined downcutting and upbuilding hypothesis is that it provides a long period of time for the formation of huge canyons such as those in the Bering Sea and in the Bahamas. This hypothesis seems decidely preferable to the idea so frequently expressed that all submarine canyons have been excavated by turbidity currents during the low sea-level stages of the Pleistocene. Discovery of old filled canyons below the floor of present canyons adds impetus to

1. Shepard, F.P.S—*Submarine Geology*, 1973, p. 328.

the newer hypothesis and suggests that canyon cutting may have alternated with fill at different episodes".[1]

The Continental Rise

At the base of the continental slope there is a gently sloping apron of sediments which merges into the deep-sea floor. This is called the continental rise. Its depth is usually bewteen 4,000 to 5,000 metres below sea level and its slope is only about 1^0 but rises upward to 6^0. It consists essentially of submarine fans which are very similar to the piedment alluvial plains found on land. This great pile of sediments laps upon to the continental slope between 1200-2000 m below sea level. The volume of the sediments is so great that they have been derived not merely from the outer shelf and upper slope, but mostly from the land probably since the Cretaceous times. Erosion of the continents would lead to pouring down of sand, silt and other sediments down the continental slope by turbidity currents and this sediment in turn has produced the great fans by deposition at its base. These sediments have been found to be as thick as 1.6 km and they extend outward for about 600 km from the base of the continental slope. The sediment fan is cut by winding and levee bounded valleys. The fan valleys are seaward continuation of submarine canyons. Their cross-profiles are either V-shaped or trough like. Their depth rarely exceeds 100 metres, and the bordering levees built above the fans resemble natural levees of rivers on land. Most of the fan valleys divide into distributaries but they rarely have tributaries. The levee walls are locally steep but not rocky and are cut into unconsolidated fan sediments. The levees alongside the fan valleys were probably formed by the reduction of speed of the turbidity currents carrying sediment down the valley as they overflew.

The continental rise covers nearly 5.3% of the total ocean surface area. They are more typical of stable continental margins and are well developed along the margins of the Atlantic coasts and the Indian ocean and are rare along the Pacific margins. Good examples exist off the east coast of N. America and in the Indian ocean. Extensive development of submarine fans has taken place below the deltas of the Ganga and the Indus.

---

1. Shepard, F.P.—Ibid, p.334.

According to the plate tectonic hypothesis, most of the Pacific continental slopes are associated with collision zones and downbending of the great ocean trenches, along with the subduction of the margin of ocean plates under the continents. The base of these slopes is generally steeper than the upper portions because of the sinking of the trenches.

Morphology of the Deep Ocean Floor

Though the deep ocean floor has a fairly uniform structure, it is characterized by a considerable variety of relief which is comparable in diversity to relief features found on land. On land an important cause of this diversity is erosion. But on the deep ocean floor sedimentation is far more important than erosion. The variety of relief features on the surface of the deep ocean floor can be explained primarily by reference to volcanism and diastrophism. The surface of deep ocean floor is, in fact, highly irregular and complex, and besides level plains, plateaus, mountain ranges, high peaks and deep trenches are found. From the morphological point of view, it is, however, possible to divide the deep ocean floor into three distinct physiographic provinces :—

(1) Abyssal plains or the ocean floor plain—a very flat surface found in many of the deep ocean basins.
(2) Mid-ocean ridges,
(3) Deep sea trenches and Island arcs.

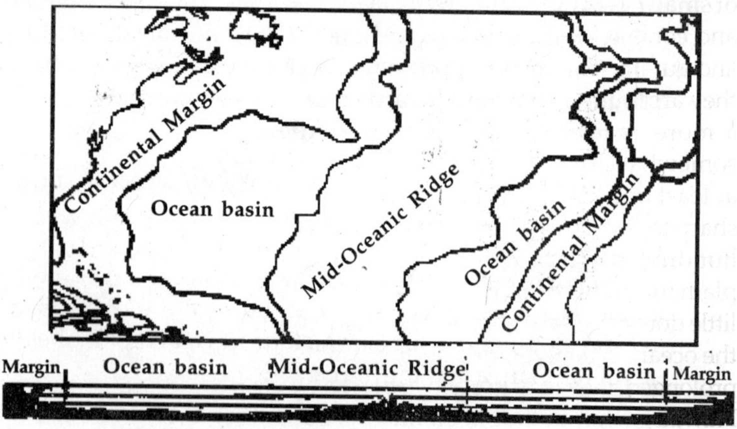

Fig. 213—Physiographic Provinces of the North Atlantic Ocean.

Abyssal Plains or Ocean Basins

The abyssal plains usually lie at the base of continental rise and extend up to the mid-ocean ridges. They are found on 41.8% of the total aea of the ocean. They are most extensive in the Indian ocean where they occupy 49.2% of the total area. On account of the relatively large area under continental shelf and the mid-ocean ridges, only about 38% of the Atlantic ocean is covered by abyssal plains. The average depth of the abyssal plains is 3,000 to 6,000 m (approximately 10,000 to 20,000 ft). They are for the most part flat and featureless having a gradient of less than 1:1,000 which is the critical botom slope across which sediment can be transported by suspension flow. Gradients flatter than this must give rise to sedimentation and it has been suggested that suspension flows provide a suitable mechanism for the very flat surfaces of abyssal plains. They are usually covered by a thick layer of the fine sediments, half a km to one km deep. It has been suggested that the original relief has been buried under sediments brought by turbidity currents. The sediments consist of clay, silt and sand brought by rivers which are mixed with pelagic oozes derived from marine organisms and volcanic ash. The composition of the oozes varies with depth. Calcium carbonate predominates in upper strata. Below this silica becomes more important, while at the bottom and in deeper waters red clay (brown clay) is mainly found.

The surface of the abyssal plains is often diversified by a number of small positive features such as abyssal hills, sea-mounts, guyots and coral atoll, most of which are geneticaly related. Sea mounts and guyots are particularly numerous in the Pacific ocean. ,although they are found in the other oceans as well. A *sea-mount* is defined as a more or less isolated elevation on the ocean floor, which is somewhat circular or elliptical in shape and has a vertical relief of at least one km (3279 ft). It is characterized by steep slopes and a sharp top with relatively limited summit area. It usually lies several hundred metres below sea-level but rises steeply from the abyssal plain to heights of 2,700 to 3,700 metres (9,000 to 12,000 ft). There is little doubt that sea mounts originated as volcanoes erupting beneath the ocean. Wherever the volcanic acivity was particularly severe or prolonged, they may rise above the sea-level to form volcanic islands. Hawaii islands in the Pacific and the Canaries island in the Atlantic ocean are examples of this. But the term sea-mount is specially used for steep and isolated elevations on the ocean floor with a local relief of 1 km or more and lying below sea-level. Their slopes tend to be steeper than land volcanoes owing to the more rapid solidification of lava in sea water. In the Pacific ocean some sea-mounts have flat tops. These are known as *guyots or table mounts*. The only difference between sea mounts and guyots is that while the former has pointed top, the latter is flat-topped. It is believed that the guyots are truncated volcanic cones which rose above sea-level at some stage in their development and whose top has been cut by subaerial and marine erosion. Their present depth of 1,500 m or more below sea level is the result of subsequent subsidence. Guyots are important in three main areas in the Pacific—1) in a north-south line running from Kamchatka to the latitude of Hawaii, 2) a group in the abyssal plain south of Alaska, and 3) a large group west of Hawaii, from the Mariana islands to the Marshall islands. In the tropical zone the guyots have provided the foundation for the growth of coral atolls. "The linear arrangement of sea-mounts and guyots may be related to elongaed zones of up-warping in the ocean floor. This process generates heat, which may help to account for the subsequent volcanic activity"[1]. All the islands, sea mounts and guyots of the Pacific ocean are associated with vulcanism, although in the atolls the volcanic material is buried beneath the thick growth of coral. Menard (1969) has shown that volcanic sea mounts increase in size with distance from the mid-ocean ridge crest, and very few rise above the ocean crust formed in the last 10 million years by sea-floor spreading. They must normally develop after this length of time. The sea mounts and guyots provide useful evidence of sea-floor spreading.

At some places on the ocean basin floor we come across areas which are several hundred metres higher than the neighbouring abyssal plains and extend over thousands of square kilometres. These are known as *Oceanic Rise*. A good example is the Bermuda Rise in the North Atlantic which contains 3 to 15 km wide and 40 to 100 m high hills. In the centre of the Bermuda Rise where the Bermuda island is located, there has been development of coral limestones (Fig. 215).

In some sections of the abyssal plains, deep-sea channels called *mid-ocean canyons* are found. These are either continuation of the fan valleys of the continental rise on to the abyssal plain or some

1. King, C.A.M.—op. cit., p. 180.

form of erosion has cut the gentle slope of these plains, developing winding channels with relatively steep walls and flat floors 1 to 8 km wide. They have a general trend which is more parallel to the distant coasts than at right angles to them. One of the best known is

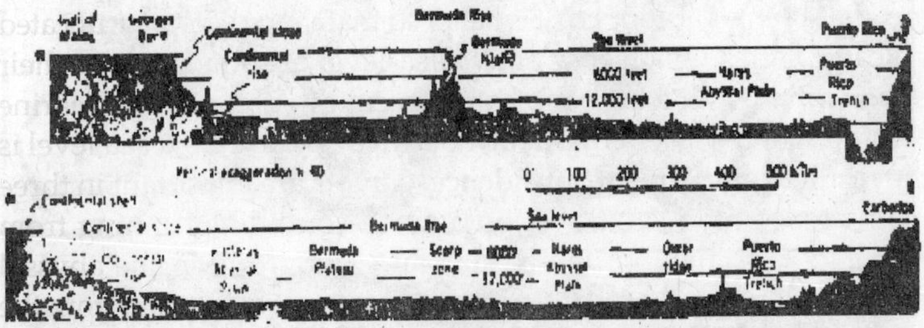

Fig. 214—Two Cross-sections of the floor of the North Atlantic Ocean. (After Heezen)

the Mid-Atlantic channel extending down south from the Baffin Bay, around New Foundland Rise and then down the Atlantic Basin to the Nares Deep, sloping continuously to the south. It has steep walls and a flat floor at least 2 km wide. The north-east Pacific has a large number of deep sea channels in the Aleutian abyssal plain, some of which appear to be extensions of glacial troughs of the shelves along the coast of British Columbia and south-east Alaska.

Mid-Ocean Ridges

Of all the physiographic features found on the deep ocean floor, the mid-ocean ridges are the most impressive and the most significant. They cover nearly one-third (32.7%) of the total surface area of the oceans. They are generally centrally placed in the ocean and run across their entire length. The mountains are bordered by plateaus and rises of varying height. They are comparable in size to the Rockies and Andes mountain systems and their total length taking all the oceans together is nearly 70,000 km. The principal ridges are about 1,200 to 3,000 km (400 to 1,000 mi) wide, and their average relief is 1,500 to 3,000 metres (5,000 to 10,000 ft). Their summits usually lie 1,500 to 2,400 metres (5 to 8 thousand ft) below sea-level, but occasionally they rise above sea-level, such as in the Azores where the peaks attain a height of 3,600 to 6,200 metres (10,000 to 20,000 ft) from their base on the ocean floor.

Though generally centrally placed in the oceans, the exact

Fig. 215—Geomorphology of the ocean floor showing especially the mid-ocean ridges, abyssal plains, and the continental shelf and slope.

alignment of the mid-ocean ridges varies with the different shapes of the oceans. The Atlantic ocean is elongate with roughly parallel margins. The Pacific margin is along a great circle and covers almost half the globe. The Indian ocean has an isosceles triangular shape with irregular boundaries. The mid-Atlantic ridge runs centrally down the entire length of the Atlantic ocean starting from the Arctic ocean in the north and is continued south of Africa into the Indian ocean and beyond into the Pacific ocean south of Australia. In the Indian ocean the ridge has an inverted Y shape. One arm of the inverted Y, the south-west Indian Ocean Ridge extends around Africa and connects with the Mid-Atlantic Ridge, while the other arm, the south-east Indian Ocean Ridge bends around Australia and joins the East Pacific Rise. North of the Y junction, the central Indian ridge runs north towards the Gulf of Aden and the Red sea (the Carlsberg Ridge) and connects with the Great Rift valley of East Africa. The Indian ocean has been recently intensively studied and the surveys indicate far greater complexity than was thought earlier. The east Pacific Rise is apparently a continuation of the system of mid-ocean ridges of the Atlantic and Indian oceans. It crosses the South Pacific and turns north to run into the Gulf of California but before doing so it diverges south-eastward towards the coast of Chile. It has a crest of 2 to 3 km above the deep-ocean floor and is thousands of kms wide. Its side slopes are gentler than those of the mid-Atlantic Ridge. There are no rifts in the crest that are so significant in the Mid-Atlantic Ridge. On the other hand, there are many volcanoes which rise above the flanks of the Rise. North of the Gulf of California the east-west fracture zones become more pronounced and dominant.

The mid -Atlantic Ridge occupies the central third of the ocean (Fig. 214). It is the best known of the mid-ocean ridges and has been most studied between 17^0 N and 54^0 N, where it is 800 to 1400 km wide and 2.5 to 4 km high. It rises in the form of a series of steps, and is more rugged towards the crest. At the crest there is a deep rift valley which is 1000-4000 m deep and 25-50 km wide. The central rift valley is a feature typical also of the Indian ocean ridge, particularly the Carlsberg ridge and in the Gulf of Aden and the Red Sea. The Central Indian Ocean Ridge has no continuous central rift valley but numerous parallel ridges and troughs, at least in the portion of the ridge between th equator and 30^0 S which has been intensively surveyed.

The mid-ocean ridges are active, as they contain evidences of large scale faulting and vulcanicity and are the scenes of frequent seismic activity. At the crest is the central rift valley, usually 30 km wide and 2 km deep which is called the *oceanic rift* or the *mid-ocean rift*. According to the plate tectonics theory, hot magma from the earth's mantle rises up through the rift and spreads out on either side of the rift. Thus the surface of the ridge has been built up by successive lava flows and volcanic cones. The lava cools rapidly in the ocean water and becomes rounded like pillows and hence is called *pillow lava*. In the absence of fluvial or glacial erosion, the slopes of these oceanic ridges are very sharp, steep and smooth. The ridges represent zones of linear tension in the earth's crust, through which new ocean crust is formed as the marginal zones move away to be subducted in deep sea trenches and the island arc systems. Although the ridge is continuous over long distance and forms one unified system, it is frequently broken by lateral displacement along fractures at right angles to its trend. These fracture lines are in the form of *transform faults*. It is along these transform faults that newly formed crust moves away from the ridge. On account of the transform faults, the mid-ocean ridge instead of being straight, assumes a zigzag line consisting of many short segments connected by fracture zones. New oceanic crust is being constantly formed by volcanic action at the mid-ocean ridges and this activity is essentially responsible for sea floor spreading and continental drift. As the new crust spreads out from the ridge, mountains and escarpments are built in the ridge area by faulting, volcanic action and lava flow. This process can be observed in Iceland which is essentially a part of the mid-Atlantic ridge raised above sea-level. Here in the middle of the ridge is found a 45 km wide rift valley. The sides of the rift valley are steep, and along the valley there are fissures through which lava wells up and spreads over the adjacent sides.

The physical feature of the mid-ocean ridge is dependent upon the rate of lava upwelling and its spreading. Fast spreading builds broad elevations and gentle slopes such as are characteristic of the East Pacific rise. On the other hand, slow spreading is responsible for the steep concave flanks of the Mid-Atlantic ridge. As the sea-floor spreading continues and the plate becomes older, the volcanic activity also declines. Normally, volcanoes become inactive by the time the crust is 20 to 30 million years old. This is shown by the existence of guyots which are found almost entirely on the crust that is more than 30 milion years old, such as the floor of the western equatorial Pacific. Guyots are believed to be drowned ancient island volcanoes that have been submerged by the gradual sinking of the ageing crust.

The Deep-Sea Trenches and Island Arcs

The trenches are the deepest parts of the oceans and cover about 7% of total surface of the oceans. Their depth below sea level is usually between 7,500 to 10,500 metres (25,000 to 35,000 ft). They occur as linear and narrow depressions on the ocean floor and are 3,000 to 6,000 metres (10,000 to 20,000 ft) deeper than the deep sea plain. Most trenches are V-shaped with steep sides, and their floor, though often flat, is usualy free from sediments. Most trenches are about 100 km wide and 1,000 km long. But several trenches are over 2,000 km long including the Kurile - Kamchatka trench, and the longest of all the Peru-Chile trench is 5,900 km long. The Java trench is 4,500 km, and Aleution trench 3,700 km long. Generally the trenches are located along those coasts where island arcs are found in areas of active volcanoes and earthquakes or where extensive mountain ranges called mountain arcs are found on the continental margins.

Island arcs and deep-sea trenches are essentially features of the Pacific ocean, though some trenches occur in the Indian ocean and the Atlantic ocean as well. The trenches form an almost continuous ring around the Pacific ocean, while the island arcs are most typically developed in the western part of the Pacific ocean. In fact, there is a difference between the trenches found in the eastern and western Pacific ocean. In the west island arc-trench system is separated from the continental margin by a small ocean, and the trenches lie on the ocean side of the island arcs. Many of these occur in the séas marginal to the west Pacific and are parallel to the island arcs. In the eastern Pacific the Andean type of trench occurs where the depression is immediately adjacent to the land and peripheral to the ocean. The sediments associated with the two types differ in character and origin. While thick turbidities are found in the Andean type, volcanic sediments are found in the island arc type of trenches. The trenches along the west coast of North and South America are close to the land but have several gaps in their continuity. The gap between the Middle America trench (west of the Rockies) and the Peru-Chile trench (west of the Andean Cordillera) is apparently related to the Caribbean block which in turn is bordered by a trench (the Puerto Rico trench) that curves around the West Indian arc in the Atlantic. Also, the southward termination of the Peru-Chile trench is supposed to be related to the other trench in the Atlantic that borders the South Sandwich arc. The East Pacific trenches are slighty deeper than those of the Atlantic but are less deep than those of the western Pacific. The Peru-Chile trench, for example, is 8,055 m deep.

The trenches associated with the island-arc system of the western Pacific form a series of somewhat disconnected deeps extending from Kamchatka in the north to as far south as the Kermadec islands, north of Newzealand. This system contains the deepest trenches, the greatest gravity anomalies and the highest seismicity. The depth of these trenches varies roughly between 8 to over 10 thousand metres. Mariana trench is probably the deepest with a depth of 10,915 m. Mindanao or Phillippines trench is 10,030 m deep. In the extreme south Tonga is 10,800 m and Kermadec 10,057 m deep.

The similarity in the maximum depth of many of these great trenches is striking. The trenches are bordered by rows of volcanoes and arc-shaped island chains, like the Aleutians, Kuriles and Marianas. Nearly all of these occur on the concave side of the arcuate trenches. The epicentres of the great earthquakes, especialy the deep-focus earthquakes are located in the trenches. " The deepest focus earthquakes are associated with the deepest and steepest trenches. This strongly suggests that the seat of the trench-producing forces lies far below the earth's surface"[1]. These trenches have also some of the greatest negative gravity anomalies in the world. There is evidently a deficiency of mass at these places, such as would be produced if the crust had been downbuckled and was being held down against isostasy by pressure from the sides"[2]. Downbuckling is also indicated by the greater depth of the Mohorovicic discotinuity under some of the trenches.

According to plate tectonics theory the trenches are zones of collision or compression where the leading edge of one moving plate collides with the leading edge of another plate. In this state of intense compression the oceanic crust moves down under the continental crust (6 to 10 cm per year) and is subducted or consumed into the mantle. The trenches occur especially where the oceanic plate comes into contact with a continental crust section. Thus they are common in the zone where the westerly moving Pacific plate meets the Eurasian and Australian plates, and again where the easterly moving Pacific plate, east of the East Pacific Rise, meets the American plate both in the northern and southern hemispheres. The subduction of one plate below another to the depth of about 100 km or more results in the formation of trenchces and accounts for the pattern of volcanic activity and deep focus earthquakes associated with the trenches. Where both the plate margins consist of oceanic crust, volcanoes and island arcs are formed, but where the upper plate is continental, mountain systems are formed on the continental margins and volcanic and earthquake activities are important. Thus the evidence concerning the trenches and the arcs on the whole favours the new plate tectonics theory, and there is now little support for the old tectogen theory of trench formation.

---

1. Fisher, R.L. and Revell, R.—"The Trenches of the Pacific, Scientific American, offprint 814, November 1955, *Readings in the Earth Sciences*, Vol. 1, Indian Reprint, 1975, p.106.
2. Shepard, F.P.—*Submarine Geology*, 1973, p. 391

Questions

1. Examine critically the problem of the origin of submarine canyons.
2. Account for the origin and characteristics of *either* continental shelf *or* continental slope and give suitable examples.
3. Give a reasoned account of the morphology of *either* the continental margins *or* the deep ocean floor.
4. Write short notes on any *three* of the following : —

   Mid-ocean ridges, Sea-mounts and Guyots, Trenches and Island arcs, Continental Rise.

Chapter 30

Coral Reefs and Coral Islands

Introduction

The coral reefs constitute a distinctive submarine morphological feature. They are organic calcareous structures formed by accumulations of various kinds of marine organisms, the most important of which are coral polyps. The reef forms an active biological system of high productivity. The most common reef formers are the hermatypic corals with symbiotic flagellates and the calcareous coralline algae, which are found principally in the tropical seas. The symbiotic flagellates (which are plants) obtain phosphorous and nitrogen from the corals and help the corals in clacification and photosynthesis but probably do not provide food to the latter. The corals live in huge colonies and their bodies are attached to one another. They are essentially dependent on the calcium obtained from the sea, and their lower part is attached to either rock surfaces or the pile of dead corals by calcareous materials. As corals die, other corals start living on the calcareous debris of the dead corals. Through this process in course of time a solid structure or reef of pure calcium carbonate is formed which is known as the coral reef. Sometimes large waves break fragments off the front of the reef and pile coral masses above sea level to produce islands. An actively growing reef soon recovers from the effects of wave erosion and starts rebuilding its damaged frame.

In several parts of the tropical oceans there has been such a prolific growth of corals that they have been able to build extensive rocky reefs. These reefs have developed either along coasts which are themselves not built of corals or far away from land on seamounts and guyots which form their foundations. The corals cannot survive above sea level. The coral reefs, therefore, do not rise above sea-level and are found either below or up to the sea-level. It is only in places where the land has been uplifted or strong wave action has piled fragments above sea-level that the coral reefs are found above the level of waves or high tide.

Conditions for Growth of Coral Reefs

The growth of coral reefs requires specific geographical conditions, as the coral polyps and the associated reef- building organisms require the following favourable conditions for their successful growth:

(i) The hermatypic corals and the associated organisms and algae which are the most common reef builders are confined to the tropical belt, extending a little north and south of the tropics where the average temperature of the ocean water is at least 21^0 C (68^0 F). Most of the reef-forming corals require a temperature of 25^0 to 30^0 C (77^0 F to 86^0 F). They cannot survive in areas where winter temperature is below 18^0 C (64^0F) or where the maximum temperature is above 35^0 C, though such high temperatures are rarely attained in the present seas. The most favourable conditions are found in areas where the surface temperature of the ocean water is between 23^0 to 25^0 C (73^0 to 77^0 F). The colonial forms of corals are therefore principally limited to the warm tropical oceans between 30^0 N to 25^0 S.

(ii) Corals can live only in saline water, and for their proper growth the average salinity should be between 2.7 to 4.0%. Higher salinity is harmful as the percentage of carbonate of lime in the sea water which is the principal food of corals, decreases with increase in salinity. Corals cannot survive in saltless water either. Consequently, few corals are found near the mouths of rivers where the rivers bring large quantities of fresh water. Sometimes when there is heavy rainfall, the flood of fresh water, kills the corals lying above sea-level at low tide.

(iii) Corals do not usually live in waters which are deeper than 65 m (200 ft), and their vigorous growth is confined to shallow water less than 46 m deep. Examples of solitary corals are found in waters which are about 100 m deep, but the sun light cannot penetrate to such depths and adequate oxygen is not usually available for the corals.

(iv) Corals also require sediment-free clear water. The silt chokes the mouth of corals and they die. The muddy waters along the mouths of rivers have a deleterious effect on coral growth. Normally coral growth is impossible in muddy silty water, but it is difficult to say how much sediment is really fatal for corals. In the region of the Great Barrier Reef of Australia the water is not less muddy than the English Channel, still there has been abundant growth of corals in the former. Similarly, in the Fiji Islands and in the Hawaiian island of Molokai reefs exist despite the muddy water. It appears that the sediments are not harmful if they are consistently present in water, but when there is a sudden increase in the amount of sediments, the corals tend to die.

(v) Water which is disturbed by ocean waves and currents is beneficial for the corals, as this ensures an adequate supply of food to the latter. This explains why enclosed seas have very few corals, and development of coral reefs takes place most rapidly in open seas and coasts, and especially in the open headlands.

(vi) Corals do not grow above the sea level, as they cannot survive for more than an hour or two above sea level at the time of low tide. This is why the reefs are usually found below the average sea-level, and are found above sea-level only where they have been uplifted, but in such cases their surface consists of dead corals. Even where there has been no uplift, dead corals may be found on the reef surface with a layer of sand and silt above them. Living corals are found only in the depressions and along the outer edge of the reef where the water is somewhat deep and the corals continue to grow actively.

(vii) For the growth of corals in the open seas it is necessary to have submarine platforms which may act as foundations for the corals. These platforms should not be deeper than 50 fathoms (about 90 m), because the corals cannot start growing and develop their colonies on deeper surfaces. On these platforms the development of corals first takes place upwards until they reach the sea-level. Then the development takes place outwards and towards the margins. Submarine platforms for coral growth are provided by submarine volcanoes which are found in all the oceans but are particularly numerous in the Pacific ocean.

Types of Coral Reefs

It is possible to identify three principal types of coral reefs based on their structure and shape. These are : (a) fringing reefs which grow out from a land mass and are connected with it, (b) barrier reefs, which are separated from land by a lagoon, and (c) atolls or oval-shaped reefs rising from deep water and surrounding a lagoon in which there is little or no land.

Fringing Reefs

Fringing reefs are those which are formed along the coasts of continents or islands. Their surface is connected with the coast and rises up to low tide water level as a rugged and uneven platform. The outer edge of the platform slopes steeply towards the sea (Fig. 217 A). As a rule the reefs are attached to the coast, but occasionally one comes across a narrow water-filled depression in the form of a miniature lagoon between the landmass and the reef. The fringing reefs are generally not very wide, and where a river enters the sea from the land, the fringing reef becomes broken and discontinuous.

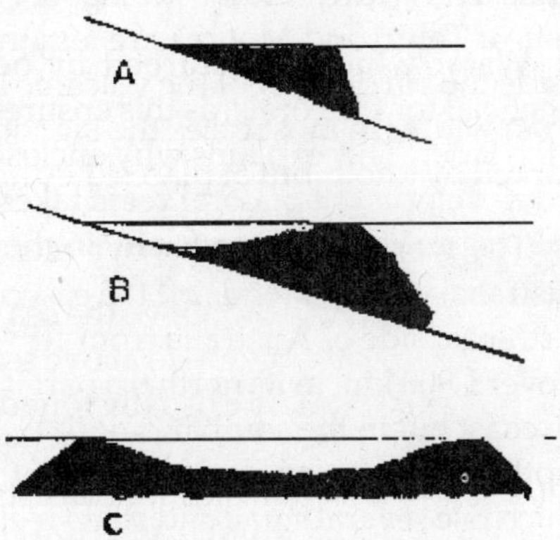

Fig. 216—Types of Coral reefs. A — Fringing reef, B — Barrier reef, C — Atoll.

Barrier Reefs

When the coral reef is situated away from the coast, and in between the coast and the reef there is a relatively wide and deep lagoon, the reef is called a barrier reef (Fig. 217 B). Barrier reefs are long, wide and extensive, and slope towards the sea at an angle of 25^0 to 45^0. The lagoons are 1/2 km to 15 km or more wide. The depth of the lagoons varies from 20 to 40 fathoms (120 to 240 ft), and sometimes isolated *coral knolls*, also called *patch reefs or pinnacles* rise from the lagoon floor to the sea-level. The barrier reefs are rarely continuous for long distances and their continuity is often broken by gaps called *passes or tidal inlets*. Through these gaps the lagoon connects with the open sea and the water brought into the lagoon by waves again returns to the sea. The rocky foundations of these reefs are generally deeper than 90 m (300 ft) which is more than the usual depth for coral growth. It is probable that there has been subsidence of the foundations after the formation of the barrier reefs.

Many tropical islands in the south-west Pacific ocean are partly surrounded by barrier reefs which are separated from the volcanic islands by a lagoon. These include Tahiti and Moorea, Samao, Truk in the Caroline islands, Bora Bora and several others. The large island of New Caledonia is almost completely surrounded by a barrier reef with a 15 km wide lagoon. The lagoons inside the barrier reefs of Tahiti and Moorea are less than 1 km in width while that inside the barrier reef of the volcanic island of Bora Bora is about 2 km in width. In most cases the lagoons are relatively shallow and marked by knolls but at places they are 45 m (150 ft) or more deep.

The world's largest barrier reef is the Great Barrier Reef of Australia, which extends off the east coast of Queensland along the north-east side of Australia from 10^0 to 24^0 S latitude for a distance of over 1,900 km. In its northern part it lies about 128 km away from the coast but in the south it is only 11 to 13 km from the coast. The depth of the lagoon is about 65 metres (200 ft) from the floor of which rise several small and steep reefs in a large variety of shapes. In fact, the Great Barrier Reef is the combination of innumerable reefs which have developed along a broad platform. In the southern portion, the reef growth is largely confined to the outer shelf, but north of 16^0 S, reefs are scattered across the shelf and form an almost continuous barrier along the outer edge. In the north their outer edge is well-marked and slopes steeply to a depth 1,000 fathoms, but in the south numerous isolated reefs with narrow and winding sea ways in between are found.

Atolls

Atolls are the most common type of coral reef. They are roughly circular or horse-shoe shaped reefs with a steep outer slope leading to deep water and a lagoon in the centre (Fig. 217 C). The average lagoon depth is 45 m, depth of 100 m or more is rare and sometimes the depth may be as little as 5 to 10 m. Most of the atolls that have been surveyed show that lagoon floor is irregular with depressions and coral knolls. Most atolls have surrounding reefs which are only roughly circular with wide variation in shape, some being elongate ellipses and others irregular-shaped masses with several protruding corners. A southeast spur is particularly common, and spurs are also usually found on the north-east. Where both are present, a relatively straight reef is found between the spurs facing the trade winds. The reefs at their outer margins are exposed only at lowest tides but further inside islands rise above much of the encircling reef. There are more islands in the south-east windward side and the reefs are also wider than the leeward side of the atolls.

Most of the world's atolls lie in the western Pacific and the neighbouring Indian ocean in the tropical zone. In the western Pacific the Funafuti atolls in the Ellice Island lying in latitude 8^0 S and longitude 177^0 E is a well known atoll which has been studied in detail by the Royal Society of England. Bikni and other atolls in the Marshall islands have also been studied by the U.S. Geological Survey. The Laccadive and the Maldive islands in the Indian ocean are composed of atolls. The atolls in the Maldive islands have numerous faros both in the lagoons and on the encircling reefs. They are also rounder in outline than most other atolls.

Characteristics of Coral Reefs

We have described three types of coral reefs, but there are some characteristics which are common to all of them. The foundation of all coral reefs consists of solid rock and not unconsolidated sand and mud. All reefs grow more actively on their outer edge than towards the land or the lagoon. This is due to several reasons. On the seaward side currents and waves bring calcium carbonate and oxygen both of which are necessary for the growth of coral polyps and algae. Towards the lagoon or the land, debris of corals get accumulated, while the water is clearer on the seaward margin and there is little accumulation of debris on account of the steep slope. Further, there is more active growth of corals on the windward than on the leeward side. On the windward side there is more active circulation of water and consequently a larger supply of food for the corals. Although corals cannot live at depths of more than 65 m (200 ft), there are several reefs which are much deeper. This is probably related to their mode of origin. On the seaward margin of all types of coral reefs, the slope in the upper part is relatively gentle (15^0 to 25^0) but becomes steep (up to 45^0) in the lower part. In the upper part this is indicative of a certain balance between coral growth and erosion by waves and currents, because both these processes are most active in the upper part. Some reefs are quite thick and their foundations are irregular, but the foundations of some other reefs are relatively flat and shallow platforms built by wave erosion and the reefs are not so thick.

Geographical Distribution of Coral Reefs

Coral reefs are limited in their geographical distribution to the tropical seas and are found almost entirely between latitudes. 30^0 N and 25^0 S. Even in this zone coral reefs are not found towards the western coast of the continents. On the other hand, we find a rich growth of reef off the eastern coasts of N. America, Africa and Australia. This is the zone of trade winds which blow off-shore on the western coasts of the continents and carry ocean water away from the coast. This results in the upwelling of cold water from underneath and the development of cold currents. This probably is the principal cause for the absence of coral growth near the western coasts, but there appear to be other causes as well, for it has been found that on the equator the temperature of oceanaic water along the western coasts is high enough for coral growth. Apart from relatively cold water, another factor which inhibits coral growth is the presence of mud or sediments in the water. Wherever the water is muddy the sediments block the mouth of the corals. Consequently on coasts where the rivers bring large quantities of sediment from the land, corals are not found. For example, we find no development of corals on the western coasts of South America on account of the presence of the cold Peru current, while in the sea west of Java and Sumatra there has been no development of coral reefs inspite of favourable temperature conditions primarily because the water is muddy.

The coral reefs are most numerous in the Pacific and the Indian oceans. In these oceans relatively shallow, warm and clear water surrounding a large number of tropical islands presents highly favourable environment for coral growth. The most extensive development has taken place in the western part of the middle Pacific ocean. The most important area of coral reef growth lies in the seas off the east coast of Australia and in the Philippines, Indonesian islands and the numerous islands lying further east. The world's largest coral reef is the Great Barrier Reef which lies off the north-east coast of Australia. In fact this is a massive assemblage of innumerable coral reefs and islands which lie on the shallow continental shelf of Queensland and which are 128 km away from the coast in the north but are only 11 km away in the middle near cape Mellville. The sea to the east of the Great Barrier Reef is called the Coral sea, but in fact it would not be an exaggeration to call the entire western section of the middle Pacific as the Coral sea. In this extensive area there are hundreds of volcanic islands along whose shores there has been development of fringing and barrier reefs, and coral islands and atolls are found in large numbers. Close by to the north lies the adjacent area of the northern Indian ocean where there has been a significant development of coral reefs and islands in the seas surrounding Indonesia, Indian Peninsula and Madagascar and off the eastern coast of Africa and in the Red sea. Apart from the examples already given, Cocos Keeling island (12^0 S and 96^0 E), the Laccadive (Minicoy and Amindive) and Maldive Islands in the Arabian sea between latitudes 0^0 to 13^0 N and some of the Islands in the Andaman and Nicobar group in the Bay of Bengal are other examples of coral reefs and islands in the Indian Ocean.

Only a few coral reefs lie outside the Indo-Pacific tropical area. In the Atlantic ocean, the seas near the West Indies including the Gulf of Mexico and the Caribbean sea have some reefs, but in the mid-Atlantic ocean there are no notable reefs with the possible exception of the reefs of Bermuda.

In the tropical seas the corals live in great colonies and build reefs. Besides these, special types of corals are found in temperate latitudes and in deep waters. They, however, do not live in colonies and are unable to build reefs. They grow like bushes and build "banks". They are found in the deep, cold and dark waters of the continental slope and the outer edge of the shelf. They live in waters which are 33 to 110 fathoms (200 to 660 ft) deep where the minimum water temperatures range between 4^0 to 15^0 C (39^0 to 59^0 F). Such deep-water coral banks or thickets are found along the eastern Atlantic coast from Norway to Cape verde islands. Similar banks are also found near the Niger Delta, the Orinico delta, Gulf of Mexico and Japan.

Coral Islands

The outer edge of the reef surface tends to be relatively high. The ocean waves break large fragments of growing corals and deposit them on the reef surface further inside. In the initial stage the reef flat behind the outer edge may contain large coral fragments, but in course of time finer fragments, sand and carbonate of lime are also deposited. By this process the edge ultimately becomes high and continuous. Under the influence of waves sand is usually deposited in heaps behind the boulder zone, and an island starts developing on the reef surface. If the wind direction is favourable, some sand is also blown from the side of the lagoon and deposited on the reef. The chief sources of sand are the fragments of corals and calcareous algae and foraminifera. Percolating water dissolves the carbonate of lime and redeposits them whereby the sand particles are cemented and become consolidated, and in this manner beach rock or beach conglomerate is formed. Since this is essentially due to wave action, such structures first develop on the windward side of the islands. Such deposits are ordinarily not continuous but occur in patches. They are, however, quite stable. Their shape, like the reef, is also arcuate. Sometimes this conglomerate occurs in steps which is probably related to sea-level changes.

In course of time the higher surface so formed tends to be extended towards the lagoon under the persistent influence of the waves. Because of the obstruction caused by the higher surface, the waves divide into two parts, and sand begins to be deposited in the relatively quieter water behind the barrier and the island starts developing. On account of the waves, it is crescentic in shape and is narrow. Several small and narrow islands may be joined together to form a relatively large island. This whole process is extremely slow and irregular, and it takes quite a long time for a full crescent-shaped island to develop. The development of the islands is to a large extent dependent upon an adequate supply of sand. Most of the sand is deposited on the windward side, but some sand is also deposited on the side of the lagoon with the result that the lagoon tends to become shallower. The currents passing through the gaps between the different islands also carry sand into the lagoon, and we find sandy shoals at the points where these pathways enter the lagoon.

Many coral islands are accumulations of sand deposited on the reef flats. In such cases they may be relatively wide and several kilometres long but they are generally low and do not rise much above the sea level. Where, however, there has been an uplift and the reef has been uplifted appreciably above sea-level, the islands may rise several hundred feet above sea level.

Seeds brought by winds and waves or by birds may result in

the growth of vegetation and the island may come to be covered by dense vegetative cover right upto the water's edge. From a distance it appears as if the trees are growing right out of water. This presents quite a striking sight in the coral seas. Growth of vegetation accelerates the process of sand deposition and adds decomposed organic matter to the soil.

Although the process described above is the most common process of the formation of coral islands, there is found a great variety in their shape and size. If the reef on which the islands are being formed is circular, a circular island or a circular line of islands may be formed. There is a lagoon in the middle of the circular atoll, and trees and plants may be found growing on the circular line of islands. Sometimes flat sand deposits are also found on isolated reefs. These are known as *cays*. Examples are to be found outside the Kingston harbour in Jamaica and in the Bay of Batavia in Java. Along the Queensland coast innumerable small islands of different shape and size of the type described above are found in the Steamer channel between the barrier reef and the mainland. Shingles of corals have been deposited by the waves on their windward side, and they are covered by mangrove forest with trees 12 to 15 metres high.

In the atoll islands the windward side is more continuous and the leeward side rather broken and fragmented. This is very clearly seen in the Funafuti atoll. This is the reason why the pathway into the lagoon lies usually behind the islands.

Theories of Coral Reef Formation

The fringing reefs are built by corals and other calcareous organisms on the bottom of shallow seas along the coasts of continents or islands. Corals start growing where the depth of water is 30 fathoms or less but they cannot grow where the water is deeper. The growth is upwards from the bottom until it reaches the sea level. Thus a platform is built, the outer edge of which does not go deeper than 30 fathoms, and it is approximately from this depth that the platform edge rises upwards. The coral growth is usually faster on the seaward side than on the landward side, as in the former the water is clearer and full of oxygen and the corals get a better food supply from the sea. Consequently, the outer edge of the reef is the first to grow up to the sea level, and a shallow waterway or a lagoon is left between the coral platform and the land.

Up to this point there is a broad unanimity of opinion regarding the mode of formation of coral reefs. The fact that the reef-building corals can grow only in shallow water indicates that virtually all reefs originated on shallow platforms. But both in the cases of the barrier reefs and the atolls their outer margin has been found to go down to such great depths that it is impossible for the reef-building corals to live and grow at those depths. This presents a complex problem and several explanations have been offered. Two major theories were proposed during the 19th century- (i) the subsidence theory of Darwin (1842) supported by Dana (1885) and Davis (1928), and (ii) the opposing theory put forward by Murray (1880), supported by Semper (1881), and Agassiz (1906). More recently Daly put forward his glacial control theory of the origin of coral reefs between 1915-48. We may broadly study the views and theories regarding the origin of barrier reefs and atolls under the following four heads:-

1. The subsidence theory of Darwin and Dana,
2. Murry's non-subsidence theory,
3. Glacial control theory of Daly,
4. The views of Davis on the coral reef problem.

Darwin's Subsidence Theory

Darwin first put forward his subsidence theory of the origin of coral reefs in 1837 and presented a revised version in 1842. Darwin proposed his hypothesis after studying only one barrier reef (Tahiti) and atoll (Keeling). It is therefore remarkable that of all the theories

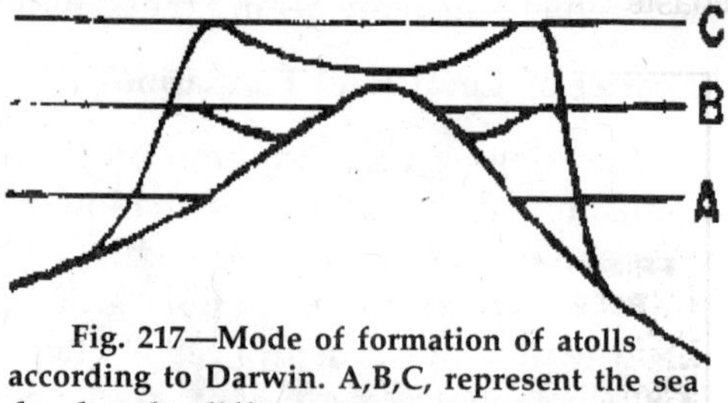

Fig. 217—Mode of formation of atolls according to Darwin. A,B,C, represent the sea level at the differnt stages of reef formation.

his theory continues to be most widely accepted. He observed that corals can live only in shallow water but the reefs formed by them go to great depths. He suggested subsidence as a possible explanation for this contradiction. The land or island along whose coast the coral reef started forming, has undergone a gradual subsidence, and as the rocky foundation of the reef subsided, the corals continued to build the reef upwards. In other words, there has been a broad correspondence or balance between the rate of subsidence and the rate of reef growth (Fig. 218).

According to this hypothesis all reefs have started as fringing reefs, and so long as there is no change in the relative level of land and sea, the reefs retain their fringing character. But when the land or the ocean floor subsides and the fringing reef is submerged, the corals start growing upwards to keep themselves alive. Their rate of growth is more towards the outer edge of the reef and less towards the coast. Consequently it is only the outer margin of the reef which is able to keep pace with the rate of subsidence, and the side nearer the coast remains submerged, resulting in the formation of a lagoon between the reef and the coast. As the subsidence proceeds, the lagoon becomes deeper and bigger, and the fringing reef changes into a barrier reef. When there is further subsidence, and if the reef is situated round an island, the island is completely drowned giving rise to a central lagoon round which there develops a ring of reef or a circular atoll (Fig. 219).

Thus according to Darwin all reefs begin as fringing reefs, and as a result of subsidence first change into barrier reefs and then into atolls. In other words, fringing reefs, barrier reefs and atolls are merely three stages in the gradual evolution of coral reefs, and atolls are the third and final form of this evolution. On the continental coasts continued subsidence will result in the conversion of the

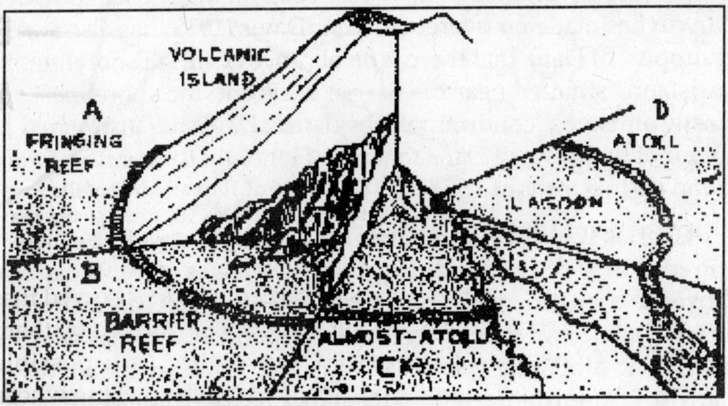

Fig. 218—Different stages (A,B,C,D,) in the formation of atoll.

fringing reefs into barrier reefs, and atolls will develop only in those areas where the subsidence resulted in the submergence of an island. In other words, subsidence is necessary for the formation of both barrier reefs and atolls, and every atoll was formerly the seat of an island which has since been drowned in the sea. Just outside the atolls the sea is deep, because here the subsidence has been so great that the central island has been completely drowned. The sea inside the lagoon is not so deep, as this was formerly occupied by an island and also continues to receive deposition of debris.

Darwin's hypothesis is remarkably simple. It requires first the formation of fringing reefs on the sides of islands, the gradual subsidence of the land with the upward growth of corals along the outer margins keeping pace with subsidence, resulting in barrier reefs, and finally when the island sank completely, the barrier reef would become an atoll. It is often not realized that Darwin did not rule out the possibility of atolls being formed in other ways. For example, he suggested that some atolls could have formed on suitable submarine craters or shallow banks or on flat, subsiding platforms and grown upwards into atolls without passing through the barrier reef stage.

Darwin's subsidence hypothesis was supported among others by Dana (1885) and Davis (1928). In fact Dana independently proposed a similar hypothesis and in the opinion of some, should get equal credit. Dana particularly drew our attention to the presence of drowned valleys and numerous embayments into the volcanic island inside barrier reefs as evidence of submergence to which Darwin has made no reference. Later Davis (1928) has also stated in support of Dana that there is an absence of cliffed shorelines in the islands situated near the barrier reefs and the shorelines are mostly embayed, confirming subsidence. Davis has also stated in favour of the views of Dana that near Honolulu the coral rocks are found up to depths of 1,178 ft.

Charles Darwin was the first to suggest that wells be drilled into an atoll to determine the origin of the masses of corals that rise above the deep ocean with no direct evidence of the basement rock on which they must have been built. A number of drillings made in the Pacific atolls between 1947 and 1965 confirm the views of Darwin. In one drilling the basement rock has been found at a depth of 4,610 ft which shows that the foundation of the atoll is a volcanic peak which rises over 3 km from the ocean floor. Earthquake studies in the Bikni atoll also indicate that the atoll is underlain by a 5,000 ft thick layer of calcareous deposits below which there is volcanic rock. It follows from this that gradual subsidence has continued here for a long time, and upward growth of corals has kept pace with the subsidence. If the subsidence had been fast, the corals would not have survived. Two holes drilled in the Midway Atoll penetrated the reef to underlying basalt—one at a depth of 157 m and the other at 384 m. The drillings indicate that the old underlying volcano was weathered and partly cut by wave erosion. After the volcano sank, it was covered by reworked clays and volcanic conglomerates, and these in turn were covered by reef limestones. The sinking of the island was interrupted by period of emergence, producing leaching of limestone and its recrystallization. H.W. Menard (1964) has suggested that possibly there was regional subsidence of the ocean floor in S.W. Pacific, as most of the atolls and guyots are found in this region. But several people differ from this view, as they feel that it will not be correct to say that there has been large scale subsidence and submergence of land and islands in the Pacific ocean or other coral areas merely on the evidence of the atolls. It will nonetheless be true to say that Darwin's subsidence theory on the whole finds confirmation from recent seismic studies and drillings in the Pacific atolls. Guyots form the base of several atolls and the coral has grown either on the portion of the flat surface of an eroded volcanic peak or which subsided at a rate at which the reef building corals could maintain their hold.

Non-Subsidence or Antecedent Platform Theories

There are some who have held that subsidence has not played a part in the formation of coral reefs and that more or less stable submarine platforms have provided the foundations for corals. This opposing view was put forward by John Murray in 1880 and was supported by Semper (1881) and Agassiz (1906). Since the reef-building corals can live only in shallow water, they conceived submarine platforms which are not deeper than 54 m. Murray's

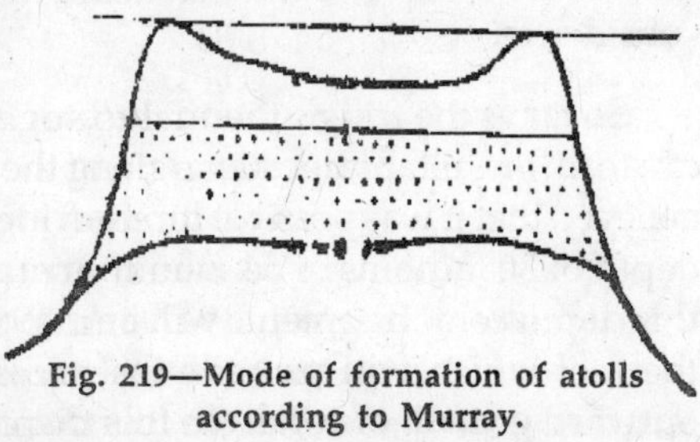

Fig. 219—Mode of formation of atolls according to Murray.

idea was that any submarine platform could be lowered by erosion or built up by deposition until it was at a suitable height for corals to grow, and atolls have been formed on the tops of those submarine plateaus and hills which rose up to the appropriate heights for coral growth. The corals then start growing upwards and the coalescence of coral colonies results in the formation of a reef. Since the corals on the outer margin grow faster, they are the first to reach the sea-level and a ring-shaped atolls is formed with a lagoon in the centre. The growth of corals on the outer margin will result in further growth of the atolls, while on the inside the lagoon will become deeper and bigger by the solution of dead corals.

According to Murray, all that is needed is the existence of several submarine hills and plateaus which are 27 to 54 m below sea level. He tried to show that even assuming that the original heights of the hills and plateaus are different, it is possible for them to attain the suitable elevation for coral growth in two ways. Most of the oceanic islands are volcanic in origin and some of them rise above the sea level. But the materials derived from volcanic action are to a large extent heaps of rocky fragments and are liable to be easily destroyed by wave action, leaving behind a shallow platform suitable for coral growth. On the other hand, there may be other volcanic peaks which rise to such heights and are below sea-level. Pelagic deposits will accumulate on them more rapidly than the deep sea, and in course of time the deposits can raise them to such height that corals start growing on them.

In other words, Murray argued that either by the erosion of volcanic hills rising above sea-level or by the deposition of sediments on those lying below sea-level, it was possible to have an adequate number of shallow submarine platforms on which the reef-building corals could grow. Thus it was possible to explain the formation of circular reefs in the middle of the ocean without reference to subsidence.

So far as the fringing and the barrier reefs are concerned, they can form on suitable locations along the continental coasts. Murray believed that it was possible for the fringing reef to extend beyond a depth of 30 fathoms or 54 m in course of time, as along the line of this depth coral fragments will be deposited as talus at the base of the reef which will provide the necessary platform for further outward growth of corals. In this manner the existence of reefs in deep water can be explained without any subsidence. But the reef above the 30 fathom depth will be formed by living corals, while at greater depths mostly coral debris will be found which will be cemented by ocean water.

According to Murray, as the reef is extended outwards the corals on the inner side die and are slowly dissolved in sea water. Thus a shallow lagoon between the coast and the inner side of the reef comes to develop, and the fringing reef changes into a barrier reef.[1]

Murry's theory has been criticized on several grounds : (1) It is difficult to accept Murry's view that the lagoon is formed by solution. The sea-water is not a good solvent. Moreover, the debris received from land erosion are deposited near the coast, therefore formation of lagoon is possible only if solution exceeds deposition. Further, has solution been so active in the lagoon, calcareous materials and coral growth would have been completely absent in it but this is not actually so. (2) Since the reef is formed on 54 m deep platforms, the depth of the lagoon has to be less than this. But quite often the lagoon is 73 to 91 m or more deep. Thus in addition to solution of coral debris, erosion of a further layer of 19 to 37 m is also necessary which seems impracticable. (3) It seems unnatural to assume that in the deep oceans where calcareous pelagic deposits are found, there is no solution and there is so much deposition of these materials that the submarine peaks are raised to within 54 m of sea level. (4) According to Murray the depth of the corals should not exceed 30 fathoms, but we do find coral deposits at greater depths. (5) It is difficult to explain the steep slope of the outer margin of the reef, as according to him the reef only upto depth of 30 fathoms has been built by living corals below which there is only talus. But the steep slope has been found to descend to much greater depths. (6) It is also doubtful if submarine platforms and reefs of the right height would be available in such large numbers. (7) The assumption that both erosion and deposition are active at the depth of 30 fathoms, does not seem logical. It is also not clear why this depth should be maximum limit of both these processes (8) Murray and his supporters have not provided any reliable evidence for rejecting the

1. Murray thought that barrier reefs had, in reality started as fringing reefs, but had grown outwards from the land, partly on the initial submarine slopes of the landmass, and partly on their own detritus. The lagoon in the rear of this reef would be hollowed out by solution. Steers, J.A.—op cit., 273-77.

subsidence theory of Darwin, and have not taken into account the geomorphic evidences available on the land adjacent to the reefs.

Agassiz who like Murray believed in non-subsidence theory, has tried to offer an alternative explanation. According to him, submarine platforms are formed by marine erosion along the coasts of islands and landmasses. This is a normal process, and in the opinion of some the continental shelves have been formed in this manner. The coral reefs which develop on these platforms are relatively less deep. He suggested that the Great Barrier Reef of Australia, the reefs of the Fiji islands and Tahiti and some other reefs have been formed in this manner . According to this theory, the reefs are built along cliffed coasts. Moreover, according to this view, there will be few or no islands in the lagoon. But in reality this is not so. In the barrier reefs of Queensland as well as New Caledonia several islands are found in the lagoons. It is true that in some islands cliffs are associated with the reef, but as a rule sea-cliff is rarely found on the inside of the barrier reef.

The non-subsidence theory was based mainly on studies conducted in the East Indies, and it has been suggested by C.A.M. King (1947) that their arguments may well apply to some of the reefs in this area and the Indian ocean, where there is some evidence of emergence.

Daly's Glacial Control Theory

Daly[1] first published his glacial-control theory of coral reef in 1915, and subsequently (1934, 1942 1948) wrote several essays explaining his view-point. Basically Daly's theory is based on the idea first suggested by A. Penck in 1894. Penck had suggested that the embayment of continental coasts and similar depth of the lagoons behind the atolls indicated that for a long time the sea level was about 100-200 m lower than at present, and the present relatively high sea level is the result of melting of ice after the Pleistocene Ice Age. Basing his ideas on the above suggestion Daly tried to analyse the influence of the changing sea level on the coral reefs..

Daly had observed in the Hawaii islands in 1909 that the reefs are narrow, and some of the Hawaiin mountains carried evidence

1. Daly, R.A.—The Glacial Control Theory of Coral Reef. *Proceed. Am. Acad. Arts. and Sci*, Vol. 51, 1915, pp. 157-251.

of past glaciation. It struck him that there is possibly some connection between temperature and reef building. The winter temperature of the seas surrounding the Hawaiin islands is slightly higher than the minimum temperature required for coral growth. He therefore concluded that in the Pleistocene Ice Age the temperature of these seas must have been considerably lower than that needed for the corals to grow. In other words, the Hawaiin reefs must have been formed after the Pleistocene. It is now generally accepted that the Pleistocene sea level was 100-150 m lower than the present sea- level. Daly assumed a fall of only 33 to 58 fathoms in the sea level. With the fall in sea-level there was also a fall in the temperature of ocean water, and it is evident that both these changes occurred on a world wide scale. Cold water and decreased sea-level would both tend to kill the corals. It is believed that even the decreased temperature was adequate for coral growth in the equatorial seas. But the decrease in sea-level exposed the coral reefs and resulted in the death of corals and the destruction of coral reefs. Further, under conditions of low sea level, not only the old coral reefs would be subjected to wave erosion but few benches and platforms would be built by coastal erosion of islands and continents. When the sea level rose again after the Ice Age ended and temperature increased, the surviving corals would start colonizing and growing on the newly formed shelves. The reefs built by these corals are post-Pleistocene. It is therefore natural that the reefs of the Hawaiin islands are small and narrow in comparison to the size of the shelves.

It can generally be stated that coral growth would be faster on the outer margins of the shelves and platforms,. The shape of the reef would therefore depend, on the shape of the platform. On the continental coasts such as in Queensland, linear reefs would be built; on erosion benches along the coasts of islands barrier reefs would be formed; and on submarine plateaus and peaks formed by erosion of the islands circular reefs and atolls would develop. The growth of corals started on the submarine platforms and as the sea-level gradually rose again with the melting of Pleistocene ice the corals continued to grow upwards. In this way the foundation of the corals became nearly 300 ft deep. With the development of the reef, the lagoon floor kept rising on account of deposition. Under the above conditions, all reefs, whether fringing, barrier or atolls, will be relatively less deep, and the depth of the lagoons of barrier reefs and atolls would be broadly similar.

As opposed to the subsidence theory of Darwin and Dana, Daly did not assume vertical movement of any kind in his theory. Nor did he assume any increase in normal erosion during the Pleistocene Ice Age. He did not rule out the possibility of local tectonic movements which might explain the difference in the depth of the lagoons but he felt that as a rule there was so much similarity in the depth of the lagoons of atolls and barrier reefs that it was more reasonable to assume crustal stability in the coral seas and platforms. Daly's theory lays stress on the fact that the Pleistocene was a period of large scale coral growth. Whatever may be the reasons for the formation of the coral reefs, it is clear that the effects of the changes in sea-level in the Pleistocene cannot be ignored in resolving this problem.[1]

Daly's theory has been criticized on the following grounds:-

(i) Some of the platforms are so wide that it does not seem likely that they could have been formed by marine erosion during the Ice Age.

(ii) The similarity in the depth of lagoons emphasized by Daly is rather overstated. Further, this depth is much greater than one would expect to result from the Pleistocene fall in sea level.

(iii) His estimate of the Pleistocene fall in sea-level is much less than the actual.

(iv) In the event of the stability of landmasses and fall in sea-level, cliffs should be formed along the coasts of continents and islands. It is probable that with the subsequent rise in sea-level, some of the cliffs would be submerged but they will not be completely destroyed. But there is complete absence of cliffed coasts in the coral seas.

---

1. "Even if Daly's hypothesis of wave-beveled platforms formed during glacial stages of low sea level is not the principal explanation of atolls and barrier reefs, undoubtedly the glacially controlled sea levels have had important effects on reef development. Some wave erosion may have occurred as the result of coral reefs being killed in the marginal zone by cooler water temperatures, but there is little evidence that the reefs in the Pacific were appreciably affected". Shepard, F.P.—*Submarine Geology*, 3rd ed. 1973, p. 363.

Views of W.M. Davis regarding Reef Formation

W.M. Davis wrote several essays on the problem of coral reef formation before the world war of 1914-18 and published his important book entitled "*The Coral Reef Problem*" in 1928.[1] Davis did not propose any new theory, but after critically reviewing the old theories he gave a new approach to the understanding of this problem and provided new evidences and arguments in support of the subsidence theory of Darwin. He particularly laid stress on the evidences available from the coasts in the study of this problem. We have seen that Dana was the first to draw our attention to this fact, but he did not adequately emphasize its significance. Davis considered that it was extremely important that practically all reef-enclosed shores show embayments, and also there is absence of cliffs in these shores. According to Davis the presence of embayments and the absence of cliffs are indicative of submergence of land, and this association confirms the subsidence theory. If the land has remained stable, not only cliffs would be formed but should continue to be present. On the other hand, if there has been subsidence, whatever cliffs were formed would be drowned. The embayments are also generally quite deep, although some have been filled with deltaic deposits. The original depth of the embayments and the inclination of the rock walls also indicate subsidence.

According to Davis, large amount of sediment is received from the land which is deposited in the lagoon situated between the reef and the land. Excessive sediment is harmful for coral growth. If it is assumed that the land was stable, then it is highly probable that lagoons would be filled up by excessive supply of land detritus and the coral growth may be destroyed. But in the event of subsidence, the sediment would go on being deposited in the subsiding lagoon with no adverse effect on reef formation.

The chief merit of the subsidence theory lies in the fact that it is comprehensive and it is able to explain the formation of reefs in different situations. The glacial-control theory is based on eustatic movements of sea level. Consequently, the rise and fall should be almost equal on all the coasts except when there has been local isostatic recovery or tectonic movement. The subsidence theory is

1. Davis, W.M.—*The Coral Reef Problem*, Am. Geog Soc. Sp. pub. No. 9, 1928.

not against the eustatic changes in sea-level and in fact includes such changes, and in addition it is applicable to those areas where there has been subsidence of land. In other words, both tectonic movements and sea-level changes are included in the subsidence theory. It is also in conformity with the suggestion of Molengraaf and Gardiner that the foundation of the reefs could also subside due to the weight of the corals. It is not at all necessary for subsidence and submergence to be continuous or for the corals to grow upwards at a consistent rate. It is possible that a period of subsidence may be followed by a period of stand-still. This unequal movement may result in the formation of a barrier reef on one side of an island, while a long period of standstill may subsequently lead to the formation of a new fringing reef on the other side of the island.

Thus Davis tried to show that Darwin's subsidence hypothesis could be applicable in a variety of situations and the form of the coasts is definitely in its favour. But it will have to be conceded that this hypothesis has not been successful in providing a satisfactory reason for uniformity in the depth of the lagoons. If some satisfactory explanation could be found, then this will further strengthen the subsidence hypothesis. At the present moment all that we can say is that the lagoon becomes shallow and relatively level on account of sedimentation and coral growth, but this does provide an answer as to why the depth of lagoons in all the coral seas is roughly equal.

It is also possible to explain the characteristics of the marginal belt coral reefs with the help of subsidence theory. Davis has divided the islands of the world in this context into three groups : (1) the islands located in the coral seas, (ii) islands which are located in relatively cold or temperate seas where corals are not found and were therefore also not found in the Pleistocence Ice Age, and (iii) two transitional zones between the coral seas and the relatively cold seas where coral growth was not possible in the Ice Age but where corals are found today. These two transitional zones are located between 25^0 and 30^0 N and 25^0 and 30^0 S latitudes. There are only a few islands in these zones, and therefore it is difficult to precisely delimit them.

The islands which are situated in the two transitional zones contain well-marked cliffs and encircling the islands are shallow banks and submarine platforms on which the coral reefs have developed. Examples of this can be seen in the north-western part of the Hawaii islands. They give the wrong impression of stability. But their shores contain many embayments which suggests instability and subsidence. Further, the shallow banks surrounding these islands are possibly old coral reefs which were destroyed during the Ice Age as a result of the lowering of sea level or the corals died when the sea temperature became low enough. As a result of the fall in sea level these reefs and the neighbouring coasts suffered erosion down to the low sea level and shallow platforms were formed. Thus the geomorphic characteristics of the coasts in the transitional zones support the subsidence hypothesis, and erosion performed by lower sea level emphasized by the glacial control hypothesis of Daly gives further strength to Darwin's hypothesis.[1]

Conclusion

The analysis of the theories regarding the origin of coral reefs highlights the complexity of the problem. The evidences available are not wholly in support of one theory, and it is difficult to arrive at a definitive conclusion. The old theories which have not taken account of the eustatic changes in sea-level are at least deficient to that extent. There is no concrete evidence to show that the submarine peaks could be raised to such great heights by pelagic deposits on them that they would become suitable for coral growth. Globigerina limestones have been found on the basement rocks of some uplifted reefs, and deep sea deposits have been found on the foundation of a reef in Solomon Island. But this is an evidence of the uplift of the sea-floor by earth movement and not of rise of the basement by deposition. It is difficult to support the view that the oceanic red clay deposits can raise parts of the ocean floor to such heights that corals can flourish on them. Similarly there is no evidence to show that solution and erosion are more important than deposition in the lagoons.

Possibly atolls and barrier reefs have been formed in different ways, and while many have been formed in areas of subsidence,

1. "In reconsidering Darwin's hypothesis in the light of the modern study of landforms, Davis has accomplished important work. He has produced many strong reasons to show that subsidence is better able to explain atolls and barrier reefs than any other postulate. His recognition of the marginal belts and the evidence they display with regard to the glacial control hypothesis is not the least important part of his thesis." Steers, J.A.—*The Unstable Earth,* Methuen, 1937, Indian Reprint, p. 290

there are others which have developed in areas which have not undergone any subsidence. But under the existing conditions, the subsidence theory is the most acceptable and receives wide support from the new evidences and arguments put forward by Davis. Daly's glacial-control theory based on rise and fall of sea-level is in a way implied in the subsidence theory. The borings made in the reefs in the different parts of the word do not fully subtantiate the subsidence theory but provide strong evidence in support of it.

Questions

1. Discuss the geographical distribution of the coral reefs with reference to the ecological conditions for their growth. Examine the effect of the last Ice Age on them.
2. Examine critically the principal theories put forward to explain the origin of barrier reefs and atolls.
3. Give a critical account of the 'subsidence' and 'glacial control' theories regarding the formation of coral reefs.
4. What theories have been advanced by Darwin and Murray to explain the origin of coral reefs? Why is Darwin's theory more acceptable?
5. "Subsidence is better able to explain atolls and barrier reefs than any other postulate". Discuss this statement.
6. Review the views of Davis on the coral reef problem and discuss the adequacy or otherwise of the glacial control theory of the origin of coral reefs.
7. Discuss the characteristics and distribution of the different types of coral reefs. How far would you agree with Darwin's view of their origin?
8. Discuss the conditions for the growth of coral reefs and their types.

Chapter 31

Review of Basic Concepts and Conflicting Views

From the preceding discussion there emerge some basic geomorphic concepts regarding the evolution of landforms. As the same time it is clear that there is a certain degree of disagreement among the scientists about some of these concepts and their relative significance. The fundamental geomorphic concepts have been admirably summarized by Thornbury in his book "*Principles of Geomorphology*" (1969) on which the present account is essentially based, but the developments during the past three decades have made it imperative to at least partially modify some of them and take note of the new dimension which has been added to the study of geomorphology by the theory of plate tectonics.

1. *The same physical processes and laws that operate today operated throughout geologic time, although not necessarily always with the same intensity as now. (Thornbury)*

This is a fundamental and most basic principle of modern geomorphology and is known as the principle of uniformitarianism. This was first propounded by James Hutton in 1785 who stated that *'the present is the key to the past'*. This principle was presented in powerful and beautiful language by Playfair in 1802, and subsequently Sir Charles Lyell gave a prominent place to it in the different editions of his famous book "*Principles of Geology*" the eleventh edition of which was published in 1872.

It would not be correct to conclude from the principle of uniformintarianism that all the physical processes have operated with equal intensity in the different periods of geological history. We know that there have been several climatic changes during geological times. Areas where today the climate is humid and fluvial processes are active, have experienced dry climatic conditions in the past. On the other hand, areas which are deserts today have had humid climates in the geological past. Glaciers were much more extensive and important during the Pleistocene than they are today. Similarly, diastrophism and volcanism have been much more active in some geological periods than they are today. But despite these differences, the rivers formed their valleys by erosion in the past exactly in the same manner in which they are doing today. Glaciers were guided in their movement and activity by the same laws during the Pleistocene as they are today. In other words, it is possible to understand the mechanism and influence of the processes operating in the past with the help of those which are operating today, and the landforms fashioned by them in the past can be easily understood. For example, by a study of the landforms and processes in the Tundra during the present times, it is possible to explain the mode of formation of relict periglacial landforms.

2. *Geological structure is a dominant control factor in the evolution of land forms and is reflected in them. (Thornbury)*

Here the term geological structure has been used in the widest sense, and includes not only rock types, folding and faulting, unconformity, mode of stratification of rocks etc., but also the hardness, permeability and solubility of rocks, the presence of joints and bedding planes, and the chemical properties of rocks. The structure influences not only the major outlines of relief but even micro-relief features and landforms bear the imprint of the rocks and their character and stratification. But when we say that particular rocks are hard or resistant and other rocks are soft or weak, , we use these terms in a relative sense with reference to the geomorphic processes, because the same rocks may be resistant in some situations but weak in other situations, and their hardness is found to differ under different climatic conditions. In other words, the geological structure in itself is a controlling factor only in a limited sense, and its influence cannot be wholly separated from the influence of processes in the evolution of landforms.

It is true that the differences in structure and nature of rocks have an important bearing on the rate of weathering and erosion of the earth's crust and as a result the various processes operate with different intensity in the different regions. But climatic conditions, local temperature and precipitation, soil humidity, vegetation cover etc. also influence the rate of activity of the processes.

On the basis of this principle, we may derive three important conclusions: —

(a) As a rule, the structural features of rocks are much older than the landforms developed on them, and in general the landforms are much younger than the structures on which they have developed.[1]

(b) The geomorphic forms can be understood only in terms of the dynamic relationship between structure and process.

(c) On account of differences in structure, local and regional differences are found in the intensity of the processes in areas of different rocks. Most of the relief features on the earth's surface owe their origin to the differential rates of operation of various processes.

Sandstone is a highly permeable or porous rock. On the countrary, shale is an impermeable rock. In sandstone areas most of the rainfall seeps underground and surface flow is less. Consequently, in such areas there are fewer rivers and the number of tributary streams is also few. The river valleys are shallow. On the contrary, in areas covered by shale surface flow is more, a river has several tributaries and the valleys are deep and well-formed.

The permeability of rocks also depends on their joints and bedding planes, and much water is able to penerate underground with their help in rocks like granites, gneisses and limestones, and then they become liable to denudation with comparative ease.

Some rocks are soluble. For example, limestone dissolves easily in water mixed with carbon-dioxide. When water penetrates underground along the joints in limestone areas, solution of underground rocks lead to the formation of underground gullies and caverns, and water carves out underground channels for its flow.

The metamorphic rocks are relatively harder and more consolidated than other kinds of rocks, and are therefore not easily liable to weathering and erosion. Slates, gneisses, quartzites, schists and such other rocks therefore form hills, mountains and high plateaus. The igneous rocks are formed inside the earth, and appear on the surface after the upper softer layer of rocks has been denuded.

---

1. Thornbury, W.D.—op. cit, p, 18.

Granite is usually found in large areas in great masses. Granites usually form domes. Domes of granite and gneisses are quite common in the Chota Nagpur plateau and in the plateau of Peninsular India generally. The edges of all such massive rocks, such as gabbro, gneiss etc. break easily and consequently the landforms produced by them are usually rounded.

3. *Geomorphic processes leave their disticntive imprint upon land forms, and each geomorphic process develops its own characteristic assemblage of land forms. (Thornbury)*

Under geomorphic processes we include all those physical and chemical processes which are responsible for the formation and evolution of landforms. These processes fall broadly into two categories—endogenetic and exogenetic. The endogenetic processes originate inside the earth's crust and include diastrophism and vulcanism. The exogenetic processes are mostly related to the atmosphere, and include weathering, mass movement and the work of rivers, underground water, glacier, wind and waves. The exogenetic processes are to some extent also subject to the influence of gravitation. The exogenetic processes continually tend to level down and reduce the irregularities produced on the earth's surface by the endogenetic forces. There is thus a continual conflict between the endogenetic and the exogenetic processes. The exogenetic processes work extremely slowly but they work unceasingly for a long period of time. On the other hand, endogenetic processes work spasmodically and after long intervals, although there are some endogenetic processes, such as mountain building or continent building movements which are slow and of long duration.

Each process gives rise to distinctive and separate landforms. Consequently, the landforms built by a particular process have identifiable and distinct characteristics, and with the help of these characteristics it is possible to differentiate the landforms produced by one process from those produced by other processes. For example floodplains, alluvial fans and deltas are the result of river action; swallow holes, underground caverns, stalactites and stalagmites are produced by underground water ; sand dunes and loess are the product of wind action; and moraines, drumlins, eskers, U-shaped valleys are clearly glacial features. Since each process produces a distinctive group of landforms, it is possible to have a genetic classification of landforms on the basis of these distinguishing characteristics. Davis realized the importance of this principle and considered this method to be best for the description of landforms. This is Davis's significant contribution to geomorphology, as prior to this the classification of landforms was done principally on morphological basis. By classifying landforms with reference to the processes responsible for their formation, their mode of formation, sequence of development and the geomorphic history can be easily understood. By keeping in mind the significance of process in the evolution of landscape, the point that is emphasized is that the evolution of landforms does not take place in a haphazard manner, and that certain landforms have a definite association with other landforms. In other words, if certain landforms exist in an area, it is possible to predict by reference to the relevant process the other probable features that are likely to be found in that area.

It is quite convenient and meaningful to study the origin of the landforms produced by each process separately but in reality the origin of most landforms in the world is the result of not one process but is the product of several processes. In other words, the different processes do not operate in isolation from one another, and as a rule, more than one processes are active at one point of time. The geomorphic processes and agents which are active in a collective and complex manner under particular climatic conditions, are collectively called a morphogenetic system. There are evidences to show that with climatic change, there have been changes in the morphogenetic system in the different periods in the past, with the result that the processes producing a particular landscape have become highly complicated.

4. *As the different erosional agents act upon the earth's surface there is produced an orderly sequence of land forms. (Thornbury)*

In his concept of the geomorphic cycle Davis emphasized the fact that there are progressive stages in the evolution of landscape and each stage is marked by distinct attributes. According to him, these stages are youth, maturity and old age at the end of which a nearly level surface called the peneplain is formed. Several geomorphologists are of the view that these stages propounded by Davis have no connection with reality and these stages cannot be identified on the landscape. The concept of peneplain has also been considered to be impractical. But most people agree with the idea of an orderly and sequential development of landforms. Thus in spite of lack of applicability in the concept of geomorphic cycle, it is difficult to find subtitute terms. Several valid objections have also been raised against the other theories proposed by Penck, King or Hack to explain the evolution of landforms. Thus it is not possible to completely rule out the use of the term geomorphic cycle, though it has to be remembered that this does not imply that geomorphic evolution is cyclic or that landscape can be easily designated as young, mature or old. Given varying structure, climate and process, the landforms produced can be different even if in each case the processes have operated over a long period of time. In fact, stage does not refer to a distinct time period in the concept of the geomorphic cycle, but has been used in the relative sense.

Related to this is the concept of partial or incomplete geomorphic cycle. In fact, in nature partial geomorphic cycle is more common than complete geomorphic cycle, because most of the earth's crust is unstable and the processes of disastrophism (destabilization) are always active on a larger or smaller scale. Before the geomorphic cycle is complete, a landscape may remain constantly either young or mature as a result of either repeated or continuous uplift, and it is highly probable that it may never be able to complete the different stages of the normal geomorphic cycle. A situation somewhat analogous to this may in fact be seen in tectonically active regions like Newzealand or California or the Himalayas.

Some people have suggested that in place of the concept of the geomorphic cycle, we should rather think in terms of evolutionary geomorphology. In other words, the entire landscape of the earth and in fact the earth itself keeps on constantly evolving and changing with the passage of time.

5. *Complexity of geomorphic evolution is more common than simplicity. (Thornbury)*

In principle the evolution of the distinctive assemblage of landforms produced by each of the processes is commonly studied separately, but in reality there is hardly a region of the earth's surface where only one process is active. A particular process may be most powerful or significant in a particular area, but quite often more than one processes operate simultaneously. For instance, although wind is the principal agent of erosion in the arid and semiarid regions, but when there is rainfall, running water too becomes an active agent of erosion in these regions. Further, the evolution of landscape is not related to one cycle of erosion but is the resultant of several erosion cycles, as there is hardly any landmass where only one cycle has progressed uninterruptedly. Consequently, besides the landforms produced in the present cycle, we come across features which were formed prior to the present cycle and they represent residual features of those formed in earlier cycles. In other words, complexity of geomorphic evolution is due to two reasons — (i) The operation of more than one process, and (ii) the existence of more than one cycle of development. Keeping these facts in mind, Horberg (1952) has divided landscapes into five principal classes:—

(i) *Simple landscapes* : These are formed by one important geomorphic process. While it is true that more than one processes are commonly responsible for the formation of a particular landscape, it has to be admitted that some landscapes are essentially the work of running water , while other landscapes are the product of solution by underground water and still others whose characteristics are essentially related to the action of winds or vulcanicity.

(ii) *Compound landscapes* : When a landscape has been formed by the combined action of more than one processes, then such a landscape has been called compound landscape. In practice compound landscapes are more common. It has therefore been said that nearly all landscapes are compound in nature. But such a generalization ignores the supremacy of one particular process in particular region, and it does not help us in understanding the origin of landscapes. The best examples of compound landscapes are to be seen in the regions affected by Pleistocene glaciation. Here we find glaciated features in higher altitudes, but at lower levels the influence of river action is evident and locally aeolian forms (especially loess deposits) are also present.

(iii) *Monocyclic landscape* : Where the landscape consists of an assemblage of landforms produced by one cycle of erosion, it is called a monocyclic landscape. It is less common than multicyclic landscapes and is generally found in newly created land surfaces such as uplifted portion of the ocean floor, the surface of volcanic cone, lava plain or plateau or areas covered with Pleistocene glacial deposits.

(iv) *Multi-cyclic landscape* : Multi-cyclic landscapes are those which have been formed by more than one cycle of erosion. The majority of the world's landscapes belong to this category. There are a variety of attributes by which the multi-cyclic character of a landscape can be identified, such as, valley in a valley, river terraces along valley sides above the present valley floor, existence of erosion surfaces, etc.

(v) *Exhumed or Resurrected landscapes* : These landscapes were formed in the geological past and then they were buried under igneous or sedimentary cover. Later as a result of prolonged denudation the outer cover of rocks is removed, and then the buried or the exhumed landscape re-appears on the surface. These are called exhumed or resurrected landscapes.

6. *Little of the earth's topography is older than Tertiary and most of it no older than Pleistocene. (Thornbury)*

Most geomorphologists are of the view that very little of the present landscape of the earth is pre-Tertiary, and the present lineaments of the landscape were carved essentially either in Pleistocene or the post-Pleistocene period. In other words, the form of the earth's surface is not old but young. Ashley (1931) has put forward this view forcefully and has shown that 90% of the earth's present landscape is post-Teritary, and possibly 99% of it is post Miocene. The Himalayas were possibly first folded during the Cretaceous, but they had not been uplifted to their present height until the Pleiocene and the development of topographic details have taken place in the Pleistocene or later.

According to this view, structure can be old, but the land forms built on them are young, and many of the erosion surfaces or old landforms are possibly exhumed forms which have been exposed on the surface after prolonged denudation of the overlying rocks.

If we accept the concept of the youthfulness of the present landscape, then it is obvious that proper interpretation of the present day landforms cannot be done without due consideration of the geological and climatic influences during the Pleistocene. Not only large areas of landmass in Eurasia and N. America were directly influenced by glaciation, but glaciation indirectly affected the outlying and the remoteareas, and there are clear evidences to show that regions where the climate is arid or semi-arid today had humid climate during those times. In addition to climatic changes, the world has witnessed extensive diastrophism from the Pliocene to the Pleistocene and even later. Pleistocene diastrophism has exerted a profound influence on the existing topography of the Pacific coast.

Some geologists do not agree with the view that present landforms of the world are young. In their view the land forms are essentially old, and their interpretation can not be isolated from geological history and tectonics, and landscape evolution can be properly understood only in the context of long time-scale of landscape evolution.[1] This view receives support from some recent and highly important concepts like sea-floor speading, plate tectonics and continental drift, as these concepts have added an altogether new dimension to the study of geomorphology. As a result a long historical perspective has become necessary in the study of landform evolution.

7. *An appreciation of world climates is necessary to a proper understanding of the varying importance of the different geomorphic processes. (Thornbury)*

Climate influences the activity of the different processes directly as well as indirectly. The direct influences are related to the amount and intensity of rainfall, intensity of evaporation, temperature and its diurnal variation, the depth and duration of frost, wind direction and velocity, climatic changes with altitude and related matters. The indirect influence makes itself felt largely through the type, distribution and density of vegetation cover. The majority of our geomorphic concepts are based on the experience of humid temperate regions, and even now knowledge about the exact working of the processes in the humid tropics, polar and arid climates is extremely limited.

Considering the important role of climate in the evolution of landscape, there has been development of climatic geomorphology and the concept of morphoclimatic systems has been propounded. These systems regulate directly as well as indirectly the disintegration of rocks, the mode of operation of gravitation, and the form and intensity of the different processes, and thus exert a

---

1. Ollier, C.D.—op.cit., pp. 299-300.

powerful influence on the evolution of landforms.[1]

Some geomorphologists are not in complete agreement with the above views. According to them the influence of present day climate is negligible on the landform of much of our present landscapes. These landscapes have experienced varying climatic conditions in their long geological history, but their influence on the landscape is not very clear and significant There are some features which are indicative of climatic conditions, for example, glacial landforms, sand dunes or coral reefs, but there are several features such as, inselberg or common fluvial features from which no definite conclusion about climate can be drawn. These geomorphologists concede that climate does influence the evolution of landforms, but they are not prepared to accept the proposition that it is possible to formulate a general theory of landscape evolution on the basis of climate.[2]

8. *Plate tectonics theory has introduced an altogether new dimension in our concept of landscape evolution.*

Plate tectonics is a comprehensive concept which has been developed after 1967. It is a great unifying principle and with its help it is possible to explain in a connected and coherent manner sea-floor speading, continental drift, geographical distribution of seismic and volcanic activities, the geomorphology of the ocean floor, geosynclines and mountain building, rift valleys etc.[3]

---

1. Morphoclimatic systems thus control the processes of landscape evolution. The vegetation cover and the climatic factor influence the stream regime and the ability of the rivers to transport debris, upon which depends their incision into uplifted regions. Vegetation and climate also control the fragmentation and alteration of rocks and the mobilisation of the debris on slopes, and so the way in which debris reaches the streams. If the debris is not too abundant or too large, it can be carried away, the rivers can incise their channels, and dissection makes progress. In the contrary case dissection is hindered. Thus there is continual interaction between morphoclimatic systems and morphogenesis. Tricart, J.—*Structural Geomorphology*, Translated by S.H. Beaver and E. Derbyshire, Longman, London, 1974, p.13.
2. "There is no doubt that climate affects landform development, but it does not provide a good basis for a general theory of landscape evolution." Ollier, C.D.—op. cit., pp. 301-02
3. This great unifying concept draws sea-floor spreading. continental drift, crustal structure and world patterns of seismic and volcanic activity together as aspects of one coherent picture. Oxburgh E.G.—"Plate tectonics, "*in Understanding the Earth*, edited by Gass, Smih and Wilson, op cit., p. 263

The plate margins are of special significance in the creation of landforms, as nearly all seismic, volcanic and tectonic activities are found concentrated near the plate margins. It is clear that there are some significant correlations between landforms and plate margins, but our knowledge about the formation of plates and the mechanism of their movement is still very incomplete. There are mountains or other landscape features which can not be directly related to plate tectonics. Further, the evidences of continental drift are all after the disintegration of pangea (i.e. Jurassic), although it is surmised that the continents must have drifted and come together more than once in the long history of the earth.

Plate tectonis is a new and powerful theory explaining and correlating many major features of the earth, but in order to understand its full significance in the origin and evolution of landscape features more detailed investigations and further in-depth studies are needed. Even so this theory has completely revolutionized our concept of landscape evolution and has added a new dimension to the analysis of landforms and has opened new vistas on the geomorphological complexities of the ocean floor and the continental margins. This theory has given an altogether new direction to geological and geomorphological thinking. The objections against the theory arise mainly from our extremely inadequate knowledge of the crust and mantle and the processes that operate in the earth's interior.

Questions

1. Give a critical account of some of the fundamental concepts in Geomorphology.
2. Explain with examples the concepts regarding the role of structure and process in the evolution of landforms.
3. "Complexity of geomorphic evolution is more common than simplicity." Discuss this statement with suitable examples.
4. "The present is the key to the past. " Discuss the significance of this statement in relation to the evolution of landforms.
5. Write short notes on *any two* of the following :-

   (a) The principle of uniformitarianism,
   (b) Influence of structure on the evolution of landforms,
   (c) Plate Tectonics and Study of landforms,
   (d) Relationship between climate and landscape evolution.

Chapter 32

Applied Geomorphology

Introduction

In recent years there has been an increasing recognition of the practical applications of geomorphic principles and the findings of geomorphological research to the understanding and solution of the problems facing mankind in his use of the natural environment. Rapid growth of population especially in the underdeveloped world has led to increasing pressure on land resources, and large scale deforestation and extension of agriculture to hilly and marginal lands have resulted in widespread and recurring man-induced catastrophies like soil erosion, landslides, floods and sedimentation. In many areas massive urbanization and various kinds of constructional and developmental projects have given rise to all kinds of environmental problems. Geomorphology can help in the identification and solution of many of these problems.

The role of applied geomorphology relates mainly to the problem of environmental resource management and consists principally of analysing and monitoring landscape forming processes in relation to surface materials and evaluating landscape changes that may arise from human interference. "The aims are to assist in the efficient discovery, assessment and wise management of the earth's finite resources, to prevent environmental deterioration and to avoid or prevent natural hazards".[1] The capability of the geomorphologist to make a worthwhile contribution has been greatly enhanced recently by quantitative study of process-form relations at the micro-level, interpretation of remote sensing imagery and air-photo analysis and his active participation in interdisciplinary research in land and project surveys and

1. Brunsden, D.- 'Geomorphology in the Service of Society', in *The Future of Geography* edited by R.J. Johnston, Methuen, London & New York, 1985, p.225

environmental management programmes. As a result agricultural scientists, engineers, geologists, hydrologists geographers, planners, policy decision makers and others are becoming increasingly aware of the importance of applied geomorphology. It is, however, customary to consult the geomorphologist only in the initial stage of investigation, though in fact he should have continuing contacts with the team involved in the planning and execution of projects.

Though some basic studies of geomorphological processes having practical applications were carried out in the first half of the 20th century, major advances really date from the 1950s. In 1954 Thornbury devoted a chapter to applied geomorphology in his *Principles of Geomorphology* and stressed the need for applying geomorphology to mineral and oil exploration, hydrological studies and engineering projects. Terrain evaluation studies which were undertaken both in developed and developing countries after the second world war primarily for military and strategic purposes yielded valuable data about land's geomorphological characteristics for assessing land potential and for providing a basis for land use planning. Rapid progress in the interpretation of aerial photographs and evaluation of remote sensing imagery has made available a wealth of information on land characteristics and assessment. In Australia the Commonwealth Scientific and Industrial Research Organization (CSIRO) developed the land-system approach in which an inventory of landform classification provided the framework for storing knowledge about the physical landscape. During the same period a comprehensive scheme for mapping land surface form, materials and processes as a basis for planning, known as geomorphological mapping, was started in Poland, Russia and other countries, its main purpose being a comprehensive analysis of the physical properties of an area with a view to assessing its natural resources or its physical limitations for development. During the 1960s and 1970s geomorphology played a leading role in the resource surveys and mapping undertaken in a number of developing countries and in otherwise unknown terrain. At the international level the International Geographical Union appointed a commission on Applied Geomorphology in 1956, followed in 1968 by the commission on (i) Man and Environment, (ii) Geomorphological Survey and Mapping, (iii) Present-day geomorphological processes, and in 1976 by the Commission on Environmental Problems and Field experiments in Geomorphology. The Unesco in its concern for man and environment has acknowledged the importance of applied geomorphology in its quarterly publication *Nature and Resources* being published since 1965 as well as in the programmes and publications of the International Hydrological Decade (IHD).

Geomorphology has diverse applications over a large area of human activity. According to Verstappen (1983) these include the following[1] :-

1. Applications in the field of earth sciences, including topographic and thematic mapping of natural resources.
2. Applications in the field of environmental studies, especially in the survey of natural hazards, landslides, avalanches, earthquakes, volcanism, land subsidence, flooding and drought.
3. Applications in the field of rural development and planning emphasizing land utilization, erosion control, conservation and river basin development.
4. Applications in the field of urbanization for urban extension, site selection or mining.
5. Applications in engineering including assessments for communication networks, river and coastal engineering.

Some geomorphologists have expressed themselves against the rise of applied geomorphology as an unhealthy sign likely to hamper basic research and lead to fewer fundamental discoveries. It is true that most applied work is within the framework of existing theories and requires routine methods and standard surveys. But there are several examples where applied research has enriched and advanced the subject. Most applied work is undertaken in relatively unknown or unusual environment which may not be otherwise possible to study scientifically, as normally finances, machinery and technology available for applied research are far superior to that available for pure research. Further, as the cost of wrong advice is very high, applied work has to be necessarily precise and

1. Verstappen, H. Th.—*Applied Geomorphology: Geomorphological Surveys for Environmental Development*, Amsterdam, Elsevier 1983. Quoted in Ibid, p. 226

thorough. A more rigorous methodology and very high standards are bound to bring positive advantages to theoretical side of the subject and enhance its capability and reputation. Applied geomorphology involves the practical testing of theories and it brings out some solid and hitherto unknown facts that can form the basis of pure research. The most important test of a theory lies in its practical application. An additional advantage is that practical application enables geomorphologists to work in close association with scientists in may related fields and thus broadens their vision and simultaneously makes other people aware of the value and principles of geomorphology. In order that geomorphologists may serve more effectively the needs of society, it is essential that they contribute their knowledge to the development process and become more involved with interdisciplinary teams and thus acquire a better understanding of the highly complex natural environmental systems.

Geomorphology and Mineral Exploration

Geomorphological knowledge can be used as an exploratory tool in the search for minerals. Geomorphic features may be helpful in locating mineral deposits in three ways :- (a) Some mineral and ore bodies have direct surface expression as landforms, (b) the relief features of an area may provide a clue to the geological structure which has favoured the deposition of particular minerals, (c) the geomorphic history of an area may provide an insight into the physical conditions under which the minerals were deposited or enriched.[1]

Surface Expression of Minerals

Some ore bodies have direct surface expression either as relief features or as structural features or as outcrops of ore or residual minerals. If the geomorphological expression of a specific ore body can be identified for a particular area, it may become relatively easy to locate similar ore bodies. In such cases aerial photographs may help in the recognition of sites worthy of detailed field examination.

The lead-zinc lode at Broken Hill, Australia is marked by a conspicuous ridge. Massive outcrops of hematite and magnetite

---

1. Thornbury, W.D.— *Principles of Geomorphology*, Wiley Eastern Reprint, 1984. p. 547

occur as whole hills in the rocks of the Dharwar system in the Indian Peninsula. The iron-bearing hematite-quartzite of Iron-ore series (upper Dharwar) in the Singhbum- Mayurbhanj region of Jharkhand and Orissa, one of the richest and largest iron-ore deposits in the world, stand out as sharply inclined hills about three thousand feet high. The Pre - Cambrian iron ore deposits of the Lake Superior region in U.S.A., are so much associated with hills and ridges that they are commonly referred to as the Iron Ore Range. The rich deposit of high grade iron ore at Cerro Bolivar, Venezuela was discovered by field investigation of two small mountains showing on the aerial photograph.

Other examples where mineralized areas stand out in distinct topographic relief are the bauxite carrying flat-topped isolated tablelands known as *pats* in western Chota Nagpur, where the basalt capping the plateau has been weathered to laterites, In the Chinhuahua region of Mexico as well as in part of Arizona and California, quartz veins stand out as hills and ridges because of their greater resistance to weathering and erosion than the surrounding rocks.

On the other hand, in some cases the mineralized areas may be found associated with depression or subsidence features. The coal-bearing Gondwana rocks of India are found in association with rivér valleys or down-faulted basins e.g. the Damodar basin, the Mahanadi basin and the Wardha valley etc. Calcite veins in parts of Arizona and British Columbia coincide with surface depressions. After reduction in bulk of the ore-body due to oxidation, the roof may collapse, resulting in what has been called *oxidation subsidence or mineralization slump*. Such slumps have been observed in limestone areas at Sierra Mojada in the state of Coahuila in northern Mexico and Bibee, Arizona, USA and used for locating lead-carbonate ores and copper ores respectively.

Weathering and Economic Deposits

Several economic mineral deposits, such as iron ore, bauxite, the clay minerals and some manganese and nickel ores are essentially weathering residues, the more mobile elements being leached away in solution. Weathering and erosion which are constantly at work on the earth's surface leave behind certain products which may form economic deposits. In the earlier stages of an erosion cycle either they do not form or they are removed and deposited elsewhere. In the later stage they may form on the erosion surfaces and be left there. They are therefore most commonly found on remnants of Tertiary erosional surfaces or along unconformities which represent ancient peneplain or near peneplain surfaces buried under later deposits. It was application of this principle that led to the discovery of bauxite deposits of Jamaica and Haiti during World War II.

The distinctive banded iron formation of Pre-cambrian age which supply much of the world's iron ore are residual iron deposits resulting from long periods of weathering which has leached the silica, leaving an enriched iron bearing deposit. Thus old erosion and weathering surfaces are particularly favourable sites for such deposits. When the higher grade Pre-cambrian ores have been exhausted, the laterite iron deposits of tropical and sub-tropical regions which are products of tertiary weathering may become important and search for them will require knowledge of ancient geomorphology and climates.

On limestones, where solution can remove most parts of the rock, insoluble residues accumulate. Most commonly they consist of clay and iron oxides and give rise to the red soil known as terra rossa. Sometimes, the residue may be silica exemplified by the flints of the English Chalk country used extensively in building. Most residual clays are formed either from solution of argillaceous limestone and dolomite or from weathered granite and pegmatites (feldspar rich rocks) or from decomposition of shales.

Bauxite, the ore of aluminium, is a weathering residue which forms primarily through intense leaching in tropical or subtropical regions and removal of all other materials including silica and iron oxides. Bauxite is widely found in the laterite caps of the Indian peninsula which have been formed by weathering on the summit of the basaltic hills and plateaus. Eocene appears to be the most important period of bauxite formation. Some of the richest deposits lie in high-level laterite of the Pat region of Ranchi and in Chatisgarh Madhya Pradesh, Gujarat and Maharastra where the weathering of the Deccan basalt under humid tropical conditions has resulted in hydrated oxides of aluminium and iron with small percentage of manganese, the other minerals of basalt having been leached out. The aluminium and iron oxides often mutually exclude each other.

When aluminium oxide is dominant, the ore is known as bauxite. The Pat level at which bauxite occurs is an Eocene erosion surface and detailed geomorphic study is essential for assessment of bauxite reserves in this region as well as in other parts of the Indian Peninsula. High relief does not appear to be conducive to bauxite formation. Apart from being a direct product of the weathering of aluminous rocks, bauxite may also form as the residuum of small percentatge of aluminous material in limestones and dolomites. The aluminous laterites of Jammu and Poonch associated with the nummulitic limestones of the Eocene age belong to this category. Here the laterite or bauxite is found on an old land surface of the Pre-Tertiary limestone, and marks a great erosional unconformity.[1] Apart from the large reserves of high grade bauxite, the nummulitic limestone also contains limonitic iron ore and coal measures.

Laterites are also a source of residual manganese deposits which occur as caps on the outcrops of Dharwar rocks. Limestone, slate and schist are the source rocks. Large lateritic manganese deposits occur in Singhbhum, Jabalpur, Bellary etc. on relatively low topographic surfaces. Large deposits are also known in China, Brazil and Morocco under similar conditions of rock and climate.

Weathering processes may also lead to the formation of enriched ore deposits. When ore deposits become exposed by erosion they are also weathered with the enclosing rocks. The valuable minerals are transported down the weathering profile where they may be redeposited. In this way rich ores may be made richer, lean ores more valuable and non-commercial primary material may be improved to commercial grade. This process is called *supergene enrichment*. Many important ore bodies have been formed in this way including the world's largest copper deposits.[2] Descending waters dissolve the copper-bearing minerals, and transport the copper down the profile, where it is re-precipitated. Some is deposited above the watertable as silicates, carbonates or sulphates and forms an oxidized zone. Just below the water table in the enrichment zone copper sulphides are deposited. This is the supergene sulphide zone below which there is a zone of primary

---

1. Wadia,D.N.—*Geology of India, Tata McGraw,* N. Delhi, 1981, p. 315.
2. Oller. C.D.—'Applications of weathering Studies', in *Applied Geomorphology* edited by John R. Hails, Elsevier Scientifie Publishing Company, Amsterdam, 1978, pp. 37-41

ore. The parent rock must be permeable and several deposits show the influence of the water table at the time of formation. In Cananea, Mexico, the oxidized section penetrates as long pendants into the sulphide zone, along fractures and faults which provide zones of high permeability. The thickness and economic value of the sulphide zone depend partly on the quality of the parent rock but geomorphic history of the weathering profile is also important. A high, nearly stationary water-table such as may occur in a humid area of low relief, gives rise to a thin but highly enriched sulphide zone. A deep water-table, especially if slowly falling, is conducive to a thick supergene sulphide zone. The most favourable condition for sulphide enrichment is active erosion with progressive depression of the water-table at such a rate that enrichment can keep pace with it. The enrichment profile of an ore body can be forecast by observing the landform under which it has been discovered. Deposits located beneath older landscapes should yield enriched ore at depth, whereas the deposits on the Atacama pediplain in Chile have their enriched ore close to the surface.

As the shape and position of the watertables at present and in the past are closely related to the form of the ground, studies have been conducted in the southern Atacama desert to establish the relations between enrichment zone and topography, so that topography may provide a clue for forecasting new enriched copper deposits. This study led to the suggestion that canyons could be largely ignored because they are likely to have their enriched zones eroded or oxidized while the undissected remnants of the pediplain and planation surfaces are the most promising localities for discovery of enriched ore close to the surface.[1] This also brings out how the desert geomorphologist may be able to help in the search for some minerals which are common in desert environment, such as secondarily enriched copper minerals and saline deposits. Workable salt deposits in deserts are frequently associated with areas where saline water accumulates and is evaporated, the two commonest environments being coastal lagoons and enclosed centripetal drainage basins in deserts. The geomorphologist can help in the appraisal of their resources by identifying topographically promising locations, delimitation of saline zones

1. Cook, R.U.- and Doornkamp, J.C. *Geomorphology in Environmental Management*, Claredon Press, Oxford, 1974, pp. 261-63.

on playas and closed drainage basins in which evaporation exceeds inflow for long periods and the inflow is sufficient to produce adequate quantity of salines, presence of saline materials close to the surface which provide a clue to the composition of the deposits and using geomorphological evidence to help predict occurrence of salines at depth, including lacustrine formations where salines might have been deposited.

Silver and nickel ores may be enriched by a process similar to the one described above. Under tropical and subtropical weathering. ultra-basic rocks become decomposed, and silica, magnesium and nickel are leached out. The silica is lost from the system but nickel and magnesium are precipitated at the base of weathering profile. This process occurs in areas of laterization. In the New Caledonian nickel mines the best ores are found where weathering has been most effective, and are concentrated near the base of the weathering profile laterite. The richest ores are on gentle slopes, saddles and spurs and not beneath the flat areas which shows that lateral seepage has aided the concentration. In limestone regions mineralization of lead and zinc sulphides is common. According to the latest view karst weathering causes both the formation of the cavities and the deposition of the sulphide minerals, and the ores are simply supergene sulphides. The stagnation of solutions in the water saturated zone creates a reducing environment for lead and zinc.[1]

Epigenetic Minerals

Minerals which are younger than the rocks containing them are called epigenetic minerals. They have been found to be associated with ancient erosion surfaces which now occur as unconformities covered by later deposits. Evidence of association of epigenetic minerals with unconformities has been found in several mineral districts in U.S.A. and Canada where are found such minerals as lead, nickel, manganese, copper, uranium, vanadium, barite and fluorite, ranging in age from Pre-cambrian to Tertiary.

Geomorphic Processes and Placer Deposits

Placer deposits are mechanically deposited minerals which accumulate where the speed of the medium carrying the minerals

1. Ollier, C.D.—op. cit., p. 42.

falls. Placer mineral concentrations result from specific geomorphological processes and are found in specific topographic positions. Nine types of placer deposits are recognised :- residual, colluvial, alluvial, aeolian, bajada, beach, glacial and buried and ancient placers. Residual placers are residues from the weathering of quartz veins and are usually of limited amount. Colluvial placers are produced by downslope creep of residual materials and are transitional between residual placers and alluvial placers. Alluvial placers are the product of fluvial activity and are the most important. Gold, tin, platinum and diamond are among the minerals obtained from alluvial placers. In India alluvial gold placers are found in the Subarnrekha valley in Chota Nagpur. They are also important in California (U.S.A), Australia and Newzealand. The placers of Malaya are partly alluvial and partly colluvial. About one-third of the world's platinum is derived from alluvial placers, particularly in Russia and Colombia. About one-fifth of the world's diamond comes from alluvial placers deposits, including the Vaal and Orange river districts of South Africa, Congo Republic, and Minas Geraes in Brazil. Aeolian placers are a source of gold in Australia and Mexico. Bajada placers form in the gravel deposits of a pediment and in the alluvial fans near the mountain base. Beach placers have yielded diamonds in Namaqualand, South Africa, gold in California, ilmenite and monazite from the Kerala coast in India, and zircon in India, Brazil and Australia. Glacial placers are not important but gold has been found in the sand and gravels of valley trains and moraines in parts of Colorado. The most important ancient placers are of Tertiary age. Some of these have been submerged beneath the sea, as a result of post-glacial rise of sea level and large quantities of ancient tin placers lie submerged in the drowned river valleys of Malaysia.

Among the bureid placers, those of the Eocene valleys of Sierra, Nevada, U.S.A. are quite famous. The tropical climate of Eocene favoured deep chemical weathering of the gold-bearing Jurassic batholiths and the gold was washed down into the valleys. The Eocene valleys were buried under later volcanic rocks. The younger Miocene and Pliocene valleys of the volcanic cover contain little gold except where they have cut into underlying batholith and intersected the buried Eocene valleys. It is therefore absolutely essential to identify the different valley systems and confine prospecting to the Eocene drainage channels. These channels do not coincide with the present river channels but contain the most valuable gold placer deposits. The problem of locating the richest placer deposit is essentially one of reconstructing the earty Tertiary (Eocene) drainage system. After the identification of buried channels principles of fluvial geomorphology can be applied to find out the exact location of economic placer deposits of heavy minerals. Placer metals are likely to be richest where there is slowing down of stream velocity. A knowledge of the changing gradient along the buried channel and irregularities on the bed-rock floor of the channel which would encourage deposition of the metals becomes important. Alluvial channels show a pattern of deep reaches alternating with shallow reaches in their long profile, and riffles and pools alternate in a downstream direction. A river incising into bed-rock may develop the same pattern with alternate deep and shallow reaches of bedrock on the bed rock floor of the valleys. Such a pattern would provide favourable site for concentration of heavy minerals. If the spacing of the depression is uniform, location of two such pockets would provide a guide to similar upstream and down-stream locations. Irregularity of the bedrock floor may be related to the nature of rocks, as some rocks like schists erode into pitted surface with small ridges and grooves whereas diorite usually erodes to a smooth surface which is not favourable for the settling of placer metals.

Another aspect of bed-rock valleys is the identification of what may be called "*inner channels*" i.e. relatively narrow and deep alluvium filled channels cut in bedrock of a valley floor. They occupy only a part of the floor and formed under conditions of high discharge at locations of maximum shear. These inner channels with their irregular longitudinal profile are ideal places for concentration of gold and other heavy minerals. They have also important implication for engineering geologists and construction engineers as they may suddenly find deep alluvium filled bed-rock channels in the floor of what may be considered to be a flat bedrock valley.

It is also well to remember the dynamism of river systems. There are alluvial gold and diamond concentrations for which the source cannot be found within the drainage basin. From this it may be concluded that the source has been completely eroded, but careful study of the drainage pattern should be made, because river capture may provide the explanation. That is, the mother lode may be found on the other side of the drainage divide as a result of the beheading of the stream of the weaker system and shifting of the drainage divide. An extreme example of this is the suggestion that the placer diamonds in Venezuela and British Guiana have their source in Africa and the rifting of the continents along the mid-ocean ridge has separated the source from the area of deposition. In this case it would be futile to look for the source of Venezualan diamonds elsewhere in South America.[1]

Geomorphology and Petroleum Exploration

Petroleum consists of hydrocarbons of marine organic origin. Most of the world's petroleum deposits are located in marine sediments of Mesozoic and Tertiary ages, generally along the flanks and in the less folded portions of orogenic belts, although some may occur also in Palaeozoic sediments (e.g. in Russia. and North America) or in the uplifted beds around salt domes. The oil pool is found in *reservoir rocks* as distinct from *source rocks* or sediments in which petroleum had its origin. The source beds such as clay and shale which are now compact and impervious were still unconsolidated with high proportion of seawater carrying dispersed globules of oil. As the source beds are compacted in response to pressure differences caused by earth movements, the fluids are squeezed out into more coarse and porous formations such as sands. In these permeable beds the oil can migrate for long distances but cannot escape unless the overlying rocks are fissured. In general oil migrates outwards and upwards from the source beds, passes into porous or fissured reservoir beds and collects into an oil pool wherever the structure provides a trap which impedes further migration. A dome or anticline of alternating permeable and impervious sediments makes a good trap for oil migrating towards it. Isolated open anticlines surrounded by extensive gathering ground provide much more favourable traps than more closely packed folds. The anticlinal structures are reflected in the topography and many oilfields have been discovered because of their striking topographic expression. Topographic highs are likely to be also structurally high. This is particularly true of late Tertiary and later earth movements where deformation has taken place so recently that subaerial denudation did not have enough time to destroy the effects of diastrophism. In areas of dense tropical forest,

1. Schumm, S.A.—'Applied Fluvial Geomorphology', in *Applied Geomorphology*, edited by John R. Hails, 1978, pp. 126-27.

the topography may not show, but aerial photographs through tonal differences in the vegetation can easily outline an anticlinal or domal structure.

Probably most of the domes and anticlines which have striking topographic expression have already been discovered and more subtle methods of identifying structure favourable to oil accumulation must be found. One such method is the detailed drainage analysis of a terrain as shown on aerial photographs. This involves detailed examination of the drainage pattern, drainage texture and stream pattern and identification of drainage anomalies or local departures from the normal drainage relationships. It is especially useful in regions where the topographic relief is slight and the sedimentary rocks are undeformed and nearly horizontal. In such areas one should expect dendritic pattern of drainage, and any departure from this pattern would constitute drainage anomaly which may be indicative of local departure from the regional structure. The influence of the local bedrock structures may be reflected in radial drainage or centripetal drainage, deflected drainage, a pronounced sharp meander along an otherwise straight stream course, sharp change in stream gradient, concentric arrangement of lesser tributaries and so on. These and other anomalies cannot be taken as positive evidence of the influence of local structures but they provide a basis for detailed field examination to prove or disprove the existence of a local structure.

In Russia recent folding and upwarping has been detected by geomorphological means including the recognition and mapping of upwarped planation surfaces and depositional downwarped basins. The upwarped structures provide favourable locations for development of oil and natural gas.

The study of buried landscapes is important for oil exploration as most stratigraphic traps are associated with unconformities or ancient erosion surfaces. Three types of stratigraphic traps may be particularly noted—ancient erosion surfaces which truncate permeable beds and are later sealed by impermeable deposits, buried karst topography, and buried elongate sandy deposits. There are many examples of a stratigraphic trap produced by the sealing of an old erosion surface. In the East Texas oil-field which is the most productive oil-field in the U.S.A., oil comes from the Cretaceous sandy formation. This deposit is truncated by an irregular erosion surface. Later this erosion surface was sealed by deposition of clay and chalk. The hills upon this buried erosion surface are an important source of oil. In the Digboi oilfield of Assam oil is found in an anticline in the Tipam sandstone (Miocene) and the sands of Barail series (Eocene), these being separated by an irregular unconformity. Tipam sandstone is overlain by clay and the Barails are also underlain by clay. As the Tipam sandstone is not of marine origin, it is believed that the oil has migrated from the Barail series, possibly by way of the Naga thrust. Similarly in Nahorkatiya oilfield the oil-sands occur as lenses in the Barail series as the Tertiary strata is cut through by many faults. The Barail sandstones are overlain by Tipan sandstone and both have been sealed by later clayey deposits.

In several oilfields, such as the West Texas Michigan Basin and Lima-Indiana fields in the U.S.A. carbonate rocks have been found to be important oil and gas reservoirs. In Gujarat and the Rann of Kutch the Nummulitic limestone (middle Eocene) contains many productive petroliferous horizons. Most oil yields from limestone have come from rocks that have high permeability produced by solution. It has been suggested that the West Texas oil reservoirs are mostly cavernous limestones. Usually the limestone reservoirs are associated with erosion surfaces on limestones beneath which there has been solutional enlargement of joints and bedding planes. This is likely to be most pronounced near the ancient drainage lines.

Sometimes petroliferous horizons are found in elongate sand basins which may still remain practically horizontal. These may be coastal deposits of beaches and bars, barrier islands or river channel or delta distributary channel deposits. It is essential to know the origin of these deposits to plan drilling programmes and the characteristics of specific topographic features can be used to advantage in the exploitation of their oil and gas possibilities. Barrier chains extend in smooth curves parallel to a former shoreline. They are generally discontinous or segmented by former tidal inlets. The bar segments may be productive, whereas the gaps between them may not contain any oil. On the other hand, the riverine deposits will be continuous and not broken, with uniform thickness and width, following sometimes a meandering course. Its trend is likely to be not parallel to but at right angle to the ancient shoreline.

Sometimes oil pools occupy the upper ends of tilted reservoir beds. The tilted bed may pass laterally into shale or it may be abruptly cut off by an impervious barrier, such as a fault throwing an impervious bed against the reservoir bed or a hill belonging to an ancient land surface which was unconformably buried by a later series of petroliferous strata or a salt dome which has ascended through the sedimentary strata.

Geomorphology and Engineering Projects

Most engineering projects, large or small, require knowledge of terrain characteristics, surficial materials and bed-rock conditions. Understanding of geomorphic principles and history is needed for construction of (i) highways, railways and airports that maintain the flow of goods and services, (ii) multipurpose dams and reservoirs for water supply, irrigation, power and flood-control, and (iii) levees, dykes, sea-walls etc. built to protect home sites, industrial structures, harbours and navigation. If proper evaluation of geomorphic conditions and principles is not made, such construction activities may result in different types of environmental disruption and give rise to may unforeseen problems.

"Topography obviously plays an important role in determining the most feasible highway route, although sometimes cultural features may necessitate routes not in best accord with topography".[1] Different types of terrain pose different kinds of problems. In *mountainous regions* like the Himalayas the gradient is so steep that roads have to be aligned along contours to reduce road gradient, making the roads extremely winding. Here highways normally hug the river valleys which provide important physiographic corridors and relatively easy routes of entry. In mountainous terrain the study of the drainage system and the valley slopes is essential for planning the alignment of roads. The road connecting Hardwar with Badrinath (over 7000 m) faithfully clings to the precipitous valley side slope of the Alaknanda (known as the Ganga below its confluence with the Bhagirathi at Devprayag). The Srinagar Leh highway makes full use of the physiographic corridor provided by the valleys of the Jhelum, the Dras and the Indus rivers. Even in areas of humbler relief, such as the Chota Nagpur plateau, which consists of several erosion surfaces of varying heights separated by

1. Thornbury, W.D.—*Principles of Geomorphology*, op, cit., p. 557

steep scarps, the scarps or ghats exert a marked influence on the structure, alignment, gradient, length, winding character and maintenance of the highways.

There are many cuts and fills and many bridges have to be built in areas of considerable relief. There is also the danger of landslides, earth flows and slumping. The roadside cut along a stabilized hillside slope or talus slope invites downslope mass movement of materials. By carefully analysing the state of relief, drainage, bedrock and regolith, the incidence of earthquakes, evidence of previous mass movements and man-made features, it is possible to identify land-slide prone areas and sites. The advantage of identifying those slopes which are potentially unstable or which have already failed, is that this knowledge can be used at the planning stage of an engineering project. Route alignment can be decided so as to avoid these slopes, or, if they must be crossed, then remedial or controlling measures can be planned. If such slopes are identified at the initial stage of a project, field sampling and laboratory analysis can concentrate on the potentially unstable slopes.[1] There is one primary cause of landsliding over which man can exercise no control and that is earthquakes. In the earthquake zones it has to be recognized that sites which are either themselves potentially unstable or which have unstable slopes above them should be avoided for highway or rail construction or for human settlements, dams, hydro-electric projects or other structures.

The *karst terrain* is marked by near-waterless surface with sinks and underground caverns. The sinkholes are, however, liable to flooding during heavy rains. A route over a karst plain therefore requires repeated cut and fill. Small caverns create special problems in the process of road construction. The local instability of the ground surface and the enlargement of solution caverns have also to be taken care of especially in the location and design of bridge abutments.

Glaciated lowlands present special engineering problems. A till plain is topographically most suitable for settlement or highway construction. But terminal moraines, eskers, kames or drumlins may present many topographic irregularities, and highway construction may necessitate cut and fill to avoid circuitous routes. Areas

1. Cook, R.U. and Doornkarmp. J.C.—*Geomorphology in Environmetal Management*, Claredon press, Oxford, 1974, p. 160.

formerly occupied by lakes and now covered with lacustrine deposits are unsuitable for heavy traffic roads, as the ground is likely to sink under the weight of heavy traffic. In such areas it may be necessary to remove the lacustrine material and replace it with materials that will not flow under heavy load. On highways which are designed to carry heavy traffic the nature of the soil beneath the road surface (called subgrade) is extremely important because of its influence on the sub-surface drainage. The life of a highway depends on the quality of the aggregate used in the highway and the soil texture and drainage of the subgrade. "An appreciation of the relationship of soils to varying topographic conditions and type of parent material becomes essential in modern highway construction".[1] Poor drainage in silty clay subgrades with high water-table causes cracks in the roads, while granular materials with a low water-table result in longer life and better performance. Pumping is the term used for expulsion of water from beneath the road slab through joints and cracks and is caused by poor drainage in the subgrade. A till plain topographically ideal for highway construction has poor internal drainage unless granular materials are introduced into the subgrade or surface drainage is improved. Till plains and moraines, lacustrine plains, outwash plains, eskers, kames, drumlins, wind-blown sand and loess present not only different types of topography but also different types of subgrade material, and an appreciation of these differences is essential for good highway construction. Because of the increasing use of granular materials in subgrades and fills, their availability and sources of supply have of late become extremely important. It is useful to map all such materials along a projected highway route with the help of aerial photographs and soil maps.

The *periglacial landscape* presents special engineering problems because of the presence of permafrost, strong frost action and solifluction, and an unstable mantle of poorly drained and saturated sediment mixed with gravels and boulders moving downslope. Most of Alaska and nearly one-half of Canada and Russia are underlain by permafrost. As most of the permafrost terrain is remote and inaccessible, reconnaissance surveys are usually undertaken in the first instance by analysing the available remote sensing imagery. The purpose of these surveys is to determine the nature of terrain, material and ground-ice condition and sources of constructional

1. Thurnbury. W.D.—op, cit., p. 558.

materials and to plan access routes and provisional layouts of structures and to select sites for detailed ground studies. Field studies invariably follow reconnaissance surveys and are directed towards validating the air-photograph predictions and analysing in detail the conditions at specific sites.

The engineering responses to permafrost may be either active or passive.[1] Permafrost can be ignored where the drainage is good and the ground materials are not susceptible to serious frost action and solifluction. In areas of thin, sporadic or discontinues permafrost, where the thawed ground has satisfactory bearing strength, it may be possible to remove the frozen ground by removing the insulating cover of surface vegetation, by excavation, by ground treatment, or by thawing of the ice with steam. Material not susceptible to frost action maybe placed on the ground. This approach has been called the active method. In regions of continuous permafrost it may not be practicable to eliminate the frozen ground, and effort may be made to preserve the permafrost by insulating the surface with vegetation or gravel cover, ventilating the undersides of heat-generating structures and preserving vegetation. This is known as the passive method. In either case a thorough understanding of the local permafrost conditions is essential. Areas suitable for location of building, roads etc. are those which have a thin active layer, bedrock near to the surface, thawed material with adequate bearing properties, good drainage, lack of frost-susceptible materials, and a stable site away from the possible areas of water seepage and icing. As most building construction problems are associated with the active zone, foundations are frequently fixed in the underlying permafrost. Other techniques include the creation of a trench around the building (generally filled with material not susceptible to frost action) in order to eliminate possible horizontal stresses; the raising of the floor above the ground by about a metre to allow air circulation, the insulation of the floor to restrict temperature rise beneath the floor and provision of skirting insulation between ground and floor with air vents that can be closed in summer, and opened in winter to allow free air movement.

The provision of utilities also poses special problems created by seasonal freezing and thawing and the presence of active and permafrost zones. Suitable water supply is usually difficult to find,

1. Cooke, R.U. and Doornkamp, J.C.-op. ciy., pp. 243-47

and the disposal and decomposition of seawage may be a problem. Freezing and thawing of such services creates additional problems. One ingenious solution is for all linear utilities—water, sewerage, electricity etc. to be placed in a single insulated and heated conduit (known as utilidor) that may be placed above or in the ground.

The problems associated with building roads, runways etc. are similar to those encountered in building construction. The aim is in- variably to provide a stable track across what is frequently an unstable surface. The problem may be complicated if the route passes through varied permafrost terrain. Drainage is particularly difficult, especially where the route crosses drainage lines. It is advisable to remove as much water as possible from accumulating beneath the road, usually by providing ditches at some distance or preventing infiltration. Road or rail road cuts may penetrate permafrost, lower the permafrost table, cause landslides or solifluction in the newly created active zone—hence they are to be avoided as far as possible.

It is thus clear that the study of the geomorphic processes of frost action and solifluction and the relations between these processes, ground disturbance and surface stability and the landforms that frequently reflect surface and sub-surface processes and conditions is absolutely essential for the formation of successful development plans and engineering designs in the periglacial regions.

Dams and Reservoirs and River Regulation

Geomorphology is an important ingredient in the assessment and utilization of water resources. In the construction of dams and reservoirs as well as in the alignment of canals geomorphic factors play a significant role. A good dam and reservoir site requires (i) a water- tight river basin of adequate size for storage with suitable bedrock, (ii) a narrow outlet of the basin with a foundation that will allow economical dam construction, (iii) facilities for building spillways to carry surplus water, (iv) availability of materials needed for dam construction, and (v) assurance that the nature of the terrain and ecology of the catchment will not make the life of the reservoir too short as a result to excessive silting.[1] Dam construction in a limestone terrain is likely to result in leakages and water is likely to seep into solution channels and emerge as seepages below the dam.

1. Thornbury. W.D.—op, cit., p. 563.

On the other hand, impervious bed-rock will provide a dependable and adequate reserviour. A constriction in a valley is most economical but it may not necessarily be a suitable dam site. In glaciated areas the surface topography may not give a correct picture of subsurface conditions, as buried bed-rock valleys containing sand and gravel are quite common. In such cases the dam is likely to leak and suitable bed-rock foundation for the dam may be found at much greater depths.

Both the construction of canals to transport water from surplus to deficit areas and ground water planning and management for transfer of water from underground to the surface involve many geomorphic considerations. The reciprocal linkage between terrain and construction is well illustrated in the great irrigation works of the Punjab, Haryana and U.P. (built mostly in the 19th century) where irrigation channels have been built to synchronize with gradient with utmost precision so that steep gradients which encourage erosion as well as shallow gradients which encourage siltation are avoided. Tube-wells, operated by electricity generated on the Ganga canal heads tap the huge reservoir of underground water in the aquifers below the Indo-Gangetic alluvium fed by rain in the Himalayan terai and bhabar, maintaining a proper balance between withdrawal and recuperative capacity, so that the ground water reservoir is not depleted. In Peninsular India while elaborate canal systems serve the fertile and gently sloping deltas of the Godavari, Krishna and Cauvery rivers, throughout the plateau innumerable tanks or storage reservoirs have been created by utilizing the natural variations in the terrain and by damming small streams and overland surface flow.

It is important to remember while regulating rivers that rivers are not static but constantly changing as a result of hydraulic forces acting on its bed and banks. These changes may result from natural environmental changes or from changes introduced by man's activities. Dam construction, for example, in the upstream reduces the quantity of bed material load and decreases the peak discharge of water moved through the channel. This produces scour below the dam. The river downstream may also reduce its width in response to decrease in flood peaks. Realignment and channelization of rivers is sometimes practised to reduce flood damage and to drain flood plain areas. Channelization steepens the channel as sinuosity is reduced, and this in turn increases the velocity of flow and the potential for bed and bank erosion. Further, the base level of the upstream part of the system has, in effect, been lowered and channel incision will begin. Thus although flood and drainage problems are temporarily solved, the increase in erosion upstream not only creates problems of lost land, destroyed bridges but the greatly increased sediment load will cause deposition in the channelized downstream with a return to the original flood and drainage problems. The long term perspective of the geomorphologist can be extremely valuable in evaluating the environmental consequences of channelization, upstream modification of hydraulic regimen and other supposedly beneficial changes. This is an area in which geomorphic expertise can prove extremely valuable in the preparation of environmental impact reports on the long term effect of river stabilizaion, channelization, and flow regulation etc.[1]

In planning a multi-purpose river valley project many major factors have to be considered including the geomorphology and geology of the area, its hydrological characteristics, ways of integration and co-ordination of all the various uses of water, such as for irrigation and flood control, hydroelectric power, navigation, domestic and industrial water supply, recreation, fish and wildlife. High altitude sites can be used for reservoirs and dams, with large falls available for hydro-electric power generation. Larger storage can be provided downstream in flatter terrain but above the more populated valleys. If physical conditions allow, more electric power can be generated downstream. At any stage in the system controls can be included to deal with flood conditions or the maintenance of adequate water supplies for the highly populated lowlands. In such projects the river catchment forms the natural unit for planning and the physical system is directly involved. Drainage basins form natural geomorphological units displaying organized relations between landforms and drainage and biotic life, and man's interference with the fluvial system requires a careful and rigorous assessment of the possible repercussions throughout the basin. For example, when new ground is broken for development, vegetation is removed and the ground is disturbed by construction equipment. Run-off and stream flow increase, land erosion is accelerated and

1. Schumm, S.A.—"Applied Fluvial Geomorphology", in *Applied Geomorphology* edited by John R. Hails, 1978, pp. 119-156.

the sediment load of the stream is often markedly increased. Excessive amounts of sediment induce the building up of channel bars, erosion of channel banks as a result of deposition within the channel, obstruction of flow and increased flooding, shifting configuration of the channel bottom, blanketing of bottom dwelling flora and fauna and alterations in these due to changes in light transmission and abrasive effects of sediments, as well as alteration of species of fish due to changes produced in the flora and fauna upon which the fish depend. Construction of houses and roads decreases natural infiltration and lowers the water table. Local interference with minor streams may cause these to flood or to undermine their banks. Untreated waste including sewage may be discharged into streams, causing pollution which in turn may be lethal for aquatic life and detrimental to the use of water downstream. Increase in water demands may neccessitate the building of more dams upstream or in adjacent basins, leading to further water flow control. All these changes may be studied and their control effected in the context of the physical characteristics of the catchment and steps have to be taken to constantly monitor the changes in the basins threatened by urbanization.

A major application of geomorphology lies in the prediction of flood-peaks and run-off and sediment yield characteristics of a region. In restricted areas where geology, land use and climate are uniform, it is the geomorphic variables that influence the hydrological characteristics of drainage systems.

An additional important role that the geomorphologist can play is in the development of flood-plain zoning laws which would restrict the activities permitted on flood plains. Any plan for utilization of flood plains should include the possibility of frequent flooding, as overbank flooding will occur every 1 to 2 years. In many areas hydrologic records are sufficiently long to permit calculation of flood peaks and from these discharge calculations depths of flooding can be determined. However, generally such data are not available and it is necessary to identify, based on vegetational, soil and geomorphic characteristics, those areas most susceptible to flooding. This is not an easy task and different criteria may be needed in different areas, but the fluvial geomorphologist must be involved in such determination. For example, Wolman shows how the character of the valley floor will influence flooding.

If the floor is flat, water will spread widely over the valley. If the valley floor is slightly concave, flood depths will be greater but flood will be confined to the flood plain nearest the channel.[1]

Schumn,[2] states that there are three ways in which a geomorpologist's knowledge of rivers and their behaviour can be applied to practical problems. The first is simply in the evaluation of existing conditions. For examples, with the help of aerial photographs, difference in drainage density and drainage patterns may be used to delineate specific rock types and to locate geologic structures. The second is an attempt to predict what will happen in the future both with constant and with changed environmental conditions, such as what will happen to the river as flood peaks are increased by urbanization or deforestation. This type of projection or prediction is based on an understanding of cause and effect relations in geomorphic systems. The third is an attempt to interpret the past, post-diction. This may have real significance for location of economic mineral deposits in fluvial sediments, as it is frequently necessary to determine past environmental conditions that were favourable for the concentration of heavy minerals and the formation of placer deposits. Post-diction is also a major component of archaeological quarternary investigations. River migration frequently destroys archaeological sites, or, as a result of cut-off, living sites are left at some distance from the active river channel. Unless the channel behaviour is understood, the environment of the site will not be correctly interpreted and the location of sites from early descriptions will be difficult.

Sand and Gravel and the Geomorphologist

Sand and gravel are required mainly for use in the construction and building industry. Industrial and commercial demands have been rising rapidly and there is a trend towards more specific and rigorous requirements for particular purposes. Increased demand has led in many areas to depletion of the best and most suitably located reserves. One of the most important requirements, and one which the geomorphologist can help to meet, is for precise knowledge of the nature and distribution of available reserves. There are many examples of surveys of sand and gravel resources

1. Schumm, S.A.—Ibid, p. 129.
2. Ibid, p. 121.

in which geomorphologists have made valuable contribution. One notable example is that of S.W. Wooldridge. His studies on the geomorphological history of south-east England were of great help in assessing the resources of the London region and elsewhere. For many years the main source of sand and gravel had been the river-terrace deposits, especially west of London, associated with the most recent phase in the evolution of River Thames and its tributaries. Studies of geomorpholgical evolution showed that the Thames formally followed a more northerly course and along this line and elsewhere above the terraces there were extensive fluvio-glacial and plateau deposits, though poorer in quality than the resources west of London. The older fluvio-glacial deposits require more beneficiation than the younger, better washed and sorted terrace sediments, but as the traditional sources have become exhausted, so the exploitation of the less satisfactory reserves has increased.

Suitable sand and gravel sources occur in five major geomorpholigcal contexts: (i) river terraces, channels, flood plains and alluvial cones and fans, (ii) fluvio-glacial environments in which a wide variety of materials may be laid down or carried in meltwater, (iii) coastal zone, (iv) on slopes where suitable materials may be found in talus or scree, and (v) wind-blown and residual deposits in arid and semi-arid areas. All these have distinctive topographic relationship and vary in quality and posibilities of development. Identification of the type of deposit is essential for proper evaluation of its potentialities. As sand and gravel are a bulky commodity of low value and expensive to transport, proximity of supplies to markets especially the major urban areas and major constructional projects is of utmost value. It is, however, near urban areas that the conflict of interest is most pronounced. Urban development and airport construction may themselves cover potentially workable deposits. Further, many river terrace and gravel deposits on urban fringes provide optimum location for market gardening which also benefits from proximity to urban markets. On the other hand, extraction of sand and gravel near urban areas may be objectionable on several environmental grounds, though the old sand and gravel workings may be used as suitable receptacles for urban garbage and debris and filled up pits can be used subsequently for forest, recreation, building or agriculture. In order to help resolve such conflicts, the precise nature and extent of sand and gravel reserves must be known.

Each of the principal sources of deposits has distinctive characteristics which make them more or less suitable for specific uses. The river deposits, in which there is extensive natural washing and sorting are usually preferable to glacial deposits as the former requires less processing. But flood-plain deposits are likely to contain high proportion of silt and clay between beds of sand and gravel. Mixed deposit of sand and gravel may be found on the river terraces which is a desirable quality. Alluvial fan and cone gravels are angular in shape and variable in size. Talus gravels are usually too large to be suitable for most use and are limited in extent. Wind-blown sands may be satisfactory but have no gravel in them. Residual deposits are unassorted and are likely to contain deeply weathered pebbles which are not suitable for use as aggregate in cement work. Terraced valley trains and outwash plains are usually favourable sites as they are usually extensive and have thin overburden. An esker may contain well-assorted gravel, but the amount of gravel available may be limited, because eskers are usually small and segmented. Kame deposits are poorly assorted, and a large amount of material may have to be discarded because it is either too large or too small. Coarse and angular sand such as is found in fluvio-glacial deposits is suitable for concrete, while rounded sand may be suitable for mortar. The thickness of the overburden and depth of water-table also affect the suitability of a reserve. In any case the quantity of material available at any one place should be sufficiently large for sustained production and for large capital investment needed in a modern operation.

The appraisal of these sediments with a view to their possible exploitation normally involves preliminary survey of possible source areas from available geological and topographical maps and air photographs, field survey of deposits and mapping of their areal extent and determination of depth and volume, sampling of deposits from available exposures, boreholes or trial pits and laboratory analysis of sample to determine grain size and shape characteristics, weathering characteristics, abradibility and shrinkage values.

Application of Geomorphology to Hydrology

Hydrology is the science of water, its properties and distribution over the globe. It is not often realized that water is the earth's most precious resource, for without water there can be no life. 97% of the earth's water is saline and unfit for human use and 2% is locked in glacier and ice. About 0.3% of the earth's water is groundwater, 0.01% lake water and 0.0001 % river water. Thus less than 1% of the water on the earth is availablde for drinking and this too is unevely distributed. Large areas therefore suffer from problems of drinking water especially in the developing countries.

We have already referred to some of hydrological problems of *fluvial regions* in the study of which geomorphological knowledge can be fruitfully employed and this need not be repeated.

The hydrological problems of *limestone regions* are intimately related to solutional erosion processes and landforms studied by the geomorphologist. Limestone varies greatly in its permeability characteristics. Permeability in limestones is partly primary and partly secondary or acquired. Secondary permeability results from joints and bedding planes and their solutional enlargement varies with topography, being greatest in the valleys and topographic lows. Much of the ground water in karst regions is confined to solution channels and caverns lying underground. Sometimes where mining operations have inadvertently touched and opend a large underground karst system, the mines have been suddebly flooded and karst springs dependent on it several kilometres away have been found to become dry. Such unexpected flooding occurred in a gypsum mine in Indiana, U.S.A. in 1960. Solutional openings with their increased permeability are of significance not only in present day karst terrains but also in buried karst landscapes concealed beneath later deposits. Generally dam and reservior construction is avoided on limestone terrain because of the underground leakage that may occur, but a number of reservoirs have been constructed on carbonate rock with adequate but expensive engineering precautions, while some other limestones have been found to be accpetable without additional engineering work. For example, saturated conditions with a well-defined watertable may exist in the limestone as a result of impermeable base to groundwater circulation provided by an extensive sill system and its associated metamorphic effects in modifying the permeability characteristics of overlying limestones. The major problem with limestone bedded reservoirs, even with detailed prior investigation and engineering precautions, is that it is difficult to forecast acurately whether a given dam will suffer leakage until the reservoir is filled, and if later work is required, the remedial operations may be so expensive that the reservoirs may have to be abandoned. It is therefore emphasized by workers in this field that fullest possible pre-construction survey of subterranean drainage is undertaken with particular reference to the part played by secondary permeability and the structural properties of the limestones.[1]

Where ancient terrain lies buried under later formations, the ability to recognize ancient karst levels now concealed beneath later deposits becomes important in the evaluation of water yielding possibilities of some regions, as highly cavernous limestone adjacent to ancient drainage lines may yield large quantities of water.

In the early stages of karst cycle the diversion of run-off to underground routes has just begun, but as the cycle advances the surface streams become fewer and shortlived and more and more run-off is diverted to the underground solution channels through the sinkholes. The hydrological importance of the underground sources thus increases with the advance of the karst cycle, and karst springs become the main source of water supply. The quality of water of such springs usually needs to be determined with reference to the sinkholes that feed the underground drainage system, as the risk of contamination is high if the water entering the sinkholes and emerging later in springs has not been filtered by a sandstone layer.

Wells dug in limestone terrain may yield large quantities of water if the limestone has primary permeabiltiy which has been further supplemented by secondary permeability in the form of solutional passageways. On the other hand, if the limestone is dense and compact with little primary permeability, the movement of water is essentially through the solutional openings. Under such conditions, satisfactory supply depends upon the chance intersection of some of the solutional routes through which the groundwater is moving. The quality of water moving through secondary opening needs to be checked as karst plains usually lack a filtering cover of sandstone.

In *glaciated areas* the possiblity of locating large supplies of groundwater depends upon a knowledge of the type of deposits

1. Smith, D.I.—'Applied Geomorphology and Hydrology of Karst region", in *Applied Geomorphology* eidted by John R. Hails, 1978, pp. 85-118.

which can yield large quantities of water together with an understanding of the geomorphic history of the area during glacial and pre-glacial times. Buried or exposed glacio-fluvial depostis like outwash plains, valley trains, intertill gravels are likely to prove good aquifers with large quantities of water in them. Most tills are poor aquifers because of the presence of clay, but they contain fragmental lenses of sand and gravel with enough water in them. No clue to these lenses is provided by surface topography but they are likely to be present in areas adjacent to lines of glacial drainage and aligned in the direction of glacial movement.

Buried preglacial and interglacial valleys may be rich potential sources of groundwater but their identification requires detailed knowledge of geomorphic history of the area and its preglacial topography. One of the great preglacieal drainage lines of North America with enormous potential groundwater supply lies buried under several hunded feet of glacial drift in Ohio, Indiana and Illinois and some of the cities in these states derive their water suply from wells fed by this source. In semi-arid Nebraska where water is a much valued resource, the buried valleys, already an important source of water supply, are likely to become even more important as more is known of the preglacial and glacial history of the region. Such water derived from depth has low tempreature and is much in demand for air conditioning. For locating such buried valleys thickness of drift-cover is first determined by seismic and other surveys and then maps of bed-rock topography are prepared. Where the buried valley has a fill of finer materials like till, clay and silt, the water yield is likely to be low, but where outwash sand and gravel are present, such as is likely along unimpeded drainage lines from an ice front, water yield is likely to be high. A thorough knowledge of the direction of glacial movement and lines of glacial outwash is necessary for assessing groundwater potentialities of buried pregalcial and inter-glacial valleys.[1]

Geomorphology and Coastal Zone Management

The coastal zone is deifned as "an area of variable width, which extends seaward to the edge of the continental shelf, but which has no distinct landward demarcation"[2] In general, however, the coastal

1. Thurnbury. W.D.—op, cit., p. 544-46.
2. Hails, J.R.—"Applied Geomorphology in Coastal-Zone Planning and Management," in *Applied Geomorphology* edited by him, 1978, p.317.

zone should be viewed, not in linear terms as a boundary between land and water, but as a dynamic region at the interface of land and water, limited in space but subjected to growing human pressure resulting from increasing concentration of industrial, commercial, recreational, fishing and agricultural activities. The rapid deterioration of the coastal environment through increasing shorline erosion, loss of valuable areas of natural beauty, pollution and extinction of species and serious disturbance of the fragile coastal eco-system and natural processes have alarmed nations throughout the world. Unesco has devoted a special section of its scientific programmes to 'Management of coastal and island regions" and has sponsored a number of multidisciplinary studies and pilot projects aimed at achieving a synthesis of knowledge on the complex interactions between terrestiral and marine environments and comprehensive evalutaion of the impacts of human activities.

Coastal zone management obviously requires an integrated and pluri-disciplinary approach. The geomorphologist can provide a systematic framework for the collection and evaluation of accurate and relevant data on the complex processes that interact with a wide range of coastal materials to create diverse but mutually interacting landforms and a wide range of morphological units. The character of the unit varies according to the processes and materials that form it. In order to control and improve this fragile and highly vulnerable environment, man must learn to work with and not against the powerful marine processes. Thus an understanding of erosional-depositional processes and their interaction with materials in the coastal system in different environmental settings is absolutely basic to an understanding of the nature and character of the coastal zone which in turn is fundamental to an understanding of man's impact on natural systems. The role of applied geomorphology is to provide for developers, engineers, planners and decision makers, the vital baseline data of land-sea interface against which possible consequences of subsequent management decisions, programmes and strategies may be properly evaluated and gauged.[1]

As the coastal processes are extremely variable and have high level of activity, the forms and materials within the coastal zone are likely to change rapidly both in time and space with the possible

1. Ibid, p. 362

exception of the hard-rock cliffed coasts where the changes may be slow and imperceptible. In evaluating the coastal process-response systems it is therefore important to identify distinct stretches of the coast along which specific responses to the physical processes can be identified in terms of morphology, material and processes. This has been called the land unit approach to the analysis of coastal areas. Coastal land untis can be mapped to delimit homogenous morphological units in such a way that the character of their material and process that formed them can be easily identified and tabulated, as in nearly all instances the three elements of form, materials and process will be found to be closely interrelated. But one must remember that there is usually a dynamic interaction between adjacent units, and these units may form part of a larger system consisting of a headland, a barrier spit, barrier islands, tidal intlets, lagoons, offshore submarine bars and the off-shore shelf zone, all of which interact within a complex system, stretching for a hundred kms or more along the coast.

Waves are the most important process element on most coasts. Waves are generated by the wind through transference of energy from air to water, so that waves grow in size with increasing wind strength, wind duration and fetch of open water. The direction of wave approach is also important, as it controls, at least partly, the longshore transport of beach material.

The division of the world coastlines into four major groups by Davies (1964) in terms of wave types is higly useful. The storm-wave coasts occur in high latitudes in the zones where strong winds blow and create high, steep storm waves that reach the coasts, often accompanied by strong onshore winds. Such coasts often have long stretches of rocky cliffs and wave-cut platforms. These rocky coasts provide suitable habitat for seaweeds and for lobsters and crabs and are, therefore, of great ecological and human significance. The lower latitudes are dominated by swells (longer and lower waves that have travelled far from the generating areas) created in the stormy high latitudes but converted to swell by the time they reach the coasts of the lower latitudes. These are differentiated into west-coast swells which are higher and hence have greater energy, and east-coast swells which are long and therefore considerably refracted before they reach the coast. The fourth type is low-energy wave coast which is found either where the fetch is very small, such as the Irish sea and parts of southern North sea where sea ice, or some offshore feature prevents effective waves from reaching the coast and includes the ice-protected polar coasts as well as coasts of enclosed seas. In parts of low-energy coasts reclamation of salt marsh can take place, where sheltered conditions allow the deposition of silt e.g. reclamations along the coasts of Holland and Denmark.

It is, however, important to realize that waves are a constantly varying force acting on the beach material, and vary through both time and space.[1] This means that the beach is rarely in equilibrium with the waves, more so where the tidal range is considerable, so that the waves are always acting at a differnt level on the beach. It is therefore necessary to consider the range of variability within which the processes operate. The time-scale is also important, as seasonal variability is likely to be more than day to day variability and variability between extreme storms and normal calm weather is likely to be most pronounced.

Storm waves will act diffently according to the material on which they operate. Steep storm waves breaking on solid rock may lead to disintegration of rocks by shock pressure, followed by attrition of the disintegrated blocks. Similar waves breaking on shingle will throw some of it to the backshore zone to create a backshore ridge far above the high-tide level. On sand beaches such waves are entirely destructive, and create steep beach scarp at the limit of their action and cary much sand offshore to their breakpoint where they create a submarine bar and some sand may also be carried to the deeper water offshore.

Sand is the most widespread beach material, particularly in low latitudes. It commonly consists of qurtz but it may consist of suitably sized basalt fragments, such as on the beaches of Hawaii or it may be organic, such as the coral sands of the tropical seas. Some beaches contain heavy mineral sands, including gold. The sorting action of the waves concentrates the heavy mineral in places forming rich reserves. Sand and gravel available on the beach have been widely used for construction purposes. But these are now increasingly obtained from the offshore zone in preference to the beach sediments, as the latter are considered essential for coastal

1. Cooke, R.U. and Doornkamp, J.C.—op. cit., p. 192-99

protection. Shingle is more plentiful around high latitude coasts, probably because much of it was derived from reworked glacial drift deposits which are quite widespread in coastal zone of norhtwest Europe and Britain. In England there has been considerable expansion of the dredging industry because of the rapid increase in demand for sand and gravel. The offshore deposits are partly glacial in origin, partly reworked terrace gravels and partly derived from erosion of the present day coast. Some offshore gravel banks have remained relatively stable for more than a century but during this time others have migrated parallel and normal to the coast or have changed their configuration significantly.[1] Little is known about the geological history of these banks and the processes inducing sediment movement across and/or around them. "As some of the banks are likely to be dredged for gravel in the near future, it is necessary to know more about sediment movement between them and, in turn, about the exchange of mateiral between these features and adjacent beaches. Predictive data are also urgently required on the rate, direction and amount of sediment transport, as well as the relative stability of the banks".[2] Though some preliminary studies have been undertaken, the possible effects of dredging on the incidence of coastal erosion, on seabed topography and wave refraction have yet to be researched. In Eastern Australia where there has been considerable development of sand mining for heavy minerals, geomorphologists have made some significant contributions towards an understanding of shoreline equilibrium. But on the whole shoreline processes have been inadequately reasearched, and as problems associated with instability of coastal ecosystems are increasing there is obvious concern, from the point of view of conservation and environmental management about (i) whether or not sandmining actually accelerates natural erosion and dune migration, (ii) whether existing post-mining restoration schemes are adequate when so little is known about the overall rate of coastal recession, and episodic dune instability, (iii) contribution of coastal erosion by cattle grazing on dune and beach-ridge vegetation, property development and other activities, and (iv) a lack of management programmes for landuse after mining operations have ceased.

Man can interfere with the dynamic coastal system intentionally

1. Hails, J.R.—op. cit., pp. 341-2
2. Ibid, p. 342.

or unintentionally. Intentional interference usually aims to prevent or reduce erosion, to conserve the beach or to reclaim land from the sea. Unintentional interference is the result of human use of the coast, such as for recreational activities or grazing. Measures designed for coast protection include sea-defence structures such as seawalls, breakwaters, jetties and groynes. Sea walls are designed to protect the backshore zone from direct erosion but since they are impermeable they increase the backwash and produce a destructive wave effect.

Groynes are built normal to the beach and are designed to trap beach material moving along shore. They are beneficial to the extent that they raise the level of the foreshore, but they reduce the amount of material passing down-drift and are deterimental to the areas down drift. This shows the necessity of considering the whole coastal system rather than just one unit. The up-drift units must also be considered, as the groynes will fail in their purpose if no material moves into the area from up-drift or offshore.

Breakwaters can be built either normal or parallel to the coast. Those built normal to the coast act as large groynes and trap longshore transported material. The aim is usually to prevent silting of tidal inlets or estuaries or to provide shelter for shipping on an open coast. But quite often this has serious consequences on the erosion of the down- drift beaches. Breakwaters built parallel to the shore provide shelter and sand accumulates in their shelter as waves refract and diffract around them.

In order to evaluate properly how sea defence and other man made structures affect shoreline equilibrium, it is necessary to monitor and quantify wave conditions, tidal currents and sediment movemnt in the nearshore zone." Along coasts composed of unconsolidated deposits surprisingly little is known about the rate at which sediment is supplied to beaches by cliff retreat and how it is subsequently dispersed by wave and tidal currents... Man's increasing use of the coastal zone is making it extremely difficult to determine, both qualitatively and quantitatively, the extent to which his activites are modifying or interrupting natural processess. It may take several years, for example, to analyse the consequences of preventing sediment from reaching coasts by damming rivers. These facts mean that feasibility studies and monitoring programmes must be incorporated into future inter-disciplinary research programmes.

The role of the geomorphologist in applied coastal studies is to measure and explain the overall sediment input/output changes along a coastline. These generally reflect natural or man induced variations in the flow and sediment regime of rivers, cliff retreat, offshore sediment budget and erosion of up-drift beaches. It is anticipated that such information will assist engineers to resolve local erosion problems and to build more suitable sea defences."[1] It is imperative that littoral processes are fully understood before beach restoration schemes are implemented.

Hails emphasizes that in the context of coastal zone management applied geomorphology must be concerned with quantitative and not descriptive research in order to obtain relevant and accurate data on (i) rates of natural erosion and deposition ; (ii) rates and amount of sediment transport from river catchments to the nearshore zone; (iii) variation in sediment composition and distribution offshore; (iv) sources of sand supply and shoreline equilibrium; (v) the rate of sand interchange between beaches and dune systems; (vi) the effects of constructing sea defences; (vii) sediment dispersal offshore and the effects of dredging on seabed morphology, sediments transport and wave refraction; and (viii) landform analysis including topography of the near-shore zone, the form of the continental shelf and of relict coast lines, particularly in terms of rock outcrops.[2] Such investigations provide not only the relevant baseline data required for planning systematic process studies and monitoring programmes, but also for implementing land -use schemes and establishing management guide-lines.

Terrain Classification and its Applications

Terrain classification is based on the principle that all landscape can be divided into smaller units having common physical attributes (or repeated landforms) that are different from those of adjacent areas, and the landforms in turn will consist of assemblages of still smaller landscape units such as ridge tops, midslope, valley floor etc. This way it is possible to build up an inventory of landscapes and landforms in the form of maps, diagrams and descriptive material which can be easily understood and communicated. If the landscape units have a high correlation with some economic aspects of land or landuse, then the inventory of

1. Hails, J.A.—Ibid, p. 321
2. Ibid, p. 318.

terrain information can be used as a predictive tool.[1] For example, if a certain unit of land is found to be good for pineapple or cashew nut, a study of the terrain map will show the extent of similar good pine-apple or cashew nut growing land; if a certain landscape unit is found to be prone to landslides, the whole area subject to such hazard can be demarcated from the terrain map.

Land Systems Approach

There are two principal methods of terrain classification—(i) land-system or landscape approach, and (ii) the parametric approach. Land-system mapping based on the landscape approach was developed by the Commonwealth Scientific and Industrial Research Organization in Australia (CSIRO) as a rapid method of reconnaissance survey of otherwise unmapped or poorly mapped areas. The primary aim was a classification to find out the suitability of land for agriculture. The first area was mapped in 1946 and since then many more areas have been mapped and reports published. A land-systems map defines those areas within which certain predictable combinations of surface forms and their associated soils and vegetation are likely to be found. Surface relief, land form analysis, soils and geology are the key elements in landsystem mapping. The method has been adopted by the Land Resources Division (LRD) of the Overseas Development Administration, Foreign and Commonwealth office in Britain and reports have been published for many countries in Africa (Tanzania, Nigeria, Bechunaland and several others).

Similar techniques are used by the Oxford MEXE group[2] to produce terrain classification maps mostly from air photo interpretation initially for military purposes for assessing the difficulties of movement across different types of ground. The basic unit of land classification is the *land-facet*, a piece of landscape usually with simple form and having a more or less uniform rock, soil and water regime. Each land facet is fairly homogenous and can be managed uniformly for most purposes. The land facet is equivalent to the land unit of the CSIRO system. Recurrent patterns of land facets are called *land systems* same as the land systems of

1. Ollier, C.D.—"Terrain Classification: Methods, application and principles", in *Applied Geomorphology* edited by John R. Hailes 1978, pp. 277-316
2. MEXE stands for Military Engineering Experimental Establishment. It is now known as MVEE, Military Vehicles Experimental Establishment.

CSIRO.

In CSIRO reports land systems are shown on a map covering the survey area ranging in size from 5,000 km[1] in New Guinea to 350,000 km^2 in arid Australia. The reports carry a series of block diagrames to illustrate the general character of each land system. On the block diagram separate component parts of the land system known as land units (land facets) are identified by numbers. A land unit has a simple form, occurs usually on a single rock type of superficial deposit and has broad uniformity of soils. Examples of a land unit would be an alluvial fan, a levee, a group of sand dunes, a bedrock cliff. Land units are recognized primarily on the basis of their geomorphology. A land unit may be composed of two or more *elements*. for example, a levee has a crest and side slopes, each of which forms separate elements. Some land units may show local deviations in one property or another (e.g. soils) from the general character of the unit. These local deviations are isolated as *variants* They are not distinguishable upon areal photographs but are based on field investigation. Land system boundaries tend to conicide with some geological or geomorphological boundaries.

In the Oxford MEXE system representation is by a large-size two page description for each land system. Also there are a large annotated block diagram, a locational map, a ground photograph, large stereophoto pairs with individual land facets indicated, and a tabular description of land facets. A land system description includes climate, rock, landscape, soil, vegetation, altitude and relief; a land-facet table describes form, soil, hydrology and vegetation. Surveys have been carried out in Uganda, Kenya and Swaziland in Africa. These reports are large and cumbersome, and the data could better be stored on standardized cards, or envelopes with annotated photo inside and other data on cover to ensure better portability and ease of operation.

In Russia a system similar to the CSIRO system evolved independently. A hierarchy of landform divisions is recognized in which units that are repeated in a regular pattern form a *landscape* which is equivalent to land system. Landscape in the Russian system is considered to be the surface expression of a homogenous geological-geomorphological foundation. Like CSIRO approach these 'landscapes' are also conceived as ecological units and form the basis for land use planning.

Land systems mapping approach has also been used in S. Africa for road engineering and land assessment for urban land planning. In Japan this type of mapping has been used for landuse improvement, flood prevention and population mapping, and in Illinious, U.S.A. for land use planning.[1]

Terrain evaluation was initiated in India in 1964 to meet certain military and engineering requirements. A Terrain Evaluation Cell was set up in the Directorate of Engineering, Research and Development Organization, Ministry of Defence and studies were conducted in diverse terrain through Departments of Geography and Geology in the Indian universities between 1964 to1970, in which the author was also involved. These studies have been found useful also for determining land capability for agriculture and land suitability for irrigation. The basic unit is a *facet* which has a unifrom character. Related facets are grouped together in a *pattern*. A facet may consist of subfacets and may also contain *variants* and *local forms*.

Parametric Approach

In the land systems method of terrain classification there is a certain amount of subjectivity in selecting the bundaries of land systems, land unit etc, and the selection of suitable criteria is also somewhat arbitrary. A more objective method is based on measurements of properties of form such as altitude, slope as relative relief. "The parametric approach may be defined as the division and classification of land on the basis of selected attribute value. This approach is more precise than the landscape method, and being quantitative it allows numerical comparision and is suited to mechanical and computer data handling".[2] This method is followed by those users, such as engineers, who have need for quantification, but in may sitautions it gives results which are remarkably similar to those produced by the other method. This method is considered superior by workers who do not believe in multi-purpose approach, as according to them every purpose requires its own evaluation and classification and the interpretation of aerial photographs in terms of the parametric model is the best, as it is unlikely that a scheme designed for general purposes will fulfil the requirement

1. Cook, R.U. and Doorkamp, J.C.—op, cit., p. 328
2. Ollier, C.D.—Terrain Classification : methods, applications and principles". op. cit., p. 286

(of engineers) unless it is specifically adopted for engineering purposes.[1] This range of approach i.e. landscape to parameric has probably its own advantages and limitations.

The parametric system of military terrain evaluation has been most fully developed by the U.S. and Canadian armies, mainly by the Quartermaster Research and Engineering Centre at Natick, Massachusetts (QREC), the U.S. Army Engineer Waterways Experiment Station at Vicksbur, Mississippi (USAEWES) and the Canadian Defence Research Board. The first organization is primarily concerned with aspects of the environment from which an army requires protection, and the other two in the means of exploiting the enviroment to military advantage, specifically cross-country mobility. QREC have produced terrain bibliographies of the Russian Arctic and Western Africa, environmental handbooks on different parts of the world and have also developed methods for analysing lines of sight and surface relief.[2]

We may now consider briefly the various *applications of terrain classification* or land systems mapping. These include military intelligence, engineering, agriculture, forestry and landuse, regional planning as well as academic applications.

Military Use

Lie of the land is of utmost importance in military operations. Various kinds of terrain information may be requried urgently. Sometimes information may be needed for cross-country mobility or 'going conditions' for tank, vehicles, troops, camels etc. i.e, the traversability of terrain . Military engineers particutlary need "going maps"—where can you go with jeeps? with infantry? with tanks? but a separate going map is required for every possible mode of travel. The terrain map provides a useful base for the preparation of these specific going maps. Then there are problems regarding visibility and the amount of cover afforded. It is important to know the distribution of potential vantage points and areas of concealment which provide maximum visibility over an eneny with minimum exposure to his oversight. Thirdly, it is important to know reaction of terrain to deliberate deforamation, such as fixing of tent pegs, digging of trenches, and dugouts or the strength of foundations for

1. Ibid, p. 315.
2. Mitchell, C.W.—*Terrain Evaluation*, Longman, London, 2nd imp. 1977, p.81.

gun emplacements or the construction of temporary airfields so that the ground will not cave in and be resistant to the penetration of projectiles. Finally, it is important to know the location and nature of sources of construction materials, such as gravel, sand and wood and the availability of potable water in adequate quantities without the danger of a high water-table which may flood entrenchement and make the surface soft and impassable.

It is almost impossible to get all the site data required for this great multiplicity of purposes, and terràin classification that provides a generalized first approximation is extremely useful. Moreover, speed is of great importance—a commander cannot wait till investigations are carried out, and so a terrain classification prepared in advance enables the genralized information to be availabe at short notice. All that is required is a certain amount of training or skill in interpreting the terrain maps.

Engineering

Terrain data for engineers is rather difficult to provide because of the wide scope of engineering activity and the variety of its points of contact with the terrain. Most progress has naturally been made in those branches of engineering which depend on terrain factors most directly and extensively, and in those in which terrain factors are the main determinants of the total cost. Mainly three aspects are involved—military, hydrological and highway engineering, and we have already referred to them earlier in different contexts. Highway engineering includes roads as well sa railways and airfields. The highway engineer has to minimise the distance between the fixed end points of the road.He has also to minimise gradients, and he has to avoid watercourses as far as possible to minimise expenditure on bridges, culverts etc. Then he has to avoid areas of unstable soils. And finally, he has to keep in mind the availability of materials of suitable nature and particle size along the route.

Both the landsystem and parametric approaches have been used for engineering purposes. The landsystem approach assumes that terrain can be classified into units which have engineering significance. The parametric approach presumes that it is possible to identify, measure and interpret all terrain factores which affect an engineering operation. The parametric approach works successfully for limited objectives and is not useful for general purpose work. The landsystem approach is general and has not been sufficiently tested. Both the systems presume that similar terrains have similar engineering properties.[1]

CSIRO Division of Applied Geomechanics (formerly Division of Soil Mechanics) have devised a scheme of terrain evaluation specifically for engineering purposes by breaking the country down to smaller units and providing more engineering details than the CSIRO Land Research Method. It is known as PUCE (Pattern- Unit-Component-Evaluation) Programme. It has an inbuilt engineering bias but claims to be operational without specialist intervention. On the other hand, in the Oxford method there is an attempt to produce a multi-purpose system without disciplinary bias but with classificatory criteria which can be interpreted only by specialists.

The British Road Research Laboratory has developed a method of terrain evaluation for engineering purposes which has been applied in northern Nigeria. It uses the conventional land system approach, but in addition the engineering characteristics of each land facet (land unit) are given in terms of the engineering properties of materials, the engineering resources and the suitability for road location. Properties of materials are assessed in the field in terms of dominat material, depth and internal drainage, and plasticity characteristics, compaction, bearing strength and texture are given based on laboratory measurement of samples. The engineering resources are assesed in terms of water supply and constructional materials.

Some of the important findings of this study are that (i) land classification provides a satisfactory basis for classifying and storing information about the engineering features of the ground, and is more effective than geological or soil maps in revealing the extent and nature of road making materials and road building problems; (ii) broad but useful generalization can be made about available materials, problems and construction methods at the level of the land system, the information being especially useful in the planning stages of regional road-building programmes; (iii) at the level of individual road projects, engineers can utilize land facets and land elements to store and retrieve design data; (iv) where large areas have similar engineering properties, substantial savings can be

1. Ollier, C.D.—op. cit., p. 293

effected in the design effort by using land classification. It also alerts engineers of problems likely to be faced in a particular terrain; and (v) land classification reduces the cost on soil sampling and testing by 70%

Agriculture, Forestry and Land Use

Land systems mapping has been applied more in agricultre and engineering project than in any other field of activity. The genesis and distribution of soils are dependent very much upon the factors which control landforms. Thus terrain classification usually provides a good basis for soil mapping. Two factors of soil formation— parent materials and topography are the very basis of terrain classification. Variations in soil type caused by climate, organisms or time may not always show in the terrain map, but usually there is a good correlation between landform and soil and it is basic assumption of CSIRO method that a soil or simple soil association is uniform in a land unit. Soil maps provide the physical basis for agriculture, and agriculture tends to follow the same patterns as the soil on the terrain maps. The CSIRO Reports usually include maps of land capability and forestry, along with the terrain maps. The terrain maps are extremely useful in opening up new land for agriculture, in planning new plantations, introduction of new crops or resettlement of displaced people.

Terrrain classification serves a dual purpose in forestry. Firstly, it depicts the presence or absence of forest and delineats different types of forest, and secondly, it helps in forest planning and afforestation.

"Land use capability is generally assessed by placing each type of land in a land class which not only indicates the nature and seriousness of the limitations to agricultureal land use, but also passes some sort of judgement on what would be the best type of land use with regard to these limitations. The factors taken into account which characterize the agricultural aspect of land units are altitude, erodibility, stoniness, drainage, flood hazard, inundation. permeability, soil depth, available soil moistrue, top-soil consistency, soil reaction, salinity and nutrient status".[1]

Regional Planning

The combination of soils, engineering and topographic

information included on terrain maps makes them very suitable for regional planning.[2] The information can be used at various levels of planning. At the broadest scale, it may be used to delineate areas for wilderness, forest, recreation and urban development, as has been done in Canada. In its programme for rural development also, Canada is developing a computerbased information system for the storage and manipulation of map- based land datea. In areas of intense development it can be used for location of residential, industrial, commercial, agricultural and recreational areas, and for the location of highways, green belts, waste deposal sites and other uses. Terrain classification is not being used so much in urban planning, but it can play an important role.

Intimately related to regional planning is urban and rural landscape planning. In urban areas competition for use of land is severe. Urban land includes mostly built-up areas—residential, commercial, industrial and transport, all of which generally require flat, well-drained sites with bed-rock well below the surface to avoid constructional and drainage problems. Residential use, however, has some preference for somewhat steeper gradients and less compact blocks of land. Industrial and commercial uses require larger expanses of flat land. Land used for transport including airports demands special treatment because of its peculiar demands and interaction with landscape imposed by moving traffic and the dangers of accident, noise and pollution. Rural or undeveloped terrain fulfills a variety of recreational needs. Land for sports and childern's play is usually flat and level. Individual areas may be small but are frequent where population is dense. In humid climates it has to be well drained so that it can be used soon after heavy rain. Then there is rural land suited to scientific and educational purposes—nature reserves, botanical and zoological gardens, field study and research and demonstration centres of differnt kinds. For these exceptional sites are most suitable, representing one of the two extremes—either representative of the area or else very exceptional. Thirdly, there is terrain suited for wide ranging forms of recreation—local and national parks, public gardens and memorial grounds. Recreational land should be dry and well-drained so that it does not form poodles or mud. Sites should also

1. Ibid, p. 302.
2. Ibid, p. 303

be well placed in relation to sun, wind and access routes. Under European conditions this means sheltered southerly orientation. Man-made modifiation of terrain such as rockeries are often valuable. All these point to the need for translating the information on terrain maps into 'relative landuse suitability' maps to help guide urban and rural developments.

Geomorphological Mapping

Geomorphological mapping brings into focus the spatial components of landforms and geomorphological processes. It leads of the recognition of properties that are found in association with each other or vary across an area. It is therefore a valuable aid in planning and economic development. It is a comprehensive scheme of mapping land surface form, materials and processes to provide an integrated analysis of the physical properties of an area with a view to assessing its potentials and limitations for development. Geomorphological maps contain information of considerable potential value in land-use planning, reclamation and conservation, hydro-engineering, civil engineering, planning of settlements and mining etc. The most useful geomorphological maps are purpose-oriented maps compiled in the context of a specified problem. Such a map may be based on a general geomorphological map, but has to be more explicit and clear.

There are many examples of the practical application of geomorphological mapping, mostly from Europe where the post-1950 extension of towns has led to the occupation of land previously considered unsuitable for urban development. It was realized that before development plans could be made, specific information was required concerning landform and materials to be found at these sites and the processes working upon them. These investigations provided the basis for geomorphological mapping. It was begun in Poland where it has been used since the early 1950s in the context of economic planning. Considerable use has been made of geomorphological mapping in former U.S.S.R., the Netherlands, France, Hungary, Czechoslovakia, Italy and other countries. The International Geographical Union set up a Commision on Geomorphic Survey and Mapping on the occasion of the 21st International Geographical Congress in New Delhi in December, 1968, but even before that there was a sub-commision on Geomorphological Mapping under the IGU Commission on Applied

Geomorphology. The Commission has compiled the geomorphological maps of Europe on the basis of unified key on 1:2,500,000 scale, under UNESCO'S aegis. Unesco has supported geomorphological mapping as a resource assessment technique. The Commission has also prepared a manual of detailed geomorphological mapping including methods of compilation of detailed geomorphological maps at scales 1:25,000 to 1:100,000 and unified key.

It has to be recognized that there are several types of geomorphological maps. Apart from the distinction between general or academic purpose maps and special purpose maps, there is also a distinction to be made between morphographic, morphogenetic, morphometric and morphochronologic maps. On *morphographic* maps landforms are identified by their specific names such as, drumlin, esker or planation surface and their shapes are shown. *Morphogenetic* maps give the origin and development of landforms with genetic descriptions on the map legend (e.g. sandy alluvial plain, sandy proglacial plain). Colours are used to indicate major genetic landform units. Form and material compositions are also shown because of their link with processes. *Morphometric maps* provide information about the dimensions of landforms and include such things as slope steepness and shape, heights of spcific features e.g. river terraces, amplitude of relief, drainage density etc. *Morphochronologic maps* distinguish between recent forms and inherited forms whose character may be related to past processes and possibly past climates. Age may be indicated or omitted when not known. The best system of geomorphological mapping is probably the I.T.C. system of geomorphological mapping developed by the International Institute of Aerial Survey and Earth Sciences, the Netherlands.

Inspite of the development of over three decades, there are a very few good geomorphological maps in print, outside of Europe and the former USSR. Firstly, to compile a good detailed geomorphological map much field-work is required. And secondly, in order to be really effective, it has to be printed in colour which is very expensive. To be of value to the user the maps have to be so framed and its legend so phrased that it can be readily understood by those for whom it is intended which includes a wide spectrum of clientale.

Questions

1. Account for the recent growth of applied geomorphology and examine its scope. Is it beneficial for the theoretical side of the subject?

2. Discuss with examples the application of geomorphology to mineral and oil exploration.

3. Show how geomorphology can help in the planning and execution of hydro-engineering projects and management of water resources.

4. Examine the applications of geomorphology to highway engineering and hydrological studies.

5. Show how thorough knowledge of the geomorphology of periglacial regions is essential for planning their development.

6. Examine the scope and applications of Terrain classification (Land systems mapping).

7. Write an essay on any *one* of the following :-

   (a) Geomorphology and Coastal zone management,
   (b) Applied fluvial geomorphology,
   (c) Applied Geomorphology of Karst regions.
   (d) Terrain classification and its applications.

8. Write short notes on any *two* of the following:-

   (a) Weathering and economic mineral deposits.
   (b) Geomorphic processes and placer deposits,
   (c) Sand and gravel and the Geomorphologist,
   (d) Geomorphological mapping.

References

Ahmad, E. (1972) *Coastal Geomorphology of India*, Orient Longmans, New Delhi

Ahmad, E. (1985) *Geomorphology*, Kalyani, N. Delhi-Ludhiana.

Ahmad, E and S.P. Das Gupta (eds) (1970) *Proceedings of symposium on Erosion Surfaces*, 21st International Geographical Congress, India, 1968.

Anderson, D.L. (1989) *Theory of the Earth*, Blackwell Scientific Publications, Oxford.

Bagnold, R.A. (1941) *The Physics of Blown Sand and Desert Dunes*, William Morrow, New York.

Bauling, H. (1935) *The changing sea-level', Institute of British Geographers*, Pub. No 3.

Beatly T., Brower, D.J. and Schwab, A.K. (1994) *An Introduction to Coastal zone management*, Island Press, Washington.

Beckindle, R.P. and Chorley, R.I. (1991) *The History of the study of landforms or the development of geomorphology*, Vol.3 : Historical and regional geomorphology, 1890-1950, Routledge, London.

Benn, D.I. and Evans D.J.A. (1998) *Glaciers and Glaciation*, Arnold, London.

Bennett, M.R. and Glasser, N.E. (1996) *Glacial geology : Icesheets and landforms*, John Wiley, Chichester.

Betty, J.K., Peterson, C.C. and Chakin, A. (eds) (1999) *The New Solar System*, 4th ed. Sky Publishing Corporation and Cambridge University Press, Cambridge.

Birkeland, *P.W.* (1974) *Pedology, Weathering and Geomorphological Research*, Oxford University Press,

Bloom A.L. (1978) *Geomorphology : a systematic analysis of late cenozoic land forms*, Prentice-Hall of India, New Delhi. 3rd ed. 1998, Prentice Hall, New Jersey.

Boillot, G. (1981) *Geology of Continntal Margins*, Translated by A Scarth, Longman.

Bolt, B.A. (1991) *Inside the Earth : evidence from earthquakes*, Freeman, San Francisco.

Bolt, B.A. (1993) *Earthquakes*, W.H. Freeman, N.Y.

Bose, S.C. (1972) *Geography of the Himalayas*, National Book Trust, New Delhi.

Bradshaw, M.J., A,J. Abytt, and A.P. Gelsthopre, (1978) *The Earth's Changing Surface*, ELBS.

Brown, G.C. & Mussett, A.E. (1993) *The Inaccessilte Earth*, 2nd ed. Chapman & Hall, London.

Brumbaugh, D.S. (1999) *Earthquakes : Science and Society*, Practice Hall, New Jersey.

Brunsden, D. (ed.) (1971) *'Slopes : form and process'. Inst, Br. Georgr, Special Pub. No. 3.*

Brunsden, D. (1985) *Geomorphology in the service of man; in The Future of Geography* edited by R.J. Johnston, Methuen.

Budel J. (1982) *Climatic Geomorphology*, Translated by L. Fischer and D. Busche, Princeton University Press.

Bunting, B.T. (1976) *The Geography of Soil*, Hutchinson, London,

Burrard, S.C. (1933) *A Sketch of the geology and geography of the Himalaya mountains and Tibet*, Govt. of India, New Delhi.

Butzer, K.W. (1976) *Geomorphology from the Earth*, Harper & Row.

Charlesworth, J.K. (1957) *The Quarternary Era*, Vol I & II, Edward Arnold, London.

Chatterjee, S.P. & S.P. Dasgupta, (1970) Physical Geography Vol I selected Papers, 21st Int Geog. cong., National committee for Geography Calcuta.

Carter, R.W.G. (1988) *Coastal environments*, Academic Press, London.

Chorley, R.J., A.J. Dunn, and R.P. Beckinsale, (1964) *The History of the Study of Landforms, I: Geomorphology before Davis,* Methuen and John Wiley, London & New York.

Chorley, R.J. (ed) (1969) *Water, Earth and Man,* Methuen, London.

Chorely, R.J. and B.A. Kennedy, (1971) *Physical Geography: A Systems Approach,* Prentice Hall, London.

Chorely R.J. and P. Haggett (ed) (1965), *Frontiers in Geographical Teaching,* Methuen, London.

Coates. D.R. (ed) (1976) *Geomorphology and Engineering,* Dowdery, Hutchinson & Ross, Stroudsburg, Binghamton Symposium Series No. 7.

Condie, K.C. (1976) *Plate Tectonics and Crustal Evolution,* Permagon Press, New York. 4th ed (1977) Buttersworth-Heinemann, Oxford.

Cooke, R.U. and J.C. Doornkamp (1974) *Geomorphology in Environmental Management,* Clarendon Press, Oxford, 2nd Ed., 1990, OUP.

Cook, R.U.A. Waren, and A.S. Guoudie, (1993) *Desert Geomorphology,* UCL Press, London.

Cotton, C.A. (1944) *Volcanoes as Landscape Forms,* Whitcombe and Tombs, Christchurch

Cotton, C.A. (1948) *Landscape.* 2nd ed., Whitcombe and Tombs, Welligngton.

Craig, R.G. and Craft, J.L. (eds) (1982) *Applied Geomorphology,* The Binghamton Symposia in Geomorphology, International Series, No. 11. George Allen & Unwin, London.

Dalrymple, G.B. (1991) *The Age of the Earth,* Stanford, Univ. Press.

Daly, R.A. (1915) *The Glacial Control Theory of Coral Reef,* Proc. Am. Acad Arts & Sc, vol 51.

Daly, R.A. (1934) *The Changing World of the Ice Age,* Yale Univ. Press.

Daly, R.A. (1936) 'Origin of Submarine Canyons', *Am Jour Sc.,* 231, 401-20.

Dauglas, I. (1977) *Humid Landforms*, MIT Press. Cambridge, Mass.

Davies, J.L. (1972) *Geographical Variation in Coastal Development*, Longman, Oliver and Boyd, Edinburgh.

Davis, W.M. (1899) 'The Geographical Cycle', *Geog. Jour*,. 14, 481-504.

Davis, W.M. (1909) *Geographical Essays*, Boston, Ginn (reprinted 1954, Dover).

Davis, W.M. (1924) *The Coral Reef Problem*, American Geographical Society.

Dewey, J.F. and J.M. Bird (1970)'Mountain belts and the new global tectonics. *Jour. of Geophy. Res.*, 75, 2625-47

Derbyshire, E. (ed) (1976) *Geomorphology and Climate*, Wiley, London.

Dietrich, Gunter, Kurt Kalle, W. Krauss and G. Siedler (1975) *General Oceanography : an introduction*, 2nd ed. Translated by Susanne and Hans Ulrich Roll, John Wiley, New York

Dikshit, K.R. (1981) The Western Ghats : a geomorphic view, in *New Perspectives in Geography*, P. Dayal felicitation Volume, edited by L.R. Singh, Thinker's Library, Allahabad.

Dikshit, K.R. (1990) 'Landscape Metabolism and Experimental Geomorphology; *Ind. Geog Jour.*, 65, 18-25.

Dikshit, K.R. *et al* (eds) (1994) *India : Geomorphological Diversity*, Rawat Publications, Jaipur.

Domenico, P.A. and Schwartz, E.N. (1990) *Physical and Chemical hydrogeology*, John Wiley, New York.

Duff, D (ed) (1993) *Holmes' Principles of Physical Geology*, 4th ed. Chapman & Hall, London.

Dury, G.H. (1959) *The Face of the Earth*, Penguin Books.

Dury, G.H. (ed.) (1966) *Essays in Geomorphology*, William Heinemann, London.

Dury, G.H. (1972) Some recent views on the nature, location, needs and potential of Geomorphology, *Professional Geographer*, 24, 199-202.

Embleton, C. Brunsden D., and D.K. C Jones (eds) (1978) *Geomorhology: Present Problems and future prospects*, O.U.P., London.

Embleton, C. and C.A.M. King. (1975) *Periglacial Geomorphology*, - Edward Arnold.

Embleton, C. and C.A.M. King. (1975) *Glacial Geomorphology*, Edward Arnold.

Embleton. C. and Thornes, J. (1979) *Process in Geomorphology*, Arnold Heinemann, First Indian Edition, 1982.

Englen, O.D. Von (1942) *Geomorphology*, New York, Macmilan, 7th Printing, 1960

Fairbridge, R.W. (ed.) (1968) *Encyclopedia of Geomorphology*, Reinhold, New York.

Faniran, A. and L.R.Jeje, (1983) *Humid Tropical Geomorphology*, Longman, London.

Flint, R.F. (1947) *Glacial Geology and the Pleistocene Epoch*, Wiley N.Y.

Flint, R.F. (1957) *Glacial and Pleistocene Geology*, Wiley, N.Y.

Flint, R.F. (1971) *Glacial and Quaternary Geology*, Wiley, New York.

Ford, D. C. & Williams P.W. **(1989)** *Karst Geomorphology and hydrology*, Unwin Hyman, London.

Francis. P. (1976) *Volcanoes*, Penguin Books.

Francis, P (1993) *Volcanoes ; a planetary perspective*, Oxford Univ. Press.

French, H.M. (1976) *The Periglacial Environment*, Longman. Harlow, 2nd ed (1996)

Gamow, G. (1959) *Biography of the Earth*. Viking Press.

Gamow, G. (1965) *A Planet Called Earth*, Bantam Pathfinder Ed. (Viking Ed. 1961).

Gansser, Augusto (1964) *Geology of the Himalayas*, Inter-Science Publishers, John Wiley, London.

Garner, H.F. (1974) *The Origin of Landscape*, Oxford Univ Press, New York.

Gass, I.G., P.J. Smith, and R.C.L. Wilson, (eds). (1973) *Understanding the Earth*, Artemis Press, 2nd Ed. E.L.B.S. Ed.

Geddes, A. (1960) 'The alluvial morhpology of the Indo-Gangetic Plain : its mapping and geographical significance'. *Trans. Inst Br. Geogr*, 28, 253-76.

Gerrard, A.J. (1990) *Mountain Environments : an examination of physical geography of mountains*, Belhaven Press, London.

Gillienson, D. (1996) *Caves processes, development and management*, Balckwell Scientific Pub, Oxford.

Gilluly, J, A.C. Waters, and A.O. Woodfors (1968) *Principles of Geology*, 3rd Ed, Freeman International Edition.

Goudie, Andrew *etal* (eds) (1981) *Geomorphological Techniques*, George Allen & Unwin, London.

Gregory, K.J. (1985) The *Nature of Physical Geography*, Edward Arnold London.

Gregory, K.J. and D.E. Walling (1973) *Drainage Basin Form and Process*, Edward Arnold, London.

Griggs, D. (1939) 'A theory of mountain building', *Am. Jour. of Sc*, 237, 611-50.

Guilcher A. (1958) *Coastal and Submarine Morphology*, Methuen.

Guilcher, A. (1988) *Coral reef Geomorphology*, John Wiley, New York.

Hails John, R. (ed) (1978) *Applied Geomorphology*, Elsevier Scientific Publishing Company, Amsterdam, Oxford, New York.

Hancock, Paul and Skinner, Barian J (2000) *The Oxford Companion to the Earth*, Oxford Univ. Press.

Holmes, A. (1965) *Principles of Physical Geology*, Nelson, London. E.L. B.S. Ed., 1966.

Holmes, Doris L (1978) *Holmes Principles of Physical Geology*, 3rd ed, ElBS and Nelson, Reprinted 1981.

Hutchinson, G.E. (1957) *A Treatise on Limnology*, Vol. I, Wiley, New York.

ICSSR (1983) A Survey of Research in Physical Geography, Concept Publishing co., New Delhi.

Jeanloz, R. (1983) The Earth's core. *Scientific American*, 249, 56-65

Jennings, I.N. (1985) *Karst Geomorphology*, Blackwell Scientific Pub. Oxford.

Johnson, D.W. (1919) *Shore Processes and Shoreline Development*, Wiley, New York.

Kahle, C.F. (ed.) (1974) Plate Tectonics—Assessments and Reassessment. *Am. Assoc. Pet. Geol.*, Memoir 23.

Kearney, P. and Vine, E.J. (1996) *Global Tectonics*, Blackwell Science, Oxford.

Kenneth, J.P. (1982) *Marine Geology*, Prentice Hall, Eaglewood cliffs, N.J.

King, C.A.M. (1959) *Beaches and Coasts*, Arnold.

King L.C. (1948) 'A Theory of Bornhardts', *Geog Jour*. 112, 83-7

King, L.C. (1951) *South African Scenery*, 2nd ed., Oliver and Boyd.

King, L.C. (1967) *Morphology of the Earth*, 2nd ed. Oliver and Boyd, Edinburgh.

Knighton, A.D. (1984) *Fluvial Forms and Processes*, Arnold, London.

Krishnan, M.S. (1982) *Geology of India and Burma*, 6th ed., C.B.S. Publishers, Delhi.

Kuenen, Ph. K. (1950) *Marine Geology*, Wiley

Kumar, Ravindra (1985) *Fundamentals of Historical Geology and Stratigraphy of India*, Wiley Eastern Ltd., New Delhi Reprinted 2003

Kummel, B. (1961) *History of the Earth*, Freeman and co., Indian Reprint, 1968, Eurasia Publishing House, New Delhi.

Lake, P. (1974) *Physical Geography*, Indian Rev. Ed. Prepared by S.C. Chatterjee, Macmillan.

Lall, J.S. and A.D. Moddie (eds.) (1981) *The Himalaya, Aspects of Change*, India International Centre and Oxford Univ. Press, Delhi.

Lancaster, N. (1995) *Geomorphology of Desert Dunes*, Routledge, New York.

Lapedes, D.N. (ed) (1978) *Mc Graw Hill Encyclopedia of the Geological Sciences.*

Law, B.C. (ed) (1968) *Mountains and Rivers of India*, National Committee for Geography, Calcutta. 21st International, Geographical Congress, India

Leopold, L.B., M.G. Wolman, and J.P. Miller, (1964) *Fluvial Processes in Geomorphology*, Freeman, San Francisco.

Leopold, L.B. (1994) *A view of the river*, Harod Univ. Press Cambridge, Mass.

Lerman. A., Imbsden D. and. Gat, J. (eds) (1995) *Physics and Chemistry of Lakes*, 2nd ed. Springer, Verlag, Berlin.

Linton, D. L (1963) 'The forms of glacial erosion'. *Trans. Inst. Br. Geogr. V. 33, 1-28.*

Livingstone D. and Warren, **A.** (1996) *Aeolian Geomorphology : an introduction*, Addison-Wesley, Longman, Harlow.

Lobeck, A.K. (1939) *Geromorphology*, Mc Graw Hill, New York.

Longwell. C.R., A. Knof, and R.F. Flint, (1948) *Physical Geology*, Wiley, 3rd ed.

Machatschek, (1969) *Geomorphology*, Longman, London and New York.

Mallory, B.F. and D.N. Cargo, (1979) *Physical Geology*, Mc Graw Hill, International Student Edition.

Mathur, S.M. (1994) *Physical Geology of India*, National Book Trust, New Delhi.

Mc Gill, G.J.H. (1996) The early history of the earth, *Geoscientist* 6 (1), 10-14.

Mc Kenzie, D.P. (1983) The Earth's mantle, *Scientific American* 249, 56-65.

Mc Manus J & Dack, R.W. (ed) (1993) *Geomorphology and Sedimentology of lakes and reservoirs*, John Wiley, Chichester.

Misra, R.P., S.P. Dasgupta and M.S. Masood (eds) (1972) *Proceedings of Symposium on Geomorpholgy of the Indian Shield,* 21st International Geographical Congress, India, 1968.

Misra, S.D. (1970) *Rivers of India,* National Book Trust, New Delhi.

Mitchell, C. (1973) *Terrain Evaluation,* Longman, London.

Morisawa, Marie (ed) (1981) *Fluvial Geomorphology,* The Binghamton Symposium in Geomorphology, International Series No. 4, George Allen & Unwin, London.

Morisawa, Marie M. (1968) *Streams : their dynamics and morphology,* McGraw Hill., New York.

Morisawa, Marie M. (1985) *River :Form and Process,* Longman, London and New York.

Nageswara Rao, K. (1985) Evolution of Landforms in the area betwen the Krishna and the Godawari deltas, *Ind. Geog. Jour,* 60, 30-36.

Nageswara Rao, K. and R. Vaidyanadhan (1978) 'Geomorphic Features in the Krishna Delta and its Evolution', *Proceedings of Symposium on Morphology and Evolution of Landforms,* University of Delhi, 120-130.

Nicholas, A. (1995) *The mid-ocean ridges : Mountains below sea level* Springer-verlog, Berlin.

Nunn P.D. (1994) *Oceanic Islands,* Blackwell, Oxford.

Ollier. C.D. (1969) *Weathering,* Oliver and Boyd, Edinburgh.

Ollier, C.D. (1981) *Tectonics and Landforms,* Longman.

Ollier C.D. (1991) *Ancient Landforms,* Belhaven Press, Cambridge, Mass.

Open University (1991) *Ocean chemistry and deep sea sediments,* 2nd ed. Permagon Press, Oxford.

Ordway, R.J. (1971) *Earth Science,* Van Nostrand, East-West Ed.

Owen LA. (1995) Shaping of the Himalayas, *Geog. Jour.* 117 (2)23-5.

Pardhasardhi, Y.J. and R. Vaidyanadhan (1976) 'Pediments and pediment forming processes around Coimbatore, South India, *Ind. Geog. Soc, Golden Jubilee Volume*, Ind. Geog. Jour, 252-59.

Pascoe, E. H. (1964) *A Manual of the Geology of India and Burma*, Manager of Publication, Delhi. Reprinted 1973.

Peel R. F. (1966) 'The landscape in aridity', *Trans. Inst. Br. Geogr.*, 38, 1-23.

Peltier L.C. (1950) 'The geographical cycle in periglacial regions as it is related to climatic geomorphology', *Ann. Ass. Am. Geogr.* 40, 214-36.

Penck, W. (1924) *Morhpological Analysis of Landforms*, English translation by H. Czech and K.C. Boswell, Macmillan, London 1953.

Petts, G. and I. Foster (1985) *Rivers and Landscape*. Edward Arnold, London.

Pichamuthu, C.S. (1970) *Physical Geography of India*. National Book Trust, New Delhi.

Pickard, G.L. & Emery, W.J. (1982) *Descriptive Physical Oceanography*, Permagon Press, Oxford.

Pitty, A.F. (1971) *Introduction to Geomorphology*, Barnes and Noble, New York.

Prasad, Nageshwar and L.R. Singh (1980) 'Evolution of Slopes in Barakar Basin, *Ind. Geog. Jour.*, 55, 22-28.

Price, R.J. (1973) *Glacial and Fluvioglacial Landforms*, Oliver and Boyd, Edinburgh (Longman group).

Preceedings of the Seminar on *Geomorphological Studies in India*, held at Sagar, November, 1965.

Ramchandran, Kaushalya, D.C. Joshi, P.C. Vats and Gheenalal (1990) 'Morphology & Texture of Shifting Sand in Western Rajasthan, India, *Ind Geog Jour*,. 65, 36-39.

Ray, S. (1978) *Geomorphology of India*, Firma KLM P ltd, Calcutta

Richards. K. (1982) *Rivers*, Methuen, London.

Rogers, J.J.W. (1993) *A history of the earth*, Cambridge Univ. Press.

Ruhe R.V. (1975) *Geomorphology*; Houghton Mifflin Co., Boston

Sanders, E.W. (1921) The cycle of erosion in a karst region (after Cvijic), *Geog. Rev*, 11, 593-604.

Salisbury, R.D. (1919) *Physiography*, 3rd ed. Henry Holt, New York.

Sambasiva Rao M. and R. Vaidayanadhan (1979) 'Morphology and Evolution of the Godawari Delta, India', Z. *Geomorph, N.H.*23, 243-55.

Savigear, R.A.G. (1952) 'Some observations on slope development in South Wales', *Tras. Inst. Inst. Br. Geogr*. No. 18, 31-52.

Schumm, S.A (1977) *The Fluvial System*, Wiley, New York.

Scientific American, (1975) *Reading in Earth Sciences* Vol. I, II & III Indian Reprint, Taraporevala.

Selby, M.L. (1985) *The Earth's Changing surface*, Clarendam Press, Oxford.

Sen, P.K. and Prasad, N. (2002) *An Introduction to the Geomorphology of India*, Allied Publishers, Mumbai & New Delhi.

Sharma, H.S. (ed) (1982) *Perspectives in Geomorphology*, vol I to IV Concept, New Delhi.

Sharma, V.K. (1986) *Geomorphology : Earth Surface Processes and Forms*, Tata Mc Graw Hill, New Delhi.

Sharma, R.C. & M. Vatal (1995) *Oceanography for geographers*, Chaitanya, Allahabad, 5th Ed.

Sharpe, C.F.S. (1938) *Landslides and related phenomena*, Columbia Univ Press.

Shepard, F.P. (1973) *Submarine Geology*, 3rd ed, Harper and Row, New York.

Singh, R.P. (1969) *Geomorphological Evolution of Chotanagpur Highlands, India*, National Geographical Society of India. Varanasi.

Singh, S. (ed) (1977) *Geomorphological Investigation of the Rajasthan*

Desert, Central Arid Zone Research Institute, Jodhpur, Monograph 7.

Small, R.J. (1978) *The Study of Landforms*, 2nd ed., Cambridge Univ. Press, Reprinted 1985.

Smart, W.M. (1959) *The Origin of the Earth*, Penguin Books.

Smith, D.G. (Ed-inchief) (1982) *The Cambridge Encyclopedia of Earth Sciences*, Cambridge, CUP.

Sparks, B.W. (1971) *Rocks and Relief*, Longman, London.

Sparks, B.W. (1960) *Geomorphology*, Longman, London, Second Ed., 1972.

Stacery, F.D. (1992) *Physics of the Earth*, 3rd ed. Brookfield Press, Brisbane.

Steers, J.A. (1937) *The Unstable Earth*, Methuen, Indian Edition, Kalyani Publishers.

Steers, J.A. (1953) *The Sea Coast*, Collins, New Naturalist Series.

Steers, J.A. (1963) *Coasts and Beaches*, Oliver and Boyd.

Steers, J.A. (ed) (1971) *Applied Coastal Geomorphology*, Macmillan, London.

Strahler, A.N. (1968) *The Earth Sciences*, Harper International Edition.

Strahler. A.N. (1969) *Physical Geography*, 3rd ed, Wiley.

Strahler, Alan and Strahler, Arthur. (2000) *Physical geography*, Wiley Singapore, Indian Ed., Delhi.

Sudgen, D.E.and B.S. John (1976) *Glacier and Landscape*, John Wiley, New York.

Summerfield, M.A. (1991) *Global Geomorphology*, Longman, Harlow.

Sweeting, M.W. (1972) *Karst Landforms*, Macmillan, London.

Sweeting, M., D.P. Ede and M.D. Newson (1973) Some Results and Application of Karst Hydrology : a symposium, *Geog-jour*, Vol. 139, pp 280-310.

Taylor, S.R. & Mc Lennan, S.M. (1985) *The Continental Crust, its Composition and distribution*, Blackwell Science, London.

Thomas, D.S.G. (ed) (1997) *Arid zone geomorphology ; Process, form & Change in drylands*, 2nd ed. John Wiley, Chechester.

Thomas, M.F. (1974) *Tropical Geomorphology*, Macmillan, London.

Thornbury, W.D. (1969) *Principles of Geomorphology*, 2nd ed, Wiley International Edition. Also Wiley Eastern Reprint, 1984.

Tricart, J. (1970) *Geomorphology of Cold Environments*, Translated by E. Watson and Martin, Macmillan, London.

Tricart, J. (1972) *The landforms of the Humid Tropics, Forests and Savannas*, Translated by C. Kiewiet de Jonge, Longman.

Tricart, J. (1974) *Structural Geomorphology*, English Translation by S.H. Beaver and E. Derbyshire, Longman.

Tricart J. and A. Cailleux (1972) *Introuduction to Climatic Geomorophology*, Translated by Jonge, Longman, London.

Trudgill, Stephen (1985) *Limestone Geomorphology*, Longman, London and New York.

Twidale, C.R. (1976) *Analysis of Landforms*. Wiley, New York.

Twidale, C.R. (1981) Granitic Inselbergs: Domed, Block-strewn and Castellated. *Geog. jour* Vol. 147, pp 54-71.

Twidale, C.R. (1982) *Granite Landform*, Elsevier, Amsterdam.

Van Amdel, T.H. (1994) *New Views on an Old Planet*, Cambridge, Univ. Press.

Verma, V.K. ed. (1973) *Recent Researches in Geology*, Vol. I. Hindustan Pub. Co., Delhi.

Verma, V.K. (1999) *Lecturers on Geomorphology*, Pilgrims Book Pvt. Ltd. New Delhi.

Verstappen, H. Th. (1983) *Applied Geomorphology : Geomorphological Surveys for Environmental Development*, Elsevier, Amsterdam.

Wadia, D.N. (1981) *Geology of India*, 4th Ed. Tata Mc Graw Hill, New Delhi.

Washburn, A.L. (1956) 'Classification of patterned ground and review of suggested origins', *Bull. Geol. Soc. Am*, 67, 823-66.

Washburn, A.L. (1973) *Periglacial Proceses and Environments*, Martin Press, New York, Edward Arnold, London.

Wegener, A. (1929) *The Origin of Continents and Oceans.* Trans from 4th Revised German ed. by J.G. Skerl, Methuen, London.

White, W.B. (1988) *Geomorphology and hydrology of Karst terrains.* Oxford Univ. Press, New York.

Windley, B.E. (1995) *The evolving Continents*, John Wiley Chichester.

Wooldridge, S.W. and D.L. Linton (1955) *Structure, Surface and Drainage in South-East England*, 2nd ed., Philip, London.

Wooldridge, S.W. and R.S. Morgan (1948) *The physical Basis of geography*, First Published 1937, Longmans.

Worcester, P.G. (1965) *A Textbook of Geomorphology*, Van Nostrand, 2nd ed., East-Wested., New Delhi.

Wylle, P.J. (1971) *The Dynamic Earth*, Wiley, New York.

Young, A. (1972) *Slopes*, Longman, New York, Oliver & Boyd Edinburgh.

Zakrzewska, Barbara (1967) Trends and methods in landforms geography', *Annals Ass. Am. Geogr.* 57, 128-165.

Zernitz, E.R. (1932) 'Drainage Patterns and their significance', *Jour Geol*, 40, 498-521.

Index

N

S